Ecological Studies

Analysis and Synthesis

Volume 232

Ecological Studies is Springer's premier book series treating all aspects of ecology. These volumes, either authored or edited collections, appear several times each year. They are intended to analyse and synthesize our understanding of natural and managed ecosystems and their constituent organisms and resources at different scales from the biosphere to communities, populations, individual organisms and molecular interactions. Many volumes constitute case studies illustrating and synthesizing ecological principles for an intended audience of scientists, students, environmental managers and policy experts. Recent volumes address biodiversity, global change, landscape ecology, air pollution, ecosystem analysis, microbial ecology, ecophysiology and molecular ecology.

More information about this series at http://www.springer.com/series/86

Patricia M. Glibert • Elisa Berdalet •
Michele A. Burford • Grant C. Pitcher •
Mingjiang Zhou
Editors

Global Ecology and Oceanography of Harmful Algal Blooms

Springer

Editors
Patricia M. Glibert
Horn Point Laboratory
University of Maryland Center for Environmental Science
Cambridge, Maryland, USA

Elisa Berdalet
Institute of Marine Sciences (CSIC)
Barcelona, Catalonia, Spain

Michele A. Burford
Australian Rivers Institute
Griffith University
Nathan, Queensland, Australia

Grant C. Pitcher
Fisheries Research and Development
Cape Town, South Africa

Mingjiang Zhou
Institute of Oceanology
Chinese Academy of Sciences
Qingdao, China

ISSN 0070-8356 ISSN 2196-971X (electronic)
Ecological Studies
ISBN 978-3-030-09930-5 ISBN 978-3-319-70069-4 (eBook)
https://doi.org/10.1007/978-3-319-70069-4

Printed on acid-free paper

This Springer imprint is published by the registered company Springer International Publishing AG part of Springer Nature.
The registered company address is: Gewerbestrasse 11, 6330 Cham, Switzerland

In Memoriam: Otto Ludwig Lange (1927–2017)

Otto Ludwig Lange, emeritus Professor at the University of Würzburg, Germany, died on August 14, 2017, at the age of 89 years. The death was unexpected; indeed, he was actively planning the celebration of his 90th birthday. With his death the ecological community has lost a prominent and valued member, and *Ecological Studies* no longer has its founder and promotor, to whom we owe this thriving book series.

Otto Ludwig Lange was borne in Dortmund and studied biology at the University of Göttingen, after he returned from serving in World War II and imprisonment in France. He obtained his doctorate degree at the University of Göttingen in 1952 under Prof F. Firbas on "Heat and drought tolerance of lichens in relation to their geographic distribution". Following habilitation as part of the German tenure procedure at the University of Göttingen in 1959, he moved to Darmstadt to join the group of Prof. O. Stocker as Associate Professor. He had already attained full professorship in Forest Botany and Technical Mycology in 1963, as well as becoming the Director of the Arboretum of the University of Göttingen. Four years later, in 1967, Otto Lange accepted an offer from the University of Würzburg to establish a new Chair of Plant Ecology, which incorporated management of the Botanic Gardens. Despite many attractive offers from around the world, Otto Lange decided to stay at Würzburg until his retirement.

Otto Lange was an enthusiastic and inspiring ecologist with a remarkably broad range of interests, and a main focus on the physiology of vascular and non-vascular plants under natural field conditions. He taught a very broad range of subjects, including plant taxonomy, vegetation science, and eco-physiology, and guided field courses and associated excursions on plant identification within Germany and across Europe. He put a lot of effort into teaching, raising the interest of students from their 1st Semester. He was full of ideas, and always eager to fully understand anatomy, morphology and existence of species in certain environments on a sound physiological basis. He travelled around the globe to better understand the distribution and existence of lichens and also fueling an interest in salt and drought tolerance. He was fascinated by dew fall and water uptake from humid air, as well as being the first person to demonstrate the response of vascular plants to air humidity, His experiments with *Citrullus colocynthis* became the classic demonstration of the regulation of leaf temperature via adjustments of the energy balance to avoid heat damage.

Otto Lange became one of the most highly recognized German ecologists, with his focus on the eco-physiology of plants under arid and cold conditions. From this long list of honours, the most remarkable recognition of his research was probably the naming of "Lange Peak" in the Lyttelton Range in the Admiralty Mountains of Northern Victoria Land, Antarctica. Three lichen species are named after Otto Lange (*Peltua langei, Hubbsia langei* and *Jeckelixia ottolangei*). These enduring milestones of Otto Lange's work remain amongst his chief legacies but he was also recognized in many other ways; he received the Balzan-Price of Italy, the Gottfried Wilhelm Leibnitz Price of the German Science Foundation, the "Bayerische Maximilian Orden" of the Bavarian Goverment and the Cothenius medal of the German Academy of Sciences, and the Ecological Society of America Award. *Ecological Studies* Volume 100 was rightly dedicated to Otto Lange's retirement, while special volumes in lichenological journals also celebrated his 70th and 80th birthday.

Besides his achievements in science, Otto Lange was widely recognized as a fair and supportive research organizer. He was Editor in Chief of the oldest German botanical journal *Flora*, and co-editor of many journals, including *Oecologia*, He was member of the referee board of the German Science Foundation, acted as the Dean of the Faculty in Würzburg, and a member of the academic Senate and the board of financial affairs of his university. In recognition of these administrative engagements the University honoured Otto Lange with the Bene Merenti Medal of the University of Würzburg.

The foundation of the book series *Ecological Studies* remains a major science achievement. Since the very first volume, published in 1970, this book series became the main international platform for publications of multidisciplinary projects, encompassing all vegetation zones and land uses on this globe, including aquatic systems.

Very sadly, we have lost an ingenious and original advisor. We will miss the stimulating discussions that lead to so many new insights in science and which helped many scholars cope with life during the ups and downs of their careers. He was also a personal friend, who cared passionately about the life of people in his institute and in his field of science. He will be sorely missed.

Ernst-Detlef Schulze, Editor of *Ecological Studies*

Contents

List of Abbreviations

2D	Two-dimensional
3D	Three-dimensional
ADCP	Acoustic Doppler current profiler
AE	Assimilation efficiency
AFM	Atomic force microscopy
AM	Adapting Mosaic, a Millennium Ecosystem Assessment scenario
ANF	Autotrophic nanoflagellates
ANN	Artificial neural network
A_s	Parameter of efficiency of nutrient uptake at the cell surface (transport)
ASIMUTH	Applied simulations and integrated modelling for the understanding of harmful algal blooms
ASP	Amnesic shellfish poisoning
AUV	Autonomous underwater vehicle
AZA	Azaspiracids
BFAR	Bureau of Fisheries and Aquatic Resources (Philippines)
BHAB	Benthic harmful algal bloom
BMAA	Beta-*N*-methylamino-L-alanine
BML	Bottom mixed layer
BNC	Benthic nepheloid layer
C	Carbon
CAFOs	Concentrated animal feeding operations
CAS	Chinese Academy of Sciences
CCM	Carbon concentrating mechanisms
CDC	Centers for Disease Control and Prevention
CDOM	Coloured dissolved organic matter
CEOHAB	Chinese Ecology and Oceanography of Harmful Algal Blooms Programme
CFP	Ciguatera fish poisoning

CIESM	Mediterranean Science Commission
COD	Chemical oxygen demand
COOP	Coastal Ocean Observing Panel
CRP	Core research project
CTX	Ciguatoxin
CyanoHABs	Cyanobacterial harmful algal blooms
DA	Domoic acid
DIN	Dissolved inorganic nitrogen
DIP	Dissolved inorganic phosphorus
DOM	Dissolved organic matter
DON	Dissolved organic nitrogen
DSP	Diarrheal/diarrhetic shellfish poisoning
DST	Diarrhetic shellfish toxins
DTX	*Dinophysis* toxins
DVM	Diel vertical migration
ECOHAB	Ecology and Oceanography of Harmful Algal Blooms (US national programme)
EDAB	Ecosystem disruptive algal bloom
EFSA	European Food Safety Authority
ELISA	Enzyme-linked immunosorbent assays
ENSO	El Niño Southern Oscillation
EOV	Essential ocean variable
EPS	Exo-polymeric substances
ESP	Environmental Sample Processor
EUROHAB	European Initiative on Harmful Algal Blooms
FAO	Food and Agriculture Organization of the United Nations
FISHSTAT	Fisheries and aquaculture database of the Food and Agriculture Organization of the United Nations
FLH	Fluorescence line height
FSS	Fine-scale sampler
GAM	Generalized additive model
GCM	Global climate model
GEF	Global Environmental Fund of the United Nations Environment Programme
GEOHAB	Global Ecology and Oceanography of Harmful Algal Blooms Programme
GLM	Generalized linear model
GLMM	Generalized linear mixed-effect model
GLOBEC	Global Ocean Ecosystem Dynamics
GM	Generation mechanism
G_{max}	Maximum rate of grazing
GNM	Global nutrient model
GO	Global Orchestration, a Millennium Ecosystem Assessment scenario
GO_2NE	Global Ocean Oxygen Network

GOM	Gulf of Maine
GOMTOX	Gulf of Maine toxic algal dynamics project on *Alexandrium fundyense* distributions
GOOS	Global Ocean Observing System
GYM	Gymnodimines
HAB	Harmful algal bloom
HABMAP	Harmful algal blooms map
HAEDAT	Harmful Algal Event Database
HARRNESS	Harmful Algal Research and Response National Environmental Science Strategy of the USA
HDTMA	Hexadecyl trimethyl ammonium bromide
HICO	Hyperspectral Imager for the Coastal Ocean
HNF	Heterotrophic nanoflagellates
HPLC	High-performance liquid chromatography
HTS	High-throughput sequencing
IAEA	International Atomic Energy Agency
IBM	Individual-based model
ICEP	Indicator for coastal eutrophication potential
ICES	International Council for the Exploration of the Sea
IFCB	Imaging Flow Cytobot, a submersible imaging flow cytometer
IFPRI	International Food Policy Research Institute
IMAGE-GNM	Integrated model to assess the global environment–global nutrient model
IMBeR	Integrated Marine Biosphere Research Program
IOC	Intergovernmental Oceanographic Commission
IOC/ANCA	IOC regional working group and network on harmful algae in the Caribbean and adjacent areas
IOCCG	International Ocean Colour Coordinating Group
IOC/FANSA	IOC regional working group and network on harmful algal blooms in South America
IOC/HANA	IOC network on harmful algae in North Africa
IOC/WESTPAC/HAB	IOC Subcommission for the Western Pacific
IOC WG Trends PO	IOC Working Group Trends of Phytoplankton in the Oceans
IPCC	Intergovernmental Panel on Climate Change
IPHAB	Intergovernmental Panel on Harmful Algal Blooms
IPSAP	IFREMER particle size analyser profiler
ISE	Ion-selective electrode
ISSHA	International Society for the Study of Harmful Algae
JGOFS	Joint Global Ocean Flux Study
K_s	Half saturation constant for nutrient uptake
K_μ	$K\mu$–Half saturation constant for growth
LC–HRMS	Liquid chromatography–high-resolution mass spectrometry

LC–MS	Liquid chromatography–mass spectrometry
LME	Large marine ecosystems
LOICZ	Land–Ocean Interactions in the Coastal Zone
LSU rDNA	Large subunit ribosomal DNA
MEA	Millennium Ecosystem Assessment
MAA	Mycosporine-like amino acids
MBARI	Monterey Bay Aquarium Research Institution
MEA	Microelectrode array
MERIS	Medium Resolution Imaging Spectrometer (sensor aboard the European Space Agency satellite)
MIB	Methylisoborneol, a bioactive compound of benthic HABs
μ	Growth rate
μ_{max}	Maximum rate of growth
μZ	Microzooplankton
MOST	Chinese Ministry of Science and Technology
MTX	Maitotoxin
N	Nitrogen
N2A	Neuroblastoma cell line
NANI	Net anthropogenic nitrogen input
NAO	North Atlantic oscillation
NARCCAP	North American Climate Change Assessment Program
NATO ASI	North Atlantic Treaty Organization Advanced Science Institute
NEWS	Global Nutrient Export from Watersheds models
NMA	Non-motile microalgae
NN	Neural network
NOAA	National Oceanic and Atmospheric Administration of the US
NORS	National Outbreak Reporting System
NPZ	Nitrogen–phytoplankton–zooplankton (model)
NPZBD	Nitrogen–phytoplankton–zooplankton–bacteria–detritus (model)
NR	Nitrate reductase
NSFC	National Natural Science Foundation of China
OA	Okadaic acid
OHHABS	One Health Harmful Algal Bloom System
OLCI	Ocean and Land Colour Instrument
OPD	Optical plankton discriminator
OSM	Open Science Meeting
OSTR	Ostreocin
OU	Optimal uptake, a nutrient kinetic parameter
OVT	Ovatoxin
P-CTX	Pacific ciguatoxin
P	Phosphorus
PAR	Photosynthetic active radiation
PCOM	Portuguese Coast Operational Model System

PCR	Polymerase chain reaction
PDO	Pacific Decadal Oscillation
PET	Plankton emergence trap
PFT	Plankton functional type
PhilHABs	Ecology and Oceanography of Harmful Algal Blooms in the Philippines
PICES	North Pacific Marine Science Organization
PLTX	Palytoxin
POGO	Observation of the Global Oceans
POM	Particulate organic matter
PSP	Paralytic shellfish poisoning
PST	Paralytic shellfish toxins
PTX	Pectenotoxin
PUFA	Polyunsaturated fatty acids
qPCR	Quantitative polymerase chain reaction
RBA	Receptor binding assay
$RBA_{(F)}$	Fluorescence-based receptor binding assay
RCM	Regional climate model
RLH	Reflectance line height
ROMS	Regional Ocean Modelling System
ROS	Reactive oxygen species
RTM	Reactive transport models
S	Substrate concentration
SCAMP	Self-contained autonomous microstructure profiler
SCM	Subsurface chlorophyll maximum
SCOR	Scientific Committee on Oceanographic Research
SEED	EU project entitled "Life cycle transformations among HAB species, and the environmental and physiological factors that regulate them"
SI	Similarity index
SML	Surface mixed layer
SOA	Chinese State Ocean Administration
SPARROW	Spatially Referenced Regressions on Watersheds
SPATT	Solid Phase Adsorption Toxin Tracker
SPX	Spirolides
SSC	Scientific Steering Committee
SST	Sea surface temperature
STX	Saxitoxin
SVM	Support Vector Machine
SWAT	Soil and Water Assessment Tool
SWRO	Sea water reverse osmosis (used in desalination plants)
TL	Thin layer
TMDL	Total Maximum Daily Load
UAV	Unmanned aerial vehicles

UNEP	United Nations Environment Programme
UNEP/NOWPAP	United Nations Environment Programme/Northwest Pacific Action Plan
UNESCO	United Nations Educational, Scientific, and Cultural Organization
USFDA	US Food and Drug Administration
UV	Ultraviolet
V	Specific velocity of nutrient uptake
V_{aff}	Nutrient uptake affinity, a nutrient kinetic parameter
V_{max}	Maximal nutrient uptake velocity
WG	Working group
WHO	World Health Organization
YSI	Yellow Springs Instruments (now Xylem Instruments)
YTX	Yessotoxins
ZINB	Zero-inflated negative binomial

Part I
Introduction to Harmful Algal Blooms and the GEOHAB Programme

Chapter 1
Introduction to the Global Ecology and Oceanography of Harmful Algal Blooms (GEOHAB) Synthesis

Patricia M. Glibert, Elisa Berdalet, Michele A. Burford, Grant C. Pitcher, and Mingjiang Zhou

This volume on the Global Ecology and Oceanography of Harmful Algal Blooms (GEOHAB) Programme is aimed at synthesizing the broad range of research and activities that took place during the nearly two decades of the international programme bearing this acronym. It is, however, but a snapshot into the diversity of work that took place all over the globe, promoted directly by GEOHAB and its core programmes, indirectly through its multiple Open Science Meetings or obliquely through the greater awareness of HABs and their impacts made known through various outreach activities, workshops and written materials. These impacts were, at least in part, a result of GEOHAB and related activities. Thus, the editorship and authorship throughout this book reflect the global reach of the GEOHAB Programme (Fig. 1.1).

There is no doubt that there are more harmful algal bloom (HAB) events, occurring more often, in more places and lasting longer than in decades past. It is now also well established that many of these events are the result of human activities, primarily through increased nutrient inputs—from land applications, direct sea applications (especially from aquaculture) and atmospheric deposition,

P.M. Glibert (✉)
University of Maryland Center for Environmental Science, Horn Point Laboratory, Cambridge, MD, USA
e-mail: glibert@umces.edu

E. Berdalet
Institute of Marine Sciences (CSIC), Barcelona, Catalonia, Spain

M.A. Burford
Australian Rivers Institute and School of Environment, Griffith University, Nathan, Australia

G.C. Pitcher
Fisheries Research and Development, Cape Town, South Africa

M. Zhou
Institute of Oceanology, Chinese Academy of Sciences, Qingdao, China

P.M. Glibert et al. (eds.), *Global Ecology and Oceanography of Harmful Algal Blooms*, Ecological Studies 232, https://doi.org/10.1007/978-3-319-70069-4_1

Fig. 1.1 The global reach of editors (*stars*) and contributors to this volume (*circles*), reflective of the broad network of GEOHAB

all of which have been increasing and have been altering the proportions and forms of nutrient loads and from alterations to coastal habitats. Species introductions are also occurring in ways and at rates not common in decades past, through ballast water and through transfer of materials in the aquaculture industry. As a result of these human activities, blooms are having more ecological, economic and human health impacts. As stated in the GEOHAB Science Plan (2001), "these occurrences of toxic or harmful microalgae represent a significant and seemingly expanding threat to human health, fishery resources, and marine ecosystems throughout the world".

This volume aims to capture the key focus areas of research under the GEOHAB umbrella. The central challenge for GEOHAB was to "understand the critical features and mechanisms underlying the population dynamics of HAB species in a variety of oceanographic regimes". GEOHAB fostered research that was multifaceted, multidisciplinary, international and within an oceanographic context.

Accordingly, many important HABs that occur in freshwaters were not studied in this programme. Additionally, toxins and human health effects were not a direct focus of GEOHAB research.

This book begins with chapters that introduce HABs and their impacts (Glibert et al. 2018b) and the GEOHAB Programme, its history and activities (Kudela et al. 2018). Then, two chapters highlight the important drivers of HABs in a global context, emphasizing changing landscapes and seascapes in relation to HABs (Glibert et al. 2018c) and climate change and its impacts on HABs (Wells and Karlson 2018). Understanding adaptive strategies of HABs was an important goal of the programme, and thus the next chapters focus on this topic, including physiological responses to nutrients across the spectrum of nutrient availability (Glibert et al. 2018d), mixotrophy and HABs (Flynn et al. 2018) and life cycles and HABs (Azanza et al. 2018).

Understanding HABs in different habitats and biomes was another important focus area of GEOHAB and represented the core research projects of the programme. Thus, chapters follow on HABs in stratified systems (Raine et al. 2018), fjords and coastal embayments (Roy et al. 2018), eastern boundary upwelling systems (Pitcher et al. 2018), eutrophic systems (Glibert et al. 2018a) and benthic systems (Berdalet and Tester 2018).

A particularly important regional focus of the GEOHAB Programme was Asia, motivated by the rapidly escalating extent and diversity of the HAB problem in this part of the world. This regional focus resulted in a subprogramme, GEOHAB Asia, and its successes and future steps in HAB research are described in the overview chapter by Furuya et al. (2018). Specific examples highlighting the expansion of HAB problems in China follow in the chapters by R. Yu et al. (2018) and by Liu and Zhou (2018). Looking beyond HABs along the Chinese coast, and in keeping up with the goal of comparative studies, the chapter by Goes et al. (2018) compares the expansion in HABs in the Arabian Sea and the Gulf of Thailand.

Another important objective of GEOHAB was to improve the detection and prediction of HABs by fostering the development of capabilities in these areas. Chapters thus include an overview of rapidly advancing capabilities for observation and prediction of HABs (Glibert et al. 2018e) and modelling approaches and advances on different scales (Franks 2018).

With the many advances stemming from GEOHAB-related research, there are still many important questions and HAB issues to be resolved. The final section of the book includes chapters related to the forward-looking issues that may be tackled by the next generation of GEOHAB-like research. The emerging HAB issues in freshwater are reviewed by Burford et al. (2018), and prevention and mitigation of HABs are reviewed by Z. Yu et al. (2018), focusing on applications used with increasing frequency in China and Korea. Lastly, the final chapter by Berdalet et al. (2018) introduces the next international effort on HABs, GlobalHAB, and provides an overview of its evolution from GEOHAB.

We thank all of our authors for their contributions. We thank all those who have participated in GEOHAB activities over the years, from attendance at the Open Science Meetings, contributions to GEOHAB-sponsored special issues, training workshops and other unsung activities that have helped to communicate the importance of this water quality issue and helped to advance our scientific knowledge of its ecology and oceanography. We thank Andrea Schlitzberger from Springer for accepting this volume into Springer's catalogue within the Ecological Studies Series. This effort has certainly taken the global village to accomplish. It is hoped that the global community can continue to advance our understanding of all facets of HAB events and to make effective progress in managing and mitigating the impacts of HABs on ecosystems and human health.

References

Azanza RV, Brosnahan ML, Anderson DM et al (2018) The role of life cycle characteristics in harmful algal bloom dynamics. In: Glibert PM, Berdalet E, Burford M et al (eds) Global ecology and oceanography of harmful algal blooms. Springer, Cham, pp 133–161

Berdalet E, Kudela R, Banas NS et al (2018) GlobalHAB: fostering international coordination on harmful algal bloom research in aquatic systems. In: Glibert PM, Berdalet E, Burford M et al (eds) Global ecology and oceanography of harmful algal blooms. Springer, Cham, pp 425–447

Berdalet E, Tester PA (2018) Key questions and recent research advances on harmful algal blooms in benthic systems. In: Glibert PM, Berdalet E, Burford M et al (eds) Global ecology and oceanography of harmful algal blooms. Springer, Cham, pp 261–286

Burford MA, Hamilton DP, Wood SA (2018) Emerging HAB research issues in freshwater environments. In: Glibert PM, Berdalet E, Burford M et al (eds) Global ecology and oceanography of harmful algal blooms. Springer, Cham pp 381–402

Flynn KJ, Mitra A, Glibert PM et al (2018) Mixotrophy in HABs: by whom, on whom, when, why, and what next. In: Glibert PM, Berdalet E, Burford M et al (eds) Global ecology and oceanography of harmful algal blooms. Springer, Cham pp 113–132

Franks PJS (2018) Recent advances in modelling of harmful algal blooms. In: Glibert PM, Berdalet E, Burford M et al (eds) Global ecology and oceanography of harmful algal blooms. Springer, Cham, pp 359–377

Furuya K, Iwataki M, Lim PT et al (2018) Overview of harmful algal blooms in Asia. In: Glibert PM, Berdalet E, Burford M et al (eds) Global ecology and oceanography of harmful algal blooms. Springer, Cham, pp 289–308

GEOHAB (2001) Global ecology and oceanography of harmful algal blooms, science plan. Glibert P, Pitcher G (eds) SCOR and IOC, Baltimore and Paris, 86 pp

Glibert PM, Al-Azri A, Allen JI et al (2018a) Key questions and recent research advances on harmful algal blooms in relation to nutrients and eutrophication. In: Glibert PM, Berdalet E, Burford M et al (eds) Global ecology and oceanography of harmful algal blooms. Springer, Cham, pp 229–259

Glibert PM, Berdalet E, Burford M et al (2018b) Harmful algal blooms and the importance of understanding their ecology and oceanography. In: Glibert PM, Berdalet E, Burford M et al (eds) Global ecology and oceanography of harmful algal blooms. Springer, Cham, pp 9–25

Glibert PM, Beusen AHW, Harrison JA et al (2018c) Changing land, sea- and airscapes: sources of nutrient pollution affecting habitat suitability for harmful algae. In: Glibert PM, Berdalet E, Burford M et al (eds) Global ecology and oceanography of harmful algal blooms. Springer, Cham, pp 53–76

Glibert PM, Heil CA, Wilkerson F et al (2018d) Nutrients and HABs: dynamic kinetics and flexible nutrition. In: Glibert PM, Berdalet E, Burford M et al (eds) Global ecology and oceanography of harmful algal blooms. Springer, Cham, pp 93–112

Glibert PM, Pitcher GC, Bernard S et al (2018e) Advancements and continuing challenges of emerging technologies and tools for detecting harmful algal blooms, their antecedent conditions and toxins, and applications in predictive models. In: Glibert PM, Berdalet E, Burford M et al (eds) Global ecology and oceanography of harmful algal blooms. Springer, Cham, pp 339–357

Goes JI, Gomes HR, Al-Hashimi K et al (2018) Ecological drivers of green *Noctiluca* blooms in two monsoonally-driven ecosystems. In: Glibert PM, Berdalet E, Burford M et al (eds) Global ecology and oceanography of harmful algal blooms. Springer, Cham, pp 327–336

Kudela RM, Raine R, Pitcher G et al (2018) Establishment, goals, and legacy of the Global Ecology and Oceanography of Harmful Algal Blooms (GEOHAB) Program. In: Glibert PM, Berdalet E, Burford M et al (eds) Global ecology and oceanography of harmful algal blooms. Springer, Cham, pp 27–49

Liu D, Zhou M (2018) Green tides of the Yellow Sea: massive free-floating blooms of *Ulva prolifera*. In: Glibert PM, Berdalet E, Burford M et al (eds) Global ecology and oceanography of harmful algal blooms. Springer, Cham, pp 317–326

Pitcher GC, Figueiras FG, Kudela RM et al (2018) Key questions and recent research advances on harmful algal blooms in eastern boundary upwelling systems. In: Glibert PM, Berdalet E, Burford M et al (eds) Global ecology and oceanography of harmful algal blooms. Springer, Cham, pp 205–227

Raine R, Berdalet E, Yamazaki H et al (2018) Key questions and recent research advances on harmful algal blooms in stratified systems. In: Glibert PM, Berdalet E, Burford M et al (eds) Global ecology and oceanography of harmful algal blooms. Springer, Cham, pp 165–186

Roy S, Montresor M, Cembella A (2018) Key questions and recent research advances on harmful algal blooms in fjords and coastal embayments. In: Glibert PM, Berdalet E, Burford M et al (eds) Global ecology and oceanography of harmful algal blooms. Springer, Cham, pp 187–203

Wells ML, Karlson B (2018) Harmful algal blooms in a changing ocean. In: Glibert PM, Berdalet E, Burford M et al (eds) Global ecology and oceanography of harmful algal blooms. Springer, Cham, pp 77–90

Yu R-C, Lü S-H, Liang Y-B (2018) Harmful algal blooms in the coastal waters of China. In: Glibert PM, Berdalet E, Burford M et al (eds) Global ecology and oceanography of harmful algal blooms. Springer, Cham, pp 309–316

Yu Z, Song X, Cao X et al (2018) Mitigation and control of HABs. In: Glibert PM, Berdalet E, Burford M et al (eds) Global ecology and oceanography of harmful algal blooms. Springer, Cham, pp 403–423

Chapter 2
Harmful Algal Blooms and the Importance of Understanding Their Ecology and Oceanography

Patricia M. Glibert, Elisa Berdalet, Michele A. Burford, Grant C. Pitcher, and Mingjiang Zhou

2.1 Introduction

Over the decades of the 1980s and 1990s, as the expansion in harmful algal blooms (HABs) was gaining recognition (e.g., Anderson 1989; Hallegraeff 1993; GEOHAB 1998; Smayda 2002), no longer was it sufficient to study bloom events in isolation; many countries were facing a bewildering array of impacts caused by species not previously known or recognized in those regions. The complexity of the HAB problem, its causative factors, and the impacts HABs have on the environment were becoming well characterized. The benefits of collaborative, cooperative, and comparative studies on HABs were recognized to advance the understanding of this phenomenon and to provide scientific guidance to managers. The aim of this chapter is to introduce several aspects of this complex phenomenon, and why an understanding of ecology and oceanography of HAB species and their associated events is so important. This chapter also briefly introduces the effects of global changes in nutrients and climate that are developed more fully in subsequent chapters [see Chap. 4, Glibert et al. (2018b), and Chap. 5, Wells and Karlson

P.M. Glibert (✉)
University of Maryland Center for Environmental Science, Horn Point Laboratory, Cambridge, MD, USA
e-mail: glibert@umces.edu

E. Berdalet
Institute of Marine Sciences (CSIC), Barcelona, Catalonia, Spain

M.A. Burford
Australian Rivers Institute and School of Environment, Griffith University, Nathan, Australia

G.C. Pitcher
Fisheries Research and Development, Cape Town, South Africa

M. Zhou
Institute of Oceanology, Chinese Academy of Sciences, Qingdao, China

P.M. Glibert et al. (eds.), *Global Ecology and Oceanography of Harmful Algal Blooms*, Ecological Studies 232, https://doi.org/10.1007/978-3-319-70069-4_2

(2018)], as well as a number of concepts relating to the adaptive strategies of HABs which help to explain why they are so successful in environments subject to many anthropogenic changes.

2.2 What Are Harmful Algal Blooms?

The majority of algae in marine and freshwaters are not only beneficial but also necessary to the functioning of aquatic ecosystems; they form the base of the food web, and it is this microscopic life on which all aquatic life ultimately depends for food (Glibert et al. 2005). Algae produce oxygen and also play an important role in regulating atmospheric CO_2 by sequestering it during production and transporting it to deeper waters. Yet, a comparatively small subset of the total known microscopic algal species can cause problems. This can occur when they accumulate in sufficient numbers, when they produce toxins, or when they directly or indirectly interfere with other organisms or alter the physical habitat indirectly, negatively impacting the growth of others. These are known as the harmful algae, and their associated proliferation events are referred to as HABs. HABs may be caused by the explosive growth of a single species that rapidly dominates the water column but may also be the result of highly toxic cells that do not accumulate in high numbers. Therefore, in some cases, toxic conditions can also occur when the water is clear with very low cell concentrations. The effects of HABs, detailed in the next section, are as varied as the organisms themselves.

The causative organisms, the harmful algae, were formerly called "red tides" because many were composed of dinoflagellates that in high densities coloured the water red, but blooms may also be green, yellow, or brown, depending on the type of algae present and their pigmentation (Fig. 2.1; Glibert et al. 2005). Many marine HABs are, indeed, dinoflagellates, but other classes of algae, including cyanobacteria, raphidophytes, and diatoms, have members that may form HABs under some conditions. Some HABs organisms are not algae at all, that is, they do not depend on photosynthesis for their nutrition; they obtain their nutrition exclusively through grazing and particle ingestion, while many of them have the ability to use varied sources of nutrition, mixing photosynthesis and inorganic nutrient uptake together with grazing, depending on the environmental conditions [see Chap. 7, Flynn et al. (2018)]. Other harmful algae that also are not technically "algae" are the cyanobacterial HABs, CyanoHABs, some of which have the ability to "fix" their own nitrogen (N) from the atmosphere. The term "HAB" also applies to blooms of some nontoxic micro- or macroalgae (seaweeds), which can grow out of control and cause major ecological impacts such as the displacement of indigenous species, habitat alteration, or oxygen depletion. All of these harmful causative organisms are included under the HAB species umbrella label, and thus the term "HAB" is operational and not technical, but this distinction is irrelevant in terms of human health and ecological and economic impacts. Moreover, with few exceptions, there are no formal definitions of the absolute abundance of a HAB species that determine whether it is a "bloom."

Fig. 2.1 Many HABs are produced by the accumulation of vividly coloured cells in surface water. The blooms shown here are from the Philippines (panel **a**, photo: http://taqplayer.info/philippine-red-tide); Florida, USA (panel **b**, photo: http://www.politicnote.com/army-corps-of-engineers-battles-guacamole-thick-florida-algae/); Qingdao, China (panel **c**, photo: www.sailjuice.com); Long Island, NY, USA (panel **d**, photo by C. Gobler); Sydney Harbour, Australia (panel **e**, photo: https://bioweb.uwlax.edu/bio203/f2013/bradford_andr/habitat.htm); and a freshwater lagoon in Uganda (panel **f**, photo by P. Glibert). These high biomass blooms can cause hypoxia, can contribute to toxicity of fish and shellfish, and can cause other environmental problems

2.3 How Are HABs Harmful?

HABs can be harmful in several fundamental ways. Many HABs produce toxins, some of which are among the most potent toxins known. These toxins have various vectors by which harmful effects are transferred through the food chain or to human consumers, and there is a wide range of potency (Table 2.1). The range of toxins produced by the common marine HABs includes brevetoxins, the cause of neurotoxic shellfish poisoning (NSP); saxitoxins, the cause of paralytic shellfish poisoning (PSP); okadaic acid (OA), the cause of diarrhetic shellfish poisoning (DSP); domoic acid (DA), the cause of amnesic shellfish poisoning (ASP); azaspiracid, the cause of azaspiracid shellfish poisoning (AZP); and ciguatoxins (CTX), the cause of ciguatera fish poisoning (Landsberg 2002; Glibert et al. 2005). There are no known antidotes for poisonings caused by HAB toxins. Although human poisoning events are few in the developed world due to careful monitoring of seafood products, it has been estimated that direct human poisonings in many parts of Asia from toxic algae number in the thousands annually (Yan and Zhou 2004). In Florida, USA, when *Karenia brevis* blooms occur, increases in hospitalizations have been reported due to gastrointestinal and respiratory irritation that especially affects those that are immune-compromised (Kirkpatrick et al. 2006). CyanoHABs also have a range of human effects, depending on the toxin involved. Some effects are as mild as skin rashes or irritations, while other effects include cancer, particularly liver cancer, caused by long-term exposure to toxins such as microcystins in drinking water (c.f., Backer and McGillicuddy 2006; Bláha et al. 2009, and references therein).

Toxins may kill shellfish or fish directly or may have little effect on them but may cause illness or death of people or other consumers in the food web when fish or shellfish that have accumulated the algal toxins are eaten (Landsberg 2002). Not all fish exposed to algal toxins die; some experience extreme illness. As an example, acute effects of brevetoxin (from *K. brevis*) on fish have been reported to include altered swimming behaviour (swimming in a spiral, twisting behaviour,

Table 2.1 Major syndromes, vectors, and human health symptoms associated with common HABs

Illness	Major vector	Symptoms
Amnesic shellfish poisoning (ASP)	Domoic acid from *Pseudo-nitzschia* sp. in shellfish	Short-term memory loss, vomiting, cramps
Diarrhetic shellfish poisoning (DSP)	Okadaic acid from *Dinophysis* sp. in shellfish	Diarrhoea, vomiting, cramping
Neurotoxic shellfish poisoning (NSP)	Brevetoxin from *Karenia* sp. in shellfish, aerosolized toxins	Nausea, diarrhoea, respiratory distress, eye irritation
Paralytic shellfish poisoning (PSP)	Saxitoxin from *Alexandrium* sp. and other species in shellfish	Numbness around the lips and mouth, respiratory paralysis, death
Cyanotoxin poisoning	Microcystins and other toxins from cyanobacteria in water	Skin irritation, respiratory irritation, tumour promotion, liver cancer, liver failure

loss of equilibrium), defecation and regurgitation, paralysis of the pectoral fin, curvature of the caudal fin, and convulsions (Landsberg 2002). Deaths of birds and mammals, such as dolphins, manatees, sea lions, and whales, have also been related to HABs and their toxins and are of immense public and economic concern when they do occur. One of the most well-documented effects of toxin transfer through the food web is that of California sea lions that have been shown to become intoxicated with DA resulting from *Pseudo-nitzschia* blooms, which accumulate in the northern anchovy, a common prey item for the sea lions. Seizures, complications of pregnancy, and death are among the commonly observed effects in these animals (Silvagni et al. 2005). A recent massive whale mortality event in Chile in 2015, including at least 340 primarily sei whales, has been linked to PSP toxicity from feeding near shore (Häussermann et al. 2017).

While the most familiar impact of HABs is their intoxication of shellfish, leading to contaminated seafood, some HAB toxins have also been shown to have adverse effects on very early life stages of consumers, leading to increased mortalities at the larval stage. For example, experiments conducted on embryos and larvae of freshly spawned oysters, *Crassostrea virginica* and *C. ariakensis*, exposed to the toxic dinoflagellate *Karlodinium veneficum* showed increased mortality relative to similar embryos and larvae exposed to a control diet (Glibert et al. 2007; Stoecker et al. 2008). These experiments also demonstrated important behavioural changes in pediveliger larvae when exposed to *K. veneficum* in their diet, leading them to stop swimming and to sink. Even if such effects did not immediately result in mortality, any change in behaviour may influence larval dispersal, reduce feeding and growth, and likely increase the susceptibility of larvae to predation. Related results have been documented for the HAB dinoflagellate *Prorocentrum minimum*. Larvae had poorer survival and lower settling success with only *P. minimum* in the diet, but survival improved when the proportion of *P. minimum* in their diet was reduced (Luckenbach et al. 1993; Wikfors and Smolowitz 1995). In Australia, the related HAB species, *P. rhathymum*, has been associated with mortalities of spat of the Japanese or Pacific oyster, *C. gigas* (Pearce et al. 2005).

The other fundamental way in which HABs are harmful is through high biomass accumulation that may lead to environmental damage, including hypoxia, anoxia, and shading of submerged vegetation, each of which, in turn, can lead to a multitude of negative environmental consequences. These are the blooms that cause visible water discoloration (Fig. 2.1). Such high biomass blooms can cause hypoxia and anoxia, and the global occurrences of such "dead zones" are expanding (Diaz and Rosenberg 2008).

Many types of harmful effects occur from HABs that are not directly toxic. Some HABs have physical structures, such as spines, that can lodge in fish gills and can cause irritation and eventual suffocation. Some HABs have more subtle effects on the ecosystem or effects that are more difficult to quantify in the wild. Blooms that are known to disrupt ecosystem function but which may or may not necessarily be toxic are considered ecosystem disruptive algal blooms (EDABs; Sunda et al. 2006). Blooms of picoplankton, such as the pelagophyte *Aureococcus anophagefferens* (brown tide) and the CyanoHAB *Synechococcus* sp., which can be sustained for long periods of time once established, can have multiple negative ecosystem effects that, in turn, may help to sustain these blooms. The positive feedbacks of reduced grazing and/or bottom shading contribute to the availability of nutrients for these blooms (Sunda et al. 2006). In Laguna Madre, Texas, which experienced a nearly decade-long bloom of the EDAB species *Aureoumbra lagunensis*, the density of protozoan grazers was found to be greatly reduced during blooms, and it was suggested that a thick polysaccharide layer around the cells may make it difficult for the protozoa to feed (Buskey and Stockwell 1993; Buskey et al. 2001). Allelopathic chemicals may also play an important role in maintaining EDAB species (Sunda et al. 2006; Granéli et al. 2008).

In addition to ecological and human health impacts of HABs, the economic impacts can also be very large. The overall economic impact of HABs is difficult to determine, but the costs range from the direct expenses of public health and medical care for those affected to losses in commercial and recreational fishing and in tourism-related activities from water quality deterioration. Single HAB fish kill events in Korea and shellfish kill events in China have been estimated to have cost from $1 to $100 million and $300 million, respectively, in lost fish or shellfish in aquaculture settings, while in Japan such events have been estimated to have resulted in losses of fish worth more than $300 million (GEOHAB 2010; Trainer and Yoshida 2014). Loss of revenue from shellfish harvesting area closures also takes enormous economic toll on communities affected. For example, in 2003, it was estimated that more than $6 million in revenue from sport fishing and tourism due to fish kills were lost in Texas due to a bloom of a small flagellate (*Prymnesium parvum*) in inland rivers and reservoirs. Yet another economic cost in some regions is the effect of HABs on seawater reverse osmosis (SWRO) desalination plants (e.g., Villacorte et al. 2015). In this case, the main concern is biofouling of the RO membranes, which can cause plants to shut down, resulting in a loss of an important utility service in regions where this is a major source of freshwater. In addition, where recurrent HABs are documented, another economic cost of HABs is that of sustained environmental monitoring to protect human health or industries such as aquaculture. Monitoring is one of the most effective ways to prevent human exposure to HAB-related toxins.

2.4 Where Do HABs Occur?

HABs are found in all parts of the world, in all types of waters. While marine HABs are emphasized herein, the trends in freshwater HABs are equally concerning. Several examples demonstrate the global expansion of HABs. In the Gulf of Maine, occurrences of the PSP-causative dinoflagellate *Alexandrium fundyense* and PSP toxicity were rare prior to about 1970 but have been frequent in the past three decades. In fact, PSP occurrences have increased globally in this same time frame (e.g., Anderson 1989; Hallegraeff 1993). PSP toxicity has increased in proportion to the growth of the human population in Puget Sound, Washington

State (Trainer et al. 2003). The bloom-forming dinoflagellate *P. minimum* is also now documented to cover most coasts throughout the world and has been documented to have expanded in concert with escalating eutrophication (Heil et al. 2005; Glibert et al. 2008, 2012). Based on analyses of frustules preserved in cores, blooms of the diatom *Pseudo-nitzschia* spp. in the Gulf of Mexico were also rare prior to the 1950s, but have increased significantly in abundance and frequency since then, concomitant with increases in nutrient loading (Parsons et al. 2002). In Europe and Asia, there are also more blooms with more toxic effects than in previous decades (Granéli et al. 1999). *Cochlodinium* (*Margalefidinium*) *polykrikoides* is among the many species of dinoflagellates which are increasingly adversely affecting fisheries, tourism, and economies around the world (e.g., Yuki and Yoshimatsu 1989; Guzman et al. 1990; Kim et al. 1999, 2004; Lee 2008; Tomas and Smayda 2008; Mulholland et al. 2009; Howard et al. 2012; Al-Azri et al. 2014). While previously most commonly observed in tropical systems (Steidinger and Tangen 1997), *C. polykrikoides* is now increasingly observed in temperate systems (e.g., Kudela et al. 2008; Mulholland et al. 2009). As such, *C. polykrikoides* appears to be similar to many other harmful dinoflagellates undergoing global expansion (e.g., Heil et al. 2005; Glibert et al. 2005, 2008; Harrison et al. 2011).

One of the areas where HAB expansion has been particularly pronounced has been in the coastal waters of Asia, as shown, for example, in increasing observations of "red tides" in Chinese coastal waters [GEOHAB 2010; see also Chap. 14, Furuya et al. (2018), and Chap. 15, Yu et al. (2018)]. Microalgae are not the only HABs increasing. In 2008, a bloom of the macroalgal species *Ulva prolifera* (*Enteromorpha prolifera* or sea lettuce) occurred at the venue of the Olympic Games sailing competition, almost blanketing the water with filamentous scum [Hu et al. 2010; Huo et al. 2013; see also Chap. 16, Liu and Zhou (2018)]. Blooms of this magnitude in this region had not previously been observed. One of the features of this species and its blooms is that it tends to float, making detection from remote sensing (satellite imaging) feasible. It is from such approaches that the scale of these blooms and their change over time can be estimated (Hu et al. 2010). A 10-year record of images of the region shows that prior to 2007, the area covered by these green tides was $<21\ km^2$. In 2008 the scale of the bloom was $>1900\ km^2$, and in 2009 it was $1600\ km^2$ (Hu et al. 2010).

Similarly, in the past several decades, a massive expansion of dinoflagellate blooms has occurred in the Arabian Sea and Sea of Oman (formerly Gulf of Oman) (Parab et al. 2006; Al-Azri et al. 2007; Gomes et al. 2008; Harrison et al. 2011). These blooms were not observed during the Joint Global Ocean Flux Study (JGOFS) cruises of the 1990s (Gomes et al. 2008) but have expanded considerably, consistent with climate changes and atmospheric warming (Goes et al. 2005) and nutrient loading (Harrison et al. 2011; Goes and Gomes 2016). These recent blooms have been dominated by the heterotrophic dinoflagellate *Noctiluca scintillans* [Parab et al. 2006; Al-Azri et al. 2007; Gomes et al. 2008; Harrison et al. 2011; see also Chap. 17, Goes et al. (2018)]. These examples represent just a sampling of the bloom events that are occurring more often and in more places.

2.5 Why Are HABs Expanding?

Although some of the factors contributing to regional and global expansion are natural, such as biological species dispersal, many others are considered to be a result of human activities. Increases in nutrient loading, changes in agriculture and aquaculture practices, overfishing, ballast water discharge, and global climate change may all be important in the global increase in HABs.

By far, the greatest change in the past several decades has been the rate and composition of nutrient loading [see also Chaps. 4, 12, Glibert et al. (2018a, b)]. Population growth and development (with associated large sewage discharges) and the production of food, both crop (with the associated expanding synthetic fertilizer use) and animal production systems (with their associated waste), have resulted in increased runoff from land to both fresh and marine waters. The production and consumption of energy also result in increased atmospheric inputs from NO_x emissions, which can then lead to increased N deposition. These nutrient sources have led to widespread coastal eutrophication throughout Europe, the USA, and Asia.

Another significant alteration in nutrient loading to the coastal zone in some regions comes from the increase in aquaculture activities [see also Chap. 4, Glibert et al. (2018b)]. These industries have altered ecosystems through input of feed and faeces, only a small percentage of which is incorporated in food biomass (Bouwman et al. 2011, 2013a, b). China's expansion of aquaculture has been especially great, and the release of nutrients from aquaculture in some provinces of China exceeds 20% of that from riverine export (Bouwman et al. 2013a). Global models of the impacts of finfish, shellfish, and aquatic plant aquaculture (Bouwman et al. 2011, 2013b) have estimated the alteration of nutrient cycling and particularly in the case of finfish, where external feeding is required, have demonstrated the magnitude of increased nutrient loads. Even mollusc production, which does not depend on exogenous nutrients to be supported, when intensive, can become point sources of nutrients as particulate nutrients are assimilated and excreted in faeces and pseudofaeces.

Fish and shellfish harvesting has also altered ecosystems, leading to changes in food chains/food webs. The exploitation of natural fish stocks has, in some cases, led to a decrease in the control of HAB species by removal of the primary grazers through trophic cascade effects. As an example, altered food webs and trophic interactions, through overfishing and the invasion of the predatory ctenophore *Mnemiopsis leidyi*, have been suggested as possible factors leading to increased HABs in the Black Sea (Lancelot et al. 2002).

Climate controls many of the fundamental parameters regulating algal growth, including water temperature, nutrients, and light, and thus can be expected to result in changes in the species composition, trophic structure, and function of marine ecosystems [Wells et al. 2015; see also Chap. 5, Wells and Karlson (2018)]. Warmer temperatures may contribute to increased growth rates and to range expansion of particular species. The available data, along with modelling approaches, suggest shifts in HAB species and the timing of their outbreaks related to increases in mean water temperature (e.g., Tester et al. 2010; Moore et al. 2008, 2015; Glibert et al. 2014). Temperatures also affect the community of organisms within which the HAB may live, including bacteria, viruses, competing phytoplankton taxa, and grazers (Wells et al. 2015 and references therein). Toxicity of many HABs also increases with warming, but this is not the case in all HABs (Davis et al. 2009; Fu et al. 2012, and references therein). The combination of temperature, together with elevated pCO_2 and altered nutrient ratios, appears to be especially potent in terms of toxicity of some HABs. For some species, higher toxicity associated with warming may be associated with slower growth rates (e.g., Ogata et al. 1989; Lewis et al. 1993), but high growth rates are not necessary for HABs to thrive [see Chap. 7, Flynn et al. (2018)].

Higher temperatures are promotive of increased water column stability and increased thermal stratification. This can favour cyanobacterial species that are buoyant because they can float to the surface (e.g., Walsby 1975; Visser et al. 2016, and references therein). Moreover, at higher temperatures water can become less viscous, buoyant cyanobacteria can float faster, and conversely sinking diatoms will sink faster (e.g., O'Neil et al. 2012 and references therein).

Long-term changes in, or intensification of, climate forces such as monsoons or interannual oscillations, such as those related to the El Niño-Southern Oscillation (ENSO), or longer-term cycles, such as North Atlantic Oscillation (NAO) and the Pacific Decadal Oscillation (PDO), can also alter conditions for HABs. In the northern Iberian Peninsula, the abundance of the harmful dinoflagellate *Gymnodinium catenatum* was high during the mid-1980s when there was a transition from downwelling-favourable conditions to upwelling-favourable conditions following a shift in the North Atlantic Oscillation (NAO) index (Alvarez-Salgado et al. 2003). In late 2013, and continuing through much of 2016, anomalously warm water developed in the northeastern Pacific Ocean (e.g., Bond et al. 2015; Freeland and Whitney 2015), a feature associated with the unusually strong El Niño event and the Pacific Decadal Oscillation (PDO), and an exceptionally large bloom of *Pseudo-nitzschia* developed along the USA west coast, lasting months and causing extensive ecological and economic harm (McCabe et al. 2016).

2.6 Why the Need for Advancing Knowledge of HAB Ecology and Oceanography?

Factors leading to the establishment of a bloom usually include environmental, chemical, as well as physical dynamics and may also include changes in trophodynamic interactions. In addition, various physiological adaptive strategies may lead to disproportionate success of a particular species. For example, blooms may result from excystment of resting cells during a restricted set of suitable conditions, transport of cells from a source region where blooms are already established, and enhanced growth due to unusual climatic or hydrographic conditions.

Once a bloom is initiated, physical processes controlling bloom transport are of paramount importance. Coastal currents driven by wind, buoyancy, or other factors can transport blooms hundreds or even thousands of kilometres along the coast, often from one management area to another. Understanding the physical dynamics underlying these transport pathways is essential to effective management and mitigation of HAB effects. A population's range and biomass are affected by physical controls such as long-distance transport, accumulation of biomass in response to water flows and swimming behaviour, and the maintenance of suitable environmental conditions (including temperature and salinity, stratification, irradiance, and nutrient supply). Thus, physical forcing, nutrient supply, and the behaviour of organisms all interact to determine the timing, location, and ultimate biomass achieved by a bloom, as well as its impacts.

Physical processes that are likely to influence the population dynamics of HAB species are operative over a broad range of spatial and temporal scales. The retentive nature of some semi-enclosed coastal systems, such as estuaries and fjords, can produce long residence times leading to prolonged suitable periods for cells to thrive [Cembella et al. 2005; see also Chap. 4, Glibert et al. (2018b)]. A linkage has been demonstrated, for example, between tidally generated fronts and the sites of massive blooms of the toxic dinoflagellate *Gyrodinium aureolum* (*Karenia mikimotoi*) in the North Sea (Holligan 1979). The typical pattern is that of a high surface concentration of cells at the frontal convergence, contiguous with a subsurface chlorophyll maximum which follows the sloping interface between the two water masses beneath the stratified side of the front. The signature of the chlorophyll maximum, sometimes visible as a "red tide," may be 1–30 km wide. Chlorophyll concentrations are generally lower and much more uniform on the well-mixed side of the front. The timing and duration of upwelling is another physical feature that plays an important role in many blooms, as in the case of the Benguela upwelling system, where HABs accumulate subsurface as stratification increases during the upwelling season and move onshore as upwelling relaxes (Kudela et al. 2005).

The importance of small-scale physical processes in HAB development is observed in the layering of the physical, chemical, and biological environment in stratified coastal systems. Off the French coast, for example, a thin layer of dinoflagellates, including the HAB species *Dinophysis* cf. *acuminata*, has been observed in the region of the thermocline (Gentien et al. 2005). The same pattern is found for *Dinophysis norvegica* in the Baltic Sea, where a 1–2-m-thick layer with up to 80,000 cells L^{-1} is usually situated between 20 and 25 m depth, where light is $<1\%$ (Gisselson et al. 2002).

Studies of species responses at the ecosystem level are essential if we are to understand the population dynamics of HABs. The specific growth rate of a species is determined by many metabolic processes, including photosynthesis and nutrient uptake and assimilation, all of which are under both environmental and genetic control. The net population growth of a species is controlled by external environmental factors, including nutrients, physical transport, grazing, and other community interactions.

HAB species do not typically appear to have higher intrinsic growth rates than many other phytoplanktons. Some HABs may actually have lower growth rates than non-HAB species, such as diatoms, but succeed when loss processes are also reduced. In fact, some HABs are actually slower growing than their non-HAB counterparts. They succeed due to the negative impact they can have on competitors or their grazers. Interactions between HAB species and other algae may be an important survival strategy for some HABs. For example, it has long been argued that production of allelopathic exudates allows some harmful species to outcompete co-occurring phytoplankton (e.g., Smayda 1998; Granéli and Johansson 2003). Some species may even use cyst formation as a survival strategy from strong allelochemicals produced by another HAB species (Fistarol et al. 2004). It may well be that some of the HAB species that form thin, subsurface, or surface layers of cells at extraordinary densities do so because this allows them to change the ambient water chemistry and light penetration in a manner that deters grazing, inhibits co-occurring algal species, or facilitates sexuality and gamete encounter rate.

Reductions in grazer abundance can also play a key role in bloom development. This might result from physical factors or behavioural strategies, which lead to spatial separation of harmful algal species and grazers. Local reductions in grazer abundances may be in direct response to HABs (e.g., avoidance or mortality induced by the HABs, Granéli and Johansson 2003) or in response to the effects of past HAB events on grazer populations. The response of zooplankton and other grazers to toxic algae is often species-specific in terms of behavioural responses and toxin susceptibility. Moreover, prey quality can vary, characterized in terms of lipid and protein content and also in terms of elemental stoichiometry. The net effect of the stoichiometric regulation of consumers together with different nutrient requirements of phytoplankton is that not only are grazers affected by food quality, but they, in turn, affect food quality by altering the composition of nutrients available to them. This process of stoichiometric interaction is seen at all levels of the food web, leading to a complex interaction of stoichiometry and trophic cascades and potential for shifts in dominant bloom species as not only is the rate of assimilation efficiency altered, but so too is the rate of regeneration, in turn disproportionately promoting those species that can use the altered nutrient forms (Mitra and Flynn 2005). It is also apparent that prey quality has strong effects on the success of mixotrophs as well (e.g., Lundgren et al. 2016; Lin et al. 2017).

A survival and growth strategy that is important to many HAB species is a complex life cycle, or a life cycle involving resting or benthic stages, such as spores or cysts [see also Chap. 8, Azanza et al. (2018)]. These life cycle stages provide a recurrent seed source or inoculum for planktonic populations, and this characteristic may be a critical factor in determining not only the geographic distribution of species but possibly their eventual abundance as well.

Moving higher in the food web, zooplankton impaired by ingesting harmful algae may be more susceptible to predation and thus may become an important vector for transferring toxins in the pelagic food web. Alternatively, zooplankton faecal pellets may be important sources of toxin to benthic communities. Herbivorous fish can accumulate and transfer toxins and even cause mass mortalities of the marine birds that consume them. During food web transfers, toxins may be bioaccumulated, excreted, degraded, or structurally modified. In order to understand the effects of HAB species on the marine food web, both the direct and the indirect, more subtle pathways by which toxins are transferred and transformed and the differential susceptibility of marine organisms to these toxins must be characterized. In some cases HABs may harm, but not kill, other members of the food web, leaving them stressed and therefore more susceptible to other sources of mortality, such as infection by pathogens (Burkholder and Glasgow 1997; Glibert et al. 2002).

The above represent some well-established examples of the effects of various processes on the distribution of harmful algae. Much has been accomplished in understanding HAB ecology and oceanography, but there are however many examples where the physical and/or biological processes common to particular ecosystems are poorly characterized and understood, and therefore their influence on HAB population dynamics and their harmful properties remain uncertain.

2.7 Conclusions and the Role of GEOHAB

In summary, the HAB problem is significant and growing worldwide and poses a major threat to public health, to ecosystem health, as well as to fisheries and economic development. The HAB problem and its impacts are diverse as are the causes and underlying mechanisms controlling the blooms. Winds, tides, currents, fronts, and other environmental features can create discrete patches or streaks of cells at all scales. A full understanding of the many biological, chemical, and physical processes that underlie HABs will continue to be a challenge, given the many different species and hydrographic systems involved.

HABs are a serious and growing problem in the global coastal ocean—one that requires the interplay of all oceanographic disciplines, as well as others such as public health and resource management. Through recognition of the diversity of these interactions, much progress has been made towards understanding the causes and impacts of these events. Nevertheless, there is much work to be done to translate such knowledge in effective management activities that will reduce the severity of such outbreaks or their impacts on ecosystem and human health in future years. The GlobalHAB Programme, the next generation of GEOHAB [see Chap. 22, Berdalet et al. (2018)], will be important in advancing the science and management of HABs on a global scale.

Acknowledgments This chapter is a contribution of the University of Maryland Center for Environmental Science (No. 5403).

References

Al-Azri A, Al-Hashmi K, Goes J et al (2007) Seasonality of the bloom-forming heterotrophic dinoflagellates *Noctiluca scintillans* in the Gulf of Oman in relation to environmental conditions. Int J Oceans Oceanogr 2(1):51–60

Al-Azri A, Piontkovski SA, Al-Hashmi KA et al (2014) Mesoscale and nutrient conditions associated with the massive 2008 *Cochlodinium polykrikoides* bloom in the Sea of Oman/ Arabian Gulf. Estuar Coasts 37:325–338

Alvarez-Salgado XA, Figueiras FG, Perez FF et al (2003) The Portugal coastal counter current of NW Spain: new insights on its biogeochemical variability. Prog Oceanogr 56:281–321

Anderson DM (1989) Toxic algal blooms and red tides: a global perspective. In: Okaichi T, Anderson DM, Nemoto T (eds) Red tides: biology, environmental science and toxicology. Elsevier, New York, pp 11–16

Azanza RV, Brosnahan ML, Anderson DM et al (2018) The role of life cycle characteristics in harmful algal bloom dynamics. In: Glibert PM, Berdalet E, Burford M et al (eds) Global ecology and oceanography of harmful algal blooms. Springer, Cham, pp 131–161

Backer LC, McGillicuddy DJ Jr (2006) Harmful algal blooms at the interface between coastal oceanography and human health. Oceanography 19(2):94–106

Berdalet E, Kudela RM, Banas NS et al (2018) GlobalHAB: fostering international coordination on harmful algal bloom research in aquatic systems. In: Glibert PM, Berdalet E, Burford M et al (eds) Global ecology and oceanography of harmful algal blooms. Springer, Cham, pp 425–447

Bláha L, Babica P, Maršálek B (2009) Toxins produced in cyanobacterial water blooms – toxicity and risks. Interdiscip Toxicol 2(2):36–41. https://doi.org/10.2478/v10102-009-0006-2LC

Bond NA, Cronin MF, Freeland H et al (2015) Causes and impacts of the 2014 warm anomaly in the NE Pacific. Geophys Res Lett 42:3414–3420. https://doi.org/10.1002/2015GL063306

Bouwman AF, Beusen AHW, Glibert PM et al (2013a) Mariculture: significant and expanding cause of coastal nutrient enrichment. Environ Res Lett 8(044026):5. https://doi.org/10.1088/1748-9326/8/4/044026

Bouwman AF, Beusen AHW, Overbeek CC et al (2013b) Hindcasts and future projections of global inland and coastal nitrogen and phosphorus loads due to finfish aquaculture. Rev Fish Sci 21:112–158

Bouwman AF, Pawlowski M, Liu C et al (2011) Global hindcasts and future projections of coastal nitrogen and phosphorus loads due to shellfish and seaweed aquaculture. Rev Fish Sci 19:331–357

Burkholder JM, Glasgow HB Jr (1997) *Pfiesteria piscicida* and other *Pfiesteria*-like dinoflagellates: behavior, impacts, and environmental controls. Limnol Oceanogr 42:1052–1075

Buskey EJ, Liu H, Collumb C et al (2001) The decline and recovery of a persistent Texas brown tide algal bloom in the Laguna Madre (Texas, USA). Estuaries 24:337–346

Buskey EJ, Stockwell DA (1993) Effects of a persistent "brown tide" on zooplankton populations in the Laguna Madre of South Texas, toxic phytoplankton blooms in the sea. In: Smayda TJ, Shimizu Y (eds) Proceedings of the fifth international conference on toxic marine phytoplankton. Elsevier, Amsterdam, pp 659–666

Cembella AD, Ibarra DA, Diogene J et al (2005) Harmful algal blooms and their assessment in fjords and coastal embayments. Oceanography 18(2):158–171

Davis TW, Berry DL, Boyer GL et al (2009) The effects of temperature and nutrients on the growth and dynamics of toxic and non-toxic strains of *Microcystis* during cyanobacteria blooms. Harmful Algae 8:715–725

Diaz RJ, Rosenberg R (2008) Spreading dead zones and consequences for marine ecosystems. Science 321:926–929. https://doi.org/10.1126/science.1156401

Fistarol GO, Legrand C, Selander E et al (2004) Allelopathy in *Alexandrium* spp.: effect on a natural plankton community and on algal monocultures. Aquat Microb Ecol 35:45–56

Flynn KJ, Mitra A, Glibert PM et al (2018) Mixotrophy by HABs: by whom, on whom, when, why and what next. In: Glibert PM, Berdalet E, Burford M et al (eds) Global ecology and oceanography of harmful algal blooms. Springer, Cham, pp 113–132

Freeland H, Whitney F (2015) Unusual warming in the Gulf of Alaska. North Pac Mar Organ (PICES) Press 22:51–52

Fu M, Wang Z, Pu X et al (2012) Changes in nutrient concentrations and N:P:Si ratios and their possible impacts on the Huanghai Sea ecosystem. Acta Oceanol Sinica 31:101–112

Furuya K, Iwataki M, Lim PT et al (2018) Overview of harmful algal blooms in Asia. In: Glibert PM, Berdalet E, Burford M et al (eds) Global ecology and oceanography of harmful algal blooms. Springer, Cham, pp 289–308

Gentien P, Donaghay P, Yamazaki H et al (2005) Harmful algal blooms in stratified systems. Oceanography 18(2):172–183

GEOHAB (1998) Global ecology and oceanography of harmful algal blooms: a plan for co-ordinated scientific research and co-operation to develop international capabilities for assessment, prediction and mitigation. Cullen, J (ed) Asian Natural Environmental Science Center, The University of Tokyo. 43 pp

GEOHAB (2010) GEOHAB Asia, global ecology and oceanography of harmful algal blooms in Asia: a regional comparative programme, Furuya K, Glibert PM, Zhou M et al (eds) IOC and SCOR, Baltimore and Paris, 69 pp

Gisselson L-Å, Carlsson P, Granéli E et al (2002) *Dinophysis* blooms in the deep euphotic zone of the Baltic Sea: do they grow in the dark? Harmful Algae 1:401–418

Glibert PM, Al-Azri A, Allen JI et al (2018a) Key questions and recent research advances on harmful algal blooms in relation to nutrients and eutrophication. In: Glibert PM, Berdalet E, Burford M et al (eds) Global ecology and oceanography of harmful algal blooms. Springer, Cham, pp 229–259

Glibert PM, Alexander J, Meritt DW et al (2007) Harmful algae pose additional challenges for oyster restoration: impacts of the harmful algae *Karlodinium veneficum* and *Prorocentrum minimum* on early life stages of the oysters *Crassostrea virginica* and *Crassostrea ariakensis*. J Shellfish Res 26:919–925

Glibert PM, Beusen AHW, Harrison JA et al (2018b) Changing land-, sea- and airscapes: sources of nutrient pollution affecting habitat suitability for harmful algae. In: Glibert PM, Berdalet E, Burford M et al (eds) Global ecology and oceanography of harmful algal blooms. Springer, Cham, pp 53–76

Glibert PM, Burkholder JM, Kana TM (2012) Recent advances in understanding of relationships between nutrient availability, forms and stoichiometry and the biogeographical distribution, ecophysiology, and food web effects of pelagic and benthic *Prorocentrum* spp. Harmful Algae 14:231–259

Glibert PM, Landsberg J, Evans J et al (2002) A fish kill of massive proportion in Kuwait Bay, Arabian Gulf, 2001: the roles of infectious bacteria, harmful algae, and eutrophication. Harmful Algae 1:1–17

Glibert PM, Maranger R, Sobota DJ et al (2014) The Haber-Bosch–harmful algal bloom (HB-HAB) link. Environ Res Lett 9:105001 (13 p). https://doi.org/10.1088/1748-9326/9/10/105001

Glibert PM, Mayorga E, Seitzinger S (2008) *Prorocentrum minimum* tracks anthropogenic nitrogen and phosphorus inputs on a global basis: application of spatially explicit nutrient export models. Harmful Algae 8:33–38

Glibert PM, Seitzinger S, Heil CA et al (2005) The role of eutrophication in the global proliferation of harmful algal blooms: new perspectives and new approaches. Oceanography 18(2):198–209

Goes JI, Gomes HR (2016) An ecosystem in transition: the emergence of mixotrophy in the Arabian Sea. In: Glibert PM, Kana TM (eds) Aquatic microbial ecology and biogeochemistry: a dual perspective. Springer International Pub., Geneva, pp 155–183

Goes JI, Gomes HR, Al-Hashimi K et al (2018) Ecological drivers of green *Noctiluca* blooms in two monsoonally-driven ecosystems. In: Glibert PM, Berdalet E, Burford M et al (eds) Global ecology and oceanography of harmful algal blooms. Springer, Cham, pp 327–336

Goes JI, Thoppil PG, Gomes HR et al (2005) Warming of the Eurasian land mass is making the Arabian Sea more productive. Science 308:545–547

Gomes HR, Goes JI, Matondkar P et al (2008) Blooms of *Noctiluca miliaris* in the Arabian Sea—an in situ and satellite study. Deep-Sea Res 55(II):751–765

Granéli E, Carlsson P, Tester P et al (1999) Effects of N:P:Si-ratios and zooplankton grazing on phytoplankton communities in the northern Adriatic Sea. I. Nutrients, phytoplankton, biomass, and polysaccharide production. Aquat Microb Ecol 18:37–54

Granéli E, Johansson N (2003) Effects of the toxic haptophyte *Prymnesium parvum* on the survival and feeding of a ciliate: the influence of different nutrient conditions. Mar Ecol Prog Ser 254: 49–56

Granéli E, Weberg M, Salomon PS (2008) Harmful algal blooms of allelopathic microalgal species: the role of eutrophication. Harmful Algae 8:94–102

Guzman HM, Cortes J, Gleynn PW et al (1990) Coral mortality associated with dinoflagellate blooms in the eastern Pacific (Costa Rica and Panama). Mar Ecol Prog Ser 60:299–303

Hallegraeff GM (1993) A review of harmful algal blooms and their apparent global increase. Phycologia 32:79–99

Harrison PJ, Furuya K, Glibert PM et al (2011) Geographical distribution of red and green *Noctiluca scintillans*. Chin J Oceanol Limnol 29:807–883

Häussermann V, Gutstein CS, Bedington M et al (2017) Largest baleen whale mass mortality during strong El Niño event is likely related to harmful toxic algal bloom. PeerJ 5:e3123. https://doi.org/10.7717/peerj.3123

Heil CA, Glibert PM, Fan C (2005) *Prorocentrum minimum* (Pavillard) Schiller – a review of a harmful algal bloom species of growing world-wide importance. Harmful Algae 4:449–470

Holligan PM (1979) Dinoflagellate blooms associated with tidal fronts around the British Isles. In: Taylor DL, Seliger HH (eds) Toxic dinoflagellate blooms. Elsevier North Holland, New York, pp 249–256

Howard MDA, Jones AC, Schnetzer A et al (2012) Quantitative real-time polymerase chain reaction for *Cochlodinium fulvescens* (Dinophyceae), a harmful dinoflagellate from California coastal waters. J Phycol 48:384–393

Hu CM, Li D, Chen C et al (2010) On the recurrent *Ulva prolifera* blooms in the Yellow Sea and East China Sea. J Geophys Res 115:C05017. https://doi.org/10.1029/2009JC005561

Huo Y, Zhang J, Chen L et al (2013) Green algae blooms caused by *Ulva prolifera* in the southern Yellow Sea: identification of the original bloom location and evaluation of biological processes occurring during the early northward floating period. Limnol Oceanogr 58:2206–2218

Kim CS, Lee SG, Kim HG et al (1999) Reactive oxygen species as causative agents in the ichthyotoxicity of the red tide dinoflagellate *Cochlodinium polykrikoides*. J Plankton Res 21: 2105–2115

Kim DI, Matsuyama Y, Nagasoe S et al (2004) Effects of temperature, salinity and irradiance on the growth of the harmful red tide dinoflagellate *Cochlodinium po*lykrikoides Margalef (Dinophyceae). J Plankton Res 26(1):1–66

Kirkpatrick B, Fleming LE, Backer LC et al (2006) Environmental exposures to Florida red tides: effects on emergency room respiratory diagnoses admissions. Harmful Algae 5:526–533

Kudela R, Pitcher G, Probyn T et al (2005) Harmful algal blooms in coastal upwelling systems. Oceanography 18(2):184–197

Kudela RM, Ryan JP, Blakeley MD et al (2008) Linking the physiology and ecology of *Cochlodinium* to better understand harmful algal bloom events: a comparative approach. Harmful Algae 7:278–292

Lancelot C, Martin JM, Panin N et al (2002) The North-western Black Sea: a pilot site to understand the complex interaction between human activities and the coastal environment. Estuar Coastal Shelf Sci 54:279–283

Landsberg JH (2002) The effects of harmful algal blooms on aquatic organisms. Rev Fish Sci 10:113–390

Lee DK (2008) *Cochlodinium polykrikoides* blooms and eco-physical conditions in the South Sea of Korea. Harmful Algae 7:318–323

Lewis NI, Bates SS, McLachlan JL et al (1993) Temperature effects on growth, domoic acid production, and morphology of the diatom *Nitzschia-pungens f. multiseries*. In: Smayda TJ, Shimizu Y (eds) Toxic phytoplankton blooms in the sea. Elsevier Science Publications B V, Amsterdam, pp 601–606

Lin C-H, Accoroni S, Glibert PM (2017) *Karlodinium veneficum* feeding responses and effects on larvae of the eastern oyster *Crassostrea virginica* under variable nitrogen:phosphorus stoichiometry. Aquat Microb Ecol 79:101–114

Liu D, Zhou M (2018) Green tides of the Yellow Sea: massive free-floating blooms of *Ulva prolifera*. In: Glibert PM, Berdalet E, Burford M et al (eds) Global ecology and oceanography of harmful algal blooms. Springer, Cham, pp. 317–326

Luckenbach MW, Sellner KG, Shumway SE, Greene K (1993) Effects of 2 bloom-forming dinoflagellates, *Prorocentrum minimum* and *Gyrodinium uncatenum*, on the growth and survival of the eastern oyster, *Crassostrea virginica* (Gmelin 1791). J Shellfsh Res 12:411–415

Lundgren V, Glibert PM, Granéli E et al (2016) Metabolic and physiological changes in *Prymnesium parvum* when grown under, and grazing on, prey of variable nitrogen:phosphorus stoichiometry. Harmful Algae 55:1–12

McCabe RM, Hickey BM, Kudela RM et al (2016) An unprecedented coastwide toxic algal bloom linked to anomalous ocean conditions. Geophys Res Lett. https://doi.org/10.1002/2016GL070023

Mitra A, Flynn F (2005) Predator-prey interactions: is "ecological stoichiometry" sufficient when good food goes bad? J Plankton Res 27:393–399

Moore SK, Johnstone JA, Banasand NS et al (2015) Present-day and future climate pathways affecting *Alexandrium* blooms in Puget Sound, WA, U.S. Harmful Algae 48:1–15

Moore SK, Trainer VL, Mantua NJ et al (2008) Impacts of climate variability and future change on harmful algal blooms and human health. Environ Health 7:S4. https://doi.org/10.1186/1476-069X-7-S2-S4

Mulholland MR, Morse RE, Boneillo GE et al (2009) Understanding causes and impacts of the dinoflagellate, *Cochlodinium polykrikoides*, blooms in the Chesapeake Bay. Estuaries 32: 734–747

O'Neil JM, Davis TW, Burford MA et al (2012) The rise of harmful cyanobacteria blooms: the potential roles of eutrophication and climate change. Harmful Algae 14:313–334

Ogata T, Sato S, Kodama M (1989) Paralytic shellfish toxins not associated with dinoflagellates. Toxicon 27:1241–1244

Parab SG, Matondkar SGP, Gomes H et al (2006) Monsoon driven changes in phytoplankton populations in the eastern Arabian Sea as revealed by microscopy and HPLC pigment analysis. Cont Shelf Res 26:2538–2558

Parsons ML, Dortch Q, Turner RE (2002) Sedimentological evidence of an increase in *Pseudo-nitzschia* (Bacillariophyceae) abundance in response to coastal eutrophication. Limnol Oceanogr 47:551–558

Pearce I, Handlinger JH, Hallegraeff GH (2005) Histopathology in Pacific oyster (*Crassostrea gigas*) spat caused by the dinoflagellate *Prorocentrum rhathymum*. Harmful Algae 4:61–74

Silvagni PA, Lowenstine LJ, Spraker T et al (2005) Pathology of domoic acid toxicity in California sea lions (*Zalophus californianus*). Vet Pathol 42:184–191

Smayda TJ (1998) Ecophysiology and bloom dynamics of *Heterosigma akashiwo* (Raphidophyceae). In: Anderson DM, Cembella AD, Hallegraeff GM (eds) Physiological ecology of harmful algal blooms, NATO ASI series, vol G41. Springer, Berlin, pp 113–131

Smayda TJ (2002) Adaptive ecology, growth strategies and the global bloom expansion of dinoflagellates. J Oceanogr 58:281–294

Steidinger KA, Tangen K (1997) Dinoflagellates. In: Tomas C (ed) Identifying marine phytoplankton. Academic Press, New York, pp 387–584

Stoecker DK, Adolf JE, Place AR et al (2008) Effects of the dinoflagellates *Karlodinium veneficum* and *Prorocentrum minimum* on early life history stages of the eastern oyster (*Crassostrea virginica*). Mar Biol 154:81–90

Sunda WG, Granéli E, Gobler CJ (2006) Positive feedback and the development and persistence of ecosystem disruptive algal blooms. J Phycol 42:963–974

Tester PA, Nau AW, Feldman RL et al (2010) Ciguatera fish poisoning and sea surface temperatures in the Caribbean Sea. Toxicon 56:698–710

Tomas CR, Smayda TJ (2008) Red tide blooms of *Cochlodinium polykrikoides* in a coastal cove. Harmful Algae 7:308–317

Trainer VL, Yoshida T (eds) (2014) Proceedings of the workshop on economic impacts of harmful algal blooms on fisheries and aquaculture. PICES Scientific Reports 47, p 85

Trainer V, Eberhart B-TL, Wekell JC et al (2003) Paralytic shellfish toxins in Puget Sound, Washington State. J Shellfish Res 22:213–223

Villacorte LO, Tabatabai SAA, Dhakal N et al (2015) Algal blooms: an emerging threat to seawater reverse osmosis desalination. Desalin Water Treat 55. https://doi.org/10.1080/19443994.2014.940649

Visser PM, Verspagen JMH, Sandrini G et al (2016) How rising CO_2 and global warming may stimulate harmful cyanobacterial blooms. Harmful Algae 54:145–159

Walsby T (1975) Gas vesicles. Annu Rev Plant Phys 26:427–439

Wells ML, Karlson B (2018) Harmful algal blooms in a changing ocean. In: Glibert PM, Berdalet E, Burford M et al (eds) Global ecology and oceanography of harmful algal blooms. Springer, Cham, pp 77–90

Wells ML, Trainer VL, Smayda TJ et al (2015) Harmful algal blooms and climate change: learning from the past and present to forecast the future. Harmful Algae 49:68–93

Wikfors GH, Smolowitz RM (1995) Experimental and histological studies of 4 life-history stages of the Eastern oyster, *Crassostrea virginica*, exposed to a cultured strain of the dinoflagellate *Prorocentrum minimum*. Biol Bull 188:313–328

Yan T, Zhou MJ (2004) Environmental and health effects associated with harmful algal bloom and marine algal toxins in China. Biomed Environ Sci 17:165–176

Yu R-C, Lü S-H, Liang Y-B (2018) Harmful algal blooms in the coastal waters of China. In: Glibert PM, Berdalet E, Burford M et al (eds) Global ecology and oceanography of harmful algal blooms. Springer, Cham, pp 309–316

Yuki K, Yoshimatsu S (1989) Two fish-killing of *Cochlodinium* from Harina-Nada, Seto Inland Sea Japan. In: Okaichi T, Anderson DM, Nemoto T (eds) Red tides: biology, environmental science, and toxicology. Elsevier, New York, pp 451–454

Chapter 3
Establishment, Goals, and Legacy of the Global Ecology and Oceanography of Harmful Algal Blooms (GEOHAB) Programme

Raphael M. Kudela, Robin Raine, Grant C. Pitcher, Patrick Gentien, Elisa Berdalet, Henrik Enevoldsen, and Ed Urban

3.1 Introduction

Single-celled organisms called phytoplankton and cyanobacteria (often called blue-green algae), together with macroalgae, support healthy aquatic ecosystems by forming the base of the food web. This broad group of organisms fix carbon and produce oxygen through photosynthesis, providing approximately 50% of the oxygen on the planet. A small subset of these organisms can cause harm to aquatic ecosystems and to humans by producing water-borne toxins that can accumulate in the food web to produce toxic seafood. Whether toxic or not, these same organisms

Patrick Gentien was deceased at the time of publication.

R.M. Kudela (✉)
Ocean Sciences Department, University of California Santa Cruz, Santa Cruz, CA, USA
e-mail: kudela@ucsc.edu

R. Raine
Earth and Ocean Sciences, National University of Ireland, Galway, Ireland

G.C. Pitcher
Fisheries Research and Development, Cape Town, South Africa

P. Gentien
IFREMER, Brest, France

E. Berdalet
Institute of Marine Sciences (CSIC), Passeig Marítim de la Barceloneta, Barcelona, Catalonia, Spain

H. Enevoldsen
Intergovernmental Oceanographic Commission of UNESCO, IOC Science and Communication Centre on Harmful Algae, University of Copenhagen, Copenhagen, Denmark

E. Urban
Scientific Committee on Oceanic Research, Robinson Hall, University of Delaware, Newark, DE, USA

P.M. Glibert et al. (eds.), *Global Ecology and Oceanography of Harmful Algal Blooms*, Ecological Studies 232, https://doi.org/10.1007/978-3-319-70069-4_3

can also form "blooms" (high biomass of cells) causing harm or disrupting aquatic ecosystems. Ecosystem damaged by high-biomass blooms may include, for instance, disruption of food webs, fish killing by gill damage, or the development of low-oxygen "dead-zones" after bloom degradation. When these events discolour the water, they are often referred to colloquially as red tides (regardless of colour), whether they produce toxins or not and often whether they are harmful or not. Regardless of whether these algae and cyanobacteria are toxic and whether they form low or high biomass, they are collectively referred to as "harmful algal blooms" (HABs) when they exhibit negative consequences for human, ecosystem, or wildlife health or socioeconomic status of the environment (Hallegraeff 1993; Glibert et al. 2005; Anderson et al. 2012; Kudela et al. 2015). The definition of a HAB is therefore a societal concept rather than a scientific definition.

HABs occur in marine, brackish, and freshwater aquatic systems, but they most dramatically affect human activities and are most obvious when they occur in coastal and inland waters. HABs have been documented from the Arctic to Antarctic, even in sparsely settled coastal regions and in offshore areas, and are emerging as a serious threat in freshwater systems such as the US Great Lakes, drinking water reservoirs, and coastal watersheds that deliver freshwater HAB cells and toxins to the coastal ocean. Thus, HABs are a serious and emerging threat throughout the world's freshwater and marine systems.

3.2 History of GEOHAB

While HAB events have been documented throughout human history, coordinated research on HABs began in earnest in 1974 with the First International Conference on toxic dinoflagellate blooms. This meeting was provoked by a massive paralytic shellfish poisoning (PSP) outbreak in 1972 in the western Gulf of Maine. The success of this first conference, which was primarily a US meeting with some international participants, later developed into a series of international conferences on harmful marine phytoplankton. At the fourth meeting of this group in 1989, it was recommended, "that international research efforts be undertaken to evaluate the possibility of global expansion of algal blooms and man's involvement in this phenomenon" (Granéli et al. 1990). This ultimately led to the request by Member States of the Intergovernmental Oceanographic Commission (IOC) for the formation of an international programme on HABs.

At the same time, the Scientific Committee on Oceanic Research (SCOR) and IOC Working Group (WG 97) on the *Physiological Ecology of Harmful Algal Blooms* also played an important role in focusing and starting to develop consensus in the HAB research community. The IOC Harmful Algal Bloom Programme was first initiated at a workshop sponsored by IOC and SCOR in 1991, followed shortly thereafter in 1992 by the establishment of the IOC Intergovernmental Panel on Harmful Algal Blooms (IOC/IPHAB), with a goal "to foster the effective management of, and scientific research on, harmful algal blooms in order to understand their causes, predict their occurrences, and mitigate their effects". At the fourth IPHAB meeting, IOC and SCOR jointly agreed to develop an international programme for the advancement of scientific research on HABs. This led to a workshop in Havreholm, Denmark, in 1998 chaired by Professor John Cullen. Attended by 37 scientists from 20 countries, this workshop led to the formation in 1998 of the Global Ecology and Oceanography of Harmful Algal Blooms (GEOHAB) Programme, cosponsored by IOC and SCOR. This activity included wide support from the international community through the member states of IOC/IPHAB but also with participation and coordination of other international programmes, including the Joint Global Ocean Flux Study (JGOFS), Land-Ocean Interactions in the Coastal Zone (LOICZ) Programme, and Global Ocean Ecosystem Dynamics (GLOBEC) Programme (GEOHAB 1998), and with the science goals building on a successful NATO ASI meeting convened by SCOR WG 97 (Anderson et al. 1998).

GEOHAB acknowledged from its inception that the HAB problem is global in nature, and that from a scientific perspective, there was incomplete information about the biological, chemical, and physical factors regulating HAB dynamics and impacts (GEOHAB 2001). Thus GEOHAB adopted a mission statement to "foster international co-operative research on HABs in ecosystem types sharing common features, comparing the key species involved and the oceanographic processes that influence their population dynamics" and a goal of "Improv(ing) prediction of HABs by determining the ecological and oceanographic mechanisms underlying their population dynamics, integrating biological, chemical, and physical studies supported by enhanced observation and modelling systems".

From foundation, a core component of the GEOHAB Programme has been to champion the comparative approach; indeed, the comparative approach is referenced within both the mission and goal statements. Mayr (1982) identified experimental approaches and comparative approaches as "the two great methods of science" (see also Anderson et al. 2005), and the organizers of GEOHAB recognized that the ecology and oceanography of HABs can best be understood through the study of causative organisms and affected systems in relation to comparable species and systems, from the cellular to the ecosystem level. The comparative approach lends itself to a coordinated, international programme (i.e. GEOHAB) where naturally occurring temporal and spatial variations in existing conditions and phenomena substitute for controlled experimental treatments. This approach directly addresses the fundamental issue that ecosystem-scale biological systems generally, and marine systems specifically, are not amenable to experimental control due to the size and complexity of these systems.

Following the formation of GEOHAB at the Havreholm meeting, it was recognized that a key requirement for an international, comparative approach to scientific understanding of HABs was the definition of the systems and phenomena to be compared. These requirements were formally documented by the first GEOHAB Scientific Steering Committee (SSC), chaired by Patrick Gentien, and published as the GEOHAB Science Plan in 2001 (GEOHAB 2001), the first of many GEOHAB reports. The Science Plan outlined five programme elements to serve as a

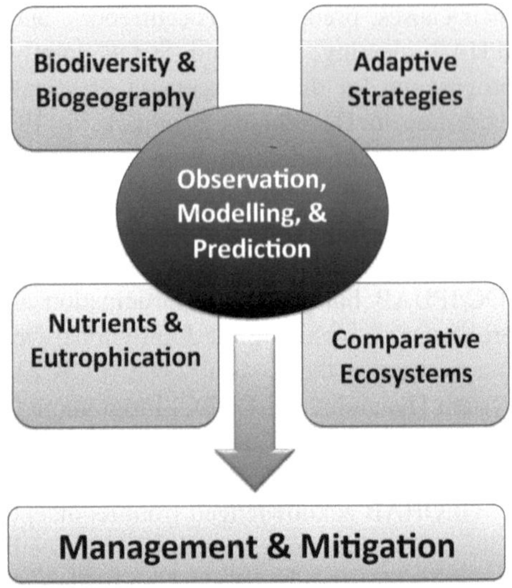

Fig. 3.1 The goals of GEOHAB were advanced through the linkages between the programme elements

framework to guide priorities and research and to define the scientific boundaries to the GEOHAB Programme (Fig. 3.1). These programme elements included *Biodiversity and Biogeography*, *Nutrients and Eutrophication*, *Adaptive Strategies*, *Comparative Ecosystems*, *and Observation, Modelling, and Prediction.* Within each programme element, there were specific objectives and questions, outlined below:

- Programme Element 1, Biodiversity and Biogeography, had an overall objective to determine the relationships among distributions of HAB species, biodiversity, and environmental change.
- Programme Element 2, Nutrients and Eutrophication, focused on determining the significance of eutrophication and nutrient transformation pathways to HAB population dynamics.
- Programme Element 3, Adaptive Strategies, attempted to define the particular characteristics and adaptations of HAB species that determine when and where they occur and produce harmful effects.
- Programme Element 4, Comparative Ecosystems, identified mechanisms underlying HAB population and community dynamics across ecosystem types through comparative studies.
- Programme Element 5, Observation, Modelling, and Prediction, sought to improve the detection and prediction of HABs by developing capabilities in observation and modelling.

It was envisioned that these elements, when combined, would provide the scientific basis for improved management and mitigation of HAB impacts and that GEOHAB would interact and coordinate with related programmes at the regional, national, and international levels. This specifically included, in the Science Plan, the Global Ocean Observing System (GOOS), LOICZ, and GLOBEC at the international level, as well as regional programmes including the Ecology and Oceanography of Harmful Algal Blooms (ECOHAB) Programme in the United States, and the European Initiative on Harmful Algal Blooms (EUROHAB) Programme in Europe (EUROHAB 1999, 2002). These partnerships have evolved through time and remain central to GEOHAB's success (Fig. 3.2). The GEOHAB Science Plan was both ambitious and forward looking. It served to steer the GEOHAB Programme from publication, in 2001, through to its conclusion in 2016, and the Science Plan also was used as the basis for the GlobalHAB Programme which succeeded GEOHAB in 2016 [Berdalet et al. 2017a; see also Chap. 22, Berdalet et al. (2018)].

GEOHAB adopted and published an Implementation Plan in 2003 (GEOHAB 2003). Implementation efforts are a major challenge for any international coordinating programme and must lay out a strategy to develop the high-level vision, mission, and goals of the programme. As described in the GEOHAB Implementation Plan, "this plan for implementation of the programme serves as an invitation to the broad scientific community to participate. Scientists working in physical, chemical and/or biological disciplines, or other fields related to harmful algal research, including the development of relevant instrumentation and models, are encouraged to contribute to this programme. Active participation by the widest international representation of the research community will be essential to ensure the success of GEOHAB". This invitation provided the basis for GEOHAB

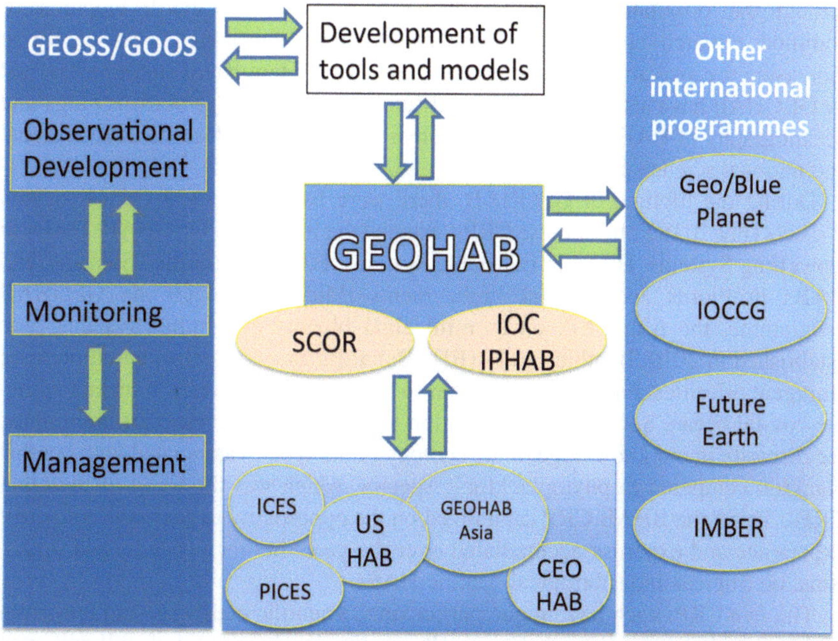

Fig. 3.2 GEOHAB cooperated and coordinated with relevant national, regional, and international programmes

activities with an emphasis on open, community participation and consensus building by international scientists focused on HAB research. A challenge that GEOHAB faced throughout its lifetime was that, in contrast to other international programmes such as GLOBEC (Turner et al. 2013), GEOHAB did not establish an International Programme Office. This led to more emphasis on support and coordination from IOC and SCOR and more reliance on the SSC and Core Research Projects (CRPs), which were developed within GEOHAB as the primary mechanism for coordinating international research.

The CRPs directly addressed Programme Element 4 of the Science Plan (comparative research) and comprised oceanographic field studies conducted in, and application of models to, comparable ecosystems. The CRPs were supported by identification of relevant organisms and measurements of the physical, chemical, and biological processes that control their population dynamics (GEOHAB 2001). A major objective of Core Research was the integration achieved by the application of coupled biological/chemical/physical models to HAB dynamics in geographically distinct ecosystems sharing common features. CRPs were one aspect of the Implementation Plan and were complemented by Targeted Research and Regional or National Projects (GEOHAB 2003). A primary distinction was that CRPs were a direct extension of the GEOHAB Programme, while Targeted and Regional/National Research were affiliated programmes under the leadership, guidance, and funding of those affiliated programmes.

GEOHAB provided useful direction to HAB scientists over the past decade through the formulation of common research objectives and the development of common research approaches. These efforts have contributed significantly to increasing the availability of comparable data. The system-based studies promoted by GEOHAB have also been useful in advancing our knowledge of the ecology and oceanography of HABs and the comparative approach and added value to regional and national science plans.

During the lifetime of GEOHAB, there were five operational Core Research Projects (Fig. 3.3). Four of these CRPs spanned the lifetime of GEOHAB: HABs in Upwelling Systems, HABs in Eutrophic Systems, HABs in Stratified Systems, and HABs in Fjords and Coastal Embayments. Midway through the GEOHAB Programme, the fifth CRP, on harmful algal blooms in Benthic Systems, was established in 2010. Benthic HABs (BHABs) arose as a core effort based on three emergent priorities. First, BHABs appear to be increasing in area and impact within the coastal ocean. Second, better attention to the BHAB phenomenon highlighted the disparity in our knowledge of the ecology of the species associated with benthic plankton compared to phytoplanktonic species, which was the focus of the other CRPs. Third, the BHAB CRP capitalized on the opportunity to share the collective experience and expertise accumulated over the years in other systems and organisms, on tropical and subtropical benthic HABs.

The five CRPs each had unique participants, scientific questions, and outcomes (Table 3.1), but a common feature to all was the development with full and open involvement from the international scientific community and with an emphasis on the comparative approach as defined in the Science Plan (GEOHAB 2001). The

Fig. 3.3 Five Core Research Projects provided the backbone for GEOHAB scientific efforts

major goals and accomplishments of each of the five CRPs are described below and expanded upon in more detail in the following chapters of this volume. Each CRP was developed through a “bottom-up” process through which the international HAB community was invited to an Open Science Meeting to help define priority research questions.

Table 3.1 Workshops, Open Science Meetings, and training sessions organized by GEOHAB

Event	Year	Location	Reference	Special issues
Joint IOC/SCOR Workshop	1998	Havreholm, Denmark	GEOHAB (1998)	
CRP: HABs in Upwelling Systems Open Science Meeting	2003	Lisbon, Portugal	GEOHAB (2005)	*Oceanography* Volume 18(2), 2005
HABWatch Workshop	2003	Villefrance-sur-mer, France	Babin et al. (2005, 2008)	*Oceanography* Volume 18(2), 2005
CRP: HABs in Fjords and Coastal Embayments Open Science Meeting	2004	Viña del Mar, Chile	GEOHAB (2010a)	*Oceanography* Volume 18(2), 2005
CRP: HABs and Stratification Open Science Meeting	2005	Paris, France	GEOHAB (2008)	*Oceanography* Volume 18(2), 2005
CRP: HABs in Eutrophic Systems Open Science Meeting	2005	Baltimore, MD, USA	GEOHAB (2006)	*Oceanography* Volume 18(2), 2005; *Harmful Algae*, Volume 8(1), 2008
GEOHAB Asia Open Science Meeting	2007	Tokyo, Japan	GEOHAB (2010b)	
GEOHAB Asia Open Science Meeting	2008	Nha Trang, Vietnam	GEOHAB (2010b)	
CRP: HABs in Eutrophic Systems, 2nd Open Science Meeting	2009	Beijing, China	Glibert et al. (2011)	*Chinese Journal of Oceanology and Limnology* Volume 29(4), 2011
GEOHAB Modelling Workshop	2009	Galway, Ireland	GEOHAB (2011)	*Journal of Marine Systems,* Volume 83 (3–4), 2011
IOCCG/GEOHAB Working Group on Harmful Algal Blooms and Ocean Colour	2010, 2011	Schulphoek, South Africa and Milan, Italy	IOCCG Monograph (pending)	
International Workshop on Fish-killing Marine Algae	2011	Oslo, Norway		
International Congress on *Ostreopsis* Development	2011	Villefrance-sur-mer, France	Zingone et al. (2012)	
CRP: HABs in Benthic Systems Open Science Meeting	2012	Honolulu, HI, USA	GEOHAB (2012)	
CRP: HABs in Fjords and Coastal Embayments 2nd Open Science Meeting	2012	Victoria, Canada	GEOHAB (2013a)	
GEOHAB International Training Workshop on the Ecology and Taxonomy of Benthic Marine Dinoflagellates	2012	Johor Island, Malaysia		

(continued)

Table 3.1 (continued)

Event	Year	Location	Reference	Special issues
CRP: HABs and Stratification, 2nd Open Science Meeting	2013	Monterey Bay, CA, USA	GEOHAB (2013b)	*Deep-Sea Research II*, Volume 100, 2014
GEOHAB Synthesis Open Science Meeting	2013	Paris, France	GEOHAB (2014)	*Oceanography*, Volume 30(1), 2017
ICES/PICES/IOC-GEOHAB Workshop on Harmful Algae Blooms in a Changing World	2013	Friday Harbor, WA, USA	ICES (2013), Wells et al. (2015)	
Harmful Algal Blooms and Climate Change	2015	Gothenburg, Sweden		

3.3 HABs in Upwelling Systems

The Upwelling CRP Science Plan was developed at an Open Science Meeting (OSM) held in 2004 (GEOHAB 2005). This CRP focused on the oceanographic processes that influence HAB population dynamics and community interactions within upwelling systems, focusing primarily on eastern boundary currents. A primary goal was to assess the extent to which HAB species within these systems respond similarly. Equally important was identification of upwelling systems that have dissimilar HAB species or groupings, to identify and understand the unique adaptations employed within each system. The CRP identified eight core questions related to upwelling systems:

- Are there definable adaptive strategies that characterize HAB species in upwelling systems?
- What seeding strategies persist within upwelling regions and are they consistent among regions?
- How do small-scale physical processes affect HAB growth and dispersion in upwelling systems?
- How do nutrient supply, type, and ratios determine HAB population dynamics in upwelling systems?
- What is the role of genetic predisposition versus environmental conditions in toxin production in different upwelling systems within a given genus or species?
- How does coastal morphology and bathymetry affect HAB dynamics in upwelling systems?
- What is the relative importance of cross-shelf and alongshore advection in different upwelling systems for HABs?
- Are climate indicators predictive of HAB events in upwelling systems?

The Upwelling CRP focused primarily on comparison of systems using existing data and field efforts, since a truly comparative field programme across multiple ecosystems requires a concerted national/international effort such as was employed by the Joint Global Ocean Flux Study (JGOFS; Karl et al. 2001). This CRP led to numerous publications, including a special issue of *Progress in Oceanography* (Pitcher and Pillar 2010) highlighting progress and still-needed effort related to the eight core questions of this CRP. Subsequent achievements of the CRP are updated in Chap. 11 (Pitcher et al. 2018).

One key success of the Upwelling CRP has been the recognition that physical forcing in upwelling systems strongly dominates and that physical models with passive tracers representing HABs provide significant understanding of these systems. Physical transport models have provided a very useful first step in predicting the movement of blooms through advection. Successful prediction of HABs is a possible outcome of the integration of real-time data into model systems as a component of operational forecasting of the ocean. This has most recently been achieved in the California Current System (Anderson et al. 2016) and can most readily be accomplished in upwelling systems, thus taking a critical step towards fulfillment of GEOHAB goals.

3.4 HABs in Eutrophic Systems

The Eutrophic Systems CRP Science Plan (GEOHAB 2006) was developed at an OSM held in 2005. This CRP focused on all aspects of HABs in eutrophic systems. An important consequence of eutrophication, or nutrient pollution, is the increased prevalence of HABs primarily through the formation of high-biomass events, and this was one aspect of HAB research that this CRP addressed. These events can lead to fish kills, accumulation and trophic transfer of algal toxins, hypoxia or anoxia caused by oxygen depletion, and altered trophic interactions through direct stimulation of HABs by nutrients. Importantly, eutrophication can also lead to detrimental impacts through more subtle alterations of ecosystem function such as through change in nutrient forms and ratios and selection for or against certain types of algae. This CRP laid out a roadmap for international research, focusing on the following questions:

- Are there clusters or specific types of HAB species that are indicative of global nutrient increases?
- To what extent do residence time and other physical processes impact the relationship between nutrient loading and HAB proliferation?
- How do feedbacks and interactions between nutrients and the planktonic, microbial food webs impact HABs and their detrimental effects?
- Do anthropogenic alterations of the food web, including overfishing and aquaculture activities, synergistically interact with nutrients to favour HABs?
- How do anthropogenic changes in land use, agricultural use of fertilizer, NO_x emissions from vehicles, and global changes in land cover affect the delivery of nutrients to coastal waters and the resulting incidence of HABs? How do the stoichiometry and quality of these nutrient sources regulate the biological responses favouring HABs?

- Do climate change and climate variability have impacts on ecosystems that augment the impacts of eutrophication in the formation of HABs?

Members of the CRP focused on establishing collaborative and comparative regional and national research programmes with eutrophication as a core component, collaborating through GEOHAB to establish an international Core Research Project. Examples include the CEOHAB in China, the GEOHAB Asia Programme, and the US Harmful Algal Research and Response National Environmental Strategy (HARRNESS 2005). Over the lifetime of GEOHAB, this CRP conducted two Open Science Meetings (Baltimore, USA and Beijing, China), produced two special issues on HABs and Eutrophication in *Harmful Algae* and the *Chinese Journal of Oceanology and Limnology* (Glibert et al. 2008a, 2011), and led to publication of a position paper describing the negative consequences of urea fertilization for enhanced carbon sequestration and fisheries production (c.f. Glibert et al. 2008b).

The conceptual scientific and management understanding of the links between nutrients and HABs has historically been based upon the somewhat simplistic notion that more nutrients lead directly to higher algal biomass. This GEOHAB CRP championed the view, based on results utilizing the comparative approach, that HAB species can be directly and/or indirectly stimulated by nutrient over-enrichment, and that chronic, subtle effects related to nutrient forms, ratios, and loads can be equally or even more important than the obvious, acute effects (Glibert et al. 2011; Glibert and Burford 2017). This has led to a fundamental reevaluation of HABs and nutrients, focusing on these often-subtle interactions rather than direct, acute responses. Accomplishments of this CRP are summarized in Chap. 12 (Glibert et al. 2018a).

3.5 HABs in Stratified Systems

The CRP on HABs in Stratified Environments held an OSM in 2005. The resulting CRP Science Plan (GEOHAB 2008) provided a cross-cutting focus on understanding the small-scale hydrographic features occurring in any environmental system, including upwelling, fjords, and semi-enclosed coastal areas. This CRP emphasized that these processes occur both with and without eutrophication; as such, questions posed by this CRP are relevant to the other GEOHAB CRPs. Particular emphasis was placed on thin layers, where many harmful phytoplankton organisms can thrive, on rheology dynamics, and on technological and modelling challenges required to resolve biological and physical interactions operating at small physical scales that influence HAB dynamics. The overall objective of this CRP was to determine the factors underlying the development of communities related to HABs in subsurface microlayers and the real-time dispersion of these microlayers as a function of turbulent and advective regimes. Several key questions were identified:

- What are the turbulence length scales relevant to harmful phytoplankton and the formation of thin layers?
- What are the main processes controlling the population evolution of a given species, and how does their ranking vary over the time course of a bloom?
- How can we quantify modifications in turbulence by phytoplankton through changes in the viscosity of its physical environment?
- What nutritional opportunities do thin layers provide to phytoplankton, especially to the species selected within thin layers?
- Are allelopathy and "chemical warfare" at work in situ within thin layers?

A strong emphasis on development of technology and methods for sampling and modelling HAB organisms in stratified systems emerged based on the unique nature of the habitat associated with these events, which typically involve formation of subsurface thin layers. A significant accomplishment from this CRP was a workshop focusing on these technological and scientific challenges (Monterey Bay, CA, 2012; GEOHAB 2013b; Berdalet et al. 2014) and subsequent publication of a special issue of *Deep-Sea Research II* (Raine et al. 2014). A main outcome of the CRP was the recognition that thin layers are important in the initiation and development of HABs, as populations develop inside retentive physical structures that can then be transported towards the coast or from the subsurface to the surface, where negative impacts occur. Accomplishments of this CRP are summarized in Chap. 9 (Raine et al. 2018).

3.6 HABs in Fjords and Coastal Embayments

The Fjords and Coastal Embayments CRP resulted from an OSM held in 2004. This CRP (GEOHAB 2010a) focused on understanding the ecology and oceanography of harmful algal species in coastal systems partially surrounded by land and thus affected by land-sea interactions operating at spatial scales much smaller than in open coastal or upwelling systems. These regions are characterized by shallow depths and proximity to land and represent a small fraction of the world ocean, but they are of considerable importance because of the ever-increasing human population utilizing the nearshore and the typically elevated productivity associated with fjords and coastal embayments. HABs frequently occur in these environments, leading to negative outcomes on human health, tourism, aquaculture, and commercial fisheries. To address these issues, seven key questions were defined in this CRP plan:

- Are there definable adaptive strategies that characterize HAB species in confined and semi-confined systems?
- What is the importance of life history transitions and cyst distribution in bloom initiation and maintenance—endogenous seed beds versus exogenous introduction?

- How do physical dispersion and aggregation processes within a semi-confined basin affect HAB growth and distribution?
- What is the relative contribution of nutrient flux and supply ratios to HAB dynamics in eutrophic versus non-eutrophic coastal embayments?
- What is the importance of spatial scale and retention time in the expression and effects of allelochemicals/toxins in semi-confined systems?
- How do embayment morphology, bathymetry, and hydrodynamic flux affect HAB dynamics?
- Are the effects of anthropogenic activities (e.g. aquaculture) and global climate change on HAB dynamics magnified in enclosed and semi-enclosed embayments?

This CRP held a second OSM (Victoria, Canada) in 2012, on "Progress in interpreting Life History and Growth Dynamics of Harmful Algal Blooms in Fjords and Coastal Environments". This meeting focused on four themes: (a) life history of HAB species; (b) allelochemical interactions; (c) genetic diversity; and (d) transport and mixing of blooms in small-scale, mesoscale, and semi-confined systems (GEOHAB 2013a). Significant accomplishments from this CRP related primarily to the life histories of HAB species, including resistance and sexual reproduction phases. This was made possible by advances in culturing, microscopy (e.g. nuclei staining and visualization), and molecular tools (e.g. Bravo et al. 2014) and in situ observation techniques (e.g. Brosnahan et al. 2015). Another significant accomplishment was recognition that species-specific strategies on cyst germination and encystment coupled to particular water circulation patterns could explain the importance of cysts on the recurrence of blooms in certain retentive areas (e.g. Onda et al. 2014) but not in others (e.g. Estrada et al. 2010).

Perhaps one of the most highly visible outcomes from participants within this CRP was the physical/biological numerical model of the population dynamics of *A. fundyense* in the Gulf of Maine, supported at a regional and national level by multiple agencies, including the US ECOHAB Programme. Continued fieldwork provided clear evidence of the interannual variability of cyst beds in the region and also showed links between cyst distribution patterns and the intensity and extension of blooms in the years immediately before and after cyst mapping surveys (McGillicuddy et al. 2011). Recent efforts suggest that it may be possible to empirically forecast the geographic extent of the forthcoming bloom based on cyst abundances (Anderson et al. 2014), a major objective of the GEOHAB Science and Implementation Plans. A more complete overview of the accomplishments of this CRP is given in Chap. 10 (Roy et al. 2018).

3.7 HABs in Benthic Systems

The newest CRP, followed from an OSM in 2010 (GEOHAB 2012), was the only CRP added to GEOHAB after development of the *Implementation Plan*. The need for this CRP emerged in part from the recognition that more frequent harmful events and increased biogeographic extension associated with benthic microalgae would benefit from a global comparative approach focusing on ecology, toxicology, and trophic interactions of the primary causative organisms, the dinoflagellates *Gambierdiscus* and *Ostreopsis*. This CRP developed an approach to facilitate coordination among researchers to identify knowledge gaps and build scientific understanding by addressing five key questions:

- What is the biogeography and biodiversity of BHABs and the relationships among distributions of BHAB species, biodiversity, and environmental changes?
- What are the relationships between eutrophication and nutrient transformation pathways and BHAB population dynamics?
- Are there particular characteristics and adaptations of BHAB species that determine when and where they occur and produce harmful effects?
- Are there mechanisms underlying BHAB population and community dynamics across ecosystem types that are revealed through comparative studies?
- What new observation and modelling approaches are available to help in the detection and prediction of BHABs?

As with all of GEOHAB, the establishment of an international programme that can coordinate information and data sharing of different science communities, many of which are geographically separated, was a primary objective of the BHAB CRP. Discoveries over the last 5–7 years have resulted in a considerable increase in publications and have brought many new researchers into the field. Although GEOHAB made progress on many of the research questions listed above, many still remain to be fully explored through future research. The impacts of benthic HABs as a group are one of the most difficult types of HABs to mitigate, and the demand for improved approaches and methods is widespread. Benthic HABs are associated with both high-value tourism (e.g. coral reef systems) and high-impact consequences such as ciguatera fish poisoning (CFP). While these implementation activities were established under GEOHAB, this effort continues under the GlobalHAB Programme (Berdalet et al. 2017a, b). Accomplishments of this CRP are further summarized in Chap. 13 (Berdalet and Tester 2018).

3.8 GEOHAB Targeted, Regional, and National Research

When GEOHAB was initiated, the SSC was well aware that the CRPs focused on specific objectives within the larger Science Plan and that further efforts were required to fully realize the GEOHAB Mission and Goal. Coordination with related and complementary programmes was also critical to the successful development of a coordinated international programme, as called out in the original workshop (GEOHAB 1998). Therefore, two other research activities were explicitly included. *Targeted Research* comprised numerous projects that were affiliated with GEOHAB, or were aligned with the goals of the CRPs, but were not directly overseen or coordinated by GEOHAB. *Regional and National Research* activities were identified as relevant to the GEOHAB Science Plan but were coordinated by regional or national organizations independent of the GEOHAB SSC. These projects and programmes ranged from narrowly focused efforts such as the US programme "Dynamics of *Alexandrium fundyense* Distributions in the Gulf of Maine: An observational and Modelling Study of Nearshore and Offshore Shellfish Toxicity, Vertical Toxin Flux, and Bloom Dynamics in a Complex Shelf Sea", known as GOMTOX, to large, national programmes such as PhilHABs (Ecology and Oceanography of Harmful Algal Blooms in the Philippines) and CEOHAB (Chinese Ecology and Oceanography of Harmful Algal Blooms). The mission and goals of GEOHAB were also articulated and implemented at a regional level through development of the EU-US Scientific Initiative on Harmful Algal Blooms (European Commission 2003) and through development of a coordinated research effort between the US ECOHAB Programme and the EU EUROHAB Programme. A total of eight regional/national programmes were endorsed by GEOHAB. During the lifetime of GEOHAB, other National Research Programmes were championed but were ultimately unsuccessful, including the proposed Canadian GEOHAB Programme in 2001 (Martin 2002) and the proposed Baltic GEOHAB Implementation Plan (ICES 2007).

A significant accomplishment facilitated by GEOHAB was the establishment of a regional plan for HAB research focus and cooperation in Asia. Two regional meetings (Tokyo, Japan, 2007 and Nha Trang, Vietnam, 2008) were held in coordination with IOC/WESTPAC and the IOC HABViet Programme [see also Chap. 14, Furuya et al. (2018)]. These meetings culminated in the publication of the GEOHAB Asia Regional Comparative Programme (GEOHAB 2010b) report, documenting these efforts. The report developed an overview of Asian HAB problems and provided a framework for guiding priorities for research modelled after the five programme elements of GEOHAB. The ecological and economic impacts of HABs in Asia are profound and increasing; the need to develop a comparative, international programme on HABs in Asia was therefore particularly compelling. Asia is characterized by (1) the highest production of aquaculture fish and shellfish globally and thus the greatest impacts from HABs on these resources, (2) a high diversity of harmful syndromes and causative organisms, (3) an apparent trend of increasing HABs throughout the region, and (4) an increasing trend towards HABs linked to regional eutrophication. GEOHAB Asia provided and still provides a template for regional coordination on HAB research to address these issues and has led to further partnerships under both GEOHAB and now GlobalHAB, as well as through ongoing efforts by IOC/WESTPAC and other regional organizations such as the North Pacific Marine Science Organization (PICES), a partner in the new GlobalHAB Programme (Berdalet et al. 2017a; Chap. 22, Berdalet et al. 2018). An overview of the important HAB issues in Asia is given in Chap. 14, Furuya et al. (2018).

3.9 Cross-Cutting and Framework Activities

The GEOHAB Implementation Plan called for framework activities in addition to research. These framework activities focused on data management, measurement protocols and quality control, capacity building, modelling, and coordination with other international programmes. An early and highly successful example of a GEOHAB framework activity was the "Workshop on real-time coastal observing systems for ecosystem dynamics and harmful algal blooms" or HABWatch, held in 2003 at the Observatoire Océanologique and Citadelle of Villefrance-sur-mer, France. The idea for the workshop, initially emerged from the Working Group on Harmful Algal Bloom Dynamics of the International Council for the Exploration of the Sea (ICES), was supported by GEOHAB and the Coastal Ocean Observing Panel (COOP) of the Global Ocean Observing System (GOOS) and was initiated with major support from the European Commission and with financial support from multiple national and international agencies. This workshop provided participants with both theory relevant to understanding the basic principles of real-time observation and modelling tools and hands-on tutorials to facilitate the acceptance and use of these tools. The workshop included 89 participants from all over the world; HABWatch has had far-reaching impacts. The lectures and tutorials are still available via the Internet (http://HABWatch.obs-vlfr.fr) and the workshop culminated in the publication of a manual within the UNESCO Oceanographic Methodology Series, "Real-time coastal observing systems for Marine ecosystem dynamics and Harmful Algal Blooms: Theory, instrumentation, and modelling" (Babin et al. 2008), with contributions from 58 authors. Additionally, a high-level summary was published as part of a GEOHAB special issue in *Oceanography* (Babin et al. 2005). Advances in real-time observational and other tools are also described more fully in Chap. 18 (Glibert et al. 2018b).

GEOHAB ran another very successful workshop and training activity focusing on modelling in 2009 in Galway, Ireland: "GEOHAB modelling: Linking Observations to Predictions" (GEOHAB 2011). As with HABWatch, the workshop emphasized training young investigators with hands-on activities, facilitating developing strategies for using observations and models to address the GEOHAB CRP science questions, international coordination, and furthering the goals of the GEOHAB Science and Implementation Plans. The workshop used the CRPs as an organizational framework and focused specifically on the science questions articulated by the eutrophic, stratified, and upwelling systems CRPs. A significant outcome beyond the workshop itself was the publication of a special issue of the *Journal of Marine Systems* (McGillicuddy et al. 2011).

More recently, a series of workshops on the relationship between HABs and climate change (Wells et al. 2015) have been jointly supported by GEOHAB, GlobalHAB, ICES, and PICES. While climate change was acknowledged in the GEOHAB Science and Implementation Plans, it was not a primary focus of the various CRPs, in large part because it was felt that there was insufficient information about the link(s) and causative relationships between HABs and climate change during the lifetime of GEOHAB. As with the Benthic HAB activities, the focus on climate change has transpired as new information, recognition of knowledge gaps, and a clearly defined role for international coordination has emerged from the scientific community. The first workshop was the "Workshop on Harmful Algae Blooms in a Changing World" (ICES 2013). Outcomes from this workshop included assessment of the current state of knowledge (Wells et al. 2015), description of two types of "sentinel" sites for observing change based on long-term observations, and planning for a larger Open Science Meeting.

The follow-on symposium took place in Gothenburg, Sweden, in 2015. Participants developed several urgent recommendations on research priorities related to HABs and climate change [see Chap. 5, Wells and Karlson (2018)]. These include focusing more intensive research on key organisms; continuing to develop ecological and forecast models; strengthening linkages among global, national, and regional observation programmes to establish the recommended HAB sentinel sites; and developing of a "best practices" manual for assessment of climate change linkages to HAB events and organisms. This is now being championed by GlobalHAB, with plans for a manual to standardize common methods and field data relevant to understanding the interaction of HABs and climate change (Berdalet et al. 2017a).

Moving forward, GlobalHAB will also interact with the IOC-UNESCO working group Global Ocean Oxygen Network (GO_2NE), recently formed to address concerns relating to declining oxygen concentrations in the ocean. The working group was formed in response to a need for a joint global analysis of ocean deoxygenation and its impacts. Formal collaboration between GlobalHAB and GO_2NE will serve investigation of the likely global increase in episodic oxygen depletion associated with increasing HABs in the coastal environment. The creation of an international network of ocean oxygen scientists was created specifically to benefit investigation of ocean deoxygenation in response to climate change and eutrophication and represented a clear opportunity for GlobalHAB to interact effectively with other international scientific programmes.

3.10 GEOHAB Legacies

While there are many ways to gauge the success and impact of international research programmes, one of the most direct methods is to document the outcomes of these activities within the scientific literature. Multiple special issues, as well as individual publications acknowledging GEOHAB, were produced during the lifetime of the GEOHAB Programme, complementing the GEOHAB report series. GEOHAB also directly coordinated research activities, cosponsored other activities, and provided endorsement for targeted, national, and regional programmes aligned with programme objectives. Many authors identified GEOHAB as a contributor to their research efforts, even though there was no requirement for acknowledgement of GEOHAB in publications. The SSC tracked the GEOHAB

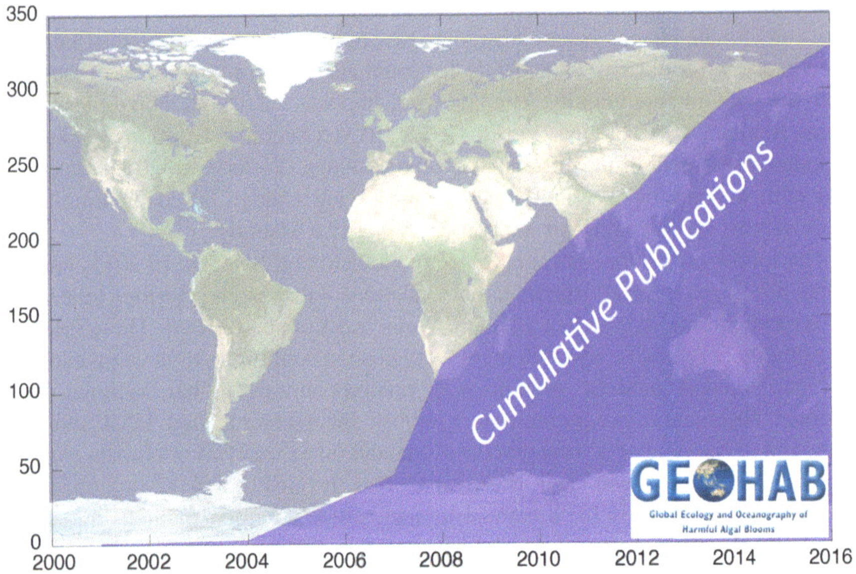

Fig. 3.4 GEOHAB is directly acknowledged in, or central to, 330 publications in 75 journals. Expanding this to include all publications that reference GEOHAB, there were more than 1000 publications as of 2016

documents, acknowledgements in other peer-reviewed manuscripts, and publications from key-endorsed projects and special journal issues sponsored by GEOHAB. At the end of 2016, there were 330 publications in 75 journals, and the GEOHAB Science and Implementation Plans were also cited 220 times. Expanding this analysis to include all publications that referenced GEOHAB in the title or text, there were more than 1000 publications as of 2016 (Fig. 3.4).

GEOHAB benefited from and had a positive influence on other international programmes. As early as the first organizing meeting (GEOHAB 1998), organizations including GOOS, LOICZ, GLOBEC, and the US ECOHAB and European EUROHAB Programmes were identified as partners. GEOHAB formally partnered with GOOS as part of the HABWatch activity. The increased visibility of HABs provided by GEOHAB and other efforts has led to the inclusion of phytoplankton generally and HAB organisms specifically, as proposed Biological Core Variables within the US Integrated Ocean Observing System (National Ocean Council 2016), while the GOOS Biology and Ecosystems Panel has identified phytoplankton species, with HABs as a specific subset, as an essential ocean variable. Open Science Meetings such as the "HABs and Climate Change" symposium, cohosted by GEOHAB, ICES, and PICES, continue to define research priorities for the next decade, while the regional planning and implementation activities established both nationally and regionally contribute to the mission and goals identified by the GEOHAB Programme. The Eutrophication CRP also benefited from a formal association with SCOR Working Group No. 132 on Land-Based Nutrients and Harmful Algal Blooms.

GEOHAB has also continued to raise awareness of the general scientific consensus that impacts from HABs on public health, recreational and tourism, fishery, aquaculture, and ecosystem have increased over the past few decades and these trends are continuing. GEOHAB has long championed the need for a coordinated, international effort, published a *Scientific Summary for Policy Makers* (Kudela et al. 2015), and contributed to the *Oceans and Society: Blue Planet* initiative coordinated by the Group on Earth Observations (GEO; Bernard et al. 2014). These efforts, while more intangible than a list of publications, continue to inform and guide research activities on HABs well beyond the formal end of the GEOHAB Programme.

The GEOHAB legacy continues beyond 2016 through ongoing activities initiated by GEOHAB and now maintained by GEOHAB partners. For example, a partnership was established between GEOHAB and the International Ocean Colour Coordinating Group (IOCCG), with the anticipated production of a monograph on satellite remote sensing of HABs. While GEOHAB has ended, the international scientific community requested that a programme like GEOHAB continues into the future. This is being implemented through formal adoption of the GlobalHAB programme by IOC (through IPHAB) and SCOR and endorsed by the International Atomic Energy Agency (IAEA). The GlobalHAB Science and Implementation Plan builds on the GEOHAB Science Plan and incorporates suggestions from the international HAB community who participated in the final GEOHAB OSM in Paris (Berdalet et al. 2017a). It is also enriched by many other contributions during the first year of the programme and subsequent inputs at the International Conference on Harmful Algae in Florianópolis, Brazil (October, 2016), where the first draft of the plan was presented. GlobalHAB builds on the scientific framework provided by GEOHAB but also expands into new disciplines and addresses emerging issues that were not addressed by GEOHAB. GlobalHAB will establish linkages with broader science domains (climate, toxicology, economy, medicine, public health), GlobalHAB will emphasize social science communications, and GlobalHAB will address management priorities [see Chap. 22, Berdalet et al. (2018)].

We look forward to seeing the programmes, initiatives, and ideas championed by GEOHAB living on through ongoing international coordination of HAB research. These ongoing and new programmes, such as GlobalHAB, will benefit from and build upon the GEOHAB legacy of informed science and international coordination to continue improving the prediction of HABs by determining the ecological and oceanographic mechanisms underlying their population dynamics, integrating biological, chemical, and physical studies supported by enhanced observation and modelling systems. These efforts will also continue to evolve, incorporating new challenges and a wider scientific scope, to translate knowledge into sound policy and to inform decision-making based on comprehensive science.

Acknowledgements We thank the many participants who have contributed to GEOHAB over the lifetime of the programme and in particular the 36 Science Steering Committee members who provided their time, ideas, and enthusiasm to make GEOHAB a success. Funding, programmatic support, and guidance were provided by SCOR, IOC, and IPHAB with significant support from the US National Science Foundation to SCOR for GEOHAB (Grants OCE-0003700, OCE-0326301, OCE-0608600, OCE-0938349, and OCE-1243377), US National Oceanic and Atmospheric Administration, and the European Commission. Support for completion of the document was provided by the Ida Benson Lynn Endowed Chair in Ocean Health and NOAA ECOHAB award NA11NOS4780030 to RMK and by the OstreoRisk project (CTM2014-53818-R) to EB. This is NOAA ECOHAB contribution number 922.

References

Anderson CR, Kudela RM, Kahru M et al (2016) Initial skill assessment of the California harmful algae risk mapping (C-HARM) system. Harmful Algae 59:1–8

Anderson DM, Cembella AD, Hallegraeff G (eds) (1998) Physiological ecology of harmful algal blooms, NATO ASI series. Series G: ecological sciences, vol 41. Springer, Berlin

Anderson DM, Cembella AD, Hallegraeff GM (2012) Progress in understanding harmful algal blooms: paradigm shifts and new technologies for research, monitoring, and management. Annu Rev Mar Sci 15:143–176

Anderson DM, Keafer BA, Kleindinst JL et al (2014) *Alexandrium fundyense* cysts in the Gulf of Maine: long-term time series of abundance and distribution, and linkages to past and future blooms. Deep Sea Res Part II 103:6–26

Anderson DM, Pitcher GC, Estrada M (2005) The comparative "systems" approach to HAB research. Oceanography 18:148–157

Babin M, Cullen JJ, Roesler CS et al (2005) New approaches and technologies for observing harmful algal blooms. Oceanography 18:210–227

Babin M, Roesler CS, Cullen JJ (eds) (2008) Real-time coastal observing systems for marine ecosystem dynamics and harmful algal blooms: theory, instrumentation and modelling. UNESCO, Paris

Berdalet E, Kudela R, Banas NS et al (2018) GlobalHAB: fostering international coordination on harmful algal bloom research in aquatic systems. In: Glibert PM, Berdalet E, Burford M et al (eds) Global ecology and oceanography of harmful algal blooms. Springer, Cham, pp 425–447

Berdalet E, Kudela R, Urban E (2017a) GlobalHAB: a new program to promote international research, observations, and modeling of harmful algal blooms in aquatic systems. Oceanography 30(1):70–81. https://doi.org/10.5670/oceanog.2017.111

Berdalet E, McManus MA, Ross ON et al (2014) Understanding harmful algae in stratified systems: review of progress and future directions. Deep Sea Res Part II 101:4–20

Berdalet E, Tester PA (2018) Key questions and recent research advances on harmful algal blooms in benthic systems. In: Glibert PM, Berdalet E, Burford M et al (eds) Global ecology and oceanography of harmful algal blooms. Springer, Cham, pp 261–286

Berdalet E, Tester PA, Chinain M et al (2017b) Harmful algal blooms in benthic systems: recent progress and future research. Oceanography 30(1):36–45. https://doi.org/10.5670/oceanog.2017.108

Bernard S, Kudela R, Velo-Suárez L et al (2014) Developing global capabilities for the observation and prediction of harmful algal blooms. In: Djavidnia S, Ott M, Seeyave S (eds) Oceans and society: blue planet. Cambridge Scholars Publishing, Newcastle upon Tyne, pp 46–52

Bravo I, Figueroa RI, Fraga S (2014) Cellular and nuclear morphological variability within a single species of the toxigenic dinoflagellate genus *Gambierdiscus*: relationship to life-cycle processes. Harmful Algae 40:1–8

Brosnahan ML, Velo-Suárez L, Ralston DK et al (2015) Rapid growth and concerted sexual transitions by a bloom of the harmful dinoflagellate *Alexandrium fundyense* (Dinophyceae). Limnol Oceanogr 60:2059–2078. https://doi.org/10.1002/lno.10155

Estrada M, Solé J, Anglès S et al (2010) The role of resting cysts in *Alexandrium minutum* population dynamics. Deep Sea Res Part II 57:308–321

EUROHAB (1999) European initiative on harmful algal blooms (EUROHAB): harmful algal blooms in European marine and brackish waters. Granéli E, Codd GA, Dale B et al (eds) European Commission, Directorate General Science, Research and Development, 93 pp

EUROHAB (2002) EUROHAB science initiative Part B: research and infrastructural needs. Granéli E, Lipiatou E (eds) National European and International Programmes, Office for Official Publications of the European Communities, Luxenbourg, 142 pp

European Commission (2003) The EU-US scientific initiative on harmful algal blooms: a report from a workshop jointly funded by the European Commission – Environment and Sustainable Development Programme and the US. National Science Foundation. Luxembourg, Belgium, p 68

Furuya K, Iwataki M, Lim PT et al (2018) Overview of harmful algal blooms in Asia. In: Glibert PM, Berdalet E, Burford M et al (eds) Global ecology and oceanography of harmful algal blooms. Springer, Cham, pp 289–308

GEOHAB (1998) Global ecology and oceanography of harmful algal blooms: a plan for co-ordinated scientific research and co-operation to develop international capabilities for assessment, prediction and mitigation. Cullen J (ed) Asian Natural Environmental Science Center, The University of Tokyo, 43 pp

GEOHAB (2001) Global ecology and oceanography of harmful algal blooms, science plan. Glibert P, Pitcher G (eds) SCOR and IOC, Baltimore and Paris, 86 pp

GEOHAB (2003) Global ecology and oceanography of harmful algal blooms, implementation plan. Gentien P, Pitcher G, Cembella A et al (eds) SCOR and IOC, Baltimore and Paris, 36 pp

GEOHAB (2005) Oceanography of harmful algal blooms, GEOHAB core research project: HABs in upwelling systems. Pitcher G, Moita T, Trainer V et al. (eds) Global ecology and IOC and SCOR, Paris and Baltimore, 88 pp

GEOHAB (2006) Global ecology and oceanography of harmful algal blooms: HABs in eutrophic systems. Glibert PM (ed) SCOR and IOC, Paris and Baltimore, 74 pp

GEOHAB (2008) Global ecology and oceanography of harmful algal blooms, GEOHAB core research project: HABs in stratified systems. Gentien P, Reguera B, Yamazaki H et al. (eds) SCOR and IOC, Paris, France, and Newark, Delaware, USA, 59 pp

GEOHAB (2010a) Global ecology and oceanography of harmful algal blooms, GEOHAB core research project: HABs in fjords and coastal embayments. Cembella A, Guzmán L, Roy S, et al. (eds) IOC and SCOR, Paris, France, and Newark, Delaware, 57 pp

GEOHAB (2010b) GEOHAB Asia, global ecology and oceanography of harmful algal blooms in Asia: a regional comparative programme. Furuya K, Glibert PM, Zhou M et al (eds) IOC and SCOR, Baltimore and Paris, 69 pp

GEOHAB (2011) GEOHAB modelling: linking observations to predictions, a workshop report. McGillicuddy DJ Jr., Glibert PM, Berdalet E et al (eds) IOC and SCOR, Paris and Newark, Delaware, 85 pp

GEOHAB (2012) Global ecology and oceangraphy of harmful algal blooms, GEOHAB core respearch project: HABs in benthic systems. Berdalet E, Tester P, Zingine A (eds) IOC and SCOR, Paris and Newark, 64 pp

GEOHAB (2013a) Global ecology and oceanography of harmful algal blooms, GEOHAB core research project: HABs in fjords and coastal embayments, second open science meeting. Roy S, Pospelova V, Montresor M et al. (eds) IOC and SCOR, Paris, France and Newark, Delaware USA, 52 pp

GEOHAB (2013b) Global ecology and oceanography of harmful algal blooms, GEOHAB core research project: HABs in stratified systems, advances and challenges for understanding physical-biological interactions in HABs in stratified environments workshop report. McManus MA, Berdalet E, Ryan J, (eds) IOC and SCOR, Paris and Newark, Delaware, USA, 88 pp

GEOHAB (2014) Global ecology and oceanography of harmful algal blooms, GEOHAB synthesis. Berdalet E, Bernard S, Burford MA, et al (eds) IOC and SCOR, Paris and Newark, Delaware, USA, 78 pp

Glibert PM, Al-Azri A, Allen JI et al (2018a) Key questions and recent research advances on harmful algal blooms in relation to nutrients and eutrophication. In: Glibert PM, Berdalet E, Burford M et al (eds) Global ecology and oceanography of harmful algal blooms. Springer, Cham, pp 229–259

Glibert PM, Anderson DM, Gentien P et al (2005) The global, complex phenomena of harmful algal blooms. Oceanography 18(2):136–147

Glibert PM, Azanza R, Burford M et al (2008a) Ocean urea fertilization for carbon credits poses high ecological risks. Mar Pollut Bull 56(6):1049–1056

Glibert PM, Burford MA (2017) Globally changing nutrient loads and harmful algal blooms: recent advances, new paradigms, and continuing challenges. Oceanography 30(1):58–69

Glibert PM, Burkholder JM, Granéli E et al (2008b) Advances and insights in the complex relationships between eutrophication and HABs: preface to the special issue. Harmful Algae 8:1–2

Glibert PM, Zhou M, Zhu M et al (2011) Preface to the special issue on eutrophication and HABs: the GEOHAB approach. Chin J Oceanol Limnol 29(4):719–723

Glibert PM, Pitcher GC, Bernard S et al (2018b) Advancements and continuing challenges of emerging technologies and tools for detecting harmful algal blooms, their antecedent conditions and toxins, and applications in predictive models. In: Glibert PM, Berdalet E, Burford M et al (eds) Global ecology and oceanography of harmful algal blooms. Springer, Cham, pp 339–357

Granéli E, Sundström B, Edler L et al (eds) (1990) Toxic marine phytoplankton. Elsevier, New York

Hallegraeff GM (1993) A review of harmful algal blooms and their apparent global increase. Phycologia 32(2):79–99

HARRNESS (2005) Ramsdell JS, Anderson DM, Glibert PM (eds), Harmful Algal Research and Response: A National Environmental Science Strategy 2005–2015. Ecological Society of America, Washington DC, 96 pp

ICES (2007) Report on the ICES-IOC-SCOR working group on GEOHAB implementation for the Baltic (WGGIB), 7–9 May 2007, Helsinki, Finland. ICES CM 2007/BCC:05, p 51

ICES (2013) Report of the ICES/PICES/IOC-GEOHAB workshop on harmful algae blooms in a changing world (WKHABCW), 18–22 March 2013, Friday Harbor Laboratories, USA. ICES CM 2013/SSGHIE:01, p 13

Karl DM, Dore JE, Lukas R et al (2001) Building the long-term picture: the U.S. JGOFS time-series programs. Oceanography 14:6–17

Kudela RM, Berdalet E, Bernard S et al (2015) Harmful algal blooms. A scientific summary for policy makers. IOC/UNESCO, Paris (IOC/INF-1320)

Martin JL (2002) Development for a Canadian GEOHAB (global ecology and oceanography of harmful algal blooms) program: 2001 workshop report. Can Tech Rep Fish Aquat Sci 2400: vii + 44

Mayr E (1982) The growth of biological thought. Harvard University Press, Cambridge, MA, p 974

McGillicuddy DJ Jr, Townsend DW, He R, Keafer BA, Kleindinst JL, Li Y, Manning JP, Mountain DG, Thomas MA, Anderson DM (2011) Suppression of the 2010 *Alexandrium fundyense* bloom by changes in physical, biological, and chemical properties of the Gulf of Maine. Limnol Oceanogr 56(6):2411–2426

National Ocean Council (2016) Biological and ecosystem observations within United States waters II: a workshop report to inform priorities for the United States Integrated Ocean Observing System. National Ocean Council, Washington DC, p 32

Onda DF, Lluisma OA, Azanza RV (2014) Development, morphological characteristics and viability of temporary cysts of *Pyrodinium bahamense* var. *compressum* (Dinophyceae) in vitro. Eur J Phycol 49:265–275

Pitcher GC, Pillar S (eds) (2010) Special issue on harmful algal blooms in upwelling systems. Prog Oceanogr 85(1–2):1–136

Pitcher G, Figueiras FG, Kudela RM et al (2018) Key questions and recent research advances on harmful algal blooms in eastern boundary upwelling systems. In: Glibert PM, Berdalet E, Burford M et al (eds) Global ecology and oceanography of harmful algal blooms. Springer, Cham, pp 205–227

Raine R, Berdalet E, McManus MA et al (2014) Understanding harmful algae in stratified systems: review of progress and future directions. Deep Sea Res Part II 101:1–254

Raine R, Berdalet E, Yamazaki H et al (2018) Key questions and recent research advances in research on harmful algal blooms in stratified systems. In: Glibert PM, Berdalet E, Burford M et al (eds) Global ecology and oceanography of harmful algal blooms. Springer, Cham, pp 165–186

Roy S, Montresor M, Cembella A (2018) Key questions and recent research advances on harmful algal blooms in fjords and coastal embayments. In: Glibert PM, Berdalet E, Burford M et al (eds) Global ecology and oceanography of harmful algal blooms. Springer, Cham, pp 187–203

Turner E, Haidvogel DB, Hofmann EE et al (2013) US GLOBEC: program goals, approaches, and advances. Oceanography 26(4):12–21

Wells ML, Karlson B (2018) Harmful algal blooms in a changing ocean. In: Glibert PM, Berdalet E, Burford M et al (eds) Global ecology and oceanography of harmful algal blooms. Springer, Cham, pp 77–90

Wells ML, Trainer VL, Smayda TJ et al (2015) Harmful algal blooms and climate change: learning from the past and present to forecast the future. Harmful Algae 49:68–93

Zingone A, Berdalet E, Bienfang P et al (2012) Harmful algae in benthic systems: a GEOHAB core research program. Cryptogamie. Algologie 33(2):225–233

Part II
Global Changes and Harmful Algal Blooms

Chapter 4
Changing Land-, Sea-, and Airscapes: Sources of Nutrient Pollution Affecting Habitat Suitability for Harmful Algae

Patricia M. Glibert, Arthur H.W. Beusen, John A. Harrison, Hans H. Dürr, Alexander F. Bouwman, and Goulven G. Laruelle

4.1 Introduction

When the first workshop related to the development of a programme related to the Global Ecology and Oceanography of Harmful Algal Blooms (GEOHAB) was held in 1998, there was consensus on two important issues (GEOHAB 1998). It was recognized first that there had been "dramatic increases in the impacts of HABs" and second that there were "no clear-cut demonstrations of the influence of specific activities on the frequency, intensity, and distribution of HABs." In the nearly two decades since the publication of that document, due in sizeable part to the efforts of the GEOHAB Programme and its associated activities, there is a new understanding,

P.M. Glibert (✉)
University of Maryland Center for Environmental Science, Horn Point Laboratory, Cambridge, MD, USA
e-mail: glibert@umces.edu

A.H.W. Beusen • A.F. Bouwman
Faculty of Geosciences, Department of Earth Sciences—Geochemistry, Utrecht University, Utrecht, The Netherlands

PBL Netherlands Environmental Assessment Agency, The Hague, The Netherlands

J.A. Harrison
Washington State University, Vancouver, WA, USA

H.H. Dürr
Ecohydrology Research Group, Department of Earth and Environmental Sciences, University of Waterloo, Waterloo, ON, Canada

G.G. Laruelle
Department Geoscience, Environment and Society (DGES), Université Libre de Bruxelles, Bruxelles, Belgium

UMR 7619 Metis, Sorbonne Universités, UPMC, Univ Paris 06, CNRS, EPHE, IPSL, Paris, France

P.M. Glibert et al. (eds.), *Global Ecology and Oceanography of Harmful Algal Blooms*, Ecological Studies 232, https://doi.org/10.1007/978-3-319-70069-4_4

and considerably more direct data, on these issues. It is now unequivocally recognized that the global expansion of HABs is continuing, with increasing abundance, frequency, and geographic extent of HABs (with new species being documented in new areas). It is also well accepted that human activities, especially nutrient pollution and climate changes, are an important, if not the most important, contributor(s) to this expansion (e.g., Heisler et al. 2008; Anderson 2014; Wells et al. 2015; Glibert and Burford 2017 and references therein). Food production and wastewater discharge accompanying the expanding global population contribute the most to nutrient pollution, but these intersect with other anthropogenic effects that have influenced freshwater flow and climate. In this chapter, we take a broad, global view of some of the massive changes that have occurred in nutrient loading as well as other physical factors that affect nutrient retention and therefore habitat suitability for HABs. Our goal here is not to describe which species may benefit under which condition, but to paint the overarching picture of changes that have occurred and that are conducive to increases in habitat suitability for HABs. We also highlight several approaches to modelling nutrient export that have been brought to bear in understanding the magnitude of these trends. While this analysis does not discount the importance of global transport of species leading to new introductions of species to new areas, and the effects of climate change and associated changes in physical structure and related environmental factors that can affect HABs [e.g., Moore et al. 2008; Doney 2010; Fu et al. 2012; see also Chap. 5, Wells and Karlson (2018)], it specifically focuses on the causes and consequences of changing nutrient loads. The rationale is that cells require nutrients for growth, and the magnitude of global nutrient changes over the recent past decades bears emphasizing [Heisler et al. 2008; Glibert et al. 2005; see also Chap. 12, Glibert et al. (2018)].

This chapter begins with an overview of changes in nutrient delivery to both the coastal and open oceans, by examining changes in land-based nutrient pollution and changes in freshwater flow, changes in sea-based nutrient pollution through aquaculture development, and changes in atmospheric nutrient deposition. Both observations and models are used in these global estimates. Second, coastal typology and the retentive nature of the different coastal types are reviewed (e.g., Pitcher et al. 2010; Dürr et al. 2011; Beusen et al. 2016). Third, applying a coastal eutrophication index comparing information on the changes in nitrogen (N) and phosphorus (P) relative to silicon (Si) (Billen and Garnier 2007; Garnier et al. 2010) and various scenarios of future change, projections of algal community composition globally for 2050 are made. Fourth, other modelling approaches of the additive effects of nutrients and climate change are reviewed showing the potential for increased algal blooms for several regions of the world.

The overarching conclusion of this synthesis is that increases in nutrient pollution are a global phenomenon, as is the alteration of water flow, altering nutrient export to coastal waters. Highest export of nutrients—from land-based sources, aquaculture sources, and atmospheric deposition—occurs in Asia, but exports from Europe and the USA are also high. An increasing risk of accelerating nutrient pollution in many world regions (western North America, eastern South America, and much of Africa and Asia) is also indicated. These changes, in turn, are altering the amount, proportion, and chemical forms of nutrients that can and do support HABs. The trajectory of more HABs is likely to increase, synergistically with other global and environmental changes, including those associated with a warmer world.

This synthesis of trends relies largely on two activities with goals that were aligned with GEOHAB: (1) Working Group 132 of the Scientific Committee on Oceanic Research (SCOR) Land-Based Nutrient Pollution and HABs and (2) a project funded under the auspices of the United Nations Environment Programme (UNEP)-Global Environmental Fund (GEF), Global foundations for reducing nutrient enrichment and oxygen depletion from land-based pollution, in support of the global nutrient cycle. The latter specifically addressed the need for more quantitative nutrient analysis, particularly in developing countries, estimated the magnitude of contributions from various nutrient sources within watersheds, and quantitatively analysed relationships between nutrient sources and coastal impacts.

4.2 Land-Based Nutrient Pollution

The global changes in N, P, and Si export to marine and freshwaters have been well documented (e.g., Beusen et al. 2016; Seitzinger et al. 2005, 2010). The global export of rivers, which has been estimated to be 40–60 Tg N per year (1 Tg = 1 million tons), has increased more than twofold since the 1860s (Boyer et al. 2006; Howarth 2008; Beusen et al. 2016). Such changes are a result of both increased human population and increases in per capita demand for nutrient-intensive foods such as meat, as well as energy and water (Seitzinger et al. 2010). Human population has now exceeded 7.5 billion people. Since 2007, most of the global population now resides in urban rather than rural areas, and it is expected that by 2050, about two thirds of all people will reside in urban areas, many of which are along coastlines (United Nations 2014). Most of the world's megacities (ten million people or more) are in China, and Africa and Asia are becoming urbanized at a more rapid pace than other parts of the world. While city growth may be favourable for some services, water and sanitation services have not grown at the same rate. Exploding urban populations have massive impacts on water supplies, and in many parts of the world, sewage remains untreated or minimally treated.

Global population growth requires growth in agriculture and animal production. In order to support this growth in agriculture, the global manufacture of N-based fertilizers has increased from <10 million metric tons N year^{-1} in 1950 to >150 million metric tons year^{-1} in 2013, with 85% of all chemical fertilizers having been produced since 1985 (Fig. 4.1a, d; Howarth 2008; Glibert et al. 2014b and references therein). In contrast to the enormous expansion in the global use of chemical N fertilizers, the use of P-based fertilizers has shown a much smaller increase, at a rate only about a third that of N (Fig. 4.1b, e; Sutton et al. 2013; Glibert et al. 2014b; Beusen et al. 2016; Bouwman et al. 2017), leading to a global

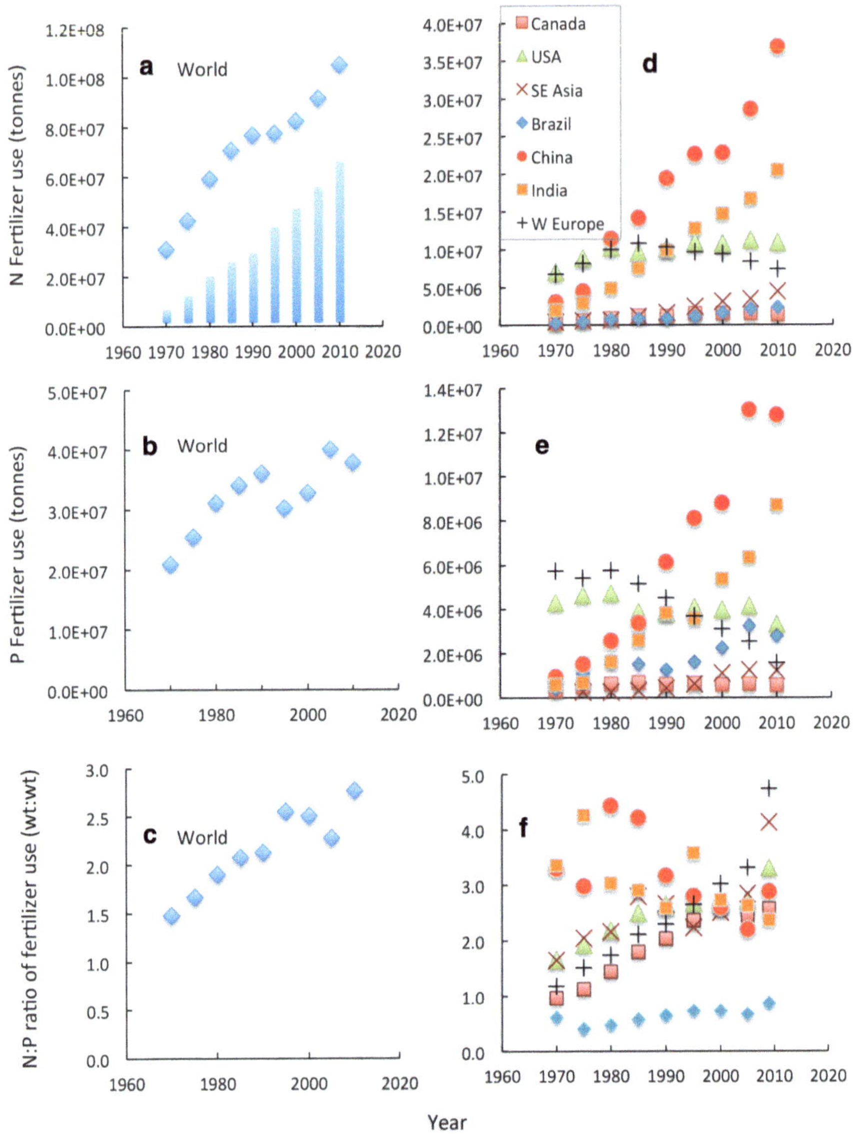

Fig. 4.1 N and P (as P_2O_5) fertilizer use and change in N:P ratio of fertilizer use by weight for the world (panels **a**, **b**, and **c**, respectively) and for selected countries or regions (panels **b**, **d**, and **e**, respectively). Superimposed on the world N use graph (**a**) is the fraction of N use as urea (*bars*). Total N and P data are from FAO (2012a, b) and data are the average of the 3 preceding years for each 5-year period; urea data are from Constant and Sheldrick (1992) through 1990 and estimated at 3.8% growth per year thereafter, comparable to urea data reported from IFA (2014). Figure reproduced from Glibert et al. (2014b) under creative commons licence

increase in N:P (Fig. 4.1c, f). Unlike N, which is fixed from the atmosphere via natural and anthropogenic processes (i.e., the Haber-Bosch process), P must be mined from the earth. Of these two major agricultural nutrients, only 10–30% actually reaches human consumers (Galloway et al. 2002; Houlton et al. 2013), and more than half is lost to the environment in direct runoff and atmospheric volatilization/eventual deposition (Galloway et al. 2014). In the case of N, a single molecule of fixed N can have multiple impacts on water, air, and soil and ecosystems (Galloway and Cowling 2002). Thus, accompanying this agricultural intensification has been an acceleration of the global N cycle, a trend projected to intensify further in the coming decades (e.g., Smil 2001; Glibert et al. 2006, 2014b; Beusen et al. 2015, 2016; Bouwman et al. 2009, 2017).

Nearly 60% of all N fertilizer now used worldwide is in the form of urea [CO $(NH_2)_2$; Fig. 4.1a; Constant and Sheldrick 1992; Glibert et al. 2006, 2014b; IFA 2014]. Global urea use as a fertilizer and feed additive has increased more than 100-fold in the past four decades (Glibert et al. 2006). It is projected that from 2012 to 2017, more than 50 new urea manufacturing plants will be constructed worldwide, half of them in China (Heffer and Prud'homme 2013), contributing to a further doubling of global urea use by 2050 (Glibert et al. 2006, 2014b). Urea can be a significant contributor to both total N and to the fraction used by phytoplankton in estuarine and coastal waters (Kaufman et al. 1983; Harrison et al. 1985; Glibert et al. 1991, 2006; Kudela and Cochlan 2000; Switzer 2008), and the frequency of reports that urea may be used preferentially by many harmful species has increased in recent years (Glibert et al. 2006 and references therein). Urea also rapidly hydrolyzes to NH_4^+ in water, another important N source for phytoplankton including HABs.

Human diets are also changing, with increasing proportions of meats as well as refined sugars. Animal agriculture is expanding to meet the dietary demands of an increasing population, and increasingly animal production is carried out in large concentrated animal feeding operations (CAFOs). To understand the scale of this nutrient source, as an example, in the Cape Fear River Basin of North Carolina, USA, it is estimated that there are 5 million hogs, 16 million turkeys, and 300 million chickens produced annually, yielding 82,700 t of N and 26,000 t of P in animal waste (Mallin et al. 2015 and references therein). In China, tens of thousands of CAFOs are estimated to produce 40 times more N pollution than other types of industries, as much of the manure produced in CAFOs is directly discharged into rivers (Ellis 2008). It has been further estimated that if unchecked, these dietary trends will, by 2050, be a major contributor to greenhouse gas emissions and global land clearing (Tilman and Clark 2014).

Spatially explicit global maps of dissolved river N and P export and dominant sources have been derived using a number of different models, including the Global Nutrient Export from WaterSheds (Global NEWS) models (Seitzinger et al. 2005; Dumont et al. 2005; Harrison et al. 2005a, b), and next-generation models, including the Integrated Model to Assess the Global Environment-Global Nutrient Model (IMAGE-GNM; Beusen et al. 2015, 2016). These models have documented not only large differences between regions and over time globally, but differences in nutrient export by element as well (Fig. 4.2a, b). The nutrient source terms that are

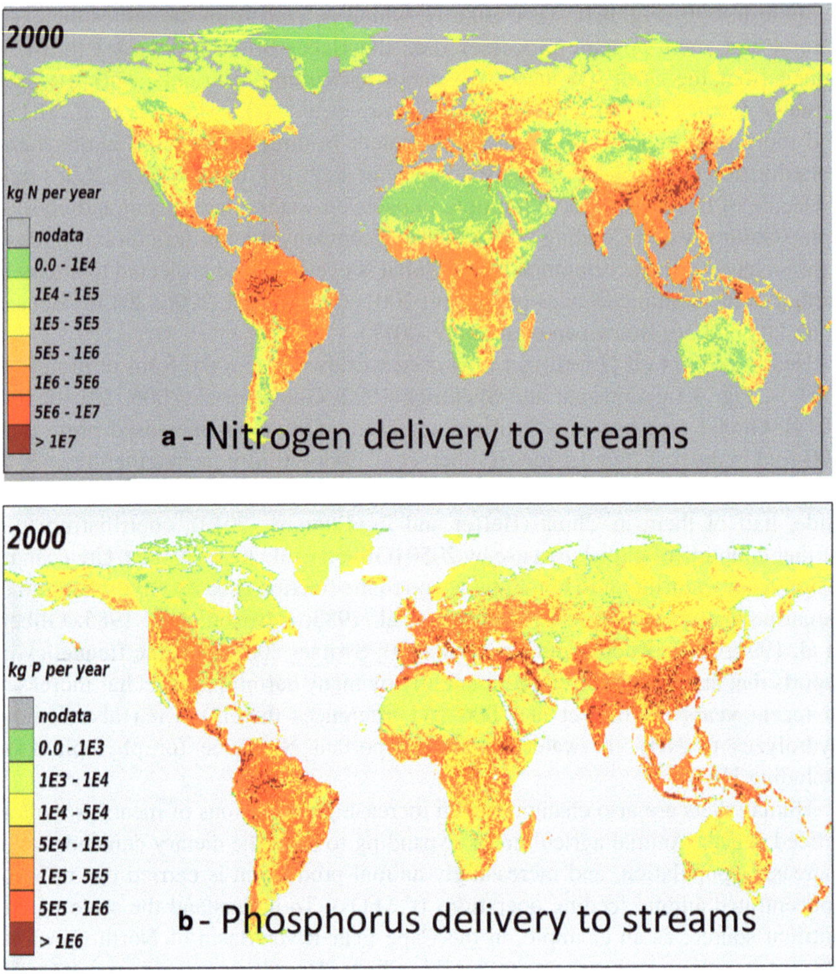

Fig. 4.2 Estimates of N (panel **a**) and P (panel **b**) delivery to streams for the year 2000 based on the Integrated Model to Assess the Global Environment-Global Nutrient Model as described in text. Figures reproduced from Supplemental Material of Beusen et al. (2016) under Creative Commons licence

considered in the Global NEWS models include natural sources such as N_2 fixation and P weathering and anthropogenic sources (nonpoint inputs from fertilizer by crop type, N_2 fixation by crops, atmospheric N deposition, and manure by animal species; point sources from sewage, as estimated by human population and treatment level; Seitzinger et al. 2005, 2010). The models also account for hydrological and physical factors including water runoff, precipitation intensity, land use and slope, as well as in-water removal processes such as dams and reservoirs and consumptive water use, and ultimately estimate nutrient export at the river mouths.

These models have shown that throughout much of Asia, Western Europe, and Eastern North America, fertilizer and manure together constitute the largest source of dissolved inorganic N (DIN) export (Dumont et al. 2005); agricultural sources of DIN (which include fertilizer, animal manures, and agricultural N_2 fixation) collectively contribute about half of the total DIN exported globally (Dumont et al. 2005). On the other hand, for dissolved inorganic P (DIP), human sewage is the largest anthropogenic source throughout much of the world, and inorganic P fertilizers and manures are less significant (Harrison et al. 2005b, 2010). The IMAGE-GNM, which is a coupled hydrology–biogeochemistry model (Beusen et al. 2015, 2016), advances the estimate of flow and retention or export of N and P by not only including the factors above but also including N and P wastewater from urban areas and from aquaculture (see Sect. 4.3) and accounts for additional in-river N transformations (Beusen et al. 2015). Both the NEWS and the IMAGE-GNM models reflect the same trends in nutrient stoichiometry as those in the fertilizer data (Fig. 4.1c), namely, that N:P ratios in water draining to the world's oceans are rapidly increasing (Beusen et al. 2016).

Previously, the Global NEWS models were used to explore the relationships between the HAB species, *Prorocentrum minimum*, and nutrient export [Glibert et al. 2008; see also Chap. 12, Glibert et al. (2018)]. Areas with high DIN and DIP yields, including Eastern Asia, Western Europe, and Eastern North America, coincide with areas in which *P. minimum* blooms have been documented and/or are increasing. This spatially explicit database also allowed identification of relationships between *P. minimum* occurrence and dominant form of N, as sewage, fertilizer, manure, or fixation of atmospheric sources and the extent to which the export of dissolved organic forms of N (DON) was attributable to anthropogenic sources.

While global models and global trends provide a macroscopic perspective, at the regional to local scale additional sources of nutrients will have to be considered in future modelling efforts. For example, for the Mediterranean Sea, it has been shown that submarine groundwater discharge likely delivers as many nutrients overall as riverine inputs (Rodellas et al. 2015). On a global scale, the N for submarine groundwater discharge is 1–1.4 Tg N (Beusen et al. 2013), but the local effect is much stronger. This is only true for N. Not for P. Even more important for near-shore areas in this region might be direct discharges of specific forms of N and P from wastewater treatment plants. While the global models estimate wastewater input, the specific level of treatment can lead to widely varying nutrient inputs even if servicing the same number of people. As described in the following section, local aquaculture is another source that is increasingly recognized to contribute regionally.

4.3 Changing Seascapes

In concert with changing diets, and in parallel with the increasing production of animals in CAFOs on land, aquaculture has been developing intensively, especially in Asia. This trend is also a function of declining natural fish stocks due to overfishing. Based on data from the Food and Agriculture Organization (FAO) on the United Nations, the annual growth rate in aquaculture was about 9% over the 1970–2000 period in both fresh and marine waters and decreased to ca. 6–7% between 2000 and 2010, with over 70% of this production in developing countries, mostly in Asia (FAO 2016); aquaculture production is expected to increase further in the coming decades (IFPRI 2003; Cork et al. 2005). Global mariculture (marine aquaculture) production is now ca. 40% of total aquaculture production, and in the most recent years, ca. 75% of mariculture is shellfish production; the remainder is finfish, including high-value marine and brackish water species (e.g., salmon, bream) in intensive farming systems in cages and net pens (FAO 2016). In 2010, 60% of global crustacean and 98% molluscan production was in marine or brackish water culture.

Aquacultural finfish production takes place in a variety of systems in freshwater and marine environments with different species and production intensities. Using data for the period 1950–2010 (derived from the database FISHSTAT, FAO 2012a, b, 2016), providing annual production, per species, country, sea, and type of environment, simple nutrient budget models were developed to describe the major flows of nutrients in shellfish and fish aquaculture systems (Bouwman et al. 2011, 2013b). The model approach describes nutrients in feed inputs, feed conversion, nutrients in fish, and the outflow in the form of faeces (solid forms based on apparent digestibility) and dissolved nutrients, as well as retention and recycling in pond and integrated aquaculture systems. Global cultured production of finfish and crustacea contributed an estimated 1.7 million tons of N and 0.46 million tons of P to receiving waters during 2008 (Verdegen 2013). Within the relatively short period from 2000 to 2006, nutrient release from shellfish cultures increased by 2.5- to 3-fold, and much larger increases are predicted in nutrient contributions from shellfish cultures by 2050 (Bouwman et al. 2011). While globally these values are small compared to riverine input, on a regional basis, they can be substantial (Bouwman et al. 2013a). As further described below, where the receiving environment is especially retentive, these nutrients can be retained, recycled, and transformed.

The implications of the expansion in aquaculture with regard to HABs are several (Bouwman et al. 2013a). Toxic and fish-killing algae are commonly associated with finfish and molluscan production (Honkanen and Helminen 2000; Wang et al. 2008), while high-biomass bloom-forming algae are more commonly associated with pond production (Alonso-Rodríguez and Páez-Osuna 2003; Wang et al. 2008). Of particular concern in both shellfish and finfish mariculture is the change in the form of nutrients as a result of excretion and associated microbial remineralization. Between 7 and 12% of the dissolved N waste of finfish consists of urea, the remainder being NH_4^+ (Kaushik and Cowey 1991). As described above, there is a mounting evidence that reduced N forms (in contrast to oxidized N forms), including urea, differentially stimulate the growth of some types of HABs (Anderson et al. 2002; Berg et al. 2003; Glibert et al. 2005, 2016), or may fuel the production of algae on which mixotrophic HABs may feed (Heisler et al. 2008;

Flynn et al. 2013). Relationships between increasing prevalence of HABs and aquaculture operations are increasingly reported with associated economic impacts (GEOHAB 2010). In China, for example, about US$2.5 million in finfish revenues were lost in 2005 due to one HAB event (Li et al. 2009). Single HAB fish-kill events in Korea have been estimated to have cost from US$1 to 100 million in lost fish, while in Japan such events have been estimated to have resulted in losses of fish worth more than US$300 million (GEOHAB 2010).

4.4 Coastal Typology and Anthropogenic Changes in Water Flow: Nutrient Retention Effects

The magnitude and effect of land-based nutrient pollution depend not only on how much nutrient is exported from land sources, but how those nutrients are retained in coastal or receiving waters and how modifications of river flow may alter nutrient export. The retention of N and P is shown to vary greatly with coastal type and latitude. Dürr et al. (2011) defined four major types of near-shore coasts, including small deltas, tidal systems, lagoons, and fjords based on hydrological, lithological, and morphological criteria. These systems represent the estuarine filter, where riverine waters mix with those of the coastal ocean, and the complex interplay of physical and biogeochemical processes takes place altering the riverine nutrient delivery to the continental shelves (Regnier et al. 2013). Fjords are particularly effective in retaining nutrients with characteristic residence times as long as decades, while small deltas have a low filtering capacity. Estuaries in temperate regions of the Northern Hemisphere are particularly important as a sink for nutrients because of the prominence of relatively long residence time tidal systems. In tropical and subtropical regions, retention is much less efficient, and a large proportion of the nutrient inputs to the ocean bypass the coastal filter by discharging directly onto the shelves in large rivers. Large rivers largely bypass the near-shore filter (McKee et al., 2004), while karstic and arheic coasts act as inactive filters. Of the nutrient inputs that enter the estuarine filter, an estimated 22% of the riverine N and 24% of the riverine P are retained (Laruelle 2009; Fig. 4.3a). These values are consistent with the limited number of system scale nutrient budgets derived from either observation (Nixon et al. 1996) or models (Arndt et al. 2009; Volta et al. 2014). These retentions are complementary to those calculated for continental shelves (Sharples et al. 2017) which predict that 75% global DIN and 80% of global DIP that enter continental shelf waters reach the open ocean. These figures suggest that the removal of nutrients taking place in estuaries and continental shelves are equally effective.

An accurate quantification of removal of nutrients by the estuarine filter generally requires extensive modelling work. The development of reactive transport models (RTMs) has advanced the capture of these complex hydrological properties (Arndt et al. 2009; Regnier et al. 2013; Volta et al. 2014). However, such complex

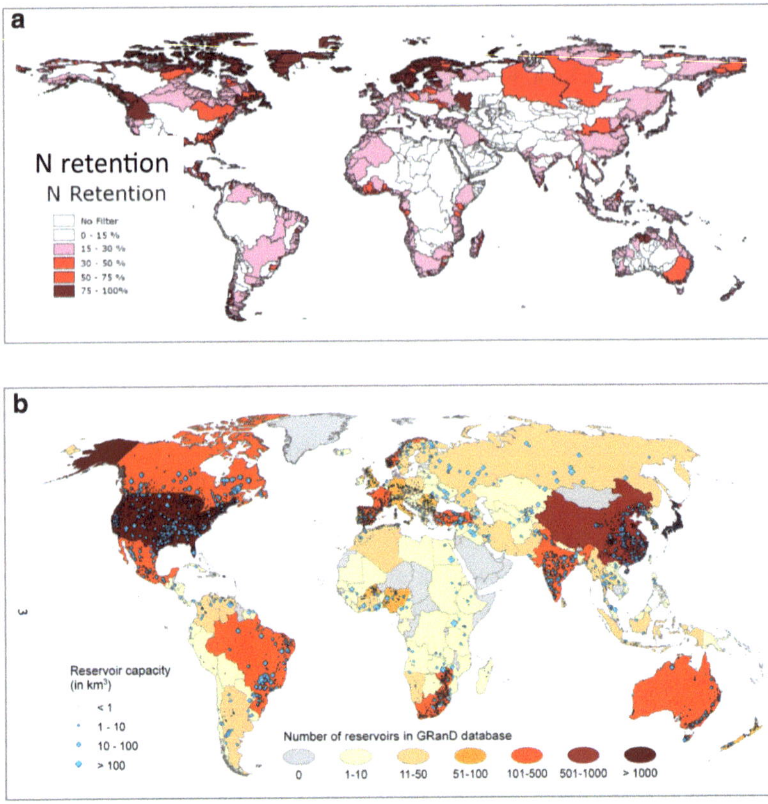

Fig. 4.3 Nitrogen retention in coastal gridcells (panel **a**). Reproduced from Laruelle (2009). Global distribution of large dams and their reservoirs, overlaid with the percentage of sediment trapped within watersheds by dams (panel **b**). Reproduced from http://www.gwsp.org/products/grand-database/global-reservoir-and-dam-grand-database-project.html (open access)

and data-intensive work can generally only be performed one system at a time, and regional applications of RTMs with less data demanding setups (Volta et al. 2014) remain in their infancy (Volta et al. 2016; Laruelle et al. 2017). As a consequence, global estimates of the efficiency of the estuarine filter in terms of nutrient retention still currently rely on up-scaling from relationships derived from a limited number of well-studied systems. Such methodology was applied by Laruelle (2009) combined with the coastal typology established by Dürr et al. (2011) to develop a global scale, spatially explicit model of the coupled N and P cycles in the estuarine zone. The model consists of a set of generic biogeochemical box models for each coastal type and was applied to estimate the estuarine retention of N and P from rivers. The model consists of a "ribbon" of cells distributed along the entire global coastline at a half-degree resolution. The model calculates the nutrient retention in each coastal cell using generic box models relying on simple empirical relationships. This typology-based modelling tool can provide an interface between spatially explicit global models of river discharge of nutrients (Seitzinger et al. 2010) and ocean global circulation models (Bernard et al. 2010, 2011; Heinze et al. 1999). The modelled retention of N and P for individual boxes compares reasonably well to data for local studies.

Coastal retention has important implications for HABs. On the one hand, in retentive regions, nutrients can be retained, recycled, and transformed, keeping the site in a nutrient-enriched condition. This can lead to enhanced phytoplankton accumulation, high-biomass HABs, and hypoxia. Coastal retention is also important in understanding impacts of aquaculture-derived nutrients, whether they remain within confined regions, in which case they are more likely to promote HABs, or whether they are likely to be flushed in a well-circulated system. Retentive zones are also important for HABs, especially those that form cysts, as the cysts accumulate and can lead to blooms whenever favourable conditions occur. Shallow, enclosed coastal lagoons can be subject to recurrent blooms once a population of HAB becomes established.

Coastal typology and features such as banks, canyons, and islands affect not only the retention of nutrients from land-based sources but also affect local circulation patterns that may transport cells or nutrients from offshore to inshore. As Pitcher et al. (2010) summarize, a rise in bottom topography can create conditions conducive to upwelling, while offshore islands may result in eddies that can influence the distribution of phytoplankton including HABs. For example, on the US West Coast, regions such as the Juan de Fuca Eddy, the Farallon Island regions, the Heceta Bank, the Monterey Bay, and the Santa Barbara Channel are all regions where *Pseudo-nitzschia* blooms are frequently observed due to their retentive nature (Hickey and Banas 2003; Pitcher et al. 2010).

In addition to effects of natural coastal typology and coastal features on nutrient retention and export, the export of nutrients to receiving waters is also affected by anthropogenic alteration of river discharge. It has been estimated that by 2030, 93% of all rivers will be affected by dams (Grill et al. 2015). Thousands of dams of a capacity of more than 1 MW have been constructed during a first construction boom in the 1950s and 1960s, and a second boom is now underway, mostly in developing economies (Grill et al. 2015; Zarfl et al. 2015; Fig. 4.3b). In fact, retention of nutrients in lakes increased >60% between 1900 and 2000 due to reservoir construction and other anthropogenic flow alterations (Beusen et al. 2016). Dam and reservoir construction and other in-river consumptive uses, as well as channelization of flow, have greatly modified the timing, magnitude, form, and stoichiometry of nutrient delivery to the coast as P is retained more efficiently than N during processing and transport in soils, groundwater, riparian zones and streams, rivers, lakes, and reservoirs (Beusen et al. 2016). For the major global dams, reactive P (total dissolved P + reactive particulate P) is more efficiently reduced (by 43%) than reactive Si (dissolved Si + reactive particulate Si) (by 21%) and total N (12%) (Maavara 2017; Maavara et al. 2014, 2015), thus increasing the N:P and N:Si ratios of riverine delivery to coastal areas. Changes in Si availability have also occurred due to sediment trapping and elemental transformations following construction of dams (e.g., Billen et al. 1999; Vörösmarty et al. 2003; Beusen et al. 2005, 2009; Syvitski et al. 2005; Harrison et al. 2012). In addition to the reductions in overall river flow that occurs from dam construction, large river systems can become fragmented preventing free movement of organisms, and severe modification of river flow alters temperature regimes and dramatically reduces sediment transport (Vörösmarty et al. 2010; Lehner et al. 2011; Liermann et al. 2012). Collectively, it is increasingly recognized that altered flow modifies not only salinity but also nutrients, in terms of total loads and proportions, and these changes can create conditions where phytoplankton assemblages change in composition. For example, retention of Si upstream following construction of the Three Gorges Dam in China has been considered to be an important factor leading to altered nutrient proportions favouring HABs in East China Sea (e.g., Zhang et al. 2015).

4.5 Changing Airscapes

In many regions, atmospheric deposition of N is not only high but increasing (e.g., Howarth 2006; Galloway et al. 2008; Duce et al. 2008). Atmospheric deposition of N reaches >700 mg N m^2 $year^{-1}$ in many regions, particularly the downwind plumes from major cities (e.g., Duce et al. 2008). Globally, emissions, mainly of ammonia (NH_3) from land, together with combustion-related emissions of nitrogen oxides (NO_x), contribute 100 Tg N $year^{-1}$ to the atmosphere (Fowler et al. 2013). This N is derived from increasing NO_x emission from fossil fuel burning and from volatilization of animal manure and other land-based fertilizer applications. Older estimates for both European and US coastal waters suggested that anthropogenic atmospheric N deposition contributes from 10 to 40% of new N loading (Jaworski et al. 1997). Recent modelling has shown that there has been a threefold increase in soluble N deposition over the land and a twofold increase over the ocean in terms of human activities since 1850 (Kanakidou et al. 2016). All forms of chemically reduced N relative to oxidized N are also increasing in atmospheric deposition, driven largely by emissions of NH_3 from agriculture. Eastern China has experienced not only high dry deposition fluxes but also has experienced the greatest increase in dry deposition fluxes over the past decade globally (Jia et al. 2016). Moreover, emissions of NH_3 are further projected to increase and to become the dominant component of N emissions with rising temperatures (Fowler et al. 2013). For HABs, this is significant, as, has been noted above, many HAB taxa may preferentially use, grow more rapidly on, or become more toxic, when the N substrate is in chemically reduced form (Glibert 2017 and references therein).

4.6 Eutrophication Potential and Global HAB Distribution

Building on the understanding that nutrient loads are changing both in total quantity and quality, there have been various efforts to relate these nutrient compositional changes to changes in phytoplankton community assemblage. One approach that has attempted to quantify the change in nutrient ratios globally is the Indicator for Coastal Eutrophication Potential (ICEP) approach (Billen and Garnier 2007). This approach estimates the total production of non-Si algal biomass that can develop in a water body in excess of that which would be required to sustain Si biomass, i.e., non-diatom versus diatom biomass (Billen and Garnier 2007).

This index has been compared to the globally available HAB data, maintained in the Intergovernmental Oceanographic Commission (IOC) HAEDAT database (Fig. 4.4a). For this analysis, each observed HAB was assigned to the corresponding Large Marine Ecosystem (LME) region. The results for the year 2000 show an agreement between positive ICEP values and observed HABs, especially in the European region, but the Asian relationship, where nutrient loading from land, sea, and air sources is especially high, was not strong (Fig. 4.4a). However, it is important to note that HAEDAT is a metadatabase under continued development containing records of harmful algal events from various parts of the world, but by virtue of differences in regional reporting, it does not yet provide a complete global perspective. The available information on individual events varies greatly from event to event and from country to country. Regions such as the ICES area (North Atlantic) since 1985 and from the PICES area (North Pacific) since 2000 are well represented, but HAB events from Asia, South America, and North Africa are generally still underrepresented in this database. For example, the vast expansion of both high-biomass HABs and toxic HABs, including those causing paralytic shellfish poisoning in Asia in the past several decades (e.g., GEOHAB 2010), is not at present well documented in HAEDAT. Undergoing efforts to enlarge HAEDAT will facilitate comparisons with the ICEP index in the future.

Furthermore, local physical and environmental conditions will, apart from the nutrient loading and element ratios used in the ICEP concept, determine the propensity of a coastal marine ecosystem to develop high-biomass algal blooms or hypoxia. This global ICEP-HAEDAT comparison, as was the case with the global projection of *P. minimum* blooms described above, is considered a first step in our understanding of the relationships between HABs and global nutrient loads and their changes. Moreover, measures such as nutrient yields and ICEP values are annual averages whereas HABs frequently are ephemeral events, and there has been no effort made to incorporate the event time scale. Thus, there may be a temporal mismatch. Notwithstanding all these limitations, on a global basis, these projections illustrate the propensity for high-biomass HABs to occur where and when dissolved N and P yields are high.

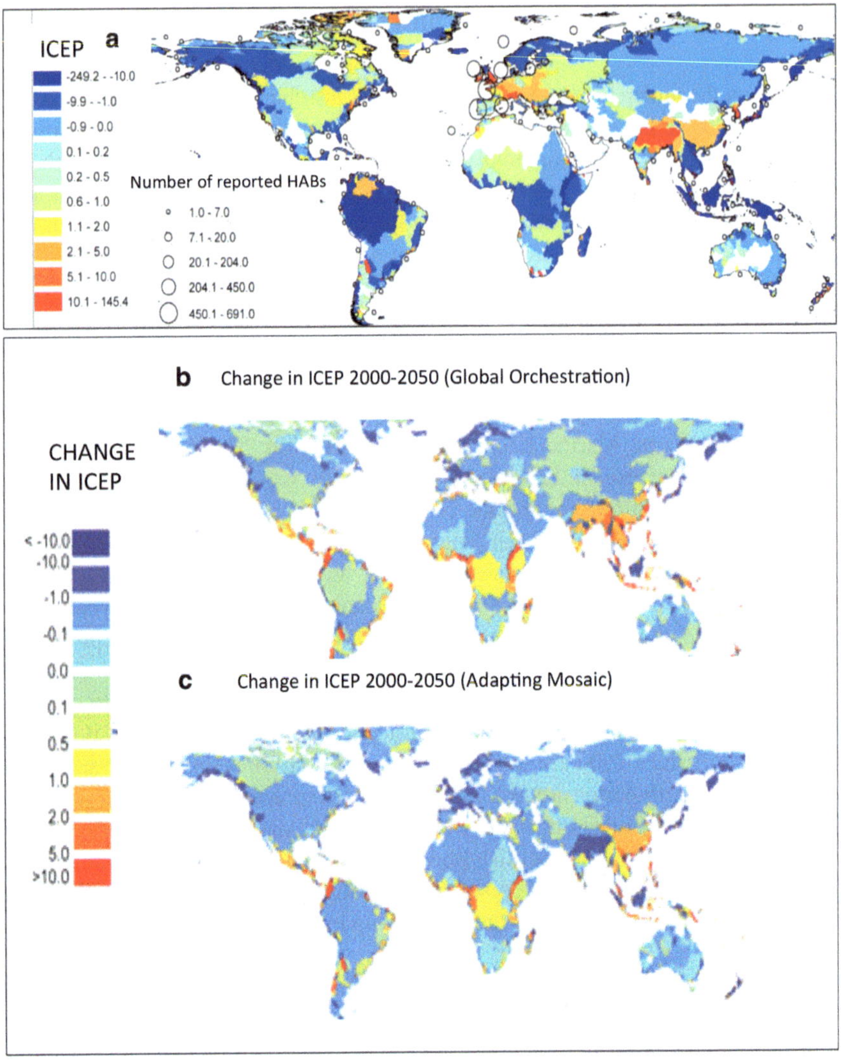

Fig. 4.4 The eutrophication index ICEP estimated globally for the year 2000 (panel **a**) and number of HABs (based on the HAEDAT data). Changes in ICEP globally estimated for the years 2000–2050 with the assumption associated with two different development scenarios, Global Orchestration (panel **b**) and Adapting Mosaic (panel **c**)

4.7 Future Projections: Millennium Ecosystem Assessment Scenarios

Future projections of eutrophication were made using a suite of assumptions described in the Millennium Ecosystem Assessment (MA) (Cork et al. 2005) and the ICEP approach. These assumptions were, in turn, based on storylines developed by the Intergovernmental Panel on Climate Change (IPCC) and translated into changes of the main anthropogenic drivers, i.e., demography, economic development, and agricultural production (Alcamo et al. 2006). Although the MA actually defines four scenarios that differ in terms of environmental management and in degree and scale of connectedness among and across country borders, here we focus on only two of these scenarios for illustration. The "Adapting Mosaic" (AM) scenario was developed assuming proactive and regional environmental management, while the "Global Orchestration" (GO) scenario assumes reactive environmental management and a trend towards globalization (Alcamo et al. 2006).

In the coming decades, coastal zones in many world regions are almost certain to see increases in river export of N and P, even accounting for increased retention by reservoirs. The MA scenarios for 2050 show major increases in N and P river export to coastal ecosystems, particularly in South and Eastern Asia and in many countries in South America and Africa. In order to guarantee food security for populations in developing countries, and to prevent land degradation or restore soil fertility, fertilizer use will have to increase in these parts of the world. As a consequence, nutrient losses by leaching, volatilization, and runoff will inevitably increase. At the same time, urbanization and lagging sewage connection and treatment of wastewater will lead to increasing nutrient discharge to surface water in developing countries. In contrast, important decreasing trends are projected in Europe and stabilization is projected in North America and Australia. Industrialized countries are assumed to reduce nutrient discharge to rivers by developing improved wastewater treatment systems and also by reducing NH_3 volatilization, leaching, and runoff by improved nutrient management.

Meanwhile, Si river export is decreasing globally as a result of eutrophication and retention in the increasing number of reservoirs in the world's river systems. The result of these simultaneous changes of N, P, and Si is an increasing ICEP value in many world regions [western part of North America, eastern part of South America, and many parts of Africa and Asia, indicating an increasing risk that severe problems associated with eutrophication may occur (Fig. 4.4b, c)]. It is worth noting that while changes in Africa were slow between 1970 and 2000, in the coming decades, changes in nutrient stoichiometry may be more significant, probably the result of the expected fast population growth and all associated societal and economic changes, such as increasing food and energy production. With increased dam construction and due to the preferential retention of P over Si and N in reservoirs, rivers that are N-limited are under greater risk of future Si limitation (Maavara 2017).

Historical data suggest that HAB risk increased considerably between 1970 and 2000. Scenario analyses for 2050 indicate that this risk will further spread (South

America, Africa) and increase in areas with current high risk (Eastern Asia). There are also large parts of the world where the HAB risk is expected to decrease as a result of higher efficiency of nutrient use in agriculture and improved wastewater treatment. This is particularly so in the AM scenario, which is a scenario with an orientation towards proactive environmental management and simple, local solutions (Fig. 4.4c).

4.8 Future Projections: Global Ecosystem Modelling Approaches

The trajectory of more HABs is only going to be additive with other global changes, such as those associated with a warmer world [Wells et al. 2015; Sinha et al. 2017; see also Chap. 5, Wells and Karlson (2018)]. Average sea surface temperatures are expected to rise as much as 5 °C over the coming century, leading to a freshening of many oceanic regions due to ice melt and altered precipitation (e.g., Moore et al. 2008; Doney 2010; Fu et al. 2012, and references therein). These changes, in turn, will alter stratification, availability of nutrients and their forms and ratios, pCO_2, and light regimes among other factors (e.g., Boyd and Doney 2003), all of which control the extent to which HABs become established, recurrent features and likely will create many "windows of opportunity" for HABs to thrive.

Two modelling approaches highlighting the effects of nutrients and climate changes as multistressors are given here. First, a suite of model projections of the effect of climate change, together with spatially explicit nutrient loads, was undertaken to estimate the potential change in HAB distribution in several regions of the globe, NW European Shelf–Baltic Sea system, NE Asia, and SE Asia (Glibert et al. 2014a). In this modelling effort projections of the effects of climate and nutrient changes on the potential for expansion of specific harmful algal genera were made by applying a coupled oceanographic-biogeochemical model (Holt et al. 2009), combined with a suite of assumed physiological "rules" for genera-specific bloom development and habitat suitability (Glibert et al. 2014a). Habitat suitability was defined by a ratio of NH_4^+:NO_3^- that exceeded 1, and an inorganic N:P ratio that was stoichiometrically imbalanced, together with genus-specific temperature and salinity criteria. Using the dinoflagellates *Prorocentrum* and *Karenia* spp. as examples, since they are globally common genera that are often associated with eutrophication, the risk of future expansion was examined in several oceanographic regions, including northwestern European Shelf–Baltic Sea system and northeastern Asia. Climate projections for "present-day" (years 1980–1990) and "future" (years 2090–2100) temperature scenarios were based on the IPCC Assessment Report on Climate Change (the "A1B" midline scenario; Solomon et al. 2007).

Model projections indicated variable habitat expansion of *Karenia* spp. and planktonic *Prorocentrum* spp. HABs under the applied assumptions of climate change (Glibert et al. 2014a; Fig. 4.5). Along the northern European coast, there

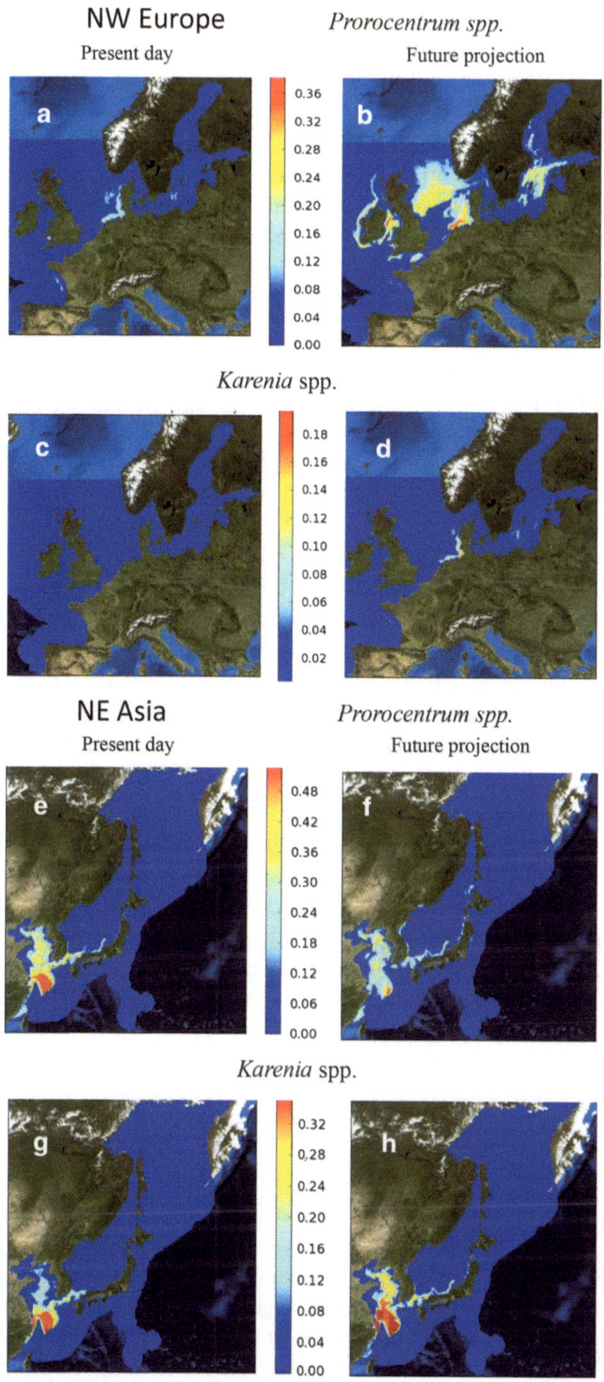

Fig. 4.5 Output of the coupled oceanographic-biogeochemical model described for NE European/ Baltic Sea region. Panels (**a**) and (**b**) depict the spatial distribution of habitat suitability (spatially explicit fraction of time of the year for which all suitable conditions were met) for *Prorocentrum*

was a considerable expansion in the number of months annually conducive to both HAB genera, but this was most notable for planktonic *Prorocentrum*. The expansion of these HABs in the future scenarios was less for the Asian coast, but there appears to be a northern geographic expansion. The projected increases in temperature and nutrient conditions suggest an even greater potential for expansion of these blooms, but the overlap of these conditions could limit the manifestation of these effects. A temporal mismatch in suitability of conditions for growth should not be viewed as evidence for limited potential for expansion. Rather, such a difference points to the possibility of expansion should there be a change in the timing of any one of the parameters. Climate forcing may alter the timing of nutrient loads relative to seasonal warming that may in turn alter the alignment of suitable conditions. Overall, the model projections described here showed the future potential expansion for these two HAB genera in two large oceanic regions, and it is foreboding. Many factors are involved, and different species may show quite different projections. Nevertheless, collectively it is clear that the expansion of HABs, exemplified by *Karenia* spp. and planktonic *Prorocentrum* spp. in some regions, is likely to continue in the coming years.

Supporting evidence for the worsening of eutrophication together with climate changes, and therefore the propensity for the worsening of HABs, comes from a recent model that explored climate change-induced precipitation changes and that showed a large potential for increases in N loading and eutrophication by the end of the century (Sinha et al. 2017). Estimates in this effort were derived from the Climate Model Intercomparison Project Phase 5 (CMIP5) models and the "business-as-usual" scenario. While the empirical model was specific to the USA, the model was applied globally by seeking regions that met specific criteria, and a number of regions with similar conditions were identified. In so doing, the Sinha et al. (2017) study reported that large portions of Asia, especially India and Eastern China, had conditions similar to some regions in the USA, including the Mississippi–Atchafalaya River Basin, and the northeast and Great Lakes regions. As discussed throughout this chapter, these regions are already global nutrient hotspots, as well as regions of frequent and/or increasing HAB occurrences, and thus these model projections serve to underscore the magnitude of the challenge of managing nutrient loads.

Fig. 4.5 (continued) spp. for present (which encompasses period from 1980 to 1990) and future conditions projected using A1B IPCC scenarios for climate change (which encompasses period from 2090 to 2100). Panels (**c**) and (**d**) are the same except for conditions suitable for *Karenia* spp. Panels (**e**)–(**h**) are the same, except for NE Asia. Reproduced and modified from Glibert et al. (2014a) with permission from the publisher

4.9 Conclusions

Recognizing the vast anthropogenic effects that nutrient pollution, harvesting and production of food (including associated fertilizer use), and altered hydrodynamics for water consumption or electricity generation are having on the globe is fundamental to understanding how these changes affect ecological function and biodiversity, including microbial biodiversity and HABs. Nutrient pollution in retentive coastal zones has fundamentally different effects than nutrient input into systems with less retentive properties. Nutrient hotspots are clear around the globe, with severe nutrient loading issues in Asia, Europe, and the USA. Scenario analysis shows that these and emerging regions around the globe will continue to face nutrient-related problems, including expanding HAB issues for decades to come. Additionally, climate change is not only altering environmental conditions, but it is altering the seasonality and timing of co-occurring suitable factors for HAB growth.

There are many opportunities to advance the understanding of HABs and environmental changes, and continued international collaborative programmes, like Global NEWS and GlobalHAB [see also Chap. 22, Berdalet et al. (2018)], will be essential to further our understanding of changes in HABs. Such advances in understanding of both HABs and the environmental conditions to which they are exposed need to be achieved across the full spectrum of scales and across the land-, sea-, and airscapes that are so rapidly changing. Multiple, co-occurring changes such as nutrient pollution, increasing reservoir capacity resulting in the increased retentiveness of rivers, and increasing global temperatures all suggest that proactive management will be required to stabilize or reduce HAB occurrences.

Acknowledgments This effort is a contribution of SCOR Working Group 132 on Land-Based Nutrient Pollution and Harmful Algal Blooms and is also contribution number 5404 from the University of Maryland Center for Environmental Science. This work received support from the Global Environment Fund, UNEP, and UNESCO-IOC funding to A. Bouwman, A. Beusen, and J. Harrison, as well as USDA Water Sustainability and Climate, USDA Earth Systems Modeling, and NSF INFEWS funding to Harrison. Laruelle is Chargé de recherches du F.R.S.-FNRS at the Université Libre de Bruxelles. Dürr received funding from the Canada Excellence Research Chair in Ecohydrology.

References

Alcamo J, Van Vuuren D, Cramer W (2006) Changes in ecosystem services and their drivers across the scenarios. In: Carpenter SR, Pingali PL, Bennett EM et al (eds) Ecosystems and human well-being: scenarios. Island Press, Washington, DC, pp 297–373

Alonso-Rodríguez R, Páez-Osuna F (2003) Nutrients, phytoplankton and harmful algal blooms in shrimp ponds: a review with special reference to the situation in the Gulf of California. Aquaculture 219:317–336

Anderson DM (2014) HABs in a changing world: a perspective on harmful algal blooms, their impacts, and research and management in a dynamic era of climatic and environmental change. In: Kim H-G, Reguera B, Hallegraeff GM et al (eds) Harmful algae 2012: Proceedings of the

15th International conference on harmful algae: October 29–November 2, 2012, CECO, Changwon, Gyeongnam, pp 3–17
Anderson DM, Glibert PM, Burkholder JM (2002) Harmful algal blooms and eutrophication: nutrient sources, composition, and consequences. Estuaries 25:704–726
Arndt S, Regnier P, Vanderborght JP (2009) Seasonally-resolved nutrient export fluxes and filtering capacities in a macrotidal estuary. J Mar Syst 78:42–58
Berdalet E, Kudela R, Banas NS et al (2018) GlobalHAB: fostering international coordination on harmful algal bloom research in aquatic systems. In: Glibert PM, Berdalet E, Burford M et al (eds) Global ecology and oceanography of harmful algal blooms. Springer, Cham, pp. 425–447
Berg GM, Balode M, Purina I et al (2003) Plankton community composition in relation to availability and uptake of oxidized and reduced nitrogen. Aquat Microb Ecol 30:263–274
Bernard CY, Dürr HH, Heinze C et al (2011) Contribution of riverine nutrients to the silicon biogeochemistry of the global ocean – a model study. Biogeosciences 8:551–564
Bernard CY, Laruelle GG, Slomp CP et al (2010) Impact of changes in river nutrient fluxes on the global marine silicon cycle: a model comparison. Biogeosciences 7:441–453
Beusen AHW, Deckers ALM, Bouwman AF et al (2005) Estimation of global river transport of sediments and associated particulate C, N and P. Global Biogeochem Cycles 19:1–17
Beusen AHW, Slomp CP, Bouwman AF (2013) Global land-ocean linkage: direct inputs of nitrogen to coastal waters via submarine groundwater discharge. Environ Res Lett 8:034035
Beusen AHW, Bouwman AF, Dürr HH et al (2009) Global patterns of dissolved silica export to the coastal zone: results from a spatially explicit global model. Global Biogeochem Cycles 23: GB0A02. https://doi.org/10.1029/2008GB003281
Beusen AHW, Bouwman AF, Van Beek LPH et al (2016) Global riverine N and P transport to ocean increased during the 20th century despite increased retention along the aquatic continuum. Biogeosciences 13:2441–2451
Beusen AHW, Van Beek LPH, Bouwman AF et al (2015) Coupling global models for hydrology and nutrient loading to simulate nitrogen and phosphorus retention in surface water – description of IMAGE–GNM and analysis of performance. Geosci Model Dev 8:4045–4067. https://doi.org/10.5194/gmd-8-4045-2015
Billen G, Garnier J (2007) River basin nutrient delivery to the coastal sea: assessing its potential to sustain new production of non-siliceous algae. Mar Chem 106:148–160. doi:10.1016.j.marchem.2006.1012.1017
Billen G, Garnier J, Deligne C et al (1999) Estimates of early-industrial inputs of nutrients to river systems: implication for coastal eutrophication. Sci Total Environ 243/244:43–52
Bouwman AF, Beusen AHW, Billen G (2009) Human alteration of the global nitrogen and phosphorus soil balances for the period 1970-2050. Global Biogeochem Cycles 23. https://doi.org/10.1029/2009GB003576
Bouwman AF, Beusen AHW, Glibert PM et al (2013a) Mariculture: significant and expanding cause of coastal nutrient enrichment. Environ Res Lett 8:044026. https://doi.org/10.1088/1748-9326/8/4/044026
Bouwman AF, Beusen AHW, Lassaletta L et al (2017) Lessons from temporal and spatial patterns in global use of N and P fertilizer on cropland. Sci Rep. https://doi.org/10.1038/srep40366
Bouwman AF, Beusen AHW, Overbeek CC et al (2013b) Hindcasts and future projections of global inland and coastal nitrogen and phosphorus loads due to finfish aquaculture. Rev Fish Sci 21:112–156. https://doi.org/10.1080/10641262.2013.790340
Bouwman AF, Pawłowski M, Liu C et al (2011) Global hindcasts and future projections of coastal nitrogen and phosphorus loads due to shellfish and seaweed aquaculture. Rev Fish Sci 19:331–357. https://doi.org/10.1080/10641262.2011.603849
Boyd PW, Doney SC (2003) The impact of climate change and feedback processes on the ocean carbon cycle. In: Fasham MJR (ed) Ocean biogeochemistry – the role of the ocean carbon cycle in global change. Springer, Berlin, pp 157–193
Boyer EW, Howarth RW, Galloway JN et al (2006) Riverine nitrogen export from the continents to the coasts. Global Biogeochem Cycles 20:GB1S91. https://doi.org/10.1029/2005GB002537
Constant KM, Sheldrick, WF (1992) World nitrogen survey. World Bank Technical paper 174, - Washington, DC

Cork S, Peterson G, Petschel-Held G (2005) Four scenarios. In: Carpenter SR, Pingali PL, Bennett EM et al (eds) Ecosystems and human well-being: scenarios, findings of the scenario working group, Millennium Ecosystem Assessment. Island Press, Washington DC, pp 223–294

Doney SC (2010) The growing human footprint on coastal and open ocean biogeochemistry. Science 328:1512–1516

Duce RA, LaRoche J, Altieri K et al (2008) Impacts of atmospheric nitrogen on the open ocean. Science 320:893–897

Dumont E, Harrison JA, Kroeze C et al (2005) Global distribution and sources of dissolved inorganic nitrogen export to the coastal zone: results from a spatially explicit, global model. Global Biogeochem Cycles 19:GB4S02

Dürr HH, Laruelle GG, Van Kempen C et al (2011) World-wide typology of near-shore coastal systems: defining the estuarine filter of river inputs to the ocean. Estuar Coasts 34:441–458. https://doi.org/10.1007/s12237-12011-19381-y

Ellis L (2008) A China environmental health project research brief: environmental health and China's concentrated animal feeding operations (CAFOs). https://www.wilsoncenter.org/sites/default/files/factory_farms_feb28.pdf

FAO (2012a) Cultured aquatic species. FAO, Rome. http://fao.stat.fao.org

FAO (2012b) FAOSTAT database collections. FAO, Rome. http://faostat.fao.org

FAO (2016) Total fishery production 1950–2014. FishStatJ – Universal software for fishery statistical time series, Rome. http://www.fao.org/fishery/statistics/software/fishstatj/en

Flynn KJ, Stoecker DK, Mitra A et al (2013) Misuse of the phytoplankton-zooplankton dichotomy: the need to assign organisms as mixotrophs within plankton functional types. J Plankton Res 35:3–11

Fowler D, Coyle M, Skiba U et al (2013) The global nitrogen cycle in the twenty-first century. Philos Trans R Soc B 368(1621):20130164. https://doi.org/10.1098/rstb.2013.0164

Fu M, Wang Z, Pu X et al (2012) Changes in nutrient concentrations and N:P:Si ratios and their possible impacts on the Huanghai Sea ecosystem. Acta Oceanol Sinica 31:101–112

Galloway JN, Cowling EB (2002) Reactive nitrogen and the world: 200 years of change. Ambio 31:64–71

Galloway JN, Cowling EB, Seitzinger SP et al (2002) Reactive nitrogen: too much of a good thing? Ambio 31:60–63

Galloway JN, Townsend AR, Erisman JW et al (2008) Transformation of the nitrogen cycle: recent trends, questions and potential solutions. Science 320:889–892

Galloway JN, Winiwarter W, Leip A et al (2014) Nitrogen footprints: past, present and future. Environ Res Lett 9:115003 (11 p). https://doi.org/10.1088/1748-9326/9/11/115003

Garnier J, Beusen AHW, Thieu V et al (2010) N:P:Si nutrient export ratios and ecological consequences in coastal seas evaluated by the ICEP approach. Global Biogeochem Cycles 24:GB0A05. https://doi.org/10.1029/2009GB003583

GEOHAB (1998) Global ecology and oceanography of harmful algal blooms: a plan for co-ordinated scientific research and co-operation to develop international capabilities for assessment, prediction and mitigation. Cullen J (ed) Asian Natural Environmental Science Center, The University of Tokyo. 43 pp

GEOHAB (2010) GEOHAB Asia, global ecology and oceanography of harmful algal blooms in Asia: a regional comparative programme. Furuya K, Glibert PM, Zhou M et al (eds) IOC and SCOR, Baltimore and Paris, 69 pp

Glibert PM (2017) Eutrophication, harmful algae and biodiversity – challenging paradigms in a world of complex nutrient changes. Mar Pollut Bull 124:591–606. https://doi.org/10.1016/j.marpolbul.2017.04.027

Glibert PM, Allen JI, Artioli Y et al (2014a) Vulnerability of coastal ecosystems to changes in harmful algal bloom distribution in response to climate change: projections based on model analysis. Glob Chang Biol 20:3845–3858. https://doi.org/10.1111/gcb.12662

Glibert PM, Al-Azri A, Allen JI et al (2018) Key questions and recent research advances on harmful algal blooms in relation to nutrients and eutrophication. In: Glibert PM, Berdalet E, Burford M et al (eds) Global ecology and oceanography of harmful algal blooms. Springer, Cham, pp 229–259

Glibert PM, Burford MA (2017) Globally changing nutrient loads and harmful algal blooms: recent advances, new paradigms, and continuing challenges. Oceanography 30(1):58–69. https://doi.org/10.5670/oceanog.2017.110
Glibert PM, Garside C, Fuhrman J et al (1991) Time- and size-dependent coupling of organic and inorganic nitrogen uptake and NH_4^+ regeneration in the plume of the Chesapeake Bay, and its regulation by large heterotrophs. Limnol Oceanogr 36:895–909
Glibert PM, Harrison JA, Heil CA et al (2006) Escalating worldwide use of urea – a global change contributing to coastal eutrophication. Biogeochemistry 77:441–463
Glibert PM, Manager R, Sobota DJ et al (2014b) The Haber-Bosch–harmful algal bloom (HB-HAB) link. Environ Res Lett 9:105001 (13 p). https://doi.org/10.1088/1748-9326/9/10/105001
Glibert PM, Mayorga E, Seitzinger S (2008) *Prorocentrum minimum* tracks anthropogenic nitrogen and phosphorus inputs on a global basis: application of spatially explicit nutrient export models. Harmful Algae 8:33–38
Glibert PM, Seitzinger S, Heil CA et al (2005) The role of eutrophication in the global proliferation of harmful algal blooms: new perspectives and new approaches. Oceanography 18:198–209
Glibert PM, Wilkerson FP, Dugdale RC et al (2016) Pluses and minuses of ammonium and nitrate uptake and assimilation by phytoplankton and implications for productivity and community composition, with emphasis on nitrogen-enriched conditions. Limnol Oceanogr 61:165–197
Grill G, Lehner B, Lumsdon AE et al (2015) An index-based framework for assessing patterns and trends in river fragmentation and flow regulation by global dams at multiple scales. Environ Res Lett 10:015001
Harrison JA, Bouwman AF, Mayorga E et al (2010) Magnitudes and sources of dissolved inorganic phosphorus inputs to surface fresh waters and the coastal zone: a new global model. Global Biogeochem Cycles 24:GB1003. https://doi.org/10.1029/2009GB003590
Harrison JA, Caraco NF, Seitzinger SP (2005a) Global patterns and sources of dissolved organic matter export to the coastal zone: results from a spatially explicit, global model. Global Biogeochem Cycles 19:GB4S04
Harrison JA, Frings P, Beusen AHW, Conley DJ, McCrackin ML (2012) Global importance, patterns, and controls of dissolved silica retention in lakes and reservoirs. Glob Biogeochem Cycles. https://doi.org/10.1029/2011GB004228
Harrison JA, Seitzinger SP, Caraco N et al (2005b) Dissolved inorganic phosphorous export to the coastal zone: results from a new, spatially explicit, global model (NEWS-SRP). Global Biogeochem Cycles 19:GB4S03
Harrison WG, Head EJH, Conover RJ et al (1985) The distribution and metabolism of urea in the eastern Canadian Arctic. Deep Sea Res 32:23–42
Heffer P, Prud'homme M (2013) Fertilizer outlook 2013-2017. In: 81st IFA annual conference, Chicago, USA, 20–22 May, 2013. www.fertilizer.org. Accessed 22 Jun 2014
Heinze C, Maier-Reimer E, Winguth AME et al (1999) A global oceanic sediment model for long-term climate studies. Global Biogeochem Cycles 13:221–250
Heisler J, Glibert PM, Burkholder JM et al (2008) Eutrophication and harmful algal blooms: a scientific consensus. Harmful Algae 8:3–13
Hickey BM, Banas NS (2003) Oceanography of the US Pacific northwest coastal ocean and estuaries with application to coastal ecology. Estuaries 26:1010–1031
Holt JT, Harle J, Proctor R et al (2009) Modelling the global coastal ocean. Phil Trans Roy Soc A: Math, Phys, Eng Sci 367:939–951
Honkanen T, Helminen H (2000) Impacts of fish farming on eutrophication: comparisons among different characteristics of ecosystem. Int Rev Hydrobiol 85:673–686
Houlton BZ, Boyer E, Finzi A et al (2013) Intentional versus unintentional nitrogen use in the United States: trends, efficiency and implications. Biogeochemistry 114:11–23
Howarth RW (2006) Atmospheric deposition and nitrogen pollution in coastal marine ecosystems. In: Visgilio G, Whitelaw DM (eds) Acid in the environment: lessons learned and future prospects. Springer, New York, NY, pp 97–116
Howarth RW (2008) Coastal nitrogen pollution: a review of sources and trends globally and regionally. Harmful Algae 8:14–20

International Fertilizer Association (IFA) (2014) IFA database. www.fertilizer.org/ifa/Home-Page/STATISTICS

International Food Policy Research Institute (IFPRI) (2003) Outlook for fish to 2020, meeting global demand. WorldFish Center, Penang, Malaysia, p 28

Jaworski NA, Howarth RW, Hetling LJ (1997) Atmospheric deposition of nitrogen oxides onto the landscape contributes to coastal eutrophication in the northeast United States. Environ Sci Technol 31:1995–2004

Jia Y, Yu G, Gao Y et al (2016) Global inorganic nitrogen dry deposition inferred from ground- and space-based measurements. Sci Rep 6:19810. https://doi.org/10.1038/srep19810

Kanakidou M, Myriokefalitakis S, Daskalakis N et al (2016) Past, present, and future atmospheric nitrogen deposition. Am Meterol Soc J. https://doi.org/10.1175/JAS-D-15-0278.1

Kaufman ZG, Lively JS, Carpenter EJ (1983) Uptake of nitrogenous nutrients by phytoplankton in a barrier island estuary: Great South Bay, New York. Estuar Coast Shelf Sci 17:483–493

Kaushik SJ, Cowey CB (1991) Dietary factors affecting nitrogen excretion by fish. In: Cowey CB, Cho CY (eds) Nutritional strategies and aquaculture waste. Proceedings of 1st international symposium on nutritional strategies in management of aquaculture waste. University of Guelph, Ontario, pp 3–19

Kudela RM, Cochlan WP (2000) Nitrogen and carbon uptake kinetics and the influence of irradiance for a red tide bloom off southern California. Aquat Microb Ecol 21:31–47

Laruelle GG (2009) Quantifying nutrient cycling and retention in coastal waster at the global scale. PhD dissertation, Geologica Ultraiectina. Mededelingen van de Faculteit Geowetenschappen Universiteit Utrecht No. 312, Utrecht University, Utrecht, p 226

Laruelle GG, Goossens N, Arndt S et al (2017) Air–water CO_2 evasion from US East Coast estuaries. Biogeosciences 14:2441–2468. https://doi.org/10.5194/bg-14-2441-2017

Lehner B, Liermann CR, Revenga C et al (2011) High-resolution mapping of the world's reservoirs and dams for sustainable river-flow management. Front Ecol Environ 9:494–502

Li J, Glibert PM, Zhou M et al (2009) Relationships between nitrogen and phosphorus forms and ratios and the development of dinoflagellate blooms in the East China Sea. Mar Ecol Prog Ser 383:11–26

Liermann CR, Nilsson C, Robertson J et al (2012) Implications of dam obstruction for global freshwater fish diversity. Bioscience 62:539–548

Maavara T (2017) Perturbations to nutrient and carbon cycles by river damming. PhD thesis, University of Waterloo, p 174

Maavara T, Dürr HH, Van Cappellen P (2014) Worldwide retention of nutrient silicon by river damming: from sparse data set to global estimate. Global Biogeochem Cycles 28:842–855

Maavara T, Parsons CT, Ridenour C et al (2015) Global phosphorus retention by river damming. Proc Natl Acad Sci USA 112:15603–15608

Mallin MA, McIver MR, Robuck AR et al (2015) Industrial swine and poultry production causes chronic nutrient and fecal microbial stream pollution. Water Air Soil Pollut 226:407. https://doi.org/10.1007/s11270-015-2669-y

McKee BA, Aller RC, Allison M et al (2004) Transport and transformation of dissolved and particulate materials on continental margins influenced by major rivers: benthic boundary layer and seabed processes. Cont Shelf Res 24:899–926

Moore SK, Trainer VL, Mantua NJ et al (2008) Impacts of climate variability and future change on harmful algal blooms and human health. Environ Health 7:S4. https://doi.org/10.1186/1476-069X-7-S2-S4

Nixon SW, Ammerman JW, Atkinson LP et al (1996) The fate of nitrogen and phosphorus at the land–sea margin of the North Atlantic Ocean. Biogeochemistry 3:141–180

Pitcher GC, Figueiras FG, Hickey BM et al (2010) The physical oceanography of upwelling systems and the development of harmful algal blooms. Prog Oceanogr 85:5–32

Regnier P, Arndt S, Goossens N et al (2013) Modelling estuarine biogeochemical dynamics: from the local to the global scale. Aquat Geochem 19:591–626

Rodellas V, Garcia-Orellana J, Masqué P et al (2015) Submarine groundwater discharge as a major source of nutrients to the Mediterranean Sea. Proc Natl Acad Sci USA 112(13): 3926–3930

Seitzinger SP, Harrison JA, Dumont E et al (2005) Sources and delivery of carbon, nitrogen, and phosphorus to the coastal zone: an overview of Global Nutrient Export from Watersheds (NEWS) models and their application. Global Biogeochem Cycles 19:GB4S01. https://doi.org/10.1029/2005gb002606

Seitzinger SP, Mayorga E, Bouwman AF et al (2010) Global river nutrient export: a scenario analysis of past and future trends. Global Biogeochem Cycles 24:GB0A08. https://doi.org/10.1029/2009GB003587

Sharples J, Middelburg JJ, Fennel K, Jickells TD (2017) What proportion of riverine nutrients reaches the open ocean? Global Biogeochem Cycles 31:39–58. https://doi.org/10.1002/2016GB005483

Sinha E, Michalak AM, Balaji V (2017) Eutrophication will increase during the 21st century as a result of prcipitaiton changes. Science 357:405–408

Smil V (2001) Enriching the Earth: Fritz Haber, Carl Bosch, and the transformation of world food. The MIT Press, Cambridge, MA, p 338

Solomon S, Qin D, Manning M et al (eds) (2007) Climate change 2007: the physical science basis. Cambridge University Press, Cambridge and New York, NY

Sutton MA, Bleeker A, Howard CM et al (2013) Our nutrient world: the challenge to produce more food and energy with less pollution. Centre for Ecology and Hydrology, Edinburgh

Switzer T (2008) Urea loading from a spring storm – Knysna estuary, South Africa. Harmful Algae 8:66–69

Syvitski JPM, Vörösmarty CJ, Kettner AJ et al (2005) Impacts on the flux of terrestrial sediment to the global ocean. Science 308:376–380

Tilman D, Clark M (2014) Global diets link environmental sustainability and human health. Nature 515:518–522. https://doi.org/10.1038/nature13959

United Nations, Department of Economic and Social Affairs, Population Division (2014) World urbanization prospects: the 2014 revision, highlights (ST/ESA/SER.A/352). https://esa.un.org/unpd/wup/publications/files/wup2014-highlights.pdf

Verdegen MCJ (2013) Nutrient discharge from aquaculture operations in function of system design and production environment. Rev Aquacult 5:158–171

Volta C, Arndt S, Savenije HHG et al (2014) C-GEM (v 1.0): a new, cost-efficient biogeochemical model for estuaries and its application to a funnel-shaped system. Geosci Model Dev 7: 1271–1295. https://doi.org/10.5194/gmd-7-1271-2014

Volta C, Laruelle GG, Regnier P (2016) Regional carbon and CO_2 budgets of North Sea tidal estuaries. Estuar Coast Shelf Sci 176:76–90

Vörösmarty CJ, Meybeck M, Fekete BE et al (2003) Anthropogenic sediment retention: major global impact from registered river impoundments. Global Planet Change 39:169–190

Vörösmarty CJ, McIntyre PB, Gessner MO et al (2010) Global threats to human water security and river biodiversity. Nature 467:555–561

Wang S, Tang D, He F et al (2008) Occurrences of harmful algal blooms (HABs) associated with ocean environments in the South China Sea. Hydrobiologia 596:79–93

Wells ML, Karlson B (2018) Harmful algal blooms in a changing ocean. In: Glibert PM, Berdalet E, Burford M et al (eds) Global ecology and oceanography of harmful algal blooms. Springer, Cham, pp 77–90

Wells ML, Trainer VL, Smayda TJ et al (2015) Harmful algal blooms and climate change: learning from the past and present to forecast the future. Harmful Algae 49:68–93

Zarfl C, Lumsdon AE, Berlekamp J et al (2015) A global boom in hydropower dam construction. Aquat Sci 77:161–170

Zhang J, Wu Y, Zhang YY (2015) Plant nutrients and trace elements from the Changjiang watersheds and East China Sea Shelf. In: Zhang J (ed) Ecological continuum from the Changjiang (Yangtze River) watersheds to the East China Sea continental margin. Springer International Publishing, Switzerland, pp 93–118

Chapter 5
Harmful Algal Blooms in a Changing Ocean

Mark L. Wells and Bengt Karlson

5.1 Introduction

There is emerging evidence that climate change already may be causing shifts in the composition of coastal and offshore marine plankton and benthic communities, and there is reasonable expectation that future climate scenario's will lead to changes in the spatial and temporal ranges of HAB species. The concern is that these changes may increase the frequency and severity of harmful algal blooms (HABs), which can be broadly categorized as toxin-producing or high-"biomass" events. But while there are many intuitive linkages, such as greater stratification leading to increased prevalence of HABs (Moore et al. 2015), these scenarios are founded on limited and often conflicting data. As a consequence, scientific evidence cannot presently establish that climate change will increase HABs, let alone what magnitude of change might be expected. More problematic, the HAB research community has been largely unprepared to rigorously address this question. International consensus is needed to help collectively guide new research strategies, a goal that will be facilitated under the umbrella of the new Global Harmful Algal Bloom (GlobalHAB) Programme.

The central challenge facing HAB research is achieving some consensus about the most efficient research and management strategies to obtain unequivocal evidence that climate will, or has, altered the frequency, distribution, or character of HABs. These strategies should include identifying meaningful hypotheses (e.g., multistressor effects) that can be tested across species and studying HABs in specific environments, such as the Arctic, where climate change impacts are

M.L. Wells (✉)
School of Marine Sciences, University of Maine, Orono, ME, USA
e-mail: mlwells@maine.edu

B. Karlson
Research & Development, Oceanography, Swedish Meteorological and Hydrological Institute, Västra Frölunda, Sweden

P.M. Glibert et al. (eds.), *Global Ecology and Oceanography of Harmful Algal Blooms*, Ecological Studies 232, https://doi.org/10.1007/978-3-319-70069-4_5

projected to be most significant (IPCC 2013). Creating synergies among these focused research strategies and observations, and assimilating these findings into ecosystem-based models, will be critical if our knowledge of HAB-climate change interactions presage actual impacts.

One of the first steps towards developing some community consensus will be to appraise what is understood about the environmental conditions that favour initiation and maintenance of different types of HAB events. This understanding is limited and fraught with structural holes, as demonstrated by our current abilities to correctly forecast HAB events, but it will serve as a springboard to focus attention on the research issues of greatest importance over the next decade. It also will help to proactively identify the fundamental parameters and research infrastructure needed to effectively hindcast current changing HAB distributions: the first step in gaining the capacity to forecast future HAB patterns in a changing climate.

The case for high-biomass HABs and climate change is more nuanced. The fundamental driver of many high-biomass HABs is well understood: elevated anthropogenic nutrient inputs [see Chaps. 4, 12, Glibert et al. (2018a, b)]. Nevertheless, climate has the potential to affect the intensity and distribution of high-biomass HAB events through, for example, changes in runoff patterns and dynamics stemming from altered precipitation regimes. The challenge will be to discriminate these effects from those associated with regional-scale shifts in human society.

A small international workshop was organized in spring 2013 at the Friday Harbor Laboratories, University of Washington, to preliminarily assess how well the current status of HAB research can serve as a foundation for projecting the effects of climate change on future HAB prevalence and character. The expertise of participants spanned the ecophysiology and nutrient acquisition of HAB species, their nutritional quality and implications for food web structure and ecosystem health, and HAB-suitable observational platforms, time series analyses, and prediction. One objective of the workshop was to produce a summary paper identifying the keystone parameters and research infrastructure needed to test the purported HAB-climate change linkage (Wells et al. 2015). The summary of those published findings that follows represents the dedicated efforts of those participants.

5.2 Direct Effects of Temperature on HABs

Temperature influences all metabolic processes, is a primary physiological constraint to species range extension, and is clearly shown to have increased in the surface mixed layer of the oceans (Beardall et al. 2009). But warming is not globally uniform, in either latitudinal or regional extents (Roemmich et al. 2012; Stocker et al. 2013), so this potential change in the physical environment will differ greatly among locations. The greatest temperature changes by far have been and will continue to be at high latitudes, but it is possible that small temperature shifts at lower latitudes might have equally significant impacts on HAB prevalence (e.g., benthic HABs). Perhaps of greater significance, though, is how temperature may expand the "windows of opportunity" for existing HAB regions, assuming that physiological limits are not exceeded during peak temperatures (i.e., late summer). However, the central question is whether temperature directly influences the competitive advantage of HAB species over non-HAB species or whether indirect temperature effects on other physical processes (e.g., increased stratification, altered winds, etc.) have the primary role in regulating potential HAB-temperature relationships.

Direct effects of temperature on phytoplankton physiology and metabolism in general are well documented (Beardall and Raven 2004; Bissenger et al. 2008; Eppley 1972; Kamykowski and McCollum 1986; Raven and Geide 1988), with recognition that the optimal and inhibitory limits differ among metabolic thresholds and species. In the case of at least dinoflagellates, increased temperatures (along with light) can stimulate, but also inhibit, germination (e.g., Anderson and Rengefors 2006; Hallegraeff et al. 1998; Itakura and Yamaguchi 2005). Temperature regimes largely constrain the biogeographical boundaries of phytoplankton species (Longhurst 1998; Okolodkov 1999, 2005), likely due in large part to their temperature tolerances. However with few exceptions (e.g., Rhodes et al. 1994), there are few data on the temperature-driven competitive interactions among HAB and non-HAB species. There also are indications that temperature is correlated directly with toxin production in some HAB species (Ogata et al. 1989), but this effect is not uniform (e.g., Lewis et al. 1993). Unusually long periods of warm and calm weather may favour cyanobacterial blooms in the Baltic Sea (e.g., Kahru et al. 2016). Such conditions were found during the ecosystem disruptive bloom of *Prymnesium polylepis* (syn. *Chrysochromulina polylepis*) in Scandinavia in 1988 (Gjosaeter et al. 2000).

There are many avenues for investigating the direct effects of temperature on the prevalence of HABs in the future ocean, but there are four leading questions that, addressed experimentally, would accelerate improved understanding. First, most studies of direct temperature effects on HAB species utilize single strains, but are these representative of the genetically diverse natural populations, or do isolates from regions that experience greater seasonal temperature variations display different resilience or competitive advantages (Boyd et al. 2013)? Second, what are the synergistic or additive effects of multistressors with temperature; i.e., will temperature effects be mitigated or exacerbated by changes in other physical parameters? Third, are there observational data that can now show HAB range expansion, or contraction, associated with documented temperature increases? Finally, will any changes in future HAB prevalence result more from changes in physiology or changes in the windows of opportunity for HABs to develop?

5.3 Direct Effects of Stratification on HABs

Climate change pressures will lead to warmer surface waters and increased precipitation and runoff in many coastal regions (IPCC 2013), both of which will lead to increased stratification of surface waters—likely more so in mid- to high-latitude regions than in tropical/subtropical waters (Behrenfeld et al. 2006). Changes in wind speed and direction will also affect stratification; altered wind speed can increase or decrease stratification. Projected increases in stratification will alter patterns of nutrient availability, in many cases leading to decreased overall nutrient supply to surface waters. In addition to altering the seasonal patterns of blooms in general, and HABs in particular (Moore et al. 2009), dinoflagellates are well adapted to exploit conditions of strong shallow nutriclines (Figueiras et al. 2006; Smayda and Trainer 2010), and this understanding serves as the precarious foundation for much of the intuitive linkage between climate change and HABs.

The fundamental uncertainty is whether increased stratification will enhance the competitive abilities of HAB organisms over non-HAB organisms, at least to the extent where toxic impacts develop in the food web. There is ample evidence that stratification favours swimmers (Peacock and Kudela 2014) and, in cases of stable stratification, the formation of thin layers that can harbour HAB organisms (e.g., Aoki et al. 2014; Farrell et al. 2012; McManus et al. 2008; Pettersson and Pozdnyakov 2013; Raine 2014; Rines et al. 2002; Ryan et al. 2010; Strom et al. 2013). Decadal patterns of increasing intensity and decreasing depth of the pycnocline in the California Current system have coincided with increased prevalence of dinoflagellates in these waters (Jester et al. 2009; Kudela et al. 2010). Widespread *Pseudo-nitzschia* blooms on the western seaboard of the USA in 2015 followed unusually intense late winter stratification from the onshore movement of anomalously warm oceanic waters (the "blob"), indicating that stratification may have helped precondition the system to support this massive event (McCabe et al. 2016). Increasing stratification also has been associated with an apparent recent emergence of *Dinophysis* species in Canada, the USA, and South Africa (Hubbart et al. 2012; Pitcher et al. 1993; Taylor et al. 2013; Trainer et al. 2013). Blooms of fish-killing flagellates have been associated with extended periods of calm weather in Scandinavia (Eckford-Soper and Daugbjerg 2016; Gjosaeter et al. 2000; Granéli and Turner 2006 and articles therein).

These apparent correlations notwithstanding, in almost all cases it is not possible to mechanistically demonstrate causation, and it is critical to remember that stratification, in the vast majority of cases, does *not* lead to HAB formation. Two important questions then are: how might increased stratification work in conjunction with other environmental parameters to favour the success of HAB species? And, how important is swimming behaviour, or the formation of thin layers (biological refugia), in facilitating HAB species success within phytoplankton assemblages? It will be equally important to ask under what scenarios increased stratification will not lead to HAB development (e.g., McGillicuddy et al. 2011). Insights to these factors would help to establish whether broadening of stratification in the future ocean would lead to the expansion of HABs.

5.4 Altered Light Field Effects on HABs

The increase in heat flux to higher latitudes over the next century is projected to cause a global increase in clouds in the mid- to high latitudes, although there remains uncertainty over these projections (IPCC 2013). Combined with trends of increased runoff that would bring higher particulate loads (Häder et al. 2010) and "brownification" (from humic substances) to coastal waters (Hansson et al. 2013), there is good reason to expect greater light attenuation in future higher latitude waters. These effects may be countered in part by shallowing of the mixed layer, but that also would lead to to increased near-surface exposure to higher-intensity photosynthetically active radiation as well as UV radiation. Despite uncertainties of the relative importance of these contrasting effects, it seems likely that light fields will be changing in much of the future ocean. Increasing levels of "brownification," or coloured dissolved organic matter (CDOM), may lead to differences in plankton community structure. For example, mixotrophs appear to have a competitive advantage in high CDOM water. Paczkowska et al. (2017) showed that in the Gulf of Bothnia, northern Baltic Sea, where CDOM concentrations are high, the proportion of mixotrophs are higher than in the southern part of the Baltic Sea, where CDOM levels are lower.

There are no known PAR requirements or tolerances that competitively separate HAB species from non-HAB species, although differing optical properties among algal taxa are well known. However, high surface irradiance can influence the production of UV absorbing toxic mycosporine-like amino acids (MAA). Although MAA are not unique to HAB species, as they are found in many non-toxic cyanobacteria and raphidophytes (Carreto and Carignan 2011; Jeffrey et al. 1999; Mohlin et al. 2012), they are presumed to contribute to the high-light tolerances of these and other HAB species.

There is equivocal evidence that high light intensities influence the "traditional" toxicity of HAB organisms. For example, saxitoxin content increases in *Alexandrium fundyense* with increasing light (Etheridge and Roesler 2005), but there is no general trend among *Alexandrium* spp. (Fu et al. 2012). In other cases, the more toxic strains of *Microcystis aeruginosa* and *Pseudo-nitzschia* spp. are observed to have greater tolerances to high light conditions. None of these or other light relationships are characterized well enough to generalize, but there are reasons to ask whether changing light fields, both increasing and decreasing maximum irradiances, will affect HAB species or alter their toxin production. In the case of tropical/subtropical benthic HABs currently prolific in high light environments, will increases in PAR/UV help to expand their range towards higher latitudes, or will light tolerance thresholds be surpassed causing contraction of their contemporary tropical distributions? Although there are a number of unresolved questions regarding light field effects, understanding how future changes in light fields may affect HAB prevalence depends upon climate models gaining better insight on the distribution and magnitude of changes in global cloud cover.

5.5 Effects of Ocean Acidification on HABs

Ocean acidification (OA) in surface waters is the consequence of atmospheric CO_2 dissolution rates into surface waters being greater than the rate of surface water exchange with the deep ocean. The end-of-century surface water pH is projected to be 0.3–0.4 units lower than present day, with most of this change occurring in high-latitude regions (IPCC 2013). These changes will affect carbon acquisition in phytoplankton, with increasing $pCO_2(aq)$ de-emphasizing the value of carbon-concentrating mechanisms (CCMs) utilized by some phytoplankton (Beardall et al. 2009; Hansen 2002; Hinga 2002). Species that contain the low-affinity Form II Rubisco (ribulose-1,5-bisphosphate carboxylase-oxygenase) and those having limited CCM capabilities, including some toxic dinoflagellates and perhaps all raphidophytes, in particular could benefit from increased $pCO_2(aq)$ (Dason and Colman 2004; Nimer et al. 1997), although findings so far have been mixed (Cho et al. 2001; Lundholm et al. 2004; Sun et al. 2011; Tatters et al. 2012).

Changes in pH also will affect other aspects of cell metabolism, including membrane potential, enzyme activity, and energy partitioning (Beardall and Raven 2004; Giordano et al. 2005). Increases in hydrogen ion activity can influence nutrient uptake, through alterations to chemical speciation or transmembrane transport, and flagellar motion, which could affect movement to favourable nutrient regimes (Hallegraeff et al. 2012). There also is evidence, though limited and conflicting, that toxin production can increase in at least some strains of *Alexandrium* at lower pH (Kremp et al. 2012), and this increased toxicity can be enhanced in *Pseudo-nitzschia* spp. under nutrient limitation (Sun et al. 2011; Tatters et al. 2012).

There are numerous unresolved questions about OA effects on HABs. For example, does increasing CO_2 availability benefit HAB species versus non-HAB species, and, if so, under what conditions might this competitive advantage affect the prevalence of HAB events? More fundamentally, are our current experimental methods adequate to enable the replication needed to gain consensus about OA effects on the toxicity and competitive interactions of HAB species, given the inherent difficulties in conducting pH manipulation experiments?

5.6 Effects of Nutrients on HABs

The majority of high-biomass HABs stem from impacts of cultural eutrophication, which will continue to be the major driving mechanism in the future ocean [see also Chaps. 4, 12, Glibert et al. (2018a, b)]. Changes in precipitation regimes are likely to diminish these HABs in some regions while exacerbating them in others. Gaining a better understanding of these changes largely waits for the adequate downscaling of global models to regional-scale predictions of environmental change. The comparative simplicity of mechanistic control for high-biomass HABs means that advances in projecting their response to climate change will be coupled to advances in downscaled climate forecasts.

Decades of HAB research has provided no clear evidence that changes in natural macronutrient supply drives a shift towards HAB species, although great progress has been made [see, however, Chaps. 4, 12, Glibert et al. (2018a, b)]. Of all nutrients, nitrogen (N) supply and forms have attracted the greatest interest, and in some cases there appear to be linkages between N supply or forms and cell toxicity (e.g., Auro and Cochlan 2013; Dyhrman and Anderson 2003; Leong et al. 2004; Levasseur et al. 1995). However, there is no overall consistent pattern among laboratory studies to suggest that specific N sources generate greater success or toxicity of HAB species—a perspective that is consistent with observations in at least eastern boundary current upwelling regimes (Kudela et al. 2010).

The cell toxicity of *Pseudo-nitzschia* spp. appears to consistently increase under N, P, or Si limited senescence (cf. reviews by Bates and Trainer 2006; Lelong et al. 2012; Trainer et al. 2012). Earlier and more rapid onset of warming in temperate waters can lead to more intense spring blooms and greater depletion of nutrients (McGillicuddy et al. 2011), which potentially could lead to more toxic *Pseudo-nitzschia* scenarios in the future.

Increased intensities or duration of stratification will decrease new nutrient inputs to surface waters, and these conditions favour flagellates (Smayda 1998). While it is clear that these changes in nutrient flux will regulate phytoplankton community development, the complexity of the response among individual species frustrates attempts to link nutrient supply to toxic HAB events. However, more detailed field observations may help refine this understanding. For example, the massive *Pseudo-nitzschia* bloom event along the western coast of N. America in 2015 may have stemmed from "preincubation" of coastal waters under low-nutrient conditions. *Pseudo-nitzschia* spp. have superior nutrient uptake capabilities relative to many centric diatoms under low-nutrient conditions (e.g., Auro and Cochlan 2013; Kudela et al. 2010; Wells 2003), so extended low-nutrient conditions may have selected for *Pseudo-nitzschia* spp. success once upwelling conditions returned (McCabe et al. 2016).

Fundamental questions remain about how climate-induced changes in nutrient fields may impact HABs. First, are the nutrient uptake kinetics and preferences of HAB species sufficiently different from non-HAB species to inform on likely competitive outcomes [see also Chap. 6, Glibert et al. (2018c)]? If so, then can improved linkages between observed nutrient fields and physical, ecological, and climate change models improve our competence in forecasting contemporary HAB patterns? Any advances in this area would add confidence to projecting how changes in nutrient fluxes may impact future HAB trends. Finally, while the vast majority of studies on nutrient/HAB interactions are restricted to the macronutrients N, P, and Si, how do changes in micronutrient and vitamin supply affect the competitive success of HAB species in coastal waters?

5.7 Grazer Effects on HABs

Most studies on HAB development focus on how environmental conditions affect the growth of HAB species. Far fewer studies have considered "top-down" controls that grazers exert on the success, or failure, of HAB species competition, and the complexity of these potential controls is daunting. The tiers of grazers include microbial pathogens, microzooplankton, copepods and other mesozooplankton, benthic invertebrates, and fish, each with unique modes of predation, prey preference, susceptibility to toxins, and ecologies (Smayda 1992). Indeed, some HAB organisms can be mixotrophic, potentially enhancing their relative abundance in the plankton community (e.g., Flynn et al. 2013; Jeong et al. 2010; see also Chap. 7, Flynn et al. 2018). Some HABs are apparently immune to grazing pressure, such as ecosystem disruptive blooms (Buskey et al. 2001), while grazing pressures are too sluggish to significantly affect other HABs (Schultz and Kjørboe 2009).

The dominant grazers on HAB species are microzooplankton (Turner 2006), and while toxic HAB prey do not appear to kill microzooplankton, their chemical defences can serve to impair grazer activity (e.g., Strom 2002). The presence of toxic prey species in some cases reduces copepod grazing rates (Engström-Ost et al. 2002; Koski et al. 1999; Schultz and Kjørboe 2009), while in others there are no effects (Engström-Ost et al. 2002). Even when prey toxins do influence copepod grazing rates, these effects are partially modulated by repeated exposures to the toxins (Colin and Dam 2002; Zheng et al. 2011).

Perhaps the greatest difficulty in assessing grazer effects on HABs in general, let alone extrapolation of this knowledge to future climate scenarios, is constructing experimental conditions that are representative of natural systems. That is, selection of appropriate prey densities, predator-prey pairings (including mixotrophs), mixed prey opportunities, and strain differences among prey and predators creates considerable logistical difficulties for designing experiments. It is possible that the current trophic-ecological paradigms may shift to altered systems that are not well represented under current environmental regimes.

Given the diversity of prey ecology and predator feeding strategies, it is unlikely that a general predator-prey interaction model will be sufficient to characterize climate change effects on grazer/HAB interactions. While modelling efforts are improving (e.g., Mitra and Flynn 2006), there remain many questions. For example, what are the best approaches to study grazing effects on HAB success? Will climate-driven changes in the coastal environment alter some or many of the known predator-prey relationships by temporally decoupling the grazer and prey life histories? To what extent will top-down control of the phytoplankton community be influenced by OA (e.g., see pH effects on sensory capacity—Lonnstedt et al. 2013), and what might this change mean for HABs in particular? Progress on understanding climate impacts on grazer/HAB species coupling will depend upon intensive study of a narrow subset of specific grazer-prey interactions under standardized future climate conditions.

5.8 General Strategies to Accelerate Understanding of Climate Change Impacts on HABs

Climate change already is causing shifts in the seasonal dynamics of production and phytoplankton community composition, but the projections of how these changes will affect the distribution, character, and intensity of HABs remains largely speculative. Some consensus is needed within the HAB research community on steps that will accelerate our understanding in both scientific and management perspectives, if our knowledge is to advance faster than any climate-induced changes in HABs. This is particularly true for high-latitude northern regions where climate change impacts are most rapid. Proactive approaches are needed that develop rigorous, testable hypotheses to guide scientists, managers, and the public on what changes are projected, and then put into place the necessary observing infrastructures (research studies, detection schemes, observing systems) to capture these critical data. Modelling or prediction of HABs under changing environmental conditions will have little hope of being accurate without the foresight to obtain the relevant data.

As a start, the fundamental limitations of single stressor or factor experiments must be recognized. Although there are many known "unknowns" on how HAB and non-HAB organisms respond to individual climate change factors, it is more relevant to ask how these species might alter their responses under multiple stressors that better represent future ocean conditions. Multistressor experiments should become the standard experimental approach. New experimental systems such as community-scale mesocosms need to be developed as tools for testing multiple stressor effects on ecosystem responses that enhance HAB species competitive success or toxicity. Effective utilization of quantitative projection models as experimental tools depends heavily on having data on these interactive outcomes.

Broad spatial and temporal observational data will be critical for informing and testing projections of climate impacts on HABs. Our current data gathering methods are largely inadequate. These programmes rely heavily on seafood safety monitoring rather than studying HAB species in their oceanographic context. The necessity for new insights to the driving mechanisms underlying HAB development helped foster the GEOHAB Programme's focus in predictive ecology and oceanography. There is a critical need to establish, or augment, long-term HAB observer sites, both in current HAB regions and more importantly in areas (e.g., high latitudes) where HABs are not currently recognized. These sites should span citizen science-based programmes involving stakeholders (Van Dolah et al. 2016), to cutting-edge automated sensor systems (Campbell et al. 2010; Preston et al. 2011), taking advantage of the local strengths in different regions (e.g., utilizing traditional "low-tech" methods such as microscopy in regions where "higher-tech" methods are less tractable). In each case, collection of a minimum range of oceanographic and meteorological data should be encouraged to help characterize environmental conditions.

Stronger linkages with existing global observing systems are needed. For example, the Global Ocean Observing System (GOOS) include HABs as part of the essential ocean variable phytoplankton biomass and diversity (see www.goosocean.org) and the GEO Task SB-01 Oceans and Society: Blue Planet initiative (Djavidnia et al. 2014) comprises ocean information systems, forecasting networks, ecosystem monitoring, and fisheries/aquaculture management, all of which are relevant to contextualizing HAB occurrences and impacts. Indeed,

HABs are specifically recognized as a task component (Bernard et al. 2014). The International Ocean Colour Coordinating Group (IOCCG) is another Programme of relevance, working towards developing remote sensing capabilities for detecting phytoplankton functional types, HABs, and water quality.

Finally, "HAB" is a scientifically arbitrary, societal-dependent definition, so understanding the full scope of climate change effects on HABs will entail ascertaining not only direct and indirect human health impacts but also the broader socioeconomic and cultural implications (Smayda 1997; Trainer and Yoshida 2014). New synergies are needed through expanding HAB research beyond HAB ecology, physiology, or direct health and economic impacts to embrace social science and economics. These broader human wellness aspects of HABs are an important part of the new IOC/SCOR-supported GlobalHAB Programme.

Acknowledgments This is NOAA ECOHAB contribution number 920.

References

Anderson DM, Rengefors K (2006) Community assembly and seasonal succession of marine dinoflagellates in a temperate estuary: the importance of life cycle events. Limnol Oceanogr 51 (2):860–873

Aoki K, Onitsuka G, Shimizu M et al (2014) Variability of factors driving spatial and temporal dispersion in river plume and *Chattonella antiqua* bloom in the Yatsushiro Sea, Japan. Mar Pollut Bull 81(1):131–139

Auro ME, Cochlan WP (2013) Nitrogen utilization and toxin production by two diatoms of the *Pseudo-nitzschia pseudodelicatissima* complex: *P. cuspidata* and *P. fryxelliana*. J Phycol 49 (1):156–169

Bates SS, Trainer VL (2006) The ecology of harmful diatoms. In: Granéli E, Turner JT (eds) Ecology of harmful algae. Springer, Berlin, pp 81–93

Beardall J, Raven JA (2004) The potential effects of global climate change on microbial photosynthesis, growth and ecology. Phycologia 43:26–40

Beardall J, Stojkovic S, Larsen S (2009) Living in a high CO_2 world: impacts of global climate change on marine phytoplankton. Plant Ecol Divers 2(2):191–205

Behrenfeld MJ, O'Malley RT, Siegel DA et al (2006) Climate-driven trends in contemporary ocean productivity. Nature 444(7120):752–755

Bernard S, Kudela R, Velo-Suárez L (2014) Developing global capabilities for the observation and prediction of harmful algal blooms. In: Djavidnia S, Cheung V, Ott M et al (eds) Oceans and society, blue planet. Cambridge Scholars Publishing, Newcastle upon Tyne, pp 46–52

Bissenger JE, Montagnes SJ, Atkinson D (2008) Predicting marine phytoplankton maximum growth rates from temperature: improving on the Eppley curve using quantile regression. Limnol Oceanogr 53:487–493

Boyd PW, Rynearson T, Armstrong EA et al (2013) Marine phytoplankton temperature versus growth responses from polar to tropical waters – outcome of a scientific community-wide study. PLoS One 8(5):e63091. https://doi.org/10.1371/journal.pone.0063091

Buskey EJ, Liu H, Collumb C, Bersano JGF (2001) The decline and recovery of a persistent Texas brown tide algal bloom in the Laguna Madre (Texas, USA). Estuaries 24:337–346

Campbell L, Olson RJ, Sosik HM et al (2010) First harmful *Dinophysis* (Dinophyceae, Dinophysiales) bloom in the US is revealed by automated imaging flow cytometry. J Phycol 46(1):66–75

Carreto JI, Carignan MO (2011) Mycosporine-like amino acids: relevant secondary metabolites. Chemical and ecological aspects. Mar Drugs 9(3):387–446

Cho ES, Kotaki Y, Park JG (2001) The comparison between toxic *Pseudo-nitzschia multiseries* (Hasle) Hasle and non-toxic *P. pungens* (Grunow) Hasle isolated from Jinhae Bay, Korea. Algae 16:275–285

Colin SP, Dam H (2002) Latitudinal differentiation in the effects of the toxic dinoflagellate *Alexandrium* spp. on the feeding and reproduction of populations of the copepod *Acartia hudsonica*. Harmful Algae 1:113–125

Dason JS, Colman B (2004) Inhibition of growth in two dinoflagellates by rapid changes in external pH. Can J Bot 82(4):515–520

Djavidnia S, Cheung V, Ott M et al (eds) (2014) Oceans and society, blue planet. Cambridge Scholars Publishing, Newcastle upon Tyne, p 285

Dyhrman ST, Anderson DM (2003) Urease activity in cultures and field populations of the toxic dinoflagellate *Alexandrium*. Limnol Oceanogr 48(2):647–655

Eckford-Soper L, Daugbjerg N (2016) The ichthyotoxic genus *Pseudochattonella* (Dictyochophyceae): distribution, toxicity, enumeration, ecological impact, succession and life history – a review. Harmful Algae 58:51–58

Engström-Ost J, Viitasalo M, Jodonasdottir S et al (2002) Calanoid copepods feed and produce eggs in the presence of toxic cyanobacteria *Nodularia spumigena*. Limnol Oceanogr 47:878–885

Eppley RW (1972) Temperature and phytoplankton growth in the sea. Fish Bull 70:1063–1085

Etheridge SM, Roesler CS (2005) Effects of temperature, irradiance, and salinity on photosynthesis, growth rates, total toxicity, and toxin composition for *Alexandrium fundyense* isolates from the Gulf of Maine. Deep Sea Res II 52:2491–2500

Farrell H, Gentien P, Fernand L et al (2012) Scales characterising a high density thin layer of *Dinophysis acuta* Ehrenberg and its transport within a coastal jet. Harmful Algae 15:36–46

Figueiras FG, Pitcher GC, Estrada M (2006) Harmful algal bloom dynamics in relation to physical processes. In: Granéli E, Turner JT (eds) Ecology of harmful algae. Springer, Berlin, pp 127–138

Flynn KJ, Stoecker DK, Mitra A et al (2013) Misuse of the phytoplankton-zooplankton dichotomy: the need to assign organisms as mixotrophs within plankton functional types. J Plankton Res 35(1):3–11

Flynn KJ, Mitra A, Glibert PM et al (2018) Mixotrophy in HABs: by whom, on whom, why, and what next. In: Glibert PM, Berdalet E, Burford M et al (eds) Global ecology and oceanography of harmful algal blooms. Springer, Cham, pp 113–132

Fu FX, Tatters AO, Hutchins DA (2012) Global change and the future of harmful algal blooms in the ocean. Mar Ecol Prog Ser 470:207–233

Giordano M, Beardall J, Raven JA (2005) CO_2 concentrating mechanisms in algae: mechanisms, environmental modulation, and evolution. Annu Rev Plant Biol 56:99–131

Gjosaeter J, Lekve K, Stenseth NC, Leinaas HP, Christie H, Dahl E, Danielssen DS, Edvardsen B, Olsgard F, Oug E, Paasche E (2000) A long-term perspective on the *Chrysochromulina* bloom on the Norwegian Skagerrak coast 1988: a catastrophe or an innocent incident? Mar Ecol Prog Ser 207:201–218

Glibert PM, Al-Azri A, Allen JI et al (2018a) Key questions and recent research advances on harmful algal blooms in relation to nutrients and eutrophication. In: Glibert PM, Berdalet E, Burford M et al (eds) Global ecology and oceanography of harmful algal blooms. Springer, Cham, pp 229–259

Glibert PM, Beusen AHW, Harrison JA et al (2018b) Changing land-, sea- and airscapes: sources of nutrient pollution affecting habitat suitability for harmful algae. In: Glibert PM, Berdalet E, Burford M et al (eds) Global ecology and oceanography of harmful algal blooms. Springer, Cham, pp 53–76

Glibert PM, Heil CA, Wilkerson F et al (2018c) Nutrients and HABs: dynamic kinetics and flexible nutrition. In: Glibert PM, Berdalet E, Burford M et al (eds) Global ecology and oceanography of harmful algal blooms. Springer, Cham, pp 93–112

Granéli E, Turner JT (eds) (2006) Ecology of harmful algae. Springer, Berlin, p 413
Häder DP, Helbling W, Williamson CE et al (2010) Effects of UV radiation on aquatic ecosystems and interactions with climate change. Photochem Photobiol Sci 10:242–260. https://doi.org/10.1039/C0PP90036B
Hallegraeff GM, Blackburn SI, Doblin MA et al (2012) Global toxicology, ecophysiology and population relationships of the chainforming PST dinoflagellate *Gymnodinium catenatum*. Harmful Algae 14:130–143
Hallegraeff GM, Marshall JA, Valentine J (1998) Short cyst-dormancy period of an Australian isolate of the toxic dinoflagellate *Alexandrium catenella*. Mar Freshw Res 49:415–420
Hansen PJ (2002) Effect of high pH on the growth and survival of marine phytoplankton: implications for species succession. Aquat Microb Ecol 28(3):279–288
Hansson L-A, Nicolle A, Granéli W et al (2013) Food-chain length alters community responses to global change in aquatic systems. Nat Clim Chang 3(3):228–233
Hinga KR (2002) Effects of pH on coastal marine phytoplankton. Mar Ecol Prog Ser 238:281–300
Hubbart B, Pitcher GC, Krock B et al (2012) Toxigenic phytoplankton and concomitant toxicity in the mussel *Choromytilus meridionalis* off the west coast of South Africa. Harmful Algae 20:30–41
Intergovermental Panel on Climate Change (IPCC) (2013) Climate change 2013: the physical science basis. In: Contribution of working group I to the fifth assessment report of the Intergovernmental Panel on Climate Change. Cambridge University Press, Cambridge and New York, NY
Itakura S, Yamaguchi M (2005) Morphological and physiological differences between the cysts of *Alexandrium catenella* and *A. tamarense* (Dinophyceae) in the Seto Inland Sea, Japan. Plankton Biol Ecol 57:85–91
Jeffrey SW, MacTavish HS, Dunlap WC et al (1999) Occurrence of UVA- and UVB-absorbing compounds in 152 species (206 strains) of marine microalgae. Mar Ecol Prog Ser 189:35–51
Jeong HJ, Yoo YD, Kim JS et al (2010) Growth, feeding and ecological roles of the mixotrophic and heterotrophic dinoflagellates in marine planktonic food webs. Ocean Sci J 45(2):65–91
Jester R, Lefebvre KA, Langlois G et al (2009) A shift in the dominant toxin-producing algal species in central California alters phycotoxins in food webs. Harmful Algae 8(2):291–298
Kahru M, Elmgren R, Savchuk OP (2016) Changing seasonality of the Baltic Sea. Biogeosciences 13(4):1009–1018
Kamykowski D, McCollum SA (1986) The temperature acclimatized swimming speed of selected marine dinoflagellates. J Plankton Res 8:275–287
Koski M, Rosenberg M, Viitasalo M et al (1999) Is *Prymnesium patelliferum* toxic for copepods? – Grazing, egg production, and egestion of the calanoid copepod *Eurytemora affinis* in mixtures of “good” and “bad” food. ICES J Mar Sci 56:131–139
Kremp A, Godhe A, Egardt J et al (2012) Intraspecific variability in the response of bloom-forming marine microalgae to changed climate conditions. Ecol Evol 2(6):1195–1207
Kudela RM, Seeyave S, Cochlan WP (2010) The role of nutrients in regulation and promotion of harmful algal blooms in upwelling systems. Prog Oceanogr 85(1–2):122–135
Lelong A, Hégaret H, Soudant P et al (2012) *Pseudo-nitzschia* (Bacillariophyceae) species, domoic acid and amnesic shellfish poisoning: revisiting previous paradigms. Phycologia 51 (2):168–216
Leong SCY, Murata A, Nagashima Y et al (2004) Variability in toxicity of the dinoflagellate *Alexandrium tamarense* in response to different nitrogen sources and concentrations. Toxicon 43(4):407–415
Levasseur M, Gamache T, St.-Pierre I et al. (1995) Does the cost of NO_3 reduction affect the production of harmful compounds by *Alexandrium excavatum*? In: Lassus, P, Arzul G, Erard E et al (eds), Harmful marine algal blooms. Technique et Documentation-Lavoisier, Intercept Ltd, pp 463–468
Lewis NI, Bates SS, McLachlan JL et al (1993) Temperature effects on growth, domoic acid production, and morphology of the diatom *Nitzschia-pungens f. multiseries*. In: Smayda TJ, Shimizu Y (eds) Toxic phytoplankton blooms in the sea, 3. Elsevier Science Publ B V, Amsterdam, pp 601–606

Longhurst A (1998) Ecological geography of the sea. Academic Press, San Diego, p 398
Lonnstedt OM, Munday PL, McCormick MI, Ferrari MCO et al (2013) Ocean acidification and responses to predators: can sensory redundancy reduce the apparent impacts of elevated CO_2 on fish? Ecol Evol 3(10):3565–3575
Lundholm N, Hansen PJ, Kotaki Y (2004) Effect of pH on growth and domoic acid production by potentially toxic diatoms of the genera *Pseudo-nitzschia* and *Nitzschia*. Mar Ecol Prog Ser 273:1–15
McCabe RM, Hickey BM, Kudela RM et al (2016) An unprecedented coastwide toxic algal bloom linked to anomalous ocean conditions. Geophys Res Lett 43(19):10366–10376
McGillicuddy DJ, Townsend DW, He R et al (2011) Suppression of the 2010 *Alexandrium fundyense* bloom by changes in physical, biological, and chemical properties of the Gulf of Maine. Limnol Oceanogr 56(6):2411–2426
McManus MA, Kudela RM, Silver MW et al (2008) Cryptic blooms: are thin layers the missing connection? Estuar Coasts 31(2):396–401
Mitra A, Flynn KJ (2006) Accounting for variation in prey selectivity by zooplankton. Ecol Model 199(1):82–92
Mohlin M, Roleda MY, Pattanaik B et al (2012) Interspecific resource competition-combined effects of radiation and nutrient limitation on two diazotrophic filamentous cyanobacteria. Microb Ecol 63(4):736–750
Moore SK, Johnstone JA, Banas NS et al (2015) Present-day and future climate pathways affecting *Alexandrium* blooms in Puget Sound, WA, USA. Harmful Algae 48:1–11
Moore SK, Mantua NJ, Hickey BM et al (2009) Recent trends in paralytic shellfish toxins in Puget Sound, relationships to climate, and capacity for prediction of toxic events. Harmful Algae 8 (3):463–477
Nimer NA, Iglesias-Rodríguez MD, Merrett MJ (1997) Bicarbonate utilization by marine phytoplankton species. J Phycol 33(4):625–631
Ogata T, Kodama M, Ishimaru T (1989) Effect of water temperature and light intensity on growth rate and toxin production of toxic dinoflagellates. In: Okaichi T, Anderson DM, Nemoto T (eds) Red tides, biology, environmental science and toxicology. Elsevier, New York, NY, pp 423–426
Okolodkov YB (1999) Species range types of recent marine dinoflagellates recorded from the Arctic. Grana 38(2–3):162–169
Okolodkov YB (2005) The global distributional patterns of toxic, bloom dinoflagellates recorded from the Eurasian Arctic. Harmful Algae 4(2):351–369
Paczkowska J, Rowe OF, Schlüter L et al (2017) Allochthonous matter: an important factor shaping the phytoplankton community in the Baltic Sea. J Plankton Res 39(1):23–34
Peacock MB, Kudela RM (2014) Evidence for active vertical migration by two dinoflagellates experiencing iron, nitrogen, and phosphorus limitation. Limnol Oceanogr 59(3):660–673
Pettersson LH, Pozdnyakov D (2013) Biology and ecology of harmful algal species. In: Pettersson LH, Pozdnyakov D (eds) Monitoring of harmful algal blooms. Springer, Berlin, pp 25–47
Pitcher GC, Horstman DA, Calder D et al (1993) The first record of diarrhetic shellfish poisoning on the South-African coast. S Afr J Sci 89(10):512–514
Preston CM, Harris A, Ryan JP et al (2011) Underwater application of quantitative PCR on an ocean mooring. PLoS One 6(8):e22522. https://doi.org/10.1371/journal.pone.0022522
Raine R (2014) A review of the biophysical interactions relevant to the promotion of HABs in stratified systems: the case study of Ireland. Deep Sea Res Part II 101:21–31
Raven JA, Geide RJ (1988) Temperature and algal growth. New Phytol 110:411–416
Rhodes LL, O'Kelly CJO, Hall JA (1994) Comparison of growth characteristics of New Zealand isolates of the prymnesiophytes *Chrysochromulina quadrikonta* and *C. camella* with those of the ichthyotoxic species *C. polylepis*. J Plankton Res 16:69–82
Rines JEB, Donaghay PL, Dekshenieks MM et al (2002) Thin layers and camouflage: hidden *Pseudo-nitzschia* spp. (Bacillariophyceae) populations in a fjord in the San Juan Islands, Washington, USA. Mar Ecol Prog Ser 225:123–137
Roemmich D, Gould WJ, Gilson J (2012) 135 years of global ocean warming between the Challenger expedition and the Argo Programme. Nat Clim Chang 2(6):425–428

Ryan JP, McManus MA, Sullivan JM (2010) Interacting physical, chemical and biological forcing of phytoplankton thin-layer variability in Monterey Bay, California. Cont Shelf Res 30(1):7–16
Schultz M, Kjørboe T (2009) Active prey selection in two pelagic copepods feeding on potentially toxic and non-toxic dinoflagellates. J Plankton Res 31:553–561
Smayda TJ (1992) Global epidemic of noxious phytoplankton blooms and food chain consequences in large ecosystems. In: Sherman K, Alexander L, Gold BD (eds) Food chains, yields, models, and management of large marine ecosystems. Westview Press, San Francisco, pp 275–307
Smayda TJ (1997) What is a bloom? A commentary. Limnol Oceanogr 42(5):1132–1136
Smayda TJ (1998) Patterns of variability characterizing marine phytoplankton, with examples from Narragansett Bay. ICES J Mar Sci 55(4):562–573
Smayda TJ, Trainer VL (2010) Dinoflagellate blooms in upwelling systems: seeding, variability, and contrasts with diatom bloom behaviour. Prog Oceanogr 85(1–2):92–107
Stocker TF, Qin D, Plattner G-K, Tignor M, Allen SK, Boschung J, Nauels A, Xia Y, Bex V, Midgley PM (2013) Climate change 2013: the physical science basis. In: IPCC (ed) Contribution of working group I to the fifth assessment report of the Intergovernmental Panel on Climate Change. Cambridge University Press, Cambridge and New York, NY, p 1535
Strom SL (2002) Novel interactions between phytoplankton and microzooplankton: their influence on the coupling between growth and grazing rates in the sea. Hydrobiologia 180:41–54
Strom SL, Harvey EL, Fredrickson KA et al (2013) Broad salinity tolerance as a refuge from predation in the harmful raphidophyte alga *Heterosigma akashiwo* (Raphidophyceae). J Phycol 49(1):20–31
Sun J, Hutchins DA, Feng Y et al (2011) Effects of changing pCO_2 and phosphate availability on domoic acid production and physiology of the marine harmful bloom diatom *Pseudo-nitzschia multiseries*. Limnol Oceanogr 56(3):829–840
Tatters AO, Fu FX, Hutchins DA (2012) High CO_2 and silicate limitation synergistically increase the toxicity of *Pseudo-nitzschia fraudulenta*. PLoS One 7(2):e32116. https://doi.org/10.1371/journal.pone.0032116
Taylor M, McIntyre L, Ritson M et al (2013) Outbreak of diarrhetic shellfish poisoning associated with mussels, British Columbia, Canada. Mar Drugs 11(5):1669–1676
Trainer V, Yoshida M (eds) (2014) Proceedings of the workshop on economic impacts of harmful algal blooms on fisheries and aquaculture. PICES Sci Rep No 47, p 85
Trainer VL, Bates SS, Lundholm N et al (2012) *Pseudo-nitzschia* physiological ecology, phylogeny, toxicity, monitoring and impacts on ecosystem health. Harmful Algae 14:271–300
Trainer VL, Moore L, Bill BD et al (2013) Diarrhetic shellfish toxins and other lipophilic toxins of human health concern in Washington State. Mar Drugs 11(6):1815–1835
Turner JT (2006) Harmful algae interactions with marine planktonic grazers. In: Granéli E, Turner JT (eds) Ecology of harmful algae. Springer, Berlin, pp 259–270
Van Dolah ER, Paolisso M, Sellner K et al (2016) Employing a socio-ecological systems approach to engage harmful algal bloom stakeholders. Aquat Ecol 50(3):577–594
Wells ML (2003) The level of iron enrichment required to initiate diatoms blooms in HNLC waters. Mar Chem 82:101–114
Wells ML, Trainer VL, Smayda et al (2015) Harmful algal blooms and climate change: learning from the past and present to forecast the future. Harmful Algae 49:68–93
Zheng Y, Dam HG, Avery DE (2011) Differential responses of populations of the copepod *Acartia hudsonica* to toxic and nutritionally insufficient food algae. Harmful Algae 10(6):723–731

Part III
Adaptive Strategies and Harmful Algal Blooms

Chapter 6
Nutrients and Harmful Algal Blooms: Dynamic Kinetics and Flexible Nutrition

Patricia M. Glibert, Cynthia A. Heil, Frances P. Wilkerson, and Richard C. Dugdale

6.1 Introduction

Increasingly, microalgae—from freshwater to marine—are bathed in nutrients coming from diverse anthropogenic activities. Whether it is fertilizer use for agriculture, animal waste, industrial pollution, or atmospheric deposition, there is no doubt that there are increasing loads of both nitrogen (N) and phosphorus (P) that are lost to the aquatic environment via multiple pathways [e.g., Howarth et al. 2002; Galloway et al. 2002; Glibert et al. 2014a; see also Chap. 4, Glibert et al. (2018b)]. Nutrient pollution is the root cause of eutrophication, and one of the important consequences of eutrophication is a change in algal communities leading to more harmful algal blooms (HABs; Anderson et al. 2002; Glibert et al. 2005; Heisler et al. 2008). The composition of nutrients making their way to aquatic systems is also changing. Accordingly, many receiving waters are now not only enriched with nutrients, but these nutrients are in proportions that differ from those of decades past—and also diverge considerably from those that have long been associated with phytoplankton growth, namely, Redfield proportions (Glibert et al. 2013; Glibert and Burford 2017). Various analyses and reviews have shown that throughout most of the world, rates of N pollution have increased more rapidly than those of P, leading to conditions of increasing N:P (e.g., Peñuelas et al. 2012; Sutton et al. 2013; Glibert et al. 2014a; Beusen et al. 2016; Bouwman et al. 2017). While this

P.M. Glibert (✉)
University of Maryland Center for Environmental Science, Horn Point Laboratory, Cambridge, MD, USA
e-mail: glibert@umces.edu

C.A. Heil
Bigelow Laboratory for Ocean Sciences, East Boothbay, ME, USA

F.P. Wilkerson • R.C. Dugdale
Romberg Tiburon Center, San Francisco State University, Tiburon, CA, USA

P.M. Glibert et al. (eds.), *Global Ecology and Oceanography of Harmful Algal Blooms*, Ecological Studies 232, https://doi.org/10.1007/978-3-319-70069-4_6

may not be the case regionally or seasonally, it is the global trajectory. Additionally, in many parts of the developed world, reductions in loads, mostly in P (e.g., in sewage effluents and laundry detergents; Litke 1999), have been undertaken as a means to reduce or control algal blooms, furthering skewing the relative proportions of N and P pollution.

The Global Ecology and Oceanography of Harmful Algal Blooms (GEOHAB) Programme had, as one of its underlying programme elements, an improved understanding of HABs in relation to nutrients and eutrophication [GEOHAB 2001, 2006; see also Chap. 3, Kudela et al. (2018) and Chap. 12, Glibert et al. (2018a)]. However, one of the most challenging questions in unravelling the effects of nutrient pollution is why a change in biodiversity, including algal diversity leading to HABs, occurs as nutrient loads change. This chapter focuses on just a few facets of this challenging question, including how classical concepts have hindered our progress and how our advancing knowledge can be applied in an understanding of how biodiversity changes with nutrients; this topic has also been addressed in detail in several other reviews (Glibert and Burford 2017; Glibert 2017).

From simple dose-response relationships, it may be apparent why increases in algal biomass may result from nutrient pollution. This is the so-called phase 1 model of eutrophication: more nutrients yield more biomass (Cloern 2001). It is far less clear, however, why the community composition may change as nutrient loads increase. There has been good progress made in understanding many important species-specific (or functional group-specific), nutrient-related adaptive strategies and physiological processes. However, there is still much to be done to bridge the conceptual divide of an understanding of how increases in biomass track increases in nutrients and an understanding of how and why biodiversity changes under nutrient-enriched conditions, including algal biodiversity leading to HABs. Applying simple, dose-response relationships to understand how HABs respond to nutrients is, quite simply, not sufficient.

One concept that has hindered understanding of biodiversity changes is the notion that the response to nutrients by specific taxa follows a fixed, invariant response, i.e., that organisms have fixed, intrinsic nutrient kinetic responses. It is this concept that has led to many comparisons of nutrient kinetics of one organism compared to another, with the idea that this may yield insight into their comparative success under different nutrient regimes. Here, we underscore the dynamic nature of nutrient responses and the fact that there is wide flexibility in physiological responses within organisms themselves as well as the variable responses between taxa. We highlight common misconceptions of kinetic constructs and promote a way forward for new models. Rather than singular entities associated with individual taxa, kinetic responses are dynamic and responsive, and such dynamic behaviour is the norm rather than the exception. Our concepts of nutrient regulation must be expanded to include an understanding of how nutrients can be regulated even in a nutrient-saturated, eutrophic environment. The ratios of externally supplied nutrients relative to internal assimilative capacity, as well as ratios of availability of one nutrient to another, are among the important factors potentially influencing biodiversity. Compounding the consideration of dynamic nutrient regulation is the consideration of the balance of autotrophy and mixotrophy. For HABs, it is not sufficient to consider how inorganic nutrients affect their growth; the contribution of particulate nutrients through phagotrophic nutrition must be recognized as the potential for mixotrophy to be synergistic, not just additive, with dissolved nutrient uptake is large.

6.2 Limiting Nutrients

The concept of limiting nutrients has long been considered to be fundamental and the evolution of thought—and controversies—surrounding the applications of the principles of the law of the minimum (Von Liebig 1855) have been well reviewed (e.g., DeBaar 1994). Decades ago, based on enrichment studies from the bioassay scale to whole lakes, the paradigm developed that phosphorus (P) was the limiting nutrient in lakes, while nitrogen (N) was the limiting nutrient in marine systems in general (Schindler 1977; Hecky and Kilham 1988; Howarth 1988), and this concept continues to be accepted as foundational in spite of the changing nutrient loads to many aquatic systems (Howarth and Marino 2006; Elser et al. 2007). In fact, seemingly new ideas and debates about multiple limiting nutrients actually date back more than a century (Nathansohn 1908, cited in DeBaar 1994).

The most common relationship describing the rate of growth as a function of nutrient availability is that of a saturating hyperbola (Fig. 6.1a). This relationship leads to the natural conclusion that if the nutrient concentration increases from a limiting value, growth should increase, and conversely, if the nutrient concentration is reduced from a level that is "saturating" (i.e., under "eutrophic" conditions) to one where concentrations become limiting, then the net effect should be a reduction in growth and ultimately biomass (Fig. 6.1b). This concept is indeed applicable for understanding biomass changes.

Importantly, however, from an ecological perspective, while biomass accumulation can be shown to increase or decrease as the concentration of the limiting nutrient changes, in natural environments, the community assemblage does not necessarily change in proportion. The most commonly cited example of this phenomenon is the Seto Inland Sea, Japan, where nutrient loads were significantly reduced following sewage upgrades. While overall numbers of blooms and their biomass declined, outbreaks of *Alexandrium tamarense* and *A. catenella* became more prevalent (Anderson et al. 2002). A similar observation was reported from the Thau Lagoon, Southern France (Collos et al. 2009).

In a recent review of lakes where P control has been implemented, based on the P limitation paradigm in lakes, Schindler et al. (2016) claim "success" due to P reduction strategies. These authors argue that reductions in N are not as effective because N_2-fixing cyanobacteria compensate for reduced external N sources. While this notion of P-only control has been well debated, importantly, there are numerous and increasing examples of non-N_2-fixing cyanobacterial blooms that are occurring in seemingly P-limited freshwaters and for which P control alone has not

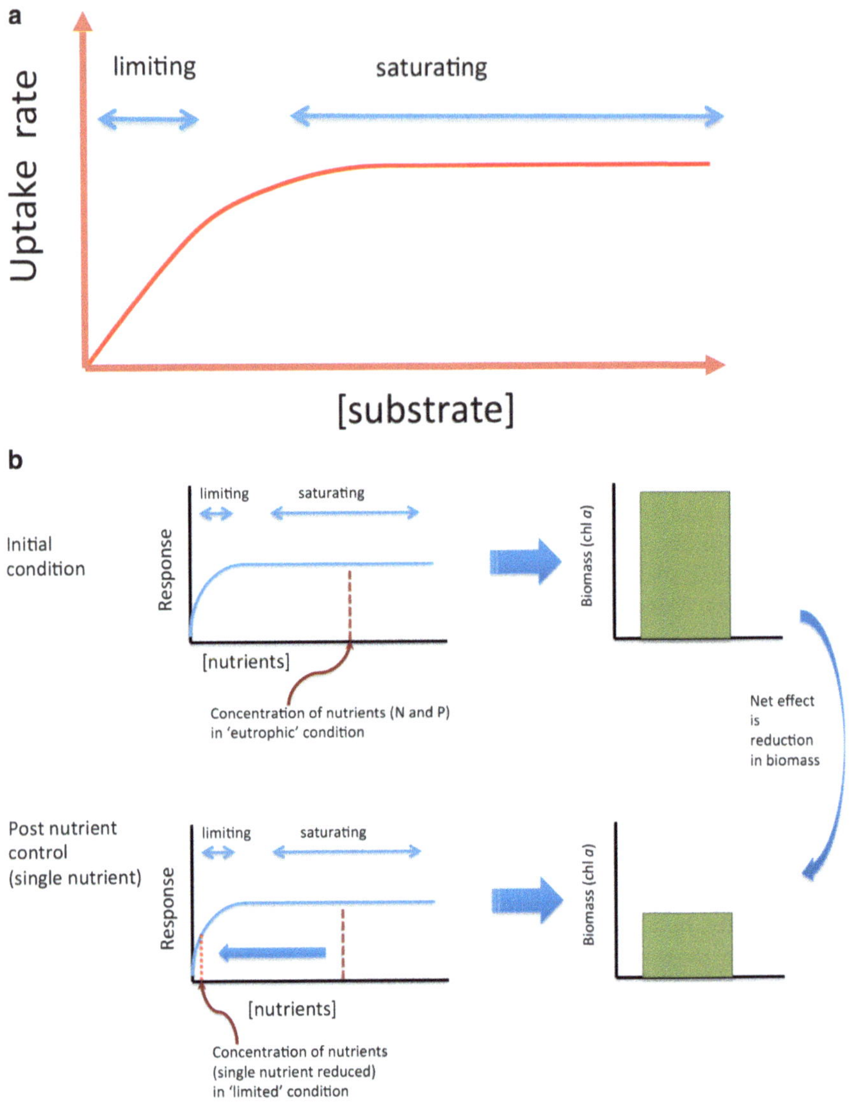

Fig. 6.1 (**a**) Classic kinetic (saturating hyperbolic) response in relation to nutrient availability. This response could depict rate of nutrient transport or growth (but such processes are only equivalent at steady state). (**b**) Hypothesized change in biomass (*right-hand panels*) in response to a reduction in the putative limiting nutrient, as shown in the kinetic response (*left-hand panels*)

been an effective control strategy. Probably the most glaring example is the case of Lake Taihu, China, where massive cyanobacterial blooms continue to exist nearly year-round even though aggressive measures have been taken to reduce P (Glibert et al. 2014a; Paerl et al. 2014). Even in the "successful" examples of

Schindler et al. (2016), only those that had cyanobacteria blooms dominated by N_2-fixing species appear to be controlled by reductions in P loads.

By further extension of the dose-response relationship, it is also often inferred that if nutrients are in excess, they do not play a regulatory role with respect to productivity or trophic structure (i.e., no further growth response in the saturated region of the response curve). If nutrients are considered to be non-limiting, then it is assumed there is another controlling factor, be it light, grazing, or a micronutrient, and therefore that factor should be the target for control efforts. Moreover, associated with the perception of lack of nutrient limitation is the subsequent assumption that their differing proportions (at saturation) will thus have no effect on metabolism and growth or community composition. This idea, particularly with respect to phytoplankton, was articulated by Reynolds (1999, p. 31) who said, "…there should be no selective effect, consequential upon different affinities of storage capabilities for a nutrient resource, that might distinguish between the potential performances of any pair of planktonic algae, so long as the resource concentrations are able to saturate the growth demand. If that is true, then the ratio between the (saturating) concentration of any of the resources also fails to exert any regulatory significance." However, as will be shown, nutrient availability and nutrient proportions—even at saturating concentration levels—have far more complicated effects than this simple perception.

6.3 Optimal Nutrients

The optimal nutrient of a species is that internal (cellular) concentration of nutrients that just meets physiological growth requirements (e.g., Klausmeier et al. 2004). Each nutrient has a given optimum, and therefore if cells use nutrients in proportion to supply, there is an optimum proportion of nutrient availability. The Redfield proportion of C:N:P of 106:16:1 is most commonly taken to be the optimal nutrient ratio for phytoplankton. Implicit in the concept of optimal nutrient ratios is the idea that a given organism or species maintains that ratio under all (or most) conditions, adjusting nutrient acquisition to just meet this physiological demand. Given that the cell—any cell—is simultaneously balancing its needs for multiple elements, theory predicts that fitness should be maximized when the cell has just what it needs and no more (Cleland and Harpole 2010 and references therein). As different functional groups and even species have different needs for N and P, for example, as the balance of these nutrients change in the environment, the success of different species should lead to new dominants (Fig. 6.2). This concept, formalized as the resource ratio theory (Tilman 1977, 1982, 1985; Smayda 1990, 1997), predicts that as the ratios of different essential elements change, the community structure will change due to competition between algae with different optimal nutrient ratios. Spanning a size range of many orders of magnitude, as well as divergent evolutionary bases of different taxa, the elemental demands of phytoplankton of different functional groups vary widely (Quigg et al. 2003; Finkel et al. 2010, 2016;

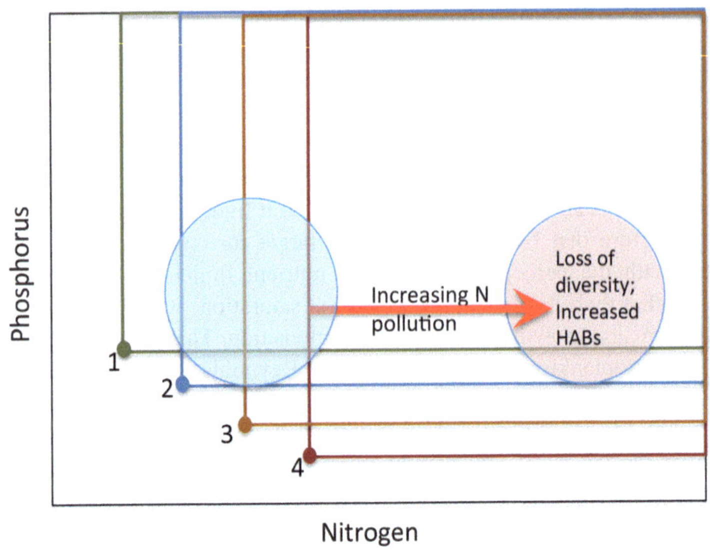

Fig. 6.2 Conceptual relationship between taxa success as nutrient proportions change. Species 1–4 each have a different optimal N/P ratio (the intersection of the *right-angled lines*). If the *blue circular* region represents historical levels of N and P supply, increasing N pollution (*red arrow*) might move the set of supply points to the *red circular* region. Under the new condition of decreased N limitation, only species 4 would persist because it is the best competitor for limiting P. Redrawn and modified from Cleland and Harpole (2010)

Hillebrand et al. 2013). In a meta-analysis of both fresh and marine studies of phytoplankton stoichiometry, Hillebrand et al. (2013) found the optimal N:P was lowest for diatoms (14.9), increased for dinoflagellates (15.1), and increased even more for cyanobacteria (25.8) and chlorophytes (27.0).

The variability in optimal ratios between different taxa is, however, further complicated by the fact that there is variability in cellular nutrient ratios even within taxa in terms of cellular nutrient ratios. Only under ideal, steady-state growth conditions do cells have an optimal nutrient balance. There is accumulating evidence that a cell's elemental stoichiometry varies considerably with growth rate and other environmental conditions (Leonardos and Geider 2004a, b; Cullen and Sherrell 2005; Glibert et al. 2013; Hillebrand et al. 2013). Different cellular N:P ratios can develop depending on the different resource supply ratios, the degree of cellular regulation, and the taxa under consideration.

In natural environments, especially in nutrient-rich or eutrophic conditions, cells are exposed to a range of nutrient availabilities and proportions. Different species have evolved many "solutions" to deal with not only limiting nutrients but also varying imbalances in the cell that may also derive from excess nutrients. As Glibert et al. (2016) reviewed, "Fundamentally there are two mechanisms to adjust imbalances: up-regulate the pathways for acquisition of the constituent in least supply, or down-regulate the cellular constituent that is in over-supply." In other words, the supplies of both "too little" (nutrient limitation) and "too much" (nutrient excess) have metabolic costs for the cell. These metabolic costs can impact different degrees of stress on different taxa, in turn leading to altered community structure over time as some taxa thrive at the expense of others. Limiting nutrients are not the only control; excess nutrients also play a role in metabolic regulation.

6.4 Dynamic Responses

Dugdale (1967) first suggested the use of the Michaelis–Menten expression (Menten and Michaelis 1913), which was developed to describe the rate of enzyme activity as a function of substrate concentration (S), to describe the rate of uptake of a limiting nutrient by phytoplankton (V). Both kinetic expressions have the same saturating hyperbolic form as the Monod (1942) expression that relates specific growth rate (μ) to substrate concentration.

$$V = V_{\max}[S/(K_s + S)] \tag{6.1}$$

$$\mu = \mu_{\max}\left[S/K_\mu + S)\right] \tag{6.2}$$

K_s and K_μ are the half-saturation constants for uptake and growth, respectively, and $V_{\max}$ and $\mu_{\max}$ are the maximal rates at saturation. The similarity of these kinetic expressions has led to much confusion and a common misconception that they are interchangeable and equivalent. There are two related problems with application of these kinetic relations, both conceptually and in model applications. First, only under conditions of steady-state growth is the rate of uptake equal to that of growth (Goldman and Glibert 1983), but under natural conditions, the assumptions of steady state are rarely met. Second, uptake kinetic relationships are, in fact, dynamic properties of both individual cells and of populations. Kinetic uptake responses change as a function of cell physiological state, temperature, and duration of exposure, among many other factors (Goldman and Glibert 1983). This is because the kinetic properties that are measured in nutrient uptake are not those of a single enzyme, but rather the summed response of carrier transport proteins, and various enzymes for assimilation, each of which can be induced, repressed, or otherwise regulated by the cellular physiological state. Single measurements and fixed kinetic values are simply a momentary image of a response that is variable. This leads to an important challenge for modellers as fixed constructs are unsuitable and indeed are considered as dysfunctional, for descriptions of algal growth under variable nutrient conditions (Goldman and Glibert 1983; Flynn 2005a, b, 2009; Kana and Glibert 2016). While it is "neither practical, not justifiable, to include all these facets of control within a model" (Flynn et al. 1999, p. 361), the challenge for physiologists is to not only describe the variability but also understand the control

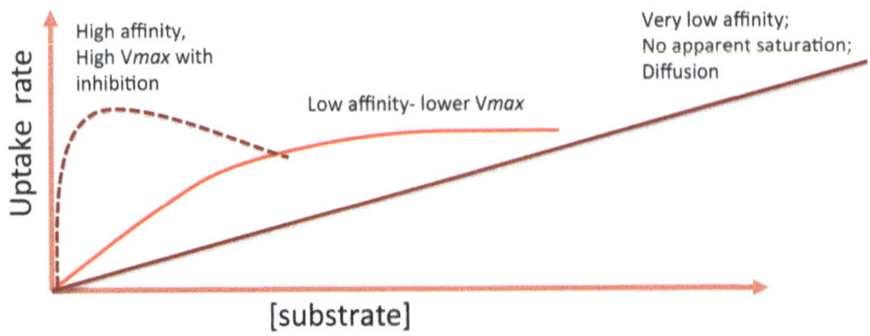

Fig. 6.3 Depiction of the range of kinetic responses that an organism can display in response to variable substrate supply. The response may be highly efficient with respect to acquisition at the low end of nutrient supply or may show inhibition of response as the substrate increases; the response may show classic saturating substrate responses; or, the response may not have characteristics of saturation and may display a linear relationship. In reality, an organism may transit through all of these response relationships depending on substrate, duration of exposure, prehistory and physiological status among other factors

of this variability with the expectation that an understanding of controls will lead to more robust general models.

A further complexity is that a saturating hyperbolic relationship is not always expressed or the most appropriate form of the response curve. Relationships between nutrient uptake and substrate concentration may be linear or biphasic or may involve an induction period (Fig. 6.3). As an example, the internal enzyme, nitrate reductase (NR), is an inducible enzyme that responds to the external concentration of NO_3^- (e.g., Packard 1973). Based on variable NR and the concept of shift-up or acceleration, in growth rates, Dugdale et al. (1990) showed that NO_3^- uptake kinetics could show linear kinetics. This phenomenon of acceleration ("shift-up") of NO_3^- uptake in the presence of NO_3^- has been well described in phytoplankton in both the classical physiological literature and more recently in molecular studies (e.g., Dugdale et al. 1981 and references therein; Allen et al. 2011 as an example). Lomas and Glibert (1999) also observed linear uptake responses for NO_3^- in nutrient-enriched natural communities, suggesting diffusion played a role at high concentrations. In fact, early studies by Falkowski (1975) have shown for several diatoms that the cell membrane-associated active carrier transporter for NO_3^-, while very efficient, does saturate and therefore diffusion may well be an important process contributing to nonsaturating NO_3^- uptake kinetics.

These complex and dynamic responses are being approached with new kinetic constructs. Rather than assuming an invariant relationship, the dynamic nature of uptake responses (and similarly, photosynthetic responses of C assimilation as a function of light as a substrate) can be thought of as a balance between control from external nutrient supply (nutrients, light) and that of the internal cellular nutrient content, a physiological trade-off (and ability to regulate) between the ability to acquire nutrients (or light) through transport and the ability to assimilate these nutrients into new biomass (Kana et al. 1997; Smith et al. 2009; Glibert et al. 2013). In physiological terms, phytoplankton alter the number of nutrient transporters,

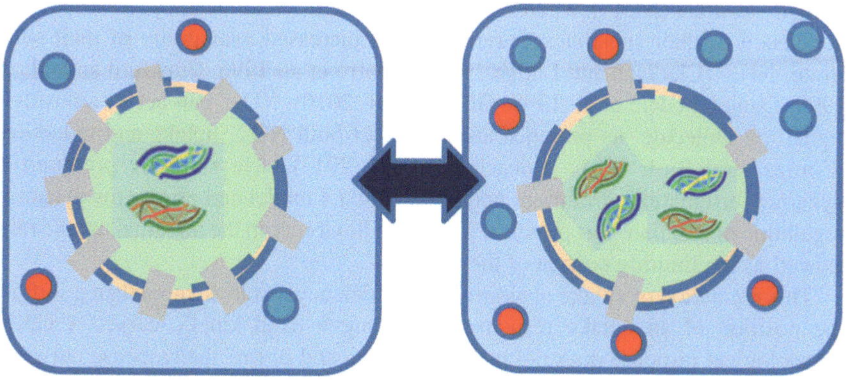

Fig. 6.4 Conceptual diagram of the relationship between acclimation to low substrate concentration (*left*) and high substrate concentration (*right*). Note that not only does the substrate concentration change, but so too does the assimilation machinery (enzymes). The important point is that cells transition between these states and are not fixed at one or the other. Redrawn and modified from Smith et al. (2009)

which determine the nutrient encounter time scale, as well as their internal enzymes, which assimilate the nutrients once encountered (Fig. 6.4; Smith et al. 2009; Pahlow and Oschlies 2013). Dugdale et al. (1981) laid the foundation for the understanding of a time-varying uptake response with a period when the uptake rate is externally controlled and a period when the uptake rate is internally controlled. While the terminology used by Dugdale et al. (1981) differs from terminology used in current modelling (and recognizing there is considerable confusion and variability in the terminology in use), the concept is quite similar: there is a dynamic regulation of uptake across the entire gradient of concentrations, and cellular physiology is continually adjusting accordingly. In contemporary terms of nutrient kinetics, this is known as an optimal uptake (OU) model (e.g., Smith et al. 2009). Such a model assumes a physiological trade-off between the efficiency of nutrient encounter at the cell surface (A_s) and the maximum rate at which a nutrient can be assimilated (V_{max}). Accordingly, rather than a Michaelis–Menten model, and instead of a half-saturation constant for uptake, K_s, the dynamic regulatory approach to kinetic modelling applies the ratio of V_{max} to A_s (e.g., (Aksnes and Egge 1991; Button et al. 2004; Smith et al. 2009). Thus, the uptake rate based on dynamic regulation, or affinity, V_{aff}, relates to substrate, S, availability as follows:

$$V_{aff} = [(V_{max}S)/((V_{max}/A_s) + S)] \tag{6.3}$$

The use of a ratio instead of fixed half-saturation constant is a way of incorporating "self-regulating" feedbacks that the cells use to upregulate the needs of the cells when resources are low or downregulate or switch off acquisition when resources are sufficient; it is a tuning knob (Kana and Glibert 2016). While the ratio may be the tuning knob, the underlying physiology may be complex and variable. As an example, in general, NO_3^- transporters are induced by the presence of their substrate (NO_3^-), whereas NH_4^+ transporters are induced by the absence or deficiency of their substrates or repressed by increased availability of their substrate, NH_4^+ (Clarkson and Luttge 1991; Navarro et al. 1996; Crawford and Glass 1998; Daniel-Vedele et al. 1998; Glibert et al. 2016). NO_3^- can act as a positive signalling molecule, its presence an inducer of both NO_3^- uptake and reduction (Coruzzi and Bush 2001). Bonachela et al. (2013) have similarly proposed a dynamic formulation of nutrient uptake in which a model cell allows for dynamic regulation of cell transport proteins, leading to flexibility in the maximal uptake rate as well as the limiting portion of the curve.

This regulation of uptake or assimilation efficiency results in more than a shift in the position of an uptake response rate along a fixed kinetic curve. A cell's physiological response does not merely "ride up and down" the response curve as substrate supply changes; the curve itself can change due to the physiological state of the cell (e.g., Glibert et al. 2013; Kana and Glibert 2016). The end result is that the shape of a kinetic curve is highly variable, with structure that may not actually fit a saturating hyperbola or may change with time and physiological state or the nutrient under consideration.

The implications of a dynamically varying, rather than fixed kinetic model, are important. On the one hand, nutrient stress can develop before nutrient availability declines below a conventionally defined half-saturation value (and bearing in mind how poorly this value is typically known), while on the other hand, regulation of nutrient uptake does not cease when availability of nutrient reaches values defined as "saturating" (Glibert et al. 2013). A dynamically regulating kinetic framework recognizes that changes in the proportions of dissolved nutrients in the environment have profound effects on cells and thus can cascade through the food web even when the availability of these elements are not in limiting proportions, with the potential of transforming ecosystems to new stable states. Across a range of substrate availability, the balance between uptake and assimilation also affects other cell properties, such as the relative proportions of ribosomes, enzyme activities, gene regulation, cellular pigmentation complement, and ultimately cell elemental composition (Glibert et al. 2013). Thus, properties such as chlorophyll *a*-to-carbon ratios (Chl:C), C:N, N:P, or other elemental properties change in response to the varying uptake rates as substrate availability changes. As the different species within a community dynamically regulate their response as nutrient availability changes, the consequence is that community composition ultimately will change.

6.5 Stoichiometry and Balancing Excess Nutrients

While most emphasis in the literature is placed on understanding how cells upregulate responses to obtain the requisite nutrients in least supply (i.e., the limiting nutrient condition), there is also emerging understanding that, especially for HABs, there are important physiological consequences when nutrients are in excess (the balance between up- and downregulation for different nutrients). How does a cell respond when there is excess nutrient beyond that which it can metabolize? There are both mechanisms for downregulation and limits to those mechanisms, and these mechanisms vary by taxa.

For HABs, this is an especially important question, as toxin production is increasingly recognized to be more prevalent when there is an excess proportion of one nutrient (especially N) relative to balanced (typically Redfield) proportions. Some cyanobacteria and dinoflagellate toxins are N-rich compounds, and thus these cells require a supply of N in order to synthesize these metabolites. The common cyanotoxins, microcystins, nodularins, cylindrospermopsins, and saxitoxins (and the same holds for many dinoflagellate toxins), have amino acid precursors, either glutamine, arginine, or leucine, and they, in turn, depend on adequate N supply for their assimilation (e.g., Oh et al. 2000; Vézie et al. 2002; Van de Waal et al. 2009, 2014; Gobler et al. 2016). Domoic acid (DA) in *Pseudo-nitzschia* also has amino acid precursors (Savage et al. 2012). Also, under conditions of elevated N:P, hemolytic activity per cell has been shown to increase by up to an order of magnitude in the prymnesiophytes *Prymnesium parvum* and *Chrysochromulina* (now *Prymnesium*) *polylepis* (Johansson and Granéli 1999), and similarly neurotoxin production has been shown to increase in the diatom *Pseudo-nitzschia multiseries* and in the dinoflagellates *Karlodinium veneficum*, *Alexandrium* sp., and *Karenia brevis* under N-enriched conditions and especially when P limitation is imposed (Granéli and Flynn 2006; Hardison et al. 2013). Toxin production may be part of the complex suite of physiological processes involved in "overflow metabolism."

In addition to toxin regulation, cells have numerous other mechanisms regulating the balance of not only acquisition and assimilation of nutrients but also all materials and energy in the cell. These differ by taxa, as well as specific environmental conditions. This is especially true when the "excess" is in the form of NH_4^+, which at high concentrations can have negative consequences for the cell. Under conditions of highly elevated NH_4^+, when the ability of the cell to balance acquisition with assimilation is exceeded, both the total N taken up and overall growth with NH_4^+ enrichment can be suppressed rather than enhanced (e.g., Dagenais-Bellefeuille and Morse 2013 and references therein, Glibert et al. 2014b). In fact, elevated NH_4^+ can be toxic for growth (e.g., Britto and Kronzucker 2002; Glibert et al. 2016 and references therein), a condition from which the cell does not easily recover. In the higher plant literature, this is known as the "NH_4^+ syndrome" (Gerendás et al. 1997; Britto and Kronzucker 2013). Relatively new data, based on higher plant studies, suggest that such growth suppression may be due, at least in part, to redox imbalances and a surplus of reductant when in excess (Escobar et al. 2006; Podgórska and Szal 2015). Increasing evidence suggests that diatoms are more susceptible to the negative effects of NH_4^+ than are many HABs, especially cyanobacteria that can thrive on high concentrations of this N form. As examples, Yoshiyama and Sharp (2006) summarized decades of data from the Delaware Bay and observed that the primary productivity rate per unit chlorophyll *a* declined exponentially with increasing NH_4^+ concentration (most of the change occurring at <10 μM NH_4^+), and similarly in the San Francisco Bay Delta, it has been suggested that suppression of growth of spring diatoms has occurred ever since NH_4^+ loading from sewage effluent increased several decades ago (Wilkerson et al. 2006; Dugdale et al. 2007) and observational and experimental evidence are confirmatory (e.g., Wilkerson et al. 2006; Parker et al. 2012). Clearly, more work is needed to understand "excess" nutrient responses; when, how, and by whom excess substrates become a "stress"; and the physiological and responses of different taxa when both nutrient and energy stresses are superimposed (Glibert et al. 2013, 2016).

6.6 Mixotrophy

Further complicating our understanding of the relationship between nutrients and HABs, and further negating the simple construct of nutrient-dose relationships based on single nutrient regulation, is the fact that for many HAB taxa, the acquisition of nutrients occurs not only in dissolved but also in particulate form [see also Chap. 7, Flynn et al. (2018)]. Indeed, understanding mixotrophy is an extreme challenge in plankton physiology because of its intrinsic complexity (Flynn and Mitra 2009; Raven et al. 2009). Planktonic mixotrophy has long been considered important in oligotrophic habitats where potentially limiting nutrients are much more concentrated in microbial prey than in the water column (Jones 1994). More recently it has been recognized that mixotrophy is also important in eutrophic systems where dissolved nutrients would seemingly be sufficient but so too are the prey on which mixotrophs graze (Burkholder et al. 2008). Most phytoflagellates appear to be capable of consuming bacteria and small eukaryotic cells (Unrein et al. 2007; Zubkov and Tarran 2008; Flynn et al. 2013). Dinoflagellates exhibit a wide range in capabilities, from those that express phagotrophy only upon extreme nutrient starvation to exclusive phagotrophic forms.

Mixotrophy can be particularly important in plankton that form toxic blooms (Burkholder et al. 2008), and this process may allow blooms to be sustained for long periods (e.g., Glibert et al. 2009). Mixotrophs also can balance their nutritional demands when inorganic nutrients are stoichiometrically imbalanced in their supply (Glibert and Burkholder 2011). There may even be growth advantages for cells in mixotrophic feeding mode. The dinoflagellates *Cochlodinium* (*Margalefidinium*) *polykrikoides* and *K. brevis* exemplify these benefits of mixotrophy. When growing as a phototroph, *C. polykrikoides* has been shown to have a growth rate of 0.17 div day^{-1} (Jeong et al. 2004). Yet, when growing as a mixotroph, with cryptophytes as its prey, the division rate of *C. polykrikoides* nearly doubles to 0.34 div day^{-1} (Jeong et al. 2004). In the case of *K. brevis*, it has been shown to graze the cyanobacterium *Synechococcus* sp., as well as cryptophytes (Jeong et al. 2005; Glibert et al. 2009). Glibert et al. (2009) found that ~1 to 80 cells of *Synechococcus* h^{-1} could be grazed by *K. brevis* with the rate varying with the predator-to-prey ratio. The rate of growth of *K. brevis* increased as the supply of

Synechococcus increased, implying that natural variability in this food source could have an effect on growth rates of the HAB in nature.

Further, there may be synergies between phototrophy and heterotrophy in mixotrophs that have important consequences for cell metabolism, especially the maintenance of cell stoichiometry. Mixotrophy may impart advantages in their ability to retain nutrients acquired from prey. These nutrients derived from feeding can help to support primary production and can supplement C acquisition through photosynthesis under conditions of light limitation (including night), although curiously some species, such as *K. veneficum*, appear to feed only during the day (Adolf et al. 2008). Even transitory benefit to C acquisition from ingestion of phototrophs (Raven et al. 2009; Jeong et al. 2010; Hansen 2011) or other mixotrophs (Minnhagen et al. 2011) may provide a growth benefit to one species over another.

The dynamic interactions between mixotroph nutrition and toxin production may also be related to cell stoichiometric (im-)balance (Glibert and Burkholder 2011), perhaps associated with the provision of secondary metabolites from photosynthetic pathways. In some instances, toxins are involved in the capture of prey and, thus, are directly related to the mixotrophic capacity of the organisms (e.g., Granéli and Flynn 2006; Stoecker et al. 2006; Adolf et al. 2007; Berge et al. 2012).

Mixotrophic protists are now known to have major roles in plankton food webs (Flynn et al. 2013; Mitra et al. 2014b), contributing substantially to N and P nutrition but importantly also to that of C. Mixotrophy also has the potential to simultaneously enhance primary production and secondary production, by converting nutrients acquired from bacterial biomass into larger particles amenable to capture by copepods and juvenile fish. Most harmful dinoflagellates, haptophytes, and raphidophytes are mixotrophs, many of which feed on their autotrophic competitors (Stoecker et al. 2006; Tillmann 1998; Jeong et al. 2010; Hansen 2011; Jeong 2011). The process of mixotrophy thus has the potential to radically change the dynamics and efficiency of both the "phytoplankton" and "microzooplankton" activities (Mitra and Flynn 2010; Flynn et al. 2013; Mitra et al. 2014a, b) and therefore overall biodiversity as nutrient loads and stoichiometry change.

6.7 Conclusions

The application of fixed kinetic models and viewpoints that are focused on singular nutrient controls with respect to nutrient limitation and fixed stoichiometry has created considerable confusion in the literature with respect to nutrient regulation of HABs. The dogma continues to be expressed that if nutrients are not limiting, then they play no regulatory effect. Yet, such concepts continue to fail when it comes to appreciation of the dynamic role of nutrient across a spectrum of availability, from limiting to excess. The oceans are changing in complex and interconnecting ways, all of which have the potential to alter food web structure and functioning—from warming and acidification to increased precipitation/runoff, increased stratification, and accelerated eutrophication—as the human population continues to increase in many regions throughout the world (Boyd and Doney 2003; Fu et al. 2012; see also Chap. 4, Glibert et al. 2018b; Chap. 5, Wells and Karlson 2018). These changes are thought to be affecting biodiversity beginning at the level of the nutrient content of individual organisms (Acquisti et al. 2009; Peñuelas et al. 2012). The potential for mixotrophy in planktonic food webs is enhanced with changing nutrient loads and ratios, as well as other environmental changes associated with changing climate. The potential for more HABs is large.

There is a need for an improved appreciation of the dynamic nature of growth and of community composition changes in relation to changes in nutrient proportions. It is time to conduct the necessary experiments to understand kinetic variability under a range of ratios. Driving systems into accentuated stoichiometric changes can perpetuate undesirable changes in community composition including HABs. The complexity of management of anthropogenic nutrient loads is well recognized, but our practice of viewing responses through the limiting-nutrient lens has hampered progress of nutrient management in relation to HABs. Controlling one nutrient may control overall biomass, but it may have unintended consequences for community structure and open the door for unwanted taxa. While there have been, and will be, successes in reducing overall chlorophyll *a* biomass, including N_2-fixing cyanobacteria, in waters following reductions in P alone (i.e., no companion reductions in N), failure to recognize that biodiversity changes as stoichiometry changes will continue to hamper progress towards HAB control. Based on our emerging knowledge of the pathways that many HABs may have for "overflow metabolism," their proliferation when only one nutrient is regulated should come as no surprise. Dynamic physiology translates into dynamic community composition. As HABs continue to increase globally, new collaborations and comparative research between systems impacted with varying nutrient loads will be helpful in advancing these physiological concepts.

Acknowledgments This is a contribution of the GEOHAB Core Research Project on HABs and Eutrophication. The authors wish to thank T. Kana, K. Flynn, and A. Mitra for many fruitful collaborations that contributed to the ideas presented herein. This is contribution number 5405 from the University of Maryland Center for Environmental Science.

References

Acquisti C, Elser JJ, Kumar S (2009) Ecological nitrogen limitation shapes the DNA composition of plant genomes. Mol Biol Evol 26:953–956

Adolf JE, Bachvaroff T, Place AR (2008) Cryptophyte abundance drives blooms of mixotrophic harmful algae: a hypothesis based on *Karlodinium veneficum* as a model system. Harmful Algae 8:119–128

Adolf JE, Krupatkina D, Bachvaroff T et al (2007) Karlotoxin mediates grazing by *Oxyrrhis marina* on strains of *Karlodinium veneficum*. Harmful Algae 6:400–412

Aksnes DL, Egge JK (1991) A theoretical model for nutrient uptake in phytoplankton. Mar Ecol Prog Ser 70:65–72

Allen AE, Dupont CL, Obornik M et al (2011) Evolution and metabolic significance of the urea cycle in photosynthetic diatoms. Nature 473:203–207

Anderson DM, Glibert PM, Burkholder JM (2002) Harmful algal blooms and eutrophication: nutrient sources, composition, and consequences. Estuaries 25(4):704–726

Berge T, Poulsen LK, Moldrup M et al (2012) Marine microalgae attack and feed on metazoans. ISME J 6:1926–1936

Beusen AHW, Bouwman AF, Van Beek LPH et al (2016) Global riverine N and P transport to ocean increased during the 20th century despite increased retention along the aquatic continuum. Biogeosciences 13:2441–2451. https://doi.org/10.5194/bg-13-2441-2016

Bonachela J, Allison SD, Martiny AC et al (2013) A model for variable phytoplankton stoichiometry based on cell protein regulation. Biogeosciences 10:4341–4356. https://doi.org/10.5194/bg-10-4341-2013

Bouwman AF, Beusen AHW, Lassaletta L et al (2017) Lessons from temporal and spatial patterns in global use of N and P fertilizer on cropland. Sci Rep. https://doi.org/10.1038/srep40366

Boyd PW, Doney SC (2003) The impact of climate change and feedback processes on the ocean carbon cycle. In: Fasham MJR (ed) Ocean biogeochemistry – the role of the ocean carbon cycle in global change. Springer, Berlin, pp 157–193

Britto DT, Kronzucker HJ (2002) NH_4^+ toxicity in higher plants: a critical review. J Plant Physiol 159:567–584

Britto DT, Kronzucker HJ (2013) Ecological significance and complexity of N-source preference in plants. Ann Bot 112:957–963

Burkholder JM, Glibert PM, Skelton HM (2008) Mixotrophy, a major mode of nutrition for harmful algal species in eutrophic waters. Harmful Algae 8:77–93. https://doi.org/10.1016/j.hal.2008.08.010

Button DK, Robertson B, Gustafson E et al (2004) The experimental and mathematical foundation of specific affinity, a cytoarchitecture-based theory for replacing the Michaelis Menten paradigm. Appl Environ Microbiol 70:5511–5521

Clarkson NM, Luttge U (1991) Mineral nutrition: inducible and repressible nutrient transport systems. Program Bot 52:61–83

Cleland EE, Harpole WS (2010) Nitrogen enrichment and plant communities. Ann N Y Acad Sci 1195:46–61. https://doi.org/10.1111/j.1749-6632.2010.05458.x

Cloern JE (2001) Our evolving conceptual model of the coastal eutrophication problem. Mar Ecol Prog Ser 210:223–253

Collos Y, Bec B, Jauzein C et al (2009) Oligotrophication and emergence of picocyanobacteria and a toxic dinoflagellate in Thau Lagoon, southern France. J Sea Res 61:68–75

Coruzzi G, Bush GR (2001) Nitrogen and carbon nutrient and metabolite signaling in plants. Plant Physiol 125:61–64

Crawford NN, Glass AND (1998) Molecular and physiological aspects of nitrate uptake in plants. Trends Plant Sci 3:389–395

Cullen JT, Sherrell RM (2005) Effects of dissolved carbon dioxide, zinc, and manganese on the cadmium to phosphorus ratio in natural phytoplankton assemblages. Limnol Oceanogr 50:1193–1204

Dagenais-Bellefeuille S, Morse D (2013) Putting the N in dinoflagellates. Front Microbiol 4:article 369 (14 p). https://doi.org/10.3389/micb.2013.00369

Daniel-Vedele F, Filleur S, Caboche M (1998) Nitrate transport: a key step in nitrate assimilation. Curr Opin Plant Biol 1:235–239

DeBaar HJW (1994) VonLiebig's law of the minimum and plankton ecology (1899-1991). Prog Oceanogr 33:347–386

Dugdale RC (1967) Nutrient limitation in the sea: dynamics, identification and significance. Limnol Oceanogr 12:685–695

Dugdale RC, Jones BH, MacIsaac JJ et al (1981) Adaptation of nutrient assimilation. Can Bull Fish Aquat Sci 210:234–250
Dugdale RC, Wilkerson FP, Hogue VE et al (2007) The role of ammonium and nitrate in spring bloom development in San Francisco Bay. Estuar Coast Shelf Sci 73:17–29
Dugdale RC, Wilkerson FP, Morel A (1990) Realization of new production in coastal upwelling areas: a means to compare relative performance. Limnol Oceanogr 35:822–829
Elser JJ, Bracken MES, Cleland EE et al (2007) Global analysis of nitrogen and phosphorus limitation of primary producers in freshwater, marine and terrestrial ecosystems. Ecol Lett 10:1135–1142
Escobar MA, Geisler DA, Rasmusser AG (2006) Reorganization of the alternative pathways of the *Arabidopsis* respiratory chain by nitrogen supply: opposing effects of ammonium and nitrate. Plant J 45:775–788
Falkowski PG (1975) Nitrate uptake in marine phytoplankton: comparison of half-saturation constants from seven species. Limnol Oceanogr 20:412–417
Finkel ZV, Beardall J, Flynn KJ et al (2010) Phytoplankton in a changing world: cell size and elemental stoichiometry. J Plankton Res 32:119–137
Finkel ZV, Follows MJ, Liefer JD et al (2016) Phylogenetic diversity in the macromolecular composition of microalgae. PLoS One 11(5):e0155977. https://doi.org/10.1371/journal.pone.0155977
Flynn KJ (2005a) Modeling marine phytoplankton growth under eutrophic conditions. J Sea Res 54:92–103
Flynn KJ (2005b) Castles built on sand: dysfunctional plankton models and the failure of the biology-modeling interface. J Plankton Res 27:1205–1210
Flynn KJ (2009) Ecological modelling in a sea of variable stoichiometry: dysfunctionality and the legacy of Redfield and Monod. Prog Oceanogr. https://doi.org/10.1016/j.pocean.2009.09.006
Flynn KJ, Mitra A (2009) Building the "perfect beast": modelling mixotrophic plankton. J Plankton Res 31:965–992
Flynn KJ, Mitra A, Glibert PM et al (2018) Mixotrophy in HABs: by whom, on whom, when, why, and what next. In: Glibert PM, Berdalet E, Burford M et al (eds) Global ecology and oceanography of harmful algal blooms. Springer, Cham, pp 113–132
Flynn KJ, Page S, Wood G et al (1999) Variations in the maximum transport rates for ammonium and nitrate in the prymnesiophyte *Emiliania huxleyi* and the raphidophyte *Heterosigma carterae*. J Plankton Res 21:355–371
Flynn KJ, Stoecker DK, Mitra A et al (2013) Misuse of the phytoplankton-zooplankton dichotomy: the need to assign organisms as mixotrophs within plankton functional types. J Plankton Res 35:3–11
Fu M, Wang Z, Pu X et al (2012) Changes in nutrient concentrations and N:P:Si ratios and their possible impacts on the Huanghai Sea ecosystem. Acta Oceanol Sinica 31:101–112
Galloway JN, Cowling EB, Seitzinger SP et al (2002) Reactive nitrogen: too much of a good thing? Ambio 31:60–63
GEOHAB (2001) Global ecology and oceanography of harmful algal blooms, science plan. Glibert P, Pitcher G (eds) SCOR and IOC, Baltimore and Paris, 86 pp
GEOHAB (2006) Global ecology and oceanography of harmful algal blooms: HABs in Eutrophic Systems. Glibert PM (ed) SCOR and IOC, Paris and Baltimore, 74 pp
Gerendás J, Zhu Z, Bendixon R et al (1997) Physiological and biochemical processes related to ammonium toxicity in higher plants. Z Pflanzenernaehr Bodenkd 160:239–251
Glibert PM (2017) Eutrophication, harmful algae and biodiversity – challenging paradigms in a world of complex nutrient changes. Mar Pollut Bull 124:591–606. https://doi.org/10.1016/j.marpolbul.2017.04.027
Glibert PM, Al-Azri A, Allen JI et al (2018a) Key questions and recent research advances on harmful algal blooms in relation to nutrients and eutrophication. In: Glibert PM, Berdalet E, Burford M et al (eds) Global ecology and oceanography of harmful algal blooms. Springer, Cham, pp 229–259

Glibert PM, Beusen AHW, Harrison JA et al (2018b) Changing land-, sea- and airscapes: sources of nutrient pollution affecting habitat suitability for harmful algae. In: Glibert PM, Berdalet E, Burford M et al (eds) Global ecology and oceanography of harmful algal blooms. Springer, Cham, pp 53–76

Glibert PM, Burford MA (2017) Globally changing nutrient loads and harmful algal blooms: recent advances, new paradigms, and continuing challenges. Oceanography 30(1):58–69. https://doi.org/10.5670/oceanog.2017.110

Glibert PM, Burkholder JM (2011) Eutrophication and HABs: strategies for nutrient uptake and growth outside the Redfield comfort zone. Chin J Oceanol Limnol 29:724–738

Glibert PM, Burkholder JM, Kana TM et al (2009) Grazing by *Karenia brevis* on *Synechococcus* enhances their growth rate and may help to sustain blooms. Aquat Microb Ecol 55:17–30

Glibert PM, Kana TM, Brown K (2013) From limitation to excess: consequences of substrate excess and stoichiometry for phytoplankton physiology, trophodynamics and biogeochemistry, and implications for modeling. J Mar Syst 125:14–28

Glibert PM, Manager R, Sobota et al (2014a) The Haber-Bosch–harmful algal bloom (HB-HAB) link. Environ Res Lett 9:105001 (13 p). https://doi.org/10.1088/1748-9326/9/10/10500

Glibert PM, Seitzinger S, Heil C et al (2005) The role of eutrophication in the global proliferation of harmful algal blooms: new perspectives and new approaches. Oceanography 18(2):198–209

Glibert PM, Wilkerson F, Dugdale RC et al (2014b) Microbial communities from San Francisco Bay Delta respond differently to oxidized and reduced nitrogen substrates – even under conditions that would otherwise suggest nitrogen sufficiency. Front Mar Sci 1:article 17. https://doi.org/10.3389/fmars.2014.00017

Glibert PM, Wilkerson F, Dugdale RC et al (2016) Pluses and minuses of ammonium and nitrate uptake and assimilation by phytoplankton and implications for productivity and community composition with emphasis on nitrogen-enriched conditions. Limnol Oceanogr 61:165–197

Gobler CJ, Christopher J, Burkholder JM et al (2016) The dual role of nitrogen supply in controlling the growth and toxicity of cyanobacterial blooms. Harmful Algae 54:87–97

Goldman JC, Glibert PM (1983) Kinetics of inorganic nitrogen uptake. In: Carpenter EJ, Capone DG (eds) Nitrogen in the marine environment. Academic Press, New York, pp 233–274

Granéli E, Flynn K (2006) Chemical and physical factors influencing toxin content. In: Granéli E, Turner JT (eds) The ecology of harmful algae. Springer, New York, NY, pp 229–241

Hansen PJ (2011) The role of photosynthesis and food uptake for the growth of marine mixotrophic dinoflagellates. J Eukaryot Microbiol 58:203–214

Hardison DR, Sunda WG, Shea D et al (2013) Increased toxicity of *Karenia brevis* during phosphate limited growth: ecological and evolutionary implications. PLoS One 8(3):e58545. https://doi.org/10.1371/journal.pone.0058545

Hecky R, Kilham P (1988) Nutrient limitation of phytoplankton in freshwater and marine environments: a review of recent evidence on the effects of enrichment. Limnol Oceanogr 33:796–822

Heisler J, Glibert PM, Burkholder JM et al (2008) Eutrophication and harmful algal blooms: a scientific consensus. Harmful Algae 8:3–13

Hillebrand H, Steinert G, Boersma M et al (2013) Goldman revisited: faster-growing phytoplankton has lower N:P and lower stoichiometric flexibility. Limnol Oceanogr 58:2076–2088

Howarth RW (1988) Nutrient limitation of net primary production in marine ecosystems. Annu Rev Ecol Syst 19:89–110

Howarth RW, Marino R (2006) Nitrogen as the limiting nutrient for eutrophication in coastal marine ecosystems: evolving views over three decades. Limnol Oceanogr 51:364–376

Howarth RW, Sharpley A, Walker D (2002) Sources of nutrient pollution to coastal waters in the United States: implications for achieving coastal water quality goals. Estuaries 25:656–676

Jeong HJ (2011) Mixotrophy in red tide algae Raphidophytes. J Eukaryot Microbiol 58:215–222

Jeong HJ, Park JY, Nho JH et al (2005) Feeding by red-tide dinoflagellates on the cyanobacterium *Synechococcus*. Aquat Microb Ecol 41:131–143

Jeong HJ, Yoo YD, Kim JS et al (2010) Growth, feeding and ecological roles of the mixotrophic and heterotrophic dinoflagellates in marine planktonic food webs. Ocean Sci J 45:65–91

Jeong HJ, Yoo YD, Kim TH et al (2004) Mixotrophy in the phototrophic harmful alga *Cochlodinium polykrikoides* (Dinophyceae): prey species, the effects of prey concentration and grazing impact. J Eukaryot Microbiol 51:563–569

Johansson N, Granéli E (1999) Influence of different nutrient conditions on cell density, chemical composition and toxicity of *Prymnesium parvum* (Haptophyta) in semi-continuous cultures. J Exp Mar Biol Ecol 239:243–258

Jones RI (1994) Mixotrophy in planktonic protists as a spectrum of nutritional strategies. Mar Microb Food Webs 8:87–96

Kana TM, Glibert PM (2016) On saturating response curves from the dual perspectives of photosynthesis and nitrogen acquisition. In: Glibert PM, Kana TM (eds) Aquatic microbial ecology and biogeochemistry: a dual perspective. Springer International Publ., Geneva, pp 93–104

Kana TM, Geider RJ, Critchley C (1997) Photosynthetic pigment regulation in microalgae by multiple environmental factors: a dynamic balance hypothesis. New Phytol 137:629–638. https://doi.org/10.1046/j.1469-8137.1997.00857.x

Klausmeier CA, Litchman E, Daufresne T et al (2004) Optimal N:P stoichiometry of phytoplankton. Nature 429:171–174

Kudela RM, Raine R, Pitcher G et al (2018) Establishment, goals, and legacy of the Global Ecology and Oceanography of Harmful Algal Blooms (GEOHAB) Program. In: Glibert PM, Berdalet E, Burford M et al (eds) Global ecology and oceanography of harmful algal blooms. Springer, Cham, pp 27–49

Leonardos N, Geider RJ (2004a) Responses of elemental and biochemical composition of *Chaetoceros muelleri* to growth under varying light and nitrate:phosphate supply ratios and their influence on critical N:P. Limnol Oceanogr 49:2105–2114

Leonardos N, Geider RJ (2004b) Effects of nitrate:phosphate supply ratio and irradiance on the C: N:P stoichiometry of *Chaetoceros muelleri*. Eur J Phycol 39:173–180

Litke DW (1999) Review of phosphorus control measures in the United States and their effects on water quality. U.S. Geological Survey Water–Resources Investigations Report 99-4007. Denver, CO

Lomas MW, Glibert PM (1999) Temperature regulation of nitrate uptake: a novel hypothesis about nitrate uptake and reduction in cool-water diatoms. Limnol Oceanogr 44:556–572

Menten L, Michaelis MI (1913) Die kinetik der invertinwirkung. Biochem Zeitung 49:333–369

Minnhagen S, Kim M, Salomon P et al (2011) Active uptake of kleptoplastids by *Dinophysis caudata* from its ciliate prey *Myrionecta rubra*. Aquat Microb Ecol 62:99–108

Mitra A, Castellani C, Gentleman W et al (2014a) Bridging the gap between marine biogeochemical and fisheries sciences; configuring the zooplankton link. Prog Oceanogr 129B:176–199

Mitra A, Flynn KJ (2010) Modelling mixotrophy in harmful algal blooms: more or less the sum of the parts? J Mar Syst 83:158–169. https://doi.org/10.1016/j.jmarsys.2010.04.006

Mitra A, Flynn KJ, Burkholder JM et al (2014b) The role of mixotrophic protists in the biological carbon pump. Biogeosciences 11:995–1005. https://doi.org/10.5194/bg-11-995-2014

Monod J (1942) Recherches sur la croissance des cultures bactériennes, 2nd edn. Hermann, Paris

Nathansohn A (1908) Uber die allgemein Produktionsbedingungen im Meere, Beiträge zur biologie des planktons, von H.H. Gran und Nathansohn. Internationale Revue der gestanten. Hydrologie 1:38–72

Navarro MT, Prieto R, Fernandez E et al (1996) Constitutive expression of nitrate reductase changes the regulation of nitrate and nitrite transporters in *Chlamydomonas reinhardtii*. Plant J 9:819–827

Oh H-M, Lee SJ, Jang M-H et al (2000) Microcystin production by *Microcystis aeruginosa* in a phosphorus-limited chemostat. Appl Environ Microbiol 66:176–179

Packard TT (1973) The light dependence of nitrate reductase in marine phytoplankton. Limnol Oceanogr 18:466–469

Paerl HW, Xu H, Hall NS et al (2014) Controlling cyanobacterial blooms in hypereutrophic Lake Taihu, China: will nitrogen reductions cause replacement of non-N_2 fixing by N_2 fixing taxa? PLoS One 9:e113123. https://doi.org/10.1371/journal.pone.0113123

Pahlow M, Oschlies A (2013) Optimal allocation backs Droop's cell-quota model. Mar Ecol Prog Ser 473:1–5

Parker AE, Hogue VE, Wilkerson FP et al (2012) Inorganic nitrogen speciation and phytoplankton growth in the high nutrient, low chlorophyll San Francisco Estuary. Estuar Coast Shelf Sci 104-105:91–101

Peñuelas J, Sardans J, Rivas-Ubach A et al (2012) The human-induced imbalance between C, N and P in Earth's life system. Glob Chang Biol 18:3–6

Podgórska A, Szal B (2015) The role of reactive oxygen species under ammonium nutrition. In: Gupta KJ, Igamberdiev AU (eds) Reactive oxygen and nitrogen species signaling and communication in plants. Springer, Heidelberg, pp 133–154

Quigg A, Finkel ZV, Irwin AJ et al (2003) The evolutionary inheritance of elemental stoichiometry in marine phytoplankton. Nature 425:291–294

Raven JA, Beardall J, Flynn KJ et al (2009) Phagotrophy in the origins of photosynthesis in eukaryotes and as a complementary mode of nutrition in phototrophs: relation to Darwin's insectivorous plants. J Exp Bot 60:3975–3987. https://doi.org/10.1093/jxb/erp282

Reynolds CS (1999) Non-determinism to probability, or N:P in the community ecology of phytoplankton. Arch Hydrobiol 146:23–35

Savage TJ, Smith GJ, Clark AT et al (2012) Condensation of the isoprenoid and amino precursors in the biosynthesis of domoic acid. Toxicon 59(1):25–33. https://doi.org/10.1016/j.toxicon.2011.10.010

Schindler DW (1977) Evolution of phosphorus limitation in lakes. Science 195:260–262

Schindler DW, Carpenter SR, Chapra SC et al (2016) Reducing phosphorus to curb lake eutrophication is a success. Environ Sci Technol. https://doi.org/10.1021/acs.est.6b02204

Smayda TJ (1990) Novel and nuisance phytoplankton blooms in the sea: evidence for a global epidemic. In: Granéli E, Sundstöm B, Edler L et al (eds) Toxic marine phytoplankton. Elsevier, New York, NY, pp 29–40

Smayda TJ (1997) Harmful algal blooms: their ecophysiology and general relevance to phytoplankton blooms in the sea. Limnol Oceanogr 42:1137–1153

Smith SL, Yamanaka Y, Pahlow M et al (2009) Optimal uptake kinetics: physiological acclimation explains the patterns of nitrate uptake by phytoplankton in the ocean. Mar Ecol Prog Ser 384:1–12

Stoecker DK, Tillmann U, Granéli E (2006) Phagotrophy in harmful algae. In: Granéli E, Turner JT (eds) The ecology of harmful algae. Springer, New York, NY, pp 177–187

Sutton MA, Bleeker A, Howard CM et al (2013) Our nutrient world: the challenge to produce more food and energy with less pollution. Centre for Ecology and Hydrology, Edinburgh

Tillmann U (1998) Phagotrophy by a plastidic haptophyte, *Prymnesium patelliferum*. Aquat Microb Ecol 14:155–160

Tilman D (1977) Resource competition between planktonic algae: an experimental and theoretical approach. Ecology 58:338–348

Tilman D (1982) Resource competition and community structure. Princeton University Press, Princeton, NJ

Tilman D (1985) The resource-ratio hypothesis of plant succession. Am Nat 125:827–852

Unrein F, Massana R, Alonso-Saez L et al (2007) Significant year-round effect of small mixotrophic flagellates on bacterioplankton in an oligotrophic coastal system. Limnol Oceanogr 52:456–469

Van de Waal DB, Smith VS, Declerck SAJ et al (2014) Stoichiometric regulation of phytoplankton toxins. Ecol Lett 17:736–742

Van de Waal DB, Verspagen JMH, Lürling M et al (2009) The ecological stoichiometry of toxins produced by harmful cyanobacteria: an experimental test of the carbon-nutrient balance hypothesis. Ecol Lett 12:1326–1335

Vézie C, Rapala J, Vaitomaa J et al (2002) Effect of nitrogen and phosphorus on growth of toxic and nontoxic *Microcystis* strains and on intracellular microcystin concentrations. Microb Ecol 43:443–454

Von Liebig, L (1855) Principles of agricultural chemistry with special reference to the late researches made in England [Reprinted in Pomeroy LR (ed) (1974) Cycles of essential elements, benchmark papers in ecology, vol I. Dowden, Hutchinson and Ross, Inc., Staussburg, PA, pp 11–28]

Wells ML, Karlson B (2018) Harmful algal blooms in a changing ocean. In: Glibert PM, Berdalet E, Burford M et al (eds) Global ecology and oceanography of harmful algal blooms. Springer, Cham, pp 77–90

Wilkerson FP, Dugdale RC, Hogue VE et al (2006) Phytoplankton blooms and nitrogen productivity in the San Francisco Bay. Estuar Coasts 29:401–416

Yoshiyama K, Sharp JH (2006) Phytoplankton response to nutrient enrichment in an urbanized estuary: apparent inhibition of primary production by overeutrophication. Limnol Oceanogr 51:424–434. https://doi.org/10.4319/lo.2006.51.1_part_2.0424

Zubkov MV, Tarran GA (2008) High bacterivory by the smallest phytoplankton in the North Atlantic Ocean. Nature 455:224–226

Chapter 7
Mixotrophy in Harmful Algal Blooms: By Whom, on Whom, When, Why, and What Next

Kevin J. Flynn, Aditee Mitra, Patricia M. Glibert, and JoAnn M. Burkholder

7.1 Introduction

The traditional view of the planktonic food web is simplistic: nutrients are consumed by phytoplankton that, in turn, support zooplankton that ultimately support fish. Historically, harmful algal species have been viewed as phototrophic phytoplankton, although with some notable exceptions. This primary producer-centric structure is the foundation of most models that have been used to explore fisheries production, biogeochemical cycling, and climate change. In recent years, however, the importance of mixotrophs increasingly has been recognized, changing scientists' view of these traditional food web interactions (Katechakis et al. 2005; Glibert and Legrand 2006; Burkholder et al. 2008; Jeong et al. 2010; Hansen 2011; Flynn et al. 2013; Mitra et al. 2016). Mixotrophy, the combination of phototrophy and heterotrophy (phago- and/or osmotrophy), enables planktonic protists traditionally labelled as "phytoplankton" or "microzooplankton" to function at multiple trophic levels. Harmful algal blooms (HABs), including both high-biomass bloom formers and toxic species, are now recognized to be not only more prevalent in waters worldwide but also typically mixotrophic (Burkholder et al. 2008). Thus, it is important that concerted efforts are taken to understand mixotrophy as an important nutritional strategy that alters food web dynamics.

K.J. Flynn (✉) • A. Mitra
Wallace Building, Swansea University, Swansea, UK
e-mail: k.j.flynn@swansea.ac.uk

P.M. Glibert
University of Maryland Center for Environmental Science, Horn Point Laboratory, Cambridge, MD, USA

J.M. Burkholder
Center for Applied Aquatic Ecology, North Carolina State University, Raleigh, NC, USA

P.M. Glibert et al. (eds.), *Global Ecology and Oceanography of Harmful Algal Blooms*, Ecological Studies 232, https://doi.org/10.1007/978-3-319-70069-4_7

The processes of mixotrophy can radically change the dynamics and efficiency of activities of the "phytoplankton" and "microzooplankton" (Flynn and Mitra 2009; Mitra and Flynn 2010; Flynn et al. 2013; Mitra et al. 2014b). Mixotrophy includes the concept of osmotrophy, the uptake of organic substrates, and phagotrophy, the ingestion of particulate material, by those organisms normally thought to be dependent on inorganic carbon (C) via photosynthesis. Osmotrophy is ubiquitous among the protists (e.g., Berman and Bronk 2003; Glibert and Legrand 2006); phagotrophy is also much more common than previously recognized and has direct implications for trophic dynamics (namely, grazing). A fundamental revision of the functional taxonomy of aquatic protists has recently been proposed in recognition of the importance of mixotrophy (Mitra et al. 2016). In this chapter we narrowly define mixotrophy as phototrophy with phagotrophy. We emphasize that mixotrophy warrants realistic descriptions within food web models; it represents a more flexible strategy in nutrient acquisition, growth, and production, with far-reaching consequences not only for growth of the mixotrophs but also for their competitors and prey. In essence, mixotrophic organisms alter the flow of nutrients through the planktonic system, potentially upgrading it biochemically and changing the nature of predator-prey interactions in a highly efficient manner. An appreciation of the importance of mixotrophy also affects which nutrient forms should be considered in water management strategies; for example, if mixotrophic HABs are promoted by events such as bacterial growth on organic substances, through ingestion of bacteria, then dissolved organics as well as the usual dissolved inorganic nutrient loads require consideration.

Within the framework of the Global Ecology and Oceanography of Harmful Algal Blooms (GEOHAB) Programme, the understanding of the role of nutrients generally, and of eutrophication more specifically, was highlighted as important areas of research needed to strengthen predictive capabilities for HABs [see Chap. 12, Glibert et al. (2018a)]. The importance of increased understanding about adaptive strategies of HABs was also recognized. When GEOHAB was conceived (GEOHAB 1998, 2001), only a few HAB species had been identified as mixotrophic, and these largely were considered curiosities. In the ensuing decades, it has been recognized that mixotrophic organisms, many of which include HAB species (e.g., Burkholder et al. 2008), are not only much more common than originally thought, but also their physiology is now known to be far more complex than could be described by simply "bolting together" descriptions of the contrasting processes of phototrophy and phagotrophy (Mitra and Flynn 2010). In addition, the central mission of GEOHAB was to strengthen the prediction of HABs through improvements in detection capability and modelling [GEOHAB 2001; see also Chap. 3, Kudela et al. (2018)]. The need for predictive models is increasingly great, particularly when considering climate change, increasing acidification, changes in nutrient input types and ratios, and increasing geographic spread of some HAB species. Yet, for the most part, one of the most important physiological functions of the dominant plankton communities—mixotrophy—is still not being explicitly included within ecosystem food web models. As aquatic ecosystems move toward an increasingly uncertain future, models describing them need to be mechanistically sound, accurately reflecting food web dynamics, and that requires incorporation of mixotrophs.

Accordingly, a working group funded under the auspices of the Leverhulme Trust of the UK was formed to address such issues. While not formally a GEOHAB endeavour, and not exclusively focused on HABs as the only types of mixotrophs, the advancements made by this working group helped to advance many important questions on nutrient physiology and adaptive strategies posed by GEOHAB. In this chapter we review some of these advancements and pose additional, still-unresolved questions regarding mixotrophic HAB species ecology.

7.2 Mixotrophy Across the Spectrum of Nutrient Supply

Planktonic mixotrophy has long been considered important in oligotrophic habitats, where potentially limiting elements are much more concentrated in microbial prey than in the water column (Jones 1994). More recently it has been recognized that mixotrophy is also important in eutrophic estuaries and coastal marine embayments, where many HAB taxa are mixotrophic (Burkholder et al. 2008). In such environments, mixotrophs respond both directly to nutrient inputs and indirectly through high abundance of bacterial and algal prey that are stimulated by the elevated nutrients. It is worth noting that in so-called eutrophic systems, light can often be limiting and hence limit newly fixed C. Thus, mixotrophy provides a mechanism to exploit alternative routes for the acquisition of C. Another important facet of mixotrophy is that all mixotrophs are not the same; Mitra et al. (2016) divide these protists into those that have a constitutive capacity for mixotrophy and those that acquire phototrophy by acquisition of photosystems from other organisms. The constitutive mixotrophs include HAB species such as small-celled toxigenic taxa (e.g., the dinoflagellate *Karlodinium* and the haptophyte *Prymnesium*), as well as non-HAB species such as the large dinoflagellate *Ceratium.* Among the non-constitutive mixotrophs, some, the plastidic ciliates, are generalists and can steal chloroplasts from a range of prey types, while others are specialists making use of plastids acquired from very few prey sources. These include the nontoxic "red tide" ciliate *Mesodinium/Myrionecta* and the toxigenic dinoflagellate *Dinophysis*, which actually obtains its plastids "3rd hand" from *Mesodinium* (Hansen et al. 2013 and references therein). Another group of non-constitutives that harbours symbiotic phototrophs includes the biogeochemically important foraminifera and radiolaria and also the HAB-forming dinoflagellate, "green" *Noctiluca* [Gomes et al. 2014; Stoecker et al. 2017; see also Chap. 17, Goes et al. (2018)]. These contrasting mixotrophic strategies impart different ecological impacts through the need and removal of prey (perhaps competitors) that could, in turn, affect the operation of other trophic pathways. Mixotrophic HABs may thus be expected to be associated with complex ecological scenarios.

Many different types of harmful microalgal species thrive in eutrophying or eutrophic estuarine and coastal marine habitats, including cyanobacteria, dinoflagellates, haptophyte flagellates, heterokontophytes (diatoms, golden flagellates,

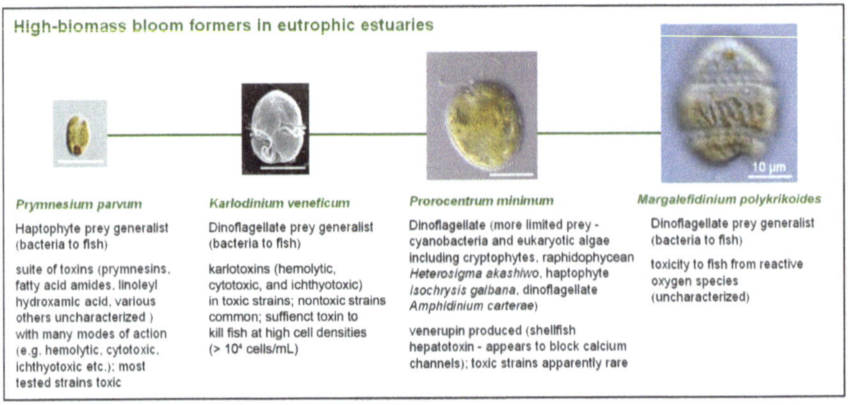

Fig. 7.1 Four examples of toxigenic mixotrophs common in eutrophic estuaries include *Prymnesium parvum*, *Karlodinium veneficum*, *Prorocentrum minimum*, and *Cochlodinium (Margalefidinium) polykrikoides*. Size comparison and other information is from Denardou-Queneherve et al. (1999), Dorantes-Aranda et al. (2010), Henrikson et al. (2010), Van Wagoner et al. (2010), Bertin et al. (2012), Glibert et al. (2012), Remmel and Hambright (2012), Johnson (2014), and references therein

brown-tide algae, and raphidophycean flagellates) (Burkholder et al. 2008; Graham et al. 2016; Fig. 7.1). In fact, the bulk of eukaryotic planktonic protists commonly classed as phototrophs is actually mixotrophic (Flynn et al. 2013; Mitra et al. 2016). Thus, it appears that most phytoflagellates (heterokontophytes, haptophytes, and cryptophytes) are capable of consuming bacteria and small eukaryotic cells (Unrein et al. 2007; Zubkov and Tarran 2008; Stukel et al. 2011). Dinoflagellates exhibit a wide range in capabilities, from those that express phagotrophy only upon extreme nutrient starvation to exclusively phagotrophic forms. Even the latter organisms may gain at least transitory inorganic C-fixing capacity from ingestion of phototrophs (Raven et al. 2009; Jeong et al. 2010; Hansen 2011) or other mixotrophs (Minnhagen et al. 2011).

Some potentially toxic species appear to have relatively poor affinities (high Ks values) for uptake of inorganic N and P substrates (DIN, dissolved inorganic N; DIP, dissolved inorganic P), and may therefore depend on mixotrophy as a means to supplement their acquisition of dissolved substrates (Smayda 1997). A contrary argument is that because of mixotrophy (phagotrophy being the ancestral state in protists—Mitra et al. 2016), these species did not need to evolve inorganic nutrient transporters with high affinity. Importantly, mixotrophy also may permit growth to be sustained or even accelerated during periods of apparent water-column nutrient deficiency or imbalance (or real nutrient deficiency for non-mixotrophic competitors) when only inorganic forms of dissolved nutrients are considered. The dinoflagellates *Cochlodinium (Margalefidinium) polykrikoides* and *Karenia brevis* exemplify these advantages of mixotrophy. When growing as a phototroph, *C. polykrikoides* has a growth rate of 0.17 div day^{-1} (Jeong et al. 2004). Yet, when growing as a mixotroph, consuming cryptophytes, the division rate of

C. polykrikoides nearly doubles to 0.34 div day^{-1} (Jeong et al. 2004). Toxigenic *Karenia brevis* has been found to graze on the cyanobacterium *Synechococcus* sp., as well as on cryptophytes (Jeong et al. 2005; Adolf et al. 2008; Glibert et al. 2009). In laboratory experiments, Jeong et al. (2005) estimated that up to 5 cells h^{-1} of *Synechococcus* were grazed by mixotrophic *K. brevis*, while Glibert et al. (2009) found that from ~1 to 80 *Synechococcus* cells h^{-1} were grazed by *K. brevis*, depending on the predator-prey ratio. The growth rate of *K. brevis* increased as the supply of *Synechococcus* increased, indicating that natural variability in this food source may affect growth rates of this HAB species in nature. In contrast, *Prymnesium parvum* (as *P. patelliferum*) and *Chrysochromulina* spp., which ingest prey under light- and nutrient-sufficient conditions, maintain similar growth rates with versus without phagotrophy (Pintner and Provasoli 1968; Larsen et al. 1993; Granéli and Carlsson 1998).

Harmful estuarine and marine algae have been reported to have significant impacts on prey populations and to contribute substantially to elemental demands of the mixotroph, illustrated by the following three examples from eutrophic Chesapeake Bay, USA. In situ grazing rates were estimated for the large dinoflagellate *Akashiwo sanguinea* (as *Gymnodinium sanguineum*; cells 60 × 45 μm, biovolume 79,000 μm^3) grazing on oligotrich ciliates (length <20 μm) (Bockstahler and Coats 1993a, b). Based on integrated station and transect averages, *A. sanguinea* removed 34% and 24%, respectively (medians 17% and 22%, respectively) of the oligotrich ciliate population daily. Daily consumption of ciliates supplied an average of 2.5% (maximum 11.6%) of the grazer C cell^{-1}. Mixotrophy also was estimated to supply 4% (maximum, 18.5%) of the grazer cellular N daily and 15% of its N requirements for asexual reproduction. Thus, it was concluded that mixotrophy may help balance the N requirements for *A. sanguinea* and thence confer an advantage over strictly phototrophic algae under N stress conditions (Bockstahler and Coats 1993a, b). Grazing by another large dinoflagellate species, *Ceratium furca* (cells 250 μm 25 μm, biovolume 3400 μm^3), was estimated in the same ecosystem using fluorescently labelled ciliate prey (Smalley and Coats 2002). Feeding rates ranged from 0 to 0.11 prey dinoflagellate^{-1} h^{-1}, and grazing by *C. furca* was estimated to remove an average of 67% of the ciliates *Strombilidium* spp. day^{-1}. Not only is such predation important for this mixotroph, but the removal of the ciliates would also have a "knock-on" effect by relieving grazing constraints upon the ciliate's prey. Phagotrophy in *Karlodinium veneficum* (reported as *Gyrodinium galatheanum* or *Karlodinium micrum*) has been the subject of intensive research focus in Chesapeake Bay. Li et al. (2000) found that *K. veneficum* became phagotrophic at suboptimal light and/or nutrient conditions, and the mean ingested prey was positively correlated with cryptophyte abundance. Adolf et al. (2008) determined that *K. veneficum* blooms in this bay were positively correlated with cryptophyte abundance. Ingestion rates among multiple strains of cultured *K. veneficum* ranged from 0 to 4 prey dinoflagellate^{-1} day^{-1}, and cultured toxic strains were capable of consuming an array of cryptophyte species (31–421 μm^3 cell^{-1}). The authors hypothesized from these data and supporting studies (e.g., Adolf et al. 2003, 2006) that cryptophyte prey abundance is a key factor that support blooms of *K. veneficum* in eutrophic habitats. This work built from a previous intensive effort with cultures to examine the balance between phototrophy and heterotrophy during mixotrophic growth of *K. veneficum* on radiolabelled cryptophyte prey (Adolf et al. 2006). Earlier work had suggested that cryptophytes are linked to blooms of other toxigenic dinoflagellates as well (Lewitus et al. 1999; Glasgow et al. 2001). Based upon such studies, it appears that mixotrophy is important in the nutrient acquisition and growth of harmful species and, therefore, should be important in the development and maintenance of their blooms in eutrophic habitats.

While the data support the premise that mixotrophic harmful algae are significant grazers and predators in eutrophic habitats, for most of these species, quantitative data continue to be lacking. Data relating laboratory information to natural field assemblages, and data to enable descriptions of various nutrition acquisition modes of mixotrophs in nature, and the implications of switching between these (such as on growth rates, vertical migration, and toxicity) have yet to be obtained. It should also be emphasized that little is known about the drivers for mixotrophy in constitutive mixotrophs, whether it is C or N (as per the studies mentioned above) or for P or growth cofactors. While rates of contribution of C or N from mixotrophy may appear minor, contributions to mixotroph well-being may be very different. For example, because the relationship between cellular-P and growth rate is strongly nonlinear (Flynn 2008), a relatively low rate of P acquisition from prey ingestion could have an impact on health and growth that extends far beyond the traditional Redfield-based expectations. For the non-constitutive mixotrophs, the ongoing role of mixotrophy is obvious; these organisms rely upon acquired phototrophy and, thus, must feed and conduct photosynthesis.

While most species are free-living, mixotrophy also occurs in some parasitic dinoflagellates with photosynthetic life stages, such as *Blastodinium* spp. (Chatton 1920; Pasternak et al. 1984), *Dissodinium psuedocalani* (Drebes 1969), and *Crepidoodinium australe* (Lom et al. 1993), within estuarine and marine waters that include eutrophic habitats. Mixotrophy is apparently operative, although not well studied, among other photosynthetic parasitic dinoflagellates as well (e.g., Gaines and Elbrächter 1987; Coats 1999). The importance of mixotrophy to overall nutrition is unknown for most of these species, but could be significant—for example, Pasternak et al. (1984) reported that the endoparasitic species acquire up to half of the energy needed for their growth through phagotrophy.

While progress is being made in understanding "who is mixotrophic and on whom," there is less confidence about questions linked to "under what conditions" and "at what rate." Thus, focus is needed to understand how the nutritional state of the mixotroph and that of its prey alter the rate of feeding and the growth response of the mixotroph. Lundgren et al. (2016) found that in studies of *Prymnesium parvum* grazing on *Rhodomonas salina*, each grown under varying N:P ratios and mixed in varying combinations, mortality rates of prey were higher when prey were N-rich, regardless of the nutritional states of *P. parvum*. Furthermore, growth of the mixotroph was higher when prey were provided in high N:P ratio conditions. Experiments by Lin et al. (2017) have shown that growth rates of initially exponential- and stationary-phase *Karlodinium veneficum* were enhanced in the presence of prey (*Rhodomonas* sp.) with reciprocal nutrient conditions. Feeding rates (measured as prey death rates) were highest for low-N:P *K. veneficum* initially growing exponentially and mixed with N-rich prey. Maximum feeding rates of low-N:P *K. veneficum* on N-rich prey during exponential growth were ~fourfold higher than feeding rates of high-N:P *K. veneficum* on N-rich prey. Both studies suggest that conditions of elevated N:P in the prey led to higher growth of the mixotroph relative to stoichiometrically balanced or low-N:P conditions in the mixotroph. As noted by Lundgren et al. (2016), it is therefore not surprising that mixotrophs are common in eutrophic waters where stoichiometrically imbalanced conditions prevail, especially high N:P ratios. These complex experiments also underscore the complexity of the natural environments inhabited by these organisms, which often are characterized by sharp gradients in salinity, light, and nutrient concentrations and ratios.

7.3 Mixotrophs and Cellular Nutrient Stoichiometry

The processes of mixotrophy have the potential to radically change the dynamics and efficiency of nutrient assimilation (Flynn and Mitra 2009; Mitra and Flynn 2010; Flynn et al. 2013; Mitra et al. 2014a, b). Mixotrophy is not simply additive or substitutional; rather, it is synergistic (Mitra and Flynn 2010; Mitra et al. 2014b). The synergism of phototrophy and phagotrophy in mixotrophs has important consequences for cell metabolism, especially the maintenance of cell stoichiometry. An advantage for mixotrophs over non-phagotrophic phytoplankton is the provision of additional nutrients (N, P, Fe) from feeding to support primary production, together with a supply of C to supplement photosynthesis under conditions of light limitation (including during the night).

However, perhaps of more importance is that mixotrophs have a stoichiometric advantage over phagotrophic heterotrophs through an ability to enhance retention of nutrients (elements and biochemicals) from their prey. With respiration and additional C losses associated with prey digestion, an organism has to release nutrients to maintain its elemental balance (e.g., C:N:P) (Sterner and Elser 2002; Mitra and Flynn 2005). An ability to decrease that stoichiometrically-linked loss would be of particular value in nutrient-impoverished waters (e.g., mid-oceans and also many temperate waters in summer). Having an additional source of organic C via photosynthesis—that is, being a mixotroph—which can balance C lost through respiration (Putt 1990; Stoecker and Michaels 1991), decreases the need to release surplus N and P. Obtaining data with sufficient temporal resolution and detail to help in the parameterisation of these events is challenging. It requires detailed understanding of both the prey and predator (mixotroph) physiologies when considered alone, and together, conducted under different conditions of light and nutrient abundance. Modelling the interactions presents parallel challenges (Flynn and Mitra 2009), especially when considering contrasts between different mixotroph functional types (Mitra et al. 2016).

7.4 Mixotrophs and the Food Web

The traditional view of the planktonic food web describes consumption of inorganic nutrients by phototrophic phytoplankton, which in turn support zooplankton and ultimately higher trophic levels. Pathways centred on bacteria provide mechanisms for nutrient recycling. This structure lies at the foundation of most models used to explore biogeochemical cycling, the functioning of the "biological pump," and the impact of climate change on these processes. Mixotrophy radically changes the traditional food web interactions, enabling primary producers to acquire nutrients directly from ingestion of prey such as bacterial and algal competitors and even from their predators.

An alternative conceptualization considers the base of this food web to be strongly supported by protist plankton communities that are mixotrophic—combining phototrophy and phagotrophy within a single cell (Flynn et al. 2013; Mitra et al. 2014b). Such a mixotroph-dominated marine food web construct considers the traditional phytoplankton-zooplankton structure (especially involving diatoms) as predominating only during short, albeit important, periods of the production cycle. These periods are characterized by pioneer species (analogous to r-selected species; Parry 1981) growing in developmental phases of ecosystems (Odum 1969), exemplified by the spring bloom in temperate waters or in upwelling waters (Fig. 7.2). They typically have enhanced illumination and inorganic nutrient availability coupled with an absence of effective grazing control (Irigoien et al. 2005), favouring rapid proliferation of strict phototrophs (e.g., diatoms and coccolithophorids during their strictly phototrophic stages; Rokitta et al. 2011). Such conditions favour the emergence of strict phagotrophs as the dominant protist predators. As these ecosystems mature, however, their nutrient regimes and particulate organic loading change, giving rise to conditions that favour mixotrophy and/or disfavour their competitors (Fig. 7.2). Therefore, during much of the planktonic production cycle, mixotrophs are abundant or dominant. The basis of the new structure suggests that the "strict" phototrophic configuration represents an optimal strategy only in immature r-selecting conditions, exemplified by the spring bloom in temperate waters, or in well-illuminated waters that are highly eutrophic (Fig. 7.2). As the nutrient conditions change (mature) due to algal growth, and with the emergence of a prey community, the assemblage transitions to a "K-selecting" environment and mixotrophy becomes selectively advantageous (Fig. 7.2). This more mature environment is dominated by heterotrophs that have phototrophic capabilities and phototrophs that have heterotrophic capabilities; the diversity of mixotrophic strategies is great (Mitra and Flynn 2010; Flynn et al. 2013) and is broadly split between constitutives and the three types of non-constitutive (acquired phototrophy) forms (Mitra et al. 2016). It is in more mature ecosystems, which often are also warmer during summer, where HABs typically develop.

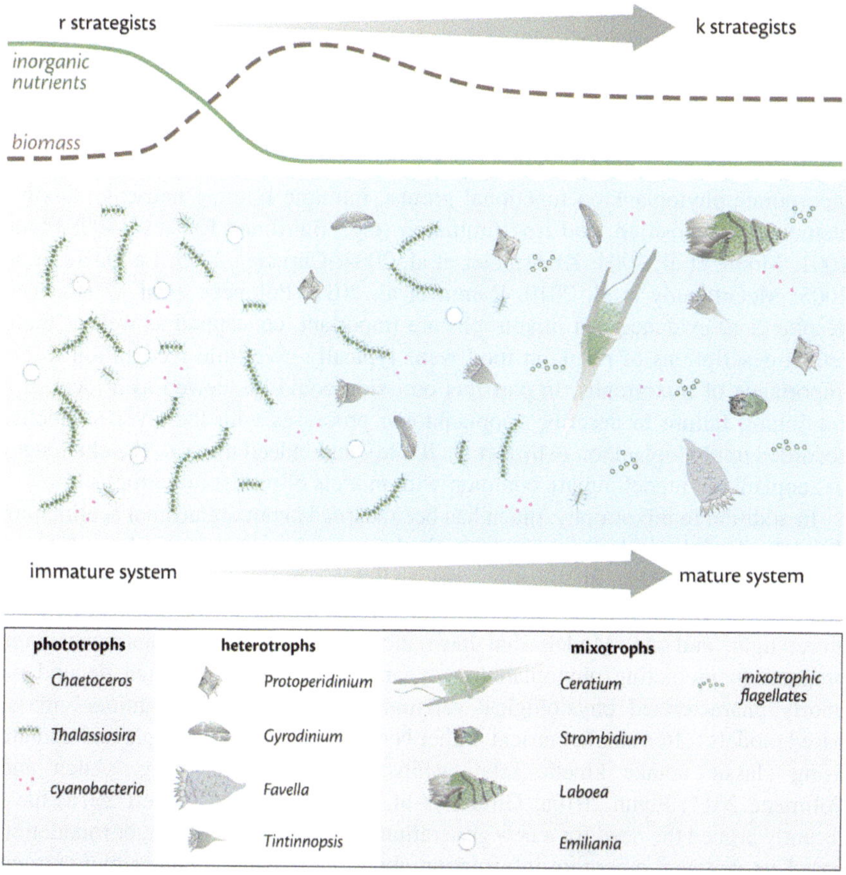

Fig. 7.2 Conceptual diagram of the changes to the planktonic food web over an annual cycle, with transitions between ecosystem states. Transitions between developmental and mature stages of the ecosystem are as indicated; early periods are optimal for strict photoautotrophs, whereas later periods (transition to the more mature state) are suboptimal for strict phototrophs and/or strict phagotrophs and more supportive for mixotrophs. Reproduced from Mitra et al. (2014b) under a creative commons licence

7.5 Inclusion of Mixotrophy in State-of-the-Art Ecosystem Modelling: The Rationale

Models, as simplifications of reality, must balance simplicity with realism and the inevitable complexity present in real biological and ecological systems. Models also depend on parameters which are not always easily measured or available and, as a result, mass fluxes, dynamics, and physiological variables are often not adequately captured. This is so even when considering the relative simplicity of the traditional "phytoplankton–zooplankton" conceptual framework. Ecosystem modelling has advanced tremendously in the past decade, with increasing sophistication in visualization of large datasets, conceptualization of complex interactions, and formulation of predictive scenarios based on model ensembles. Physiological traits of marine organisms are now being applied in models of emergent marine biodiversity (Follows et al. 2007). Newer-generation, mechanistic population dynamic models and three-dimensional ocean biogeochemical models incorporate phytoplankton functional groups, multiple limiting nutrients, flexible elemental composition, and iron limitation (e.g., Baird and Emsley 1999; Flynn 2001; Moore et al. 2004; Klausmeier et al. 2004; Christian 2005; Le Quéré et al. 2005; McGillicudy et al. 2010; Ramin et al. 2012; Polimene et al. 2015). Yet, despite clear evidence that mixotrophs are important, conceptual as well as theoretical descriptions of plankton food webs typically give little recognition to the importance of mixotrophs. In part this omission could be viewed as a facet of a continuing failure to describe zooplanktonic processes with the level of fidelity accorded to phytoplankton (Mitra et al. 2014a), and indeed many of the challenges in zooplankton modelling are common with models of protist mixotrophs.

In addition to mixotrophy, much has been learned about organismal acclimation through physiological responses to variable environments—from phytoplankton growth to competition, allelopathy, prey switching and/or prey rejection, and the relationships between these processes and various abiotic factors such as temperature, light, and pH. Models that have incorporated these findings have made progress in predicting phytoplankton dynamics, but they still often depend on poorly characterized physiological relationships. In general, traditional mass-based models (often single nutrient, either N or C) are relatively simple and operate using classic uptake kinetic relationships. Various authors (e.g., Allen and Polimene 2011; Flynn 2010a; Glibert et al. 2013; Kana and Glibert 2016) have recently argued the need for a new generation of models with a stronger foundation based on new and emerging information about the dynamic cellular biology and ecophysiology of phytoplankton [see also Chap. 6, Glibert et al. (2018c)].

In conditions under which mixotrophs dominate (and that, it turns out, includes most conditions), the minimum configuration of a model should consider C:N:P:Chl ratios. Variable elemental stoichiometric models should become the norm. Multielement descriptions, as opposed to fixed Redfield constructs, provide a basis for the development of mechanistic models that contain functional response descriptors with recognized physiological bases (Flynn 2010a; Glibert et al. 2010, 2013). Multi-element descriptions also support bioenergetic descriptions, which may be important for predicting the survival of organisms under unfavourable conditions. There are important implications of variable stoichiometry for food web dynamics (Sterner and Elser 2002; Mitra and Flynn 2005; Glibert et al. 2013). There may be preferential grazing and also alterations in flow of regenerated nutrients, which can create the potential for shifts in dominant bloom species including HABs. Therefore, it is not sufficient to consider only a single currency (C, N, or P), or even pairs of elements, in experiments or in models. Rather, it is the stoichiometric (im)balance between these that controls phytoplankton physiology, and conceptualization, parameterization, and modelling of this stoichiometric balance in mixotrophy should be prioritized in models of HABs. Multi-stressor studies are also essential; there are very few studies in which multi-stressor impacts have been studied and even fewer in which non-limiting nutrient consumptions are also measured. It is essential to conduct such studies, because what is not removed by today's generation is available for tomorrow's and vice versa. For studies of mixotrophic HABs, this challenge for experiments is extended further because, as with zooplankton modelling (Mitra et al. 2014a), there is the added complexity of sampling the prey and predator. It is also important to correctly describe the growth of the diatoms, as the only major group of protist non-mixotrophs; correctly describing C:N:P:Si in diatom populations sets the basis for nutrient availability and thence plankton succession in the summer (Flynn 2005, 2010a, b).

7.6 Including Mixotrophy in State-of-the-Art Ecosystem Modelling: An Approach

Most plankton food web models continue to be based upon derivations of the classic nutrient-phytoplankton-zooplankton (NPZ) model (Fasham et al. 1990; Plagányi 2007). Within these models, plankton functional types (PFTs) are increasingly used to aid descriptions of processes in biogeochemistry through to fisheries (e.g., end-to-end models; Rose et al. 2010). This has driven an expansion of the classic NPZ model structure (originally N-based) such that "phytoplankton" may be split into groups (e.g., the diatom PFT which require Si or non-diatom PFTs that do not), with each having its own state variables (e.g., Baretta et al. 1996; Flynn 2001; Thingstad et al. 2007). Far less emphasis has been placed on expanding the "zooplankton" component (Mitra et al. 2014a). Even given the current trend of including multiple compartments to represent different groups of PFTs, most existing descriptions of planktonic food web and ecosystem models do not explicitly include mixotrophs. The implicit assumption is that these organisms do not warrant a "box" of their own within these multicomponent ecosystem models. If considered at all, the phototrophic activity of mixotrophs is often assigned to the "phytoplankton" group, while the heterotrophic activity is assigned to the "zooplankton" component, an approach that has been called into question because it fails to capture the synergy of mixotrophic physiology (Mitra and Flynn 2010). Such assumptions stem from two misconceptions. One is that mixotrophy serves a substitutional function in providing C or a nutrient, rather than a *synergistic* function which demands a multi-nutrient description rather than the traditional "single-currency" approach (usually N-based). Secondly, there is a perceived notion that mixotrophy represents some form of inferior middle ground, such that mixotrophs only survive in specific niches (de Castro et al. 2009; Ward et al. 2011). That incorrect assumption typically is based on assumed lower growth rates of mixotrophs compared to their "specialized" counterparts. Yet, when viewed in the context that *mixotrophs dominate in mature ecosystems*, growth at slower rates (in keeping with fluxes through the ecosystem) would be a *desirable* trait rather than an inferior trait (Flynn 2009).

Recasting plankton models with inclusion of mixotrophs may be approached in many ways. Mitra et al. (2014b) extended the NPZBD (NPZ-bacteria-detritus) approach (Fasham et al. 1990; Fig. 7.3), expressing the new mixotrophic paradigm in a new model structure. A new box is added, the "M" box, which includes all protists that are mixotrophic through a combination of phototrophy and phagotrophy. This box is linked to the other boxes in reflection of the realistically wide range of physiological options; thus, mixotrophs perform phototrophy while grazing on bacteria, phytoplankton, microzooplankton, detritus, and/or other mixotrophs. Including mixotrophs within a simple NPZBD model in this way results in a substantial difference in planktonic trophic dynamics. This is because food web dynamics are not merely a summation of a series of rate processes (and kinetic curves). They are an outcome of both the quantity *and quality* of the substrate (or food) provided, not just the rate at which it is produced. The ecophysiology (nutritional quality) of the component organisms present in the ecosystem affects not only their own growth potential but also the activities of co-occurring organisms: the cellular composition of algae, for instance, has consequences for grazers and their emergent properties, with potential for a positive feedback loop generating ungrazable primary producers (Mitra and Flynn 2006; Glibert et al. 2010, 2013). Thus, for grazers, the nutritional value of the prey, not just its availability, is a key aspect of their population success (Lundgren et al. 2016; Lin et al. 2017). From the population standpoint, rate of growth of the prey population is also important.

A critical issue in modelling mixotrophs, then, is controlling the growth rate as a function of the supply of energy and nutrients presented by different modes of nutrition. The classic way of describing both phytoplankton and zooplankton growth rates in models depicts the maximum growth rate (μ_{max}) as a reflection of the maximum nutrient acquisition rate (Fasham et al. 1990). Thus, in models of N-limited phytoplankton, μ_{max} is matched to the maximum rate of N assimilation. For zooplankton, μ_{max} is a function of the maximum grazing rate (G_{max}) and constants defining assimilation efficiency (AE) and respiration (Mitra et al. 2014a). In both instances these are shortcuts that have no basis, at all, in reality. Nutrient transport in phytoplankton may exceed that required to support μ_{max} by an order of magnitude (e.g., McCarthy and Goldman 1979; Goldman and Glibert 1983; Flynn et al. 1999). Likewise, and setting aside the fact that AE is not a constant but, rather, declines with ingestion rate and food quality, the value of G_{max} greatly exceeds the grazing rate needed to match μ_{max} (Mitra et al. 2014a). For models of mixotrophs, the mismatch between such simplifying modelling approaches and reality is particularly apparent because of the interactions between phototrophy and phagotrophy. Those interactions are more complex due to the multi-nutrient basis of the metabolic processes involved. Physiological regulation occurs across the spectrum of substrate conditions, and classic "kinetic curves"

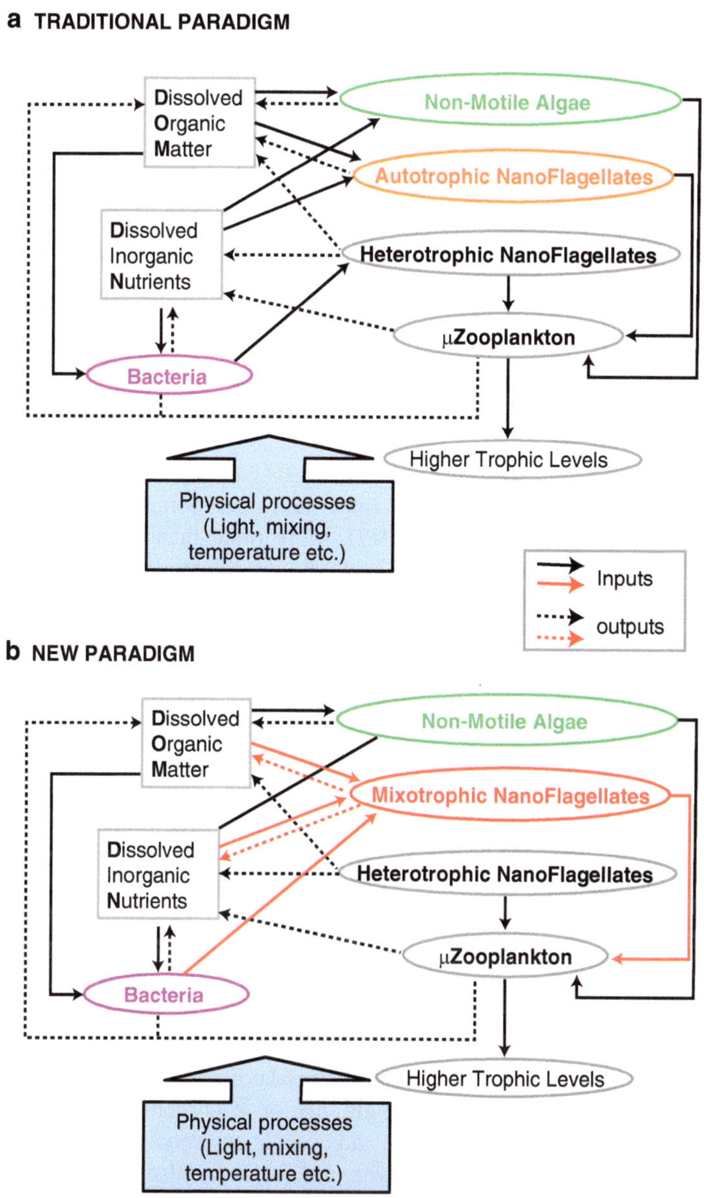

Fig. 7.3 Schematic representing two alternate modelled food web structures. In the traditional paradigm (panel **a**), the physiology of the photoautotrophic flagellates (ANFs) is similar to that of the nonmotile microalgae (NMA), and neither group is mixotrophic. Inorganic nutrients for supporting primary production are regenerated via bacteria, heterotrophic nanoflagellates (HNFs), and phagotrophic microzooplankton (μZ). In the new paradigm (panel **b**), the ANFs are replaced with mixotrophic nanoflagellates (MNFs), conforming to present understanding of protist physiology (Mitra et al. 2016). The MNFs are capable of consuming bacteria (*red-lined black arrow*) and, hence, of deriving nutrients for supporting their primary production, directly rather

cannot capture the extent of the resultant physiological regulation that organisms must attain (Glibert et al. 2013; Kana and Glibert 2016). Moreover, the often-used description of predation by reference to half-saturation abundance of prey also gives rise to kinetics that are not consistent with biological reality (Flynn and Mitra 2016). Even the most complex of the aforementioned phytoplankton models, operating within global ecosystem simulations, has been married with grossly oversimplified zooplankton models. With the emergence of mixotrophy as an important facet of plankton ecology, the zooplankton descriptions need to be upgraded not only to enhance generic realism, but because those same grazing descriptions are needed within the mixotroph descriptions themselves.

7.7 Conclusions

In summary, mixotrophic protists are important, often dominant, members of the plankton community (Leles et al. 2017). More often than not, the mixotrophic potential in these protists likely is realized. Mixotrophic microzooplankton are less susceptible to classic swings in predator-prey dynamics since, for the most part, they are not doomed to death upon exhaustion of a given type of prey; instead, they can be sustained for longer periods of time. The trophic and allometric upgrading in mixotrophy provides a route to introduce important biochemicals (notably fatty acids) into what would otherwise be microbial-loop organisms. Through mixotrophs the microbial loop becomes an effective link and not just a sink to the "classic food web," on stoichiometric and allometric grounds. Lastly and perhaps most importantly, most HABs are mixotrophs; phagotrophs exclude cyanobacteria and diatoms, but those HAB species are osmotrophs (e.g., Burkholder et al. 2008 and references therein).

The last decade or so has been a time of great advancement in our appreciation of mixotrophs—their ubiquitous characteristic, their physiological complexity, and their role in shaping trophic dynamics. Although much is known about which organisms are mixotrophs and how they feed, little is known about how the organisms modulate phototrophic and phagotrophic activities or how the flow of energy and materials through mixotrophic predator-prey interactions are altered under varying nutrient, temperature, light, pH, or pCO_2 conditions—and all these factors are rapidly changing in coastal and oceanic environments with accelerating eutrophication and climate change [Flynn et al. 2015; see also Chap. 4, Glibert et al. (2018b) and Chap. 5, Wells and Karlson (2018)]. This raises the potential for more

---

Fig. 7.3 (continued) than (for ANFs in panel **A**) via the activities of HNFs + μZ. *Dashed arrows* indicate functions contributing to nutrient pools (*blue*, inorganic; *brown*, organic). *Heavy black arrows* indicate predatory links. Reproduced from Mitra et al. (2014b) under a creative commons licence

HABs (Burkholder et al. 2008; Glibert et al. 2014). Lacking are critically needed experimental data that systematically assess how primary production, mixotrophy, and N–P–C flow vary under these changing environmental conditions, and the appropriate inclusion of these organismal processes in plankton food web models. Accurate representation of mixotrophy in next-generation models, within the context of the suite of environmental drivers likely to change, will require acquisition of substantial new experimental data that explicitly explore the connectivity between phototrophy and phagotrophy. Because of the critical role of grazing as both a fundamental feature of the zooplanktonic regulation of HAB development (as per traditional food web constructs), and as a contribution to the growth of HAB mixotrophs, enhancing models of predatory activity are of particular importance. Accurate parameterization of mixotrophy should become a high priority in water quality and fisheries models used as aids to regional and/or international policy development. The next-generation international HAB Programme, GlobalHAB [see Chap. 22, Berdalet et al. (2018)], may present opportunities for these greatly needed advancements.

Acknowledgments The authors thank the Leverhulme Trust for funding an international network (F/00 391/V), as well as the many contributors to the associated mixotrophy working groups. This is a contribution of the GEOHAB Core Research Project on HABs and Eutrophication and is number 5406 from the University of Maryland Center for Environmental Science.

References

Adolf JE, Bachvaroff T, Place AR (2008) Can cryptophytes trigger toxic *Karlodinium veneficum* blooms in eutrophic estuaries? Harmful Algae 81:19–128

Adolf JE, Stoecker DK, Harding LW Jr (2003) Autotrophic growth and photoacclimation of *Karlodinium micrum* (Dinophyceae) and *Storeatula major* (Cryptophyceae). J Phycol 39: 1101–1108

Adolf JE, Stoecker DK, Harding LW Jr (2006) The balance of autotrophy and heterotrophy during mixotrophic growth of *Karlodinium micrum*. J Plankton Res 28:737–751

Allen JI, Polimene L (2011) Linking physiology to ecology: towards a new generation of plankton models. J Plankton Res 33:989–997

Baird ME, Emsley SM (1999) Towards a mechanistic model of plankton population dynamics. J Plankton Res 21:85–126

Baretta JW, Baretta-Bekker JG, Ruardij P (1996) From EMS to ERSEM, towards generic coastal ecosystem models. Aquabiology 104:197–209

Berdalet E, Kudela R, Banas NS et al (2018) GlobalHAB: fostering international coordination on harmful algal bloom research in aquatic systems. In: Glibert PM, Berdalet E, Burford M et al (eds) Global ecology and oceanography of harmful algal blooms. Springer, Cham, pp 425–447

Berman T, Bronk DA (2003) Dissolved organic nitrogen: a dynamic participant in aquatic ecosystems. Aquat Microb Ecol 31:279–305

Bertin MJ, Zimba PV, Beauchesne KR et al (2012) Identification of toxic fatty acid amides isolated from the harmful alga *Prymnesium parvum* Carter. Harmful Algae 20:111–116

Bockstahler KR, Coats DW (1993a) Grazing of the mixotrophic dinoflagellate *Gymnodinium sanguineum* on ciliate populations of Chesapeake Bay. Mar Biol 116:477–487

Bockstahler KR, Coats DW (1993b) Spatial and temporal aspects of mixotrophy in Chesapeake Bay dinoflagellates. J Eukaryot Microbiol 40:49–60
Burkholder JM, Glibert PM, Skelton H (2008) Mixotrophy: a major mode of nutrition for harmful algal species in eutrophic waters. Harmful Algae 8:77–93
Chatton É (1920) Les Péridiniens parasites. Morphologie, reproduction, éthologie. Arch Zool Exp Gen 59:1–475
Christian JR (2005) Biogeochemical cycling in the oligotrophic ocean: Redfield and non-Redfield models. Limnol Oceanogr 50:646–657
Coats DW (1999) Parasitic life styles of marine dinoflagellates. J Eukaryot Microbiol 46:402–409
de Castro F, Gaedke U, Boenigk J (2009) Reverse evolution: driving forces behind the loss of acquired photosynthetic traits. PLoS One 4(12):e8465. https://doi.org/10.1371/journal.pone.0008465
Denardou-Queneherve A, Grzebyk D, Pouchus YF et al (1999) Toxicity of French strains of the dinoflagellate *Prorocentrum minimum* experimental and natural contaminations of mussels. Toxicon 37:1711–1719
Dorantes-Aranda JJ, García-de la Parra LM, Alonso-Rodríguez R et al (2010) Toxic effect of the harmful dinoflagellate *Cochlodinium polykrikoides* on the spotted rose snapper *Lutjanus guttatus*. Environ Toxicol 25:319–326
Drebes G (1969) *Dissodinium pseudocalani* sp. nov., ein parasitischer Dinoflagellat auf Copepodeneiern. Helgoländer Wiss Meeresunters 19:58–67
Fasham MJR, Ducklow HW, McKelvie SM (1990) A nitrogen-based model of plankton dynamics in the oceanic mixed layer. J Mar Res 48:591–639
Flynn KJ (2001) A mechanistic model for describing dynamic multi-nutrient, light, temperature interactions in phytoplankton. J Plankton Res 23:977–997
Flynn KJ (2005) Modelling marine phytoplankton growth under eutrophic conditions. J Sea Res 54:92–103
Flynn KJ (2008) Use, abuse, misconceptions and insights from quota models: the Droop cell-quota model 40 years on. Oceanogr Mar Biol Annu Rev 46:1–23
Flynn KJ (2009) Going for the slow burn: why should possession of a low maximum growth rate be advantageous for microalgae? Plant Ecol Divers 2:179–189
Flynn KJ (2010a) Ecological modelling in a sea of variable stoichiometry; dysfunctionality and the legacy of Redfield and Monod. Prog Oceanogr 84:52–65
Flynn KJ (2010b) Do external resource ratios matter? – Implications for modelling eutrophication events and controlling harmful algal blooms. J Mar Syst 83:170–180
Flynn KJ, Clark DR, Mitra A, Fabian H, Hansen PJ, Glibert PM, Wheeler GL, Stoecker DK, Blackford JC, Brownlee C (2015) Ocean acidification with (de)eutrophication will alter future phytoplankton growth and succession. Proc R Soc B 282. https://doi.org/10.1098/rspb.2014.2604
Flynn KJ, Mitra A (2009) Building the "perfect beast": modelling mixotrophic plankton. J Plankton Res 31:965–992
Flynn KJ, Mitra A (2016) Why plankton modelers should reconsider using rectangular hyperbolic (Michaelis-Menten, Monod) descriptions of predator-prey interactions. Front Mar Sci. https://doi.org/10.3389/fmars.2016.00165
Flynn KJ, Page S, Wood G et al (1999) Variations in the maximum transport rates for ammonium and nitrate in the prymnesiophyte *Emiliania huxleyi* and the raphidophyte *Heterosigma carterae*. J Plankton Res 21:355–371
Flynn KJ, Stoecker DK, Mitra A et al (2013) Misuse of the phytoplankton-zooplankton dichotomy: the need to assign organisms as mixotrophs within plankton functional types. J Plankton Res 35:3–11
Follows MJ, Dutkiewicz S, Grant S et al (2007) Emergent biogeography of microbial communities in a model ocean. Science 315:1843–1846
Gaines G, Elbrächter M (1987) Heterotrophic nutrition. In: Taylor FJR (ed) The biology of dinoflagellates. Blackwell Scientific Publications, Oxford, pp 224–268

GEOHAB (1998) Global ecology and oceanography of harmful algal blooms: a plan for co-ordinated scientific research and co-operation to develop international capabilities for assessment, prediction and mitigation. Cullen J (ed) Asian Natural Environmental Science Center, The University of Tokyo, 43 pp

GEOHAB (2001) Global ecology and oceanography of harmful algal blooms, science plan. Glibert P, Pitcher G (eds) SCOR and IOC, Baltimore and Paris, 86 pp

Glasgow HB, Burkholder JAM, Mallin MA, Deamer-Melia NJ, Reed RE (2001) Field ecology of toxic pfiesteria complex species and a conservative analysis of their role in estuarine fish kills. Environ Health Perspect 109(s5):715–730

Glibert PM, Allen JI, Artioli Y et al (2014) Vulnerability of coastal ecosystems to changes in harmful algal bloom distribution in response to climate change: projections based on model analysis. Glob Chang Biol 20:3845–3858

Glibert PM, Allen JI, Bouwman L et al (2010) Modeling of HABs and eutrophication: status, advances, challenges. J Mar Syst 83:262–275

Glibert PM, Al-Azri A, Allen JI et al (2018a) Key questions and recent research advances on harmful algal blooms in relation to nutrients and eutrophication. In: Glibert PM, Berdalet E, Burford M et al (eds) Global ecology and oceanography of harmful algal blooms. Springer, Cham, pp 229–259

Glibert PM, Beusen AHW, Harrison JA et al (2018b) Changing land, sea- and airscapes: sources of nutrient pollution affecting habitat suitability for harmful algae. In: Glibert PM, Berdalet E, Burford M et al (eds) Global ecology and oceanography of harmful algal blooms. Springer, Cham, pp 53–76

Glibert PM, Heil CA, Wilkerson F et al (2018c) Nutrients and HABs: dynamic kinetics and flexible nutrition. In: Glibert PM, Berdalet E, Burford M et al (eds) Global ecology and oceanography of harmful algal blooms. Springer, Cham, pp 93–112

Glibert PM, Burkholder JM, Kana TM et al (2009) Grazing by *Karenia brevis* on *Synechococcus* enhances their growth rate and may help to sustain blooms. Aquat Microb Ecol 55:17–30

Glibert PM, Burkholder JM, Kana TM (2012) Recent advances in understanding of relationships between nutrient availability, forms and stoichiometry and the biogeographical distribution, ecophysiology, and food web effects of pelagic and benthic *Prorocentrum* spp. Harmful Algae 14:231–259

Glibert PM, Kana TM, Brown K (2013) From limitation to excess: consequences of substrate excess and stoichiometry for phytoplankton physiology, trophodynamics and biogeochemistry, and implications for modeling. J Mar Syst 125:14–28

Glibert PM, Legrand C (2006) The diverse nutrient strategies of HABs: focus on osmotrophy. In: Granéli E, Turner J (eds) Ecology of harmful algae. Springer, New York, NY, pp 163–176

Goes JI, Gomes HR, Al-Hashimi K et al (2018) Ecological drivers of green *Noctiluca* blooms in two monsoonally driven ecosystems. In: Glibert PM, Berdalet E, Burford M et al (eds) Global ecology and oceanography of harmful algal blooms. Springer, Cham, pp 327–336

Goldman JC, Glibert PM (1983) Kinetics of inorganic nitrogen uptake. In: Carpenter EJ, Capone DG (eds) Nitrogen in the marine environment. Academic Press, New York, NY, pp 233–274

Gomes H, Goes JI, Matondkar SGP et al (2014) Massive outbreaks of *Noctiluca scintillans* blooms in the Arabian Sea due to spread of hypoxia. Nat Commun 5:4862. https://doi.org/10.1013/ncomms5862

Graham LE, Graham JM, Wilcox LW et al (2016) Algae, 3rd edn. LJLM Press, LLC, Madison, WI

Granéli E, Carlsson P (1998) The ecological significance of phagotrophy in photo- synthetic flagellates. In: Anderson DM, Cembella AD, Hallegraeff GM (eds) Physiological ecology of harmful algal blooms, NATO ASI series, vol G41. Springer, Berlin, Heidelberg, pp 539–557

Hansen PJ (2011) The role of photosynthesis and food uptake for the growth of marine mixotrophic dinoflagellates. J Eukaryot Microbiol 58:203–214

Hansen PJ, Nielsen LT, Johnson M et al (2013) Acquired phototrophy in *Mesodinium* and *Dinophysis* – a review of cellular organization, prey selectivity, nutrient uptake and bioenergetics. Harmful Algae 28:126–139

Henrikson JC, Gharfeh MS, Easton AC et al (2010) Reassessing the ichthyotoxin profile of cultured *Prymnesium parvum* (golden algae) and comparing it to samples collected from recent freshwater bloom and fish kill events in North America. Toxicon 55:1396–1404

Irigoien X, Flynn KJ, Harris RP (2005) Phytoplankton blooms: a 'loophole' in microzooplankton grazing impact? J Plankton Res 27:313–321

Jeong HJ, Park JY, Nho JH et al (2005) Feeding by red-tide dinoflagellates on the cyanobacterium *Synechococcus*. Aquat Microb Ecol 41:1331–1343

Jeong HJ, Yoo YD, Kim JS et al (2004) Mixotrophy in the phototrophic harmful alga *Cochlodinium polykrikoides* (Dinophyceae): prey species, the effects of prey concentration, and grazing impact. J Eukaryot Microbiol 51:563–569

Jeong HJ, Yoo YD, Kim JS et al (2010) Growth, feeding and ecological roles of the mixotrophic and heterotrophic dinoflagellates in marine planktonic food webs. Ocean Sci J 45:65–91

Johnson MD (2014) Inducible mixotrophy in the dinoflagellate *Prorocentrum minimum*. J Eukaryot Microbiol 64:431–434. https://doi.org/10.1111/jeu.12198

Jones R (1994) Mixotrophy in planktonic protists as a spectrum of nutritional strategies. Mar Microb Food Webs 8:87–96

Kana TM, Glibert PM (2016) On saturating response curves from the dual perspectives of photosynthesis and nitrogen acquisition. In: Glibert PM, Kana TM (eds) Aquatic microbial ecology and biogeochemistry: a dual perspective. Springer International Publishing, Geneva, pp 93–104

Katechakis A, Haseneder T, Kling R et al (2005) Mixotrophic versus photoautotrophic specialist algae as food for zooplankton: the light:nutrient hypothesis might not hold for mixotrophs. Limnol Oceanogr 50:1290–1299

Klausmeier CA, Litchman E, Daufresne T et al (2004) Optimal N:P stoichiometry of phytoplankton. Nature 429:171–174

Kudela RM, Raine R, Pitcher G et al (2018) Establishment, goals, and the legacy of the Global Ecology and Oceanography of Harmful Algal Blooms (GEOHAB) Program. In: Glibert PM, Berdalet E, Burford M et al (eds) Global ecology and oceanography of harmful algal blooms. Springer, Cham, pp 27–49

Larsen A, Eikrem W, Paasche E (1993) Growth and toxicity in *Prymnesium patelliferum* (Prymnesiophyceae) isolated from Norwegian waters. Can J Bot 71:1357–1362

Le Quéré C, Harrison SP, Prentice IC et al (2005) Ecosystem dynamics based on plankton functional types for global ocean biogeochemistry models. Glob Chang Biol 11:2016–2040

Leles SG, Mitra A, Flynn KJ et al (2017) Oceanic protists with different forms of acquired phototrophy display contrasting biogeographies and abundance. Proc R Soc B. https://doi.org/10.1098/rspb.2017.0664

Lewitus AJ, Glasgow HB, Burkholder JAM (1999) Kleptoplastidy in the Toxic Dinoflagellate *Pfiesteria Piscicida* (Dinophyceae). J Phycol 35(2):303–312

Li A, Stoecker DK, Coats DW (2000) Mixotrophy in *Gyrodinium galatheanum* (Dinophyceae): grazing responses to light intensity and inorganic nutrients. J Phycol 36:33–45

Lin C-H, Accoroni S, Glibert PM (2017) Mixotrophy in the dinoflagellate *Karlodinium veneficum* under variable nitrogen:phosphorus stoichiometry: feeding response and effects on larvae of the eastern oyster (*Crassostrea virginica*). Aquat Microb Ecol 79:101–114. https://doi.org/10.3354/ameo01823.

Lom J, Rohde K, Dykov' I (1993) *Crepidoodinium australe* n. sp., an ectocommensal dinoflagellate from the gills of *Sillago ciliata*, an estuarine fish from the New South Wales coast of Australia. Dis Aquat Organ 15:63–72

Lundgren V, Glibert PM, Granéli E et al (2016) Metabolic and physiological changes in *Prymnesium parvum* when grown under, and grazing on, prey of variable nitrogen:phosphorus stoichiometry. Harmful Algae 55:1–12

McCarthy JJ, Goldman JC (1979) Nitrogenous nutrition of marine phytoplankton in nutrient-depleted waters. Science 203:670–672

McGillicudy DJ Jr, de Young B, Doney S et al (2010) Models: tools for synthesis in international oceanographic research programs. Oceanography 23:126–139
Minnhagen SM, Kim M, Salomon P et al (2011) Active uptake of kleptoplastids by *Dinophysis caudata* from its ciliate prey *Myrionecta rubra*. Aquat Microb Ecol 62:99–108
Mitra A, Flynn KJ (2005) Predator-prey interactions: is "ecological stoichiometry" sufficient when good food goes bad? J Plankton Res 27:393–399
Mitra A, Flynn KJ (2006) Accounting for variation in prey selectivity by zooplankton. Ecol Model 199:82–92
Mitra A, Flynn KJ (2010) Modelling mixotrophy in harmful algal blooms: more or less the sum of the parts? J Mar Syst 83:158–169
Mitra A, Castellani C, Gentleman WC et al (2014a) Bridging the gap between marine biogeochemical and fisheries sciences; configuring the zooplankton link. Prog Oceanogr 129:176–199
Mitra A, Flynn KJ, Burkholder JM et al (2014b) The role of mixotrophic protists in the biological carbon pump. Biogeosciences 11:995–1005
Mitra A, Flynn KJ, Tillmann U et al (2016) Defining planktonic protist functional groups on mechanisms for energy and nutrient acquisition; incorporation of diverse mixotrophic strategies. Protist 167:106–120
Moore JK, Doney SC, Lindsay K (2004) Upper ocean ecosystem dynamics and iron cycling in a global 3-D model. Global Biogeochem Cycles 18:GB4028. https://doi.org/10.1029/2004GB002220
Odum EP (1969) The strategy of ecosystem development. Science 164:262–270
Parry GD (1981) The meaning of r- and K-selection. Oecologia 48:260–264
Pasternak AF, Arashkevich YG, Sorokin YS (1984) The role of the parasitic algal genus *Blastodinium* in the ecology of planktonic copepods. Oceanology 24:748–751
Pintner IJ, Provasoli L (1968) Heterotrophy in subdued light of 3 *Chrysochromulina* species. Bull Misaki Mar Biol Inst Kyoto Univ 12:25–31
Plagányi EE (2007) Models for an ecosystem approach to fisheries. Food and Agriculture Organization of the United Nations (FAO) Fisheries Technical Paper, No. 477. FAO, Rome
Polimene L, Mitra A, Sailley SF et al (2015) Decrease in diatom palatability contributes to bloom formation in the Western English Channel. Prog Oceanogr 137B:484–497. https://doi.org/10.1016/j.pocean.2015.04.026
Putt M (1990) Metabolism of photosynthate in the chloroplast-retaining ciliate *Laboea strobila*. Mar Ecol Prog Ser 60:271–282
Ramin M, Perhar G, Shimoda Y et al (2012) Examination of the effects of nutrient regeneration mechanisms on plankton dynamics using aquatic biogeochemical modeling. Ecol Model 240:139–155
Raven JA, Beardall J, Flynn KJ et al (2009) Phagotrophy in the origins of photosynthesis in eukaryotes and as a complementary mode of nutrition in phototrophs: relation to Darwin's insectivorous plants. J Exp Bot 60:3975–3987
Remmel EJ, Hambright KD (2012) Toxin-assisted micropredation: experimental evidence shows that contact micropredation rather than exotoxicity is the role of *Prymnesium* toxins. Ecol Lett 15:126–132
Rokitta SD, de Nooijer LJ, Trimborn S, de Vargas C, Rost B, John U (2011) Transcriptional analyses reveal differential gene expression patterns between the life cycle stages of Emiliania huxleyi (Haptophyta) and reflects specialization to different ecological niches. J Phycol 47: 829–838
Rose KA, Allen JI, Artioli Y et al (2010) End-to-end models for the analysis of marine ecosystems: challenges, issues, and next steps. Mar Coast Fish 2:115–130
Smalley GW, Coats DW (2002) Ecology of the red-tide dinoflagellate *Ceratium furca*: distribution, mixotrophy, and grazing impact on ciliate populations of Chesapeake Bay. J Eukaryot Microbiol 49:64–74
Smayda TJ (1997) Harmful algal blooms: their ecophysiological and general relevance to phytoplankton blooms in the sea. Limnol Oceanogr 42:1137–1153

Sterner RW, Elser JJ (2002) Ecological stoichiometry: the biology of elements from molecules to the biosphere. Princeton University Press, Princeton, NJ

Stoecker DK, Hansen PJ, Caron DA et al (2017) Mixotrophy in the marine plankton. Ann Rev Mar Sci 9:311–335

Stoecker DK, Michaels AE (1991) Respiration, photosynthesis and carbon metabolism in planktonic ciliates. Mar Biol 108:441–447

Stukel MR, Landry MR, Selph KE (2011) Nanoplankton mixotrophy in the eastern equatorial Pacific. Deep Sea Res Part II 58:378–386

Thingstad TF, Havskum H, Zweifel UL et al (2007) Ability of a 'minimum' microbial food web model to reproduce response patterns observed in mesocosms manipulated with N and P, glucose, and Si. J Mar Syst 64:15–34

Unrein F, Massana R, Alonso-Saez L et al (2007) Significant year-round effect of small mixotrophic flagellates on bacterioplankton in an oligotrophic coastal system. Limnol Oceanogr 52:456–469

Van Wagoner RM, Deeds JR, Tatters AO et al (2010) Structure and relative potency of several karlotoxins from *Karlodinium veneficum*. J Nat Prod 73:1360–1365

Ward BA, Dutkiewicz S, Barton AD et al (2011) Biophysical aspects of resource acquisition and competition in algal mixotrophs. Am Nat 178:98–112

Wells ML, Karlson B (2018) Harmful algal blooms in a changing ocean. In: Glibert PM, Berdalet E, Burford M et al (eds) Global ecology and oceanography of harmful algal blooms. Springer, Cham, pp 77–90

Zubkov MV, Tarran GA (2008) High bacterivory by the smallest phytoplankton in the North Atlantic Ocean. Nature 455:224–226

Chapter 8
The Role of Life Cycle Characteristics in Harmful Algal Bloom Dynamics

Rhodora V. Azanza, Michael L. Brosnahan, Donald M. Anderson, Inga Hense, and Marina Montresor

8.1 Introduction

Many phytoplankton species have multiphasic life cycles that include stages with different morphologies, behaviours, and functional roles. Studies of life cycles under controlled conditions in the laboratory, observations in species' natural environments, and model simulations have all provided valuable insights on species-specific bloom dynamics and associated impacts. This chapter presents an overview of the role that life cycles play in the bloom dynamics of representative, diverse, toxic harmful algal bloom (HAB)-forming taxa: specifically, the dinoflagellates *Pyrodinium bahamense* and *Alexandrium fundyense*, diatom species of the genus *Pseudo-nitzschia*, and the cyanobacterium *Nodularia spumigena*. These are among the most widespread harmful algal species described, causing a range of impacts from human illness and death from ingestion of potent toxins via the food chain to fish and wildlife mortalities and ecosystem disruption. There are many other HAB species that cause these and other impacts, but the four species highlighted here are selected because features of their life cycles are representative of many different HAB species that undergo sexuality or have resting stages in their

R.V. Azanza (✉)
The Marine Science Institute, University of the Philippines, Quezon City, Philippines
e-mail: rvazanza@up.edu.ph

M.L. Brosnahan • D.M. Anderson
Biology Department, Woods Hole Oceanographic Institution, Woods Hole, MA, USA

I. Hense
Institute for Hydrobiology and Fisheries Science, Center for Earth System Research and Sustainability, University of Hamburg, Hamburg, Germany

M. Montresor
Integrative Marine Ecology Department, Stazione Zoologica Anton Dohrn, Naples, Italy

P.M. Glibert et al. (eds.), *Global Ecology and Oceanography of Harmful Algal Blooms*, Ecological Studies 232, https://doi.org/10.1007/978-3-319-70069-4_8

life cycles. *Pyrodinium* and *Alexandrium* are representative of cyst-forming dinoflagellates and illustrate some of the ways that dinoflagellate life cycles are adapted to tropical and temperate habitats, respectively. *Pseudo-nitzschia* species undergo sexuality, but have no resting stage, similar to many other diatoms, and *Nodularia* is an example of a bloom-forming, toxic cyanobacteria that form akinetes as a resting stage. Consistent with the comparative approach in Global Ecology and Oceanography of Harmful Algal Blooms (GEOHAB 2001), these four species are compared and contrasted to identify common and unique strategies for survival and proliferation within varied aquatic environments. This chapter also illustrates the research conducted under the GEOHAB Programme to clarify adaptive strategies of harmful species.

8.2 Dinoflagellates: *Alexandrium fundyense* and *Pyrodinium bahamense*

Dinoflagellates represent a disproportionate fraction of all HAB species and are further overrepresented among those shown to be expanding globally in recent decades (Anderson 1989; Smayda 1997). These patterns are attributable in large part to the meroplanktonic life cycles of these organisms (including both benthic and free-swimming stages). Benthic stages, particularly diploid and haploid cysts, enable many dinoflagellates to endure short- and long-term periods of adverse conditions and also facilitate dispersal through currents and other natural mechanisms, as well as anthropogenic processes such as global shipping and aquaculture transplantations (Hallegraeff 1998). Planktonic stages are similarly adaptable as swimming cells that can migrate to exploit gradients in nutrient and light availability or, for mixotrophic and kleptoplastic species, to locate and capture prey. These same swimming behaviours facilitate dispersal within coastal current systems and contribute to patterns of cyst deposition that sustain recurrent threats to public health and ecosystems. Here, we highlight research exploring the ecology and life cycle characteristics of two marine cyst-forming species, *A. fundyense* and *P. bahamense*.

Alexandrium fundyense (Fig. 8.1a) is one of multiple species within the *Alexandrium* genus that cause outbreaks of paralytic shellfish poisoning (PSP) in temperate, subtropical, and subpolar regions around the world and is one of three toxic species within the *A. tamarense* species complex. Species boundaries within this subgroup were long delineated using morphological criteria including the presence/absence of chain formation and/or an apical pore within the 1′ Kofoidean thecal plate. However, these criteria are now known to be polyphyletic (and therefore invalid), and the formerly recognized morphospecies designations *A. tamarense*, *A. fundyense*, and *A. catenella* have been redefined or replaced using molecular characters that divide the complex into five distinct species (John et al. 2014). *A. fundyense* is the most widespread of these and is known to cause PSP on the east and west coasts of North and South America as well as in areas along the northeast Asian coast and in northern Europe (Lilly et al. 2007).

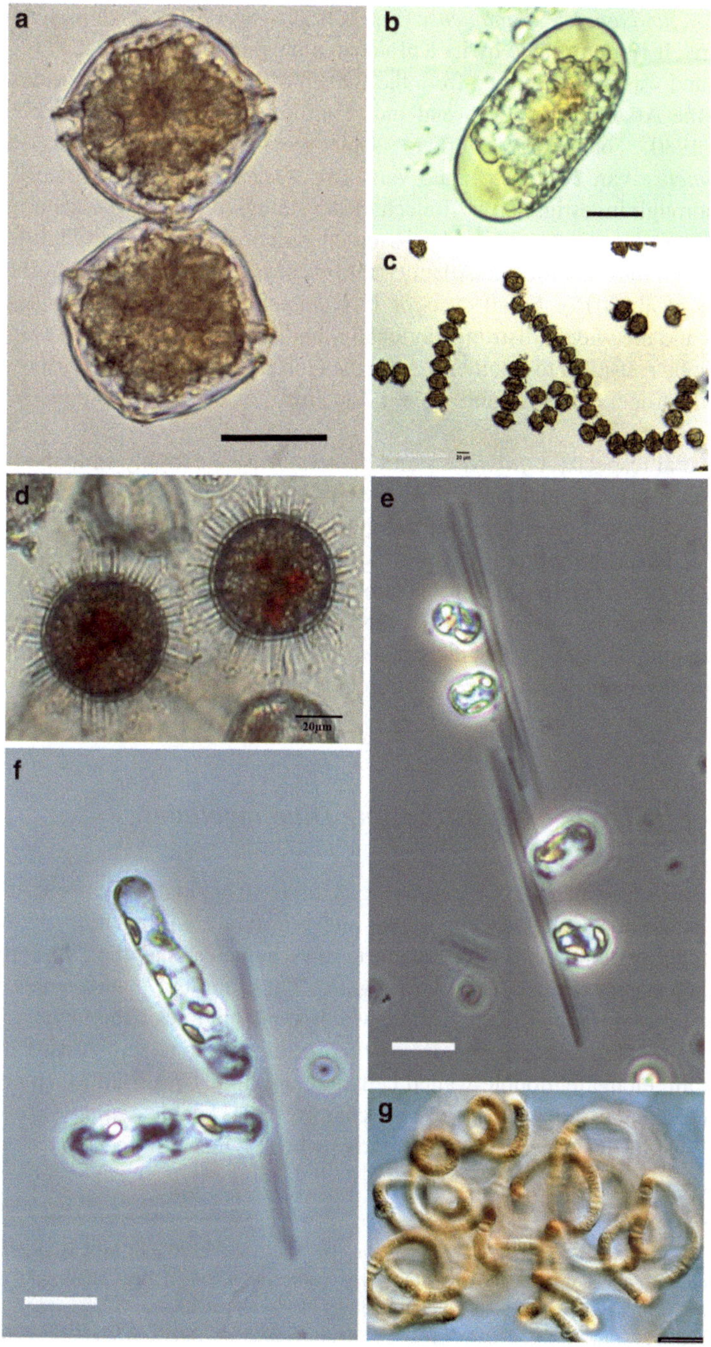

Fig. 8.1 Various life cycle stages of four HAB species. (**a**) Vegetative cell pair after division and (**b**) resting cyst of *Alexandrium fundyense*, *scale bars* = 10 μm; (**c**) vegetative cells and (**d**) resting

Like *A. fundyense*, *P. bahamense* (Fig. 8.1c) causes PSP through its synthesis of saxitoxins. It is identifiable by its Kofoidean plate pattern of (Po, Pi) 4′, 6″, 6c, 8s, 6‴, 2⁗ and was first described from the Bahamas (Plate 1906). The species occurs in both the Atlantic-Caribbean and Indo-Pacific tropics (Bohm 1931; Steidinger et al. 1980), but early workers sometimes differentiated two varieties, *P. bahamense* var. *bahamense* and var. *compressum*, on the basis of cell shape. More thorough investigation by Balech (1985) failed to detect consistent morphological differences, however, and subsequent genetic analyses showed differentiation between Indo-Pacific and Atlantic-Caribbean populations but not morphotypes (Mertens et al. 2015). Fossil cysts of *P. bahamense* are called *Polysphaeridium zoharyii* and are widely distributed globally, but living populations of resting cysts appear to be restricted to Southeast Asia, the Gulf of California, the Caribbean, and Western Europe (Amorim and Dale 1998; Furio et al. 2012; Morquecho et al. 2014).

In several respects, *A. fundyense* and *P. bahamense* are highly similar—both are thecate, cyst formers that cause PSP through the synthesis of saxitoxins—but are also distinct in their phenology and geographical distribution. *P. bahamense* forms sustained, late summer/rainy season blooms and is primarily found in tropical regions, while *A. fundyense* impacts temperate and subpolar coasts during spring, summer, and fall seasons. The aim in choosing these two species is to present ways that their unique, species-specific adaptations within a single, common life cycle have enabled them to exploit very different climate regimes.

8.2.1 Life Cycle of Cyst-Forming Dinoflagellates

Both *A. fundyense* and *P. bahamense* (Fig. 8.2a) have haplontic life cycles in which motile, haploid, vegetative cells divide and proliferate within the plankton and, then, under certain conditions transform into gametes that then fuse to form swimming zygotic cells called planozygotes. These diploid planozygotes in turn transform into immobile resting cysts, which are sometimes called hypnocysts or hypnozygotes (Pfiester and Anderson 1987). *A. fundyense* resting cysts (Fig. 8.1b) are smooth, elongate, double-walled cells that are highly resistant to temperature, mechanical, and osmotic stresses. These characteristics enable *A. fundyense* cysts to remain viable for decades, particularly when lying within anoxic habitats

Fig. 8.1 (continued) cysts of *Pyrodinium bahamense*, *scale bar* = 20 μm; (**e**) young auxospores still attached to the empty/negative mating type gametangia joined in a chain and (**f**) more auxospores connected to the gametangial frustule of *Pseudo-nitzschia* cf. *delicatissima* from Gulf of Naples, *scale bars* = 10 μm; (**g**) filaments of *Nodularia spumigena*, *scale bar* = 30 μm. Photos by D.M. Anderson (**a** and **b**), R.V. Azanza culture, photo taken by Y. Fukuyo (**c**), R.V. Azanza (**d**), Diana Sarno of Stazione Zoologica Anton Dohrn (**e** and **f**), Seija Hällfors of Finnish Environment Institute Syke (**g**)

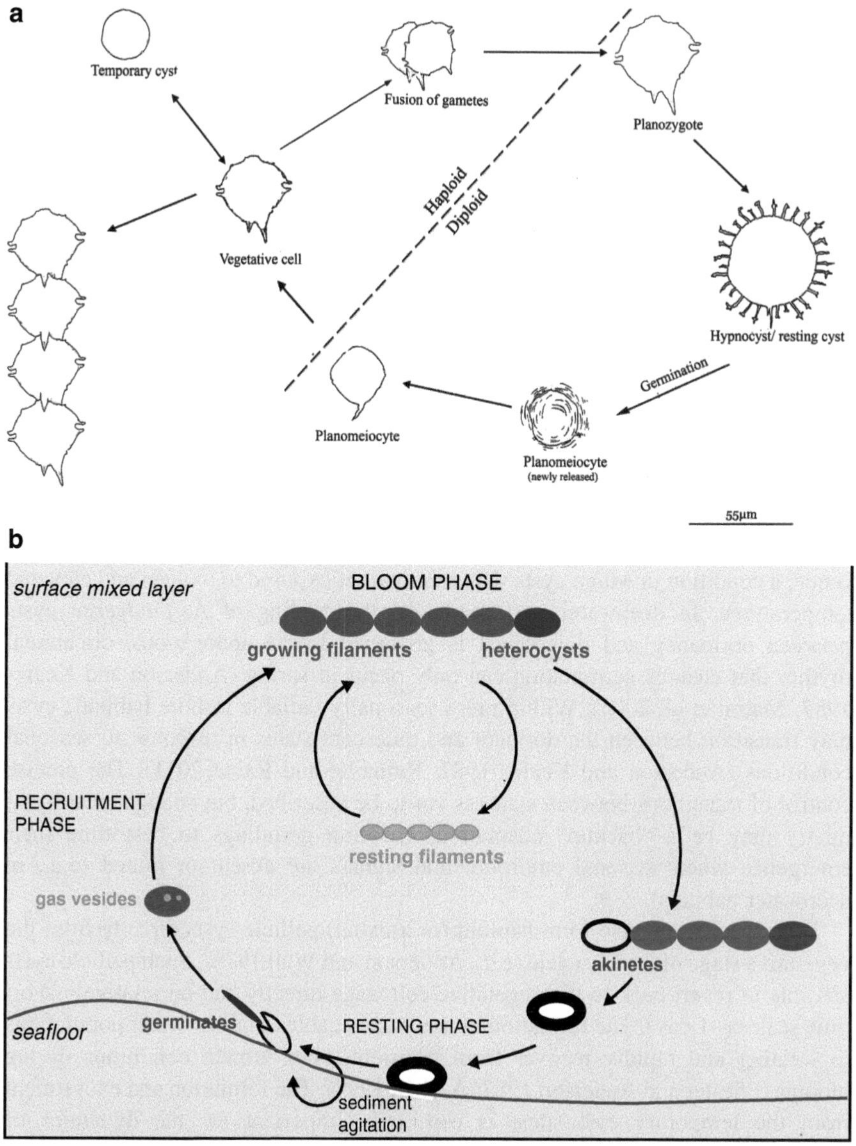

Fig. 8.2 Schematic representation of the life cycle of *Pyrodinium bahamense* (**a**), based on Azanza (1997) and *Nodularia spumigena* (**b**), modified after Hense and Beckmann (2006)

(Anderson et al. 1987; Feifel et al. 2015). The cysts' durability also facilitates their spread via a wide range of natural and anthropogenic vectors, the latter including dredging and sediment relocation, ballast water transport, and aquaculture (Hallegraeff and Bolch 1991; Anderson et al. 2012). *P. bahamense* resting cysts (Fig. 8.1d) in contrast are spherical and chorate, having trumpet-shaped processes distributed over the cyst wall surface (Wall and Dale 1969). It remains to be shown to what extent *P. bahamense* cysts remain viable when displaced or buried through various mechanisms, but these cysts similarly are able to survive long periods in anoxic marine sediments (Azanza et al. 2004; Siringan et al. 2008). Both *P. bahamense* and *A. fundyense* cysts tend to accumulate in areas that collect fine sediment, leading to formation of "seedbeds" or areas of high cyst concentration that inoculate new blooms when cysts germinate (e.g., Anderson et al. 2014a).

Before transformation back to the haploid phase of the life cycle, diploid resting cysts must undergo a maturation process that can last several weeks or months, after which they may be stimulated to germinate, giving rise to a short-lived stage called a planomeiocyte that reverts back to the haploid, vegetative stage through meiotic divisions. Despite very different temperature regimes in their habitats, cyst maturation periods for both species are similar, 1–3 months for *A. fundyense* (Anderson and Morel 1979) and 2.5–3 months for *P. bahamense* (Corrales et al. 1995). For *Alexandrium*, the maturation interval is shortest at warmer temperatures. After completion of their initial maturation, *A. fundyense* resting cysts are known to cycle between germinable and non-germinable states: dormancy, a condition in which cysts are not responsive to ambient environmental conditions, and quiescence, a condition in which cysts will germinate if exposed to oxygen and elevated temperatures. In deepwater cyst seedbeds, the cycling of *A. fundyense* cysts between dormancy and quiescence is governed by an endogenous, circannual rhythm that ensures germination can only occur in spring (Anderson and Keafer 1987; Matrai et al. 2005). Within more seasonally variable inshore habitats, cysts may transition between the dormant and quiescent states in response to seasonal conditions (Anderson and Keafer 1987; Rathaille and Raine 2011). The precise control of transitions between states is yet to be described, but endogenous rhythmicity may be a "backup" adaptation, enabling germlings to best time their emergence when seasonal environmental signals are absent or muted (e.g., in deepwater habitats).

Both species can also form haploid (or asexual) pellicle cysts directly from the vegetative stage of the life cycle (e.g., Anderson and Wall 1978). Such pellicle cysts are able to revert back to the vegetative cell stage directly and on relatively short time scales (~1 day). The formation of this stage enables shallow water populations to weather and rapidly recover from transient, poor growth conditions during blooms (Pfiester and Anderson 1987; Azanza 1997). The formation and excystment from the temporary cyst stage is especially important for the dynamics of *P. bahamense* blooms (Onda et al. 2014) and is known to play a more significant role in the dynamics of some other *Alexandrium* HABs (e.g., Bravo et al. 2010).

Further differences between *A. fundyense* and *P. bahamense* include the propensity of vegetative cells to form chains and the nature of their gamete stages. *P. bahamense* commonly forms long chains of up to 16 cells in both natural populations and in cultures (Fig. 8.1c). *A. fundyense* on the other hand is often present only as single or paired cells though can occur in chains of up to eight cells. Isolates from these chain- forming *A. fundyense* populations tend not to form chains unless stimulated by turbulence (Sullivan et al. 2003). Gametes of the two species differ in their size relative to progenitor vegetative cells. *P. bahamense* forms isogametes that are smaller than vegetative cells (20–25 μm versus 25–45 μm diameter). *A. fundyense* gametes may be similar in size to vegetative cells (30–35 μm diameter) or smaller in different populations and are more variable in size, forming both isogamous and anisogamous mating pairs (Pfiester and Anderson 1987). Both species have a heterothallic mating system (i.e., clonal daughters will not pair), and gametes fuse to form nondividing planozygotes that grow progressively larger until encystment (diameter of 50–55 μm in *P. bahamense* and >43 μm in *A. fundyense*).

8.2.2 Bloom Dynamics

8.2.2.1 Major Study Areas

Bloom dynamics are the result of a species' behaviour within its environment. Therefore, when reviewing the interactions between these species life cycles and their ecology, it is important to also consider the environments in which blooms of these species have been most heavily studied.

Pyrodinium blooms routinely affect areas throughout Southeast Asia (Furio et al. 2012 and references therein) but have been characterized most extensively in the Philippines, where the intensity of *Pyrodinium* blooms is often correlated with cyst dynamics, i.e., cyst availability/concentration, cyst bed location, cyst distribution by currents and waves, etc. (Villanoy et al. 1996; Azanza et al. 2004; Siringan et al. 2008; Azanza 2013). Sediment surveys have indicated that cysts can be found year round, especially in seedbed areas of high cyst concentrations (similar to *A. fundyense*). Increased rainfall triggers higher land-based nutrient discharge during southwest (SW) monsoon periods when temperature and salinity conditions also tend to be highly favourable for *Pyrodinium* growth and bloom development (29–32 °C and 26–30 salinity, respectively). In small embayments, such as Manila Bay and Sorsogon Bay (Philippines), there are sometimes massive, sustained blooms lasting weeks to months. Throughout the Philippines, blooms typically occur in late summer, a period that marks the start of SW monsoon when waters tend to be (at least initially) stratified. Within shallow systems like these, *Pyrodinium* often undergoes several cycles of transformation between its pellicle cyst stage and its vegetative stage during a single bloom period (Azanza 1997; Azanza et al. 2004; Onda et al. 2014). This life cycle behaviour enables *P. bahamense* to withstand shear stress associated with the high-energy SW monsoon. The bloom period ends with the beginning of northeast (NE) monsoon, when waters in the Philippines are more well mixed (Azanza and Taylor 2001; Azanza 2013).

The dynamics of *A. fundyense* blooms and associated outbreaks of PSP have been studied most intensively in the northwest Atlantic, along the coast of maritime Canada and within the Gulf of Maine (GOM), areas that span a broad continental shelf sea and support productive shellfisheries. Within this region, coastal blooms extending for hundreds of kilometres along the coast annually cause toxicity in both nearshore and offshore shellfish beds (e.g., Anderson et al. 2014b). There are also inshore populations of *A. fundyense* that cause localized blooms within embayments only a few km^2 in size (Anderson 1997). These inshore blooms have been especially valuable in revealing details of life cycle transitions due to the localized or captive nature of their blooms, which can be studied from beginning to end, sometimes with novel in situ instruments that greatly enhance the frequency and resolution of observations [e.g., Brosnahan et al. 2014, 2015; see also Chap. 18, Glibert et al. (2018)].

8.2.2.2 Cyst Distributions and Initiation of Planktonic Blooms

Study of any phytoplankton species on the geographical scale of coastal blooms is challenging because populations are patchy in space, depth, and time. In this respect, studies of cyst-forming dinoflagellates have had the advantage that significant insights into bloom ecology are possible through surveys of their immobile cysts, which are also typically concentrated in thin layers (TLs) at the sediment-water interface [see also Chap. 9, Raine et al. (2018)]. Such surveys of both *A. fundyense* and *P. bahamense* have documented large seedbeds that provide inoculums to major coastal blooms of these two species.

Within the GOM, the vast majority of *A. fundyense* cysts accumulate within two major seedbeds, one within the Bay of Fundy and the other offshore from the central Maine coast (Anderson et al. 2014a). These accumulation zones have been documented repeatedly through mapping cruises in 1997 and annually from 2004 to 2016 (Fig. 8.3a), and their persistence attributed to the combination of the region's bathymetry, sediment characteristics, currents, and annual replenishment from spring-summer blooms along the Maine coast. Still, there is significant interannual variability in total cyst abundance across these areas (up to tenfold between years) which is strongly correlated with geographic measures of bloom extent the subsequent year (e.g., the length of coastline closed due to shellfish toxicity or the southernmost latitude of shellfish closures). It is thus possible to use cyst abundance to empirically forecast the geographic extent of the forthcoming bloom (Anderson et al. 2014a). Statistical assessment of the GOM cyst distributions has further shown that many locations within the GOM harbour a relatively consistent proportion of its total cysts in a given year (Fig. 8.4). This discovery has led to development of a structure-based model of cyst distribution for estimation of the total number of cysts during interbloom periods (Solow et al. 2014). Guided by this model, annual cyst surveys could effectively allow future annual forecasts using a significantly smaller number of stations (e.g., 40% of those sampled in past surveys), greatly reducing the cost of future survey efforts. Similar approaches have great promise for other dinoflagellate species and regions but require multi-year data sets to establish common patterns of cyst abundance and distribution.

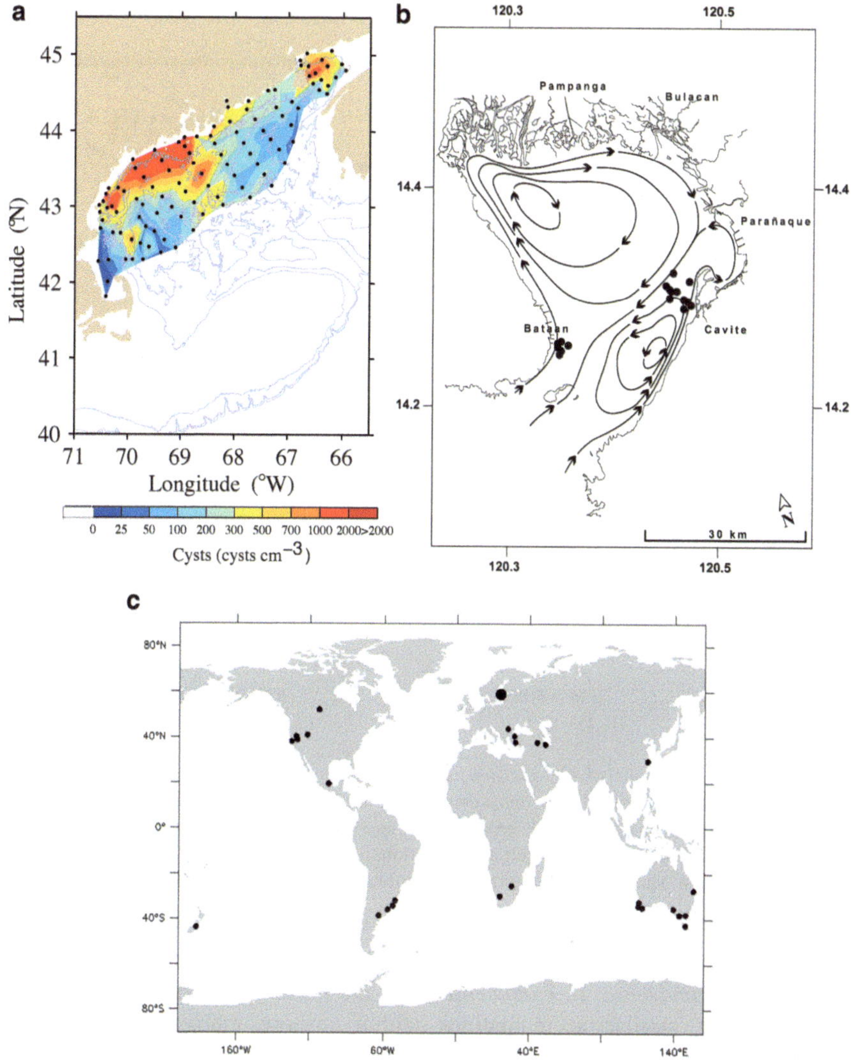

Fig. 8.3 (**a**) Map of the mean *Alexandrium fundyense* cyst abundance in the Gulf of Maine, calculated for years 2004–2009, showing the two main cyst seedbeds; *black dots* are station locations (modified from Solow et al. 2014). (**b**) Map of Manila Bay (Philippines) showing cyst beds (*black dots*), sources of *Pyrodinium bahamense* blooms, distribution of which follows two cyclonic wind-driven current systems during the southwest monsoon (redrawn from Villanoy et al. 2006). (**c**) Occurrences of *Nodularia spumigena*; please note that this map is by no means complete; only those with clear evidence for *N. spumigena* reported in peer-reviewed English language journals are depicted; most reports stem from the past 35 years; historical evidence for *N. spumigena* blooms exists for the Baltic Sea, Australia, New Zealand, as well as inland salt lakes in the USA and Canada

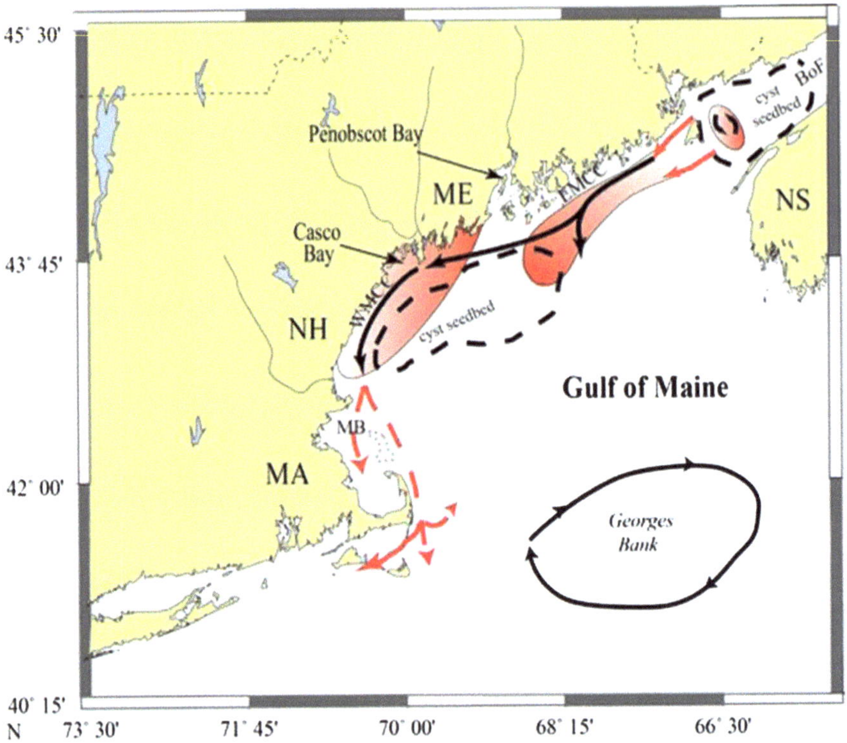

Fig. 8.4 Conceptual model of *Alexandrium fundyense* cyst and motile cell dynamics in the Gulf of Maine. Areas enclosed with *dashed lines* denote cyst seedbeds that provide inoculum cells. The *shaded areas* represent regions with growth and transport of motile cells, with *darker shading* denoting areas with the most dense cell accumulations. *Black arrows* show major currents within the eastern Gulf of Maine, and *grey arrows* show alternative current pathways. Modified from Anderson et al. (2014a)

Unlike *A. fundyense*, *P. bahamense* cysts are not known to occur in offshore seedbeds, but similar accumulations within inshore embayments are well documented. Within the Philippines, spatial and vertical surveys have documented highest concentrations of cysts in the innermost parts of Manila Bay, Sorsogon Bay, Malampaya Sound, and Juag Lagoon, areas where the sediments are mostly muddy and where mangrove stands are present (Corrales and Crisostomo 1996; Azanza et al. 2004; Siringan et al. 2008). The cysts that lie in the shallow areas where seedbeds do occur are susceptible to displacement by the high energies experienced during the NE monsoon. The newest of these displaced cysts, however, likely cannot germinate because of their obligatory maturation dormancy period of 2.5–3 months, approximately the same duration as the summer/inter-monsoon period that follows the SW monsoon, the periods when blooms typically occur and produce new cysts (Azanza 1997, 2013). Trends in *Pyrodinium* cyst and cell concentrations and a summary of physicochemical conditions in Manila Bay and

Sorsogon Bay during the different seasons have been previously reported, but relationships between cyst abundance and the timing and magnitude (if they exist) remain to be shown (Azanza 2013).

The physical resuspension of *P. bahamense* cysts contributes to the initiation of planktonic blooms, especially when resuspension events coincide with periods when *P. bahamense* is able to germinate. In the absence of such physical resuspension, germlings may also emerge directly from the near surface of the ocean bottom. Direct observation of such cyst emergence from cyst seedbeds has been difficult and remains an area for which new observational methodologies are needed. In situ cyst germination rates for *A. fundyense* have been estimated using various techniques like plankton emergence traps (PETs), but average excystment percentages are 10–20 times lower than laboratory-derived rates (Ishikawa et al. 2007). Presumably this difference reflects challenges in re-creating the chemical and physical conditions that cysts experience in sediments (e.g., a tortuous path through sediment particles for germling cells to reach the water column). Germling cells reared in the dark (similar to seedbeds at 100–150 m in the GOM) have been shown to have relatively poor survival (28% after 7 days, the estimated time of reaching the euphotic zone in a 100 m water column; Vahtera et al. 2014). Combined, the PET and germling results suggest that only a small fraction of resting cysts progress to the vegetative stage of the life cycle but also indicate a need for further study of these critical processes.

Similar to the case of *P. bahamense*, initiation of *A. fundyense* blooms may rely in part on physical resuspension of cysts. In the GOM and many other regions, substantial numbers of resting cysts occur within benthic nepheloid layers (BNLs). BNLs form through the action of near-bottom currents and are characterized by high concentrations of resuspended sediments, detritus, and organisms, sometimes in layers tens of metres thick. *A. fundyense* cyst abundance is enriched within BNLs but also quite patchy, varying spatially by three orders of magnitude, features that make surveys of BNLs for *A. fundyense* cysts challenging (Pilskaln et al. 2014). Still, cyst fluxes recorded by near-bottom sediment traps, maps of cyst abundance within surface sediment, and estimated BNL cyst inventories suggest that the residence time of cysts within sediments varies across the GOM, such that cysts in the eastern gulf have shorter residence times than those in the western gulf. The latter finding has significant implications especially for understanding the initiation of blooms that occur in the southern GOM/Georges Bank region, areas that are downcurrent from the primary GOM seedbeds within the Bay of Fundy and off the mid-Maine coast.

8.2.2.3 Bloom Development and Transport

Alexandrium fundyense blooms tend to occur in two major habitats: inshore embayments and in open coastal waters. The dynamics of blooms in these distinct habitats have been explored through development of spatially explicit models that incorporate hydrography and physical forcing in addition to biological models of germination, cell division, grazing, and other loss processes. Results from these studies have demonstrated the ways in which inshore and open coastal blooms differ with respect to the life cycle processes and behaviours that are principally responsible for their observed dynamics.

Within inshore areas like the Nauset Marsh system (Cape Cod, MA, USA), retention of bloom populations occurs behind shallow sills and other bathymetric barriers to dispersal (Anderson and Stonzenbach 1985; Crespo et al. 2011). This retentive behaviour arises from surface avoidance by vegetative stage cells during their daily vertical migrations and is reinforced by strong salinity- and temperature-driven density gradients that limit wind-driven vertical mixing and dispersal of cells. The timing of blooms and their harmful impacts is most strongly driven by temperature, which controls the rate of *A. fundyense* division. Unlike *P. bahamense* blooms in Manila Bay or coastal *A. fundyense* blooms (described below), the intensity and timing of inshore blooms are not dependent on cyst distribution or abundance (Ralston et al. 2015).

In more open, coastal systems like the GOM, cyst distributions, freshwater plumes, meteorological conditions, and circulation patterns all play significant roles in determining the intensity of harmful impacts, which are felt most strongly when expansive offshore blooms are concentrated near the coast by downwelling favourable conditions (Franks and Anderson 1992; Fauchot et al. 2008). For example, cells germinated from the two GOM cyst beds are advected in the alongshore direction from east to west in a coastal current. Growth of the vegetative cells is limited primarily by temperature from April through June throughout the GOM, whereas nutrient limitation occurs in July and August in its western areas. Thus, an observed seasonal shift in the bloom's centre of mass from west to east can be explained by changing growth conditions: early in the season, growth is more rapid in the west due to warmer temperatures there and then later more rapid in the east due to increasing nutrient limitation in the west (McGillicuddy et al. 2005).

Further explorations of GOM bloom dynamics have applied a coupled physical-biological model that is parameterized using data from several decades of observations and experiments (Stock et al. 2005; McGillicuddy et al. 2005; He et al. 2008; Li et al. 2009). The model is a nested hydrodynamic model of the GOM that includes a biological submodule. Realistic simulations of bloom development are possible using observed cyst distributions, cyst germination rates (Anderson et al. 2005), vegetative cell growth rates, and continuous real-time river flow and hydrographic data (e.g., He et al. 2008).

A large driver of the GOM model's success is the system's dependence on cyst abundance and behaviour. By virtue of their endogenous germination rhythm, the timing of bloom inoculation is well constrained, especially in comparison to more variable inshore blooms. Total cyst abundance (measured in the fall or winter) appears to be a first-order predictor of GOM bloom magnitude the following year (McGillicuddy et al. 2011). This is supported by correlations between annual cyst abundance and the geographic extent of PSP closures along the GOM coast in subsequent bloom seasons, a relationship that may be related to diminished vegetative division in these populations relative to inshore blooms (Anderson et al.

2014a). Physical and chemical processes are still likely to modulate bloom development in the GOM and may override the cyst-PSP relationship, as happened in 2010 when an anomalously warm and freshwater mass entered the GOM and likely suppressed development of that year's *A. fundyense* bloom (McGillicuddy et al. 2011).

In a similar manner, *Pyrodinium* cell and cyst movements follow wind velocity directions and cyclonic trends associated with the circulation of Manila Bay (Fig. 8.3b). Blooms form via two primary mechanisms: (1) an inoculum from cyst beds in the west that is advected along the west-northwest coast, and (2) a second inoculum from cyst beds in the southeast that spreads across the east-southeast area (Villanoy et al. 2006; Azanza 2013). Numerical models of *Pyrodinium* blooms in Manila Bay (Philippines) have been developed from physicochemical and biological data sets collected from 1994 to 2002 as well as from laboratory experiments examining *P. bahamense* behaviour (Villanoy et al. 1996, 2006; Azanza et al. 2004). Three-dimensional and spatially explicit models have been used to explore the role of *Pyrodinium* life cycle processes including reproduction, mortality, encystment, and excystment in Sorsogon Bay (Yñiguez et al. 2012). Forced by seasonally varying heat, freshwater (precipitation, runoff), and momentum fluxes (wind), temperature, salinity, nutrient, and velocity fields are simulated, which, combined with turbulent mixing and convection, affect the distribution, transport, and behaviour of *P. bahamense* cells (Villanoy et al. 2006; Azanza 2013). The model output mirrors features of *Pyrodinium* blooms in this area, which generally develop at the onset of SW monsoon and end with the onset of the NE monsoon.

8.2.2.4 Sexual Induction and Cyst Formation

Historically, investigations of resting cyst production by *A. fundyense*, *P. bahamense* (Corrales et al. 1995), and many other dinoflagellates have been culture based and have relied on nutrient starvation to induce sex. Specific limiting nutrients that induce resting cyst formation by *A. fundyense* include nitrogen (N), phosphorus (P), and iron (Anderson et al. 1984; Doucette et al. 1989). These approaches are consistent with a prevailing paradigm in the study of microbial eukaryotes that transitions from asexual reproduction to sexual fusion are only advantageous when a population's environment no longer supports continued cell division (Margulis and Sagan 1985). However, evidence that nutrient limitation induces field populations to encyst is limited.

Studies of localized, inshore *A. fundyense* blooms often report the occurrence of sexual stages (gametes and planozygotes) when measured nutrient concentrations would be expected to support continued division by vegetative cells (e.g., Anderson et al. 1983; Anglès et al. 2012; Brosnahan et al. 2015). Linkage of nutrient stress and sexual induction in offshore blooms has been similarly challenging. A simple nutrient-based induction model was able to replicate the cyst deposition zones in the Bay of Fundy and western GOM (McGillicuddy et al. 2005). However, direct linkage of sexual induction and nutrient stress was not supported in the only direct observation of encystment in this system, a remarkable red water event that occurred in 2009. Concentrations of N and P during that event should have been sufficient to support continued division (Townsend et al. 2014), but instead, the population was dominated by planozygotes, dissipated within 1 week, and was successfully linked to a major deposition of new resting cysts during a subsequent benthic survey (McGillicuddy et al. 2014).

An alternative stimulus for sexual induction may be high cell concentrations. Recent intensive investigations of *A. fundyense* blooms in the Nauset Marsh system (Cape Cod, MA USA) have shown that growing degree days, a simple temperature-based metric, can collapse interannual and spatial variability in the timing of both the development and peak of Nauset blooms (Ralston et al. 2014). This metric relies on the linear response of *A. fundyense* population growth rate to increasing temperature, a relationship that has been shown through both observed accumulation rates within the Nauset ponds and through deployments of an Imaging FlowCytobot (IFCB), a submersible imaging-in-flow cytometer that enables direct estimation of in situ division rates [Brosnahan et al. 2015; see also Chap. 18, Glibert et al. (2018)]. Nauset bloom peaks occur at accumulations of growing degree days that translate to peak concentrations of a few million cells L^{-1}, very similar to the concentrations observed during encystment by the 2009 GOM population described by McGillicuddy et al. (2014). High-frequency sampling with an IFCB has shown that conversions from vegetative stage cells to gametes occur en masse at these concentrations and further that encystment is likely the dominant process leading to the termination of these blooms (Velo-Suárez et al. 2013; Brosnahan et al. 2015; Choi et al. 2017). Thus, the success of the growing degree metric in predicting Nauset bloom peaks may arise from its skill in predicting the accumulation of cells to concentrations in excess of several million cells L^{-1}, a threshold at which blooms shift from continued division to sexual fusion. It remains to be shown whether *A. fundyense* in other environments and regions behave similarly. If so, it may be that *A. fundyense* recognize and exploit ephemeral physical and/or biological conditions in order to return to the benthic resting cyst stage rather than wait for growth limitation as an encystment cue. Still other factors known to induce encystment by other dinoflagellate species include the presence of parasites (Toth et al. 2004) and grazers (Rengefors et al. 1998). Both of these factors have the potential to decimate natural HAB populations, particularly when they reach such high cell concentrations. Future field investigations of *A. fundyense*, *P. bahamense*, and other dinoflagellate HAB species should further explore the role of these and other species-species interactions as triggers of encystment, a process that is known to also occur within populations that are far less dense than those observed during Nauset blooms and the 2009 GOM event.

In contrast to *A. fundyense*, *P. bahamense* blooms are often composed of mixtures of different life stages, i.e., vegetative cells, dividing cells, gametes, temporary cysts, and resting cysts, suggesting that vegetative division and encystment occur concurrently in these populations (Azanza 2013; Azanza et al. 2004). This conclusion is further supported by model-based studies of *Pyrodinium* blooms in Manila Bay and Sorsogon Bay for which sensitivity experiments concerning sexual responses to light, temperature, salinity, and nutrient conditions have been carried out (Villanoy et al. 2006; Yñiguez et al. 2012; Cayetano et al. 2013). Future investigations like those described for *A. fundyense* that document the relative distribution of sexual stages in time and space during *P. bahamense* blooms will likely shed further light on the role of encystment in limiting bloom intensity and promoting bloom decline. Still, the dominant factor leading to the termination of *P. bahamense* blooms in the Philippines appears to be the altered hydrographic conditions associated with the NE monsoon season. Mixed investments in cell division and encystment by these populations may reflect adaptation to this type of abrupt habitat change which is uncoupled in its timing from normal bloom progression. It is noteworthy that for both *P. bahamense* and *A. fundyense*, the onset of sexual processes does not coincide with conditions that would be expected to inhibit further vegetative division. This is in contrast to conclusions drawn from culture studies and underscores the importance of new cyst formation for the survival and persistence of these organisms.

8.3 Diatoms: *Pseudo-nitzschia* spp.

Pseudo-nitzschia are marine planktonic diatoms belonging to the raphid lineage, the most recent one in the evolutionary history of diatoms (Kooistra et al. 2007). About 35 species are presently included in this genus (Lelong et al. 2012; Teng et al. 2014; Lim et al. 2013), and up to 18 are known to produce the neurotoxin domoic acid (DA), responsible for amnesic shellfish poisoning (ASP) syndrome. Like the PSP toxins produced by *P. bahamense* and *A. fundyense*, DA is accumulated by filter feeders and thus enters marine food webs. Consumption of this toxin causes a range of human health impacts including gastrointestinal distress, neurological disorders, and, in more severe cases, coma and death (Lefebvre and Robertson 2010). It has also been linked to broader ecosystem impacts including mortality of sea lions, birds, and whales (Lefebvre et al. 2002). Blooms of *Pseudo-nitzschia* species occur worldwide in both coastal seas and open oceanic waters where closure of aquaculture installations for prolonged periods can cause severe economic losses (Trainer et al. 2012).

8.3.1 Life Cycle

Like the majority of diatoms described, *Pseudo-nitzschia* species undergo progressive cell size diminution during vegetative growth, due to the distinctive structure of the siliceous frustule and the modality of cell division (Montresor et al. 2016). To escape from this miniaturization trap, large-sized cells must be produced, and this "rejuvenation" occurs within the sexual phase. Sexual reproduction has been illustrated in various details for 14 *Pseudo-nitzschia* species (Lelong et al. 2012). The majority of *Pseudo-nitzschia* has been shown to have heterothallic mating systems and allogamous sexual reproduction. Species are dioecious, i.e., sexual fusion only occurs between cells of opposite mating type (MT+ and MT−) and vegetative clones are fully self-sterile.

One fundamental requirement for *Pseudo-nitzschia* sex to start is that cells of opposite mating type should be below a species-specific sexualization size threshold (Chepurnov et al. 2005; Scalco et al. 2014). Round, non-motile, and morphologically undistinguishable haploid gametes are produced by meiosis in separate gametangia. They are however behaviourally different, because gametes produced within the MT− gametangium are sessile, while gametes produced by the MT+ gametangium slide towards the sessile ones and conjugate, thus restoring the diploid phase (Scalco et al. 2015), which in turn transforms into an auxospore (Fig. 8.1e, f). It is within the auxospore—a stage that lacks a rigid siliceous frustule—that a cell of the maximum size is produced. Exceptions to this general scheme have been reported for *P. subcurvata* and *P. brasiliana*, which have a homothallic mating system in which gametes of opposite mating types are produced by single clonal strains (Fryxell et al. 1991; Quijano-Scheggia et al. 2009).

Results of laboratory experiments have shown that in *P. multistriata* successful sexual reproduction can only be achieved when parental strains are in the exponential growth phase; when crossing strains in the late- or post-exponential growth phase, only few gametes were detected, and auxospores and initial cells were not produced (Scalco et al. 2014). Moreover, the onset of the sexual phase in this species requires a threshold cell concentration to start, and it is accompanied by the arrest of growth of the vegetative parental cells. These observations suggest the presence of sex pheromones that mediate the attraction of opposite mating types and induce a cell cycle arrest in G_1 when cells can be committed to shift to meiosis, as has been recently shown for the benthic diatom *Seminavis robusta* (Moeys et al. 2016).

Although sexual reproduction is a fundamental phase in *Pseudo-nitzschia* life cycles, there are only two reports of sex events in the natural environment. Scarcity of reports is attributable to low sampling frequencies and limitations of the visual identifications of the sexual stages, which tend to be relatively rare and ephemeral. Sexual stages of two *Pseudo-nitzschia* species have been recorded at the LTER station MareChiara in the Bay of Naples (Mediterranean Sea), where they accounted for 9.2 and 14.3% of the total number of cells of *P. cf. delicatissima* and *P. cf. calliantha*, respectively (Sarno et al. 2010). Another massive sex event has been reported along the Washington coast (USA) involving *P. australis* and *P. pungens* (Holtermann et al. 2010). Auxospores were detected for a longer time interval (about 3 weeks) and accounted for 59% of the *P. australis* population at the end of the bloom.

8.3.2 Bloom Dynamics

A considerable number of studies have been carried out on *Pseudo-nitzschia* bloom dynamics in the last decades, showing that these species can be recorded both in coastal and offshore waters, in upwelling systems, as well as in more secluded environments like fjords and bays [Trainer et al. 2012; see also Chaps. 10, 11, Roy et al. (2018) and Pitcher et al. (2018)]. Intensive monitoring programmes coupled with the continued development of physical/biogeochemical models are delivering forecast capabilities that promise to better safeguard public health and local economies (e.g., Lane et al. 2009; Anderson et al. 2009; Giddings et al. 2014). However, the biological mechanisms leading to the formation and demise of *Pseudo-nitzschia* blooms are not yet fully understood. Like all unicellular microalgae, *Pseudo-nitzschia* species experience periodic increases in their cell number, often with a marked phenology (Fehling et al. 2006; Ruggiero et al. 2015). The inoculum for the bloom does not derive from the germination of benthic resting stages, which have never been reported in areas where *Pseudo-nitzschia* species are regular members of the phytoplankton community (e.g., McQuoid 2005; Montresor et al. 2013). However, in periods between subsequent blooms, cells can be present in the water column at very low, undetected concentrations or might accumulate at physical discontinuities within the water column, such as the pycnocline and/or subsurface TLs (Rines et al. 2002; Timmerman et al. 2014).

Blooms reflect increases in cell growth rates which demand ample supplies of nutrients and light, yet these factors do not fully explain variability in the timing of phytoplankton blooms. A complex signalling mediated by infochemicals between *P. multiseries* and a species of *Sulfitobacter* has been recently discovered (Amin et al. 2015). These bacteria can convert tryptophan produced by the diatom into indole-3-acetic acid, which in turn promotes diatom cell division. Interactions between *Pseudo-nitzschia* species and their microbial consortium may also modulate the toxicity of these species and their susceptibility to attacks by algicidal bacteria (Sison-Mangus et al. 2013). Species with different levels of toxicity also differ in their bacterial communities, with toxic *Pseudo-nitzschia* hosting lower microbial diversity than nontoxic ones. Massive sinking at the demise of blooms is a distinctive trait of diatom life cycles, and in this way toxic *Pseudo-nitzschia* transport DA to the deep ocean. Considerable amounts of toxin have been recorded in sediment traps at depths greater than 800 m, where DA enters benthic food webs through filter feeders and their predators (Sekula-Wood et al. 2009).

Changes in *Pseudo-nitzschia* species composition have been observed in sediment core samples (Lundholm et al. 2010) and at sites where long-term data are available (reviewed by Trainer et al. 2012). Though these composition changes are most often thought to reflect environmental changes, disappearances of some species may also be caused by failures to complete their sexual cycles. The consequent lack of recruitment of large cells to these populations is an understudied mechanism for species extirpations in existing time series data sets.

8.3.3 Modelling of Life Cycle Transitions

The pattern of cell size variation over time can help to infer the timing of sex in natural diatom populations. In the Gulf of Naples, *P. multistriata* blooms in summer-autumn, and a regular alternation of uni- and bimodal cell size classes has been recorded over 11 years (D'Alelio et al. 2010). Large-sized cohorts, most likely representing the product of sexual reproduction, were recorded every second year. The population progressively reduced in size and, upon reaching the cell size threshold for sex in the bloom of the second year, produced a new cohort of large cells. Life cycle dynamics and the regular pattern of cell size reduction and restitution were simulated with an individual-based model parameterized with data on growth rates, cell size reduction dynamics, and cell size threshold for sex gained from laboratory observations. The model output reproduced the coherent pattern in *P. multistriata* population dynamics observed in nature only when a seasonal variation in the rate of cell division was imposed; a rapid acceleration of growth rate during the bloom phase was in fact mirrored by cell size reduction. Further, this fine-tuned organization of its life cycle appears to be independent from more proximate environmental conditions. The model also predicted that the species would become locally extinct if sex did not occur within 4 years.

The biovolume of *Pseudo-nitzschia* cells during the life cycle can vary about one order of magnitude between the largest and the smallest cells. This has implications for growth capability (D'Alelio et al. 2009; Amato et al. 2005), toxicity, sinking rates, and selective grazing. The inclusion of life cycle traits, such as cell size reduction and restitution, in theoretical models can provide insights on the impact of different size classes on diatom population dynamics (Hense and Beckmann 2015).

8.4 Cyanobacteria: *Nodularia spumigena*

Cyanobacteria are among the most prominent HAB-producing taxa and cause a variety of negative impacts including contamination of drinking water and livestock and wildlife mortalities [see also Chap. 20, Burford et al. (2018)]. *Nodularia spumigena* Mertens ex Bornet & Flahault 1888 (Fig. 8.1g) is of particular interest because this species produces nodularin, a cyclic pentapeptide hepatotoxin that affects many higher-trophic-level organisms including humans (Rinehart et al. 1988). Recent studies suggest that *N. spumigena* also produces a specific neurotoxic amino acid, but due to methodological difficulties, its relevance is still under debate (e.g., Cox et al. 2005; Kruger et al. 2010). *N. spumigena* is present worldwide (Fig. 8.3c), predominantly in brackish waters of temperate regions. Exceptionally large blooms occur in the Baltic Sea where surface rafts of this buoyant species can cover more than 200,000 km^2 (Kahru and Elmgren 2014). Environmental factors like sea surface temperature play an important role in bloom formation, but bloom inoculum size is equally important and is strongly determined by life cycle dynamics.

8.4.1 Life Cycle

The life cycle of *N. spumigena* can be characterized by three different phases (Fig. 8.2b) that are synchronized with the seasonal cycle: a bloom phase, a resting phase, and a recruitment phase. The bloom phase occurs in summer when surface temperature is high (>20 °C), the dissolved inorganic N pool is depleted, and wind and shear stresses are weak. During this phase, filaments consisting of photosynthetic vegetative cells and specialized N-fixing, anaerobic heterocysts accumulate to form blooms. Through N fixation, heterocysts supply their neighbouring vegetative cells with N, an energetically demanding process. At the end of summer, the growth rate decreases due to deteriorating external conditions such as lower temperature, an exhausted P pool, and/or reduced solar radiation. The resting phase begins during which some filaments hibernate in the water column (Suikkanen et al. 2010) until the return of favourable growth conditions the following summer. In addition, thick-walled cells, called akinetes, are formed that can withstand unfavourable conditions for several years (e.g., Huber 1984). These cells sink to bottom sediments where they may be lost by burial or decomposition, but others will germinate, stimulated by environmental cues. Light, in particular, is assumed to trigger germination in this recruitment phase. Exposure may be mediated by wind-driven resuspension or activity of specific benthic fauna (Karlson et al. 2012). Cells may also ascend to and recolonize the surface through production of gas vesicles during late spring and early summer. If surface temperature exceeds approximately 13 °C, the cells divide and initiate the next bloom phase. While the described succession of *N. spumigena* life cycle stages is "typical" under temperate conditions, it is not known how life cycle strategies vary in different environments and geographic regions. It is also unclear whether toxin production is influenced by life cycle dynamics (though this is often speculated to be the case; e.g., Cox et al. 2005).

8.4.2 Bloom Dynamics

Several characteristics of *N. spumigena* blooms are best explained as the result of life cycle transitions. One of these is the rapid increase and decrease of blooms, most likely caused by transitions from or to resting stages (Hense and Beckmann 2006). Such transitions may further be associated with high year-to-year or interannual variability in total biomass and spatial extent of *N. spumigena* blooms (e.g., Galat et al. 1990; Hense and Burchard 2010; Kahru and Elmgren 2014). Under favourable environmental conditions, resting stages may accumulate over several years before they transform into growing stages. Such dynamics may explain substantial differences in bloom concentrations observed year to year. Life cycle transitions may also be hindered by unfavourable conditions leading to low concentrations or absence of blooms. Feedback mechanisms associated with the life cycle are therefore likely to play an important role in bloom variability; e.g., intense accumulations of growing cells may tend to produce more akinetes, leading to larger inocula during subsequent blooms (Hense 2007).

8.4.3 Dispersal and Future Distribution

The survival of resting stages for several years will principally facilitate the dispersal of *N. spumigena*. Indeed, there are reports of *N. spumigena* along bird migration routes (Stoyneva 2016). There is also evidence that *N. spumigena* can be transported in ballast water over long distances (Olenin et al. 2000), but whether colonization is successful or not will strongly depend on local environmental conditions, including interactions with endemic species. Overall, a clear statement about biogeographic expansion of *N. spumigena* over the past decades cannot be given, because a higher awareness and improved tools for detection of these cells may also explain the rise in reports of *N. spumigena*.

Continued presence and spread of *N. spumigena* depends on several factors. First, global warming is predicted to favour cyanobacteria in general and *N. spumigena* in particular (Paerl and Huisman 2009; Paerl et al. 2011; Hallegraeff 2010; O'Neil et al. 2012), because their growth optimum lies at about 25 °C (e.g., Lehtimaki et al. 1997). Second, the state of eutrophication will have consequences for the development of *N. spumigena*. If both nutrient loads and inventories, notably N and P, are reduced, the intensity of cyanobacterial blooms is likely to decline, especially if other factors like temperature remain unchanged. Third, ocean acidification might affect *N. spumigena*. Despite large uncertainties, recent studies suggest that the growth rate of this species is reduced under high pCO_2 (Czerny et al. 2009; Eichner et al. 2014). Toxin production, however, may still rise, because toxin content has been shown to increase with temperature (Lehtimaki et al. 1997). Again, life cycle dynamics will strongly determine how *N. spumigena* responds to climate change (Hense et al. 2013). Differing environmental requirements for different life cycle stages and the two-way interaction of each stage with its environment lead to a chaotic system, in the sense that specific features in such systems remain unpredictable, even if other aspects are known. One of these features may well be specific outbreaks of *N. spumigena*.

8.5 Synthesis and Recommendations

HAB-causative and other phytoplankton species are confronted with a dynamic aquatic environment and must acquire resources for growth while also guarding against diverse mortality processes in order to ensure their survival. Common stressors include variations in nutrient supply, turbulence, light availability, and temperature, as well as changing diversity of other species, including competitors and predators. Here, we have described ways in which four prominent HAB-causative organisms cope with these variable stressors through the evolution of distinctive traits within their respective life cycles. Common among these are their capability to persist in the face of extended periods of species-specific poor growth conditions between blooms. Also common is the ability to initiate transitions between life cycle phases, either through direct sensing of changing environmental conditions, anticipation of seasonal changes, high cell concentrations, or other factors that facilitate successful life cycle transitions.

Modelling studies have been a particularly effective means to identify and weigh the relative importance of different life cycle behaviours for determining the population dynamics of different HAB species in response to diverse environmental factors. Among the three meroplanktonic species reviewed here (*P. bahamense*, *A. fundyense*, and *N. spumigena*), potentially long-lived, durable benthic stages figure prominently in their overall ecology. Major seedbeds of the dinoflagellates' resting cysts and akinetes of the cyanobacterium *N. spumigena* reflect dynamic interactions between hydrography and patterns of new benthic stage production, particularly in areas where these species are well established. Moreover, accumulations of cysts can persist for years even in the absence of new depositions, imposing a continued threat of HAB impacts through future bloom inoculations. Spatial and temporal records of the resulting seedbeds are useful predictors of future bloom impacts in some cases. A prime example is the development of a HAB forecast for *A. fundyense* based on annual surveys of cyst abundance across the GOM. Short time scales for transitions between benthic and planktonic stages of *N. spumigena* and *P. bahamense* appear to be integral to their bloom dynamics and enable these species to escape short periods of adverse conditions during bloom periods. Rapid downturns in dinoflagellate populations may also be explained by production of resting cysts. The latter process may be spread over the whole bloom period in some populations or may be concentrated near the end of blooms, sometimes acting as a dominant factor leading to bloom decline.

Pseudo-nitzschia spp. differ from the other species discussed here in that no benthic stages in their life cycles are known. Persistence of these species during periods of non-optimal growth conditions is presumed to rely on low concentrations of cells that may linger or concentrate at density discontinuities in the water column. Like all other diatoms, these species also differ from *P. bahamense*, *A. fundyense*, and *N. spumigena* in their mode of division, a basic process in the life cycle of all microbes that is integral to these species' ability to form blooms. Specifically, *Pseudo-nitzschia* spp. undergo diminution as they divide and must undergo sexual fusion and auxospore formation to restore large cells to their population. This cell size-driven entry and exit from the sexual phase of its life cycle that occurs when cells are in exponential growth contrasts with a prevailing paradigm about sex in microbial eukaryotes, namely, that such recombination events are only likely to be favoured when abiotic conditions are limiting. Similarly, in many field populations of the dinoflagellates *A. fundyense* and *P. bahamense*, cells appear to enter the sexual phase of their life cycle prior to the onset of significant growth limitation. This suggests that biological factors (e.g., density-dependent mechanisms and/or sex pheromones) may play important roles in determining the timing and extent of sexual conversions. Continued research and observation of natural populations is needed to better characterize the history of HAB populations leading up to sexual events and to better evaluate the roles of endogenous and exogenous factors promoting these transitions, which are crucial for the recurrence of blooms.

An ongoing challenge for studies of life cycle dynamics is the characterization of stage behaviours and transitions between stages within natural populations. Though much progress towards characterizing stage-specific behaviours has been made through laboratory experiments, it is also clear that the conditions and rates of many fundamental life cycle processes differ substantially between naturally occurring and laboratory-reared cells. The factors contributing to these differences are numerous and include contributions from natural bacterial flora, differences in nutrient supply and light conditioning, and interactions with other co-occurring phyto- and zooplankton species and parasites—all factors which are exceedingly difficult to replicate in cultures. Accordingly, in situ observations are needed, but such efforts are still in their infancy in the context of life cycle transitions.

The obstacles inherent to in situ observations include the logistics of observing oceanic phenomena than can span large areas, difficulty in discriminating morphologically similar planktonic stages (e.g., planomeiocytes, vegetative cells, gametes, and planozygotes in many dinoflagellate species), and a basic inability to observe benthic processes without massive disruption of potentially informative physical associations (i.e., for resting cysts, akinetes, and other benthic stages). Automated and robotic instrumentation is beginning to address these challenges (e.g., Brosnahan et al. 2015), but more development and application of these systems is needed. Existing and future in situ approaches must be applied to a broader range of HAB taxa to elucidate natural growth and mortality processes and other factors that modulate the timing and intensity of blooms. Ongoing and future research on the roles and interactions of different life cycle stages with bacteria, viruses, and parasites are similarly expected to yield great insights into the ecological responses of these organisms. Finally, long-term data sets on the temporal and spatial occurrences of these and other plankton species need to be maintained and expanded to enable future insights into the roles that life cycle dynamics play in shaping the ecology of these species, especially as ecosystems are challenged by climatic shifts, severe weather, and anthropogenic influences (e.g., prolonged El Niño for *Pyrodinium* and anomalous warm/freshwater inputs affecting GOM *A. fundyense* blooms). Through these activities, more descriptive and predictive models can be generated that will further our understanding of HAB dynamics and better protect society from their impacts. Global efforts, such as may be achieved through the next-generation global interdisciplinary HAB Programme, GlobalHAB, will be important to advance this understanding and to provide the necessary advanced predictive models.

Acknowledgments The authors acknowledge the support of (1) the Department of Science and Technology (DOST), Philippines, for R.V. Azanza; (2) the National Science Foundation (OCE-0430724, OCE-0911031, and OCE-1314642) and National Institutes of Health (NIEHS-1P50-ES021923-01) through the Woods Hole Center for Oceans and Human Health, MIT Sea Grant (NA14OAR4170077), and Woods Hole Sea Grant (NA14OAR4170074, R/P-84) for M. Brosnahan and D. Anderson; (3) the national flagship programme RIcerca ITaliana per il MARE (RITMARE) for M. Montresor; and (4) the Cluster of Excellence "CliSAP" (EXC177), University of Hamburg, funded by the German Science Foundation (DFG) for I. Hense. Thanks go to J.C. Dungca-Santos for manuscript typing and Seija Hällfors (Syke) for providing the picture of *N. spumigena*.

References

Amato A, Orsini L, D'Alelio D et al (2005) Life cycle, size reduction patterns, and ultrastructure of the pennate planktonic diatom *Pseudo-nitzschia delicatissima* (Bacillariophyceae). J Phycol 41:542–556

Amin SA, Hmelo LR, van Tol HM et al (2015) Interaction and signalling between a cosmopolitan phytoplankton and associated bacteria. Nature 522:98–101

Amorim A, Dale B (1998) Distribution of cysts from toxic or potentially toxic dinoflagellates along the Portuguese coast. In: Reguera B, Blanco J, Fernandez M et al (eds) Harmful algae. Proceedings of the 8th International Conference on harmful algae, Vigo, Spain, 25–29 June 1997. Xunta de Galicia and Intergovernmental Oceanographic Commission of UNESCO, Paris, pp 64–65

Anderson CR, Siegel DA, Kudela RM et al (2009) Empirical models of toxigenic Pseudo-nitzschia blooms: potential use as a remote detection tool in the Santa Barbara Channel. Harmful Algae 8:478–492

Anderson DM (1989) Toxic algal blooms and red tides: a global perspective. In: Okaichi T, Anderson DM, Nemoto T (eds) Red tides: biology, environmental science, and toxicology. Elsevier Science Publishing Co. Inc., New York, pp 11–16

Anderson DM (1997) Bloom dynamics of toxic *Alexandrium* species in the northeastern US. Limnol Oceanogr 42:1009–1022

Anderson DM, Alpermann TJ, Cembella AD, Collos Y, Masseret E, Montresor M (2012) The globally distributed genus *Alexandrium*: Multifaceted roles in marine ecosystems and impacts on human health. Harmful Algae 14:10–35

Anderson DM, Chisholm SW, Watras CJ (1983) The importance of life cycle events in the population dynamics of *Gonyaulax tamarensis*. Mar Biol 76:179–190

Anderson DM, Keafer BA (1987) An endogenous annual clock in the toxic marine dinoflagellate *Gonyaulax tamarensis*. Nature 325:616–617

Anderson DM, Keafer BA, Kleindinst JL et al (2014a) *Alexandrium fundyense* cysts in the Gulf of Maine: long-term time series of abundance and distribution, and linkages to past and future blooms. Deep Sea Res Part II 103:6–26

Anderson DM, Kulis DM, Binder BJ (1984) Sexuality and cyst formation in the dinoflagellate *Gonyaulax tamarensis*: cyst yield in batch cultures. J Phycol 20:418–425

Anderson DM, McGillicuddy DJ Jr, DeGrasse SL et al (eds) (2014b) Harmful algae in the Gulf of Maine: oceanography, population dynamics, and toxin transfer in the food web. Deep Sea Res Part II 103:1–376

Anderson DM, Morel FMM (1979) The seeding of two red tide blooms by the germination of benthic *Gonyaulax tamarensis* hypnocysts. Estuar Coast Mar Sci 8:279–293

Anderson DM, Stonzenbach KD (1985) Selective retention of two dinoflagellates in a well-mixed estuarine embayment: The importance of diel vertical migration and surface avoidance. Mar Ecol Prog Ser 25:39–50

Anderson DM, Taylor CD, Armbrust EV (1987) The effects of darkness and anaerobiosis on dinoflagellate cyst germination. Limnol Oceanogr 32:340–351

Anderson DM, Townsend DW, McGillicuddy DJ et al (eds) (2005) The ecology and oceanography of toxic *Alexandrium fundyense* blooms in the Gulf of Maine. Deep Sea Res II 52:2365–2876

Anderson DM, Wall D (1978) Potential of benthic cysts of *Gonyaulax tamarensis* and *G. excavata* in initiating toxic dinoflagellate blooms. J Phycol 14:224–234

Anglès S, Garcés E, Hattenrath-Lehmann TK et al (2012) In situ life-cycle stages of *Alexandrium fundyense* during bloom development in Northport Harbor (New York, USA). Harmful Algae 16:20–26

Azanza RV (1997) Contributions to the understanding of the bloom dynamics of *Pyrodinium bahamense* var. *compressum*: a toxic red tide causative organism. Sci Diliman 9:1–6

Azanza RV (2013) Harmful algal blooms in tropical embayments affected by monsoons. In: Roy S, Pospelova V, Montresor M et al (eds) GEOHAB 2013. Global ecology and oceanography of harmful algal blooms, GEOHAB core research project: HABs in fjords and coastal embayments. Second open science meeting. progress in interpreting life history and growth dynamics of harmful algal blooms in fjords and coastal environments. IOC and SCOR, Paris, France and Newark, DE, USA, pp 20–23

Azanza RV, Siringan FP, San Diego-McGlone ML et al (2004) Horizontal dinoflagellate cyst distribution, sediment characteristics and benthic flux in Manila Bay, Philippines. Phycol Res 52:376–386

Azanza RV, Taylor FJRM (2001) Are *Pyrodinium* blooms in the Southeast Asian region recurring and spreading? A view at the end of the millennium. Ambio 30:356–364

Balech E (1985) A rediscription of *Pyrodinium bahamense* Plate (Dinoflagellata). Rev Paleobot Palynol 45:17–34

Bohm A (1931) Peridineen aus dem Persischen Golf und dem Golf von Oman. Arch Protistenk 74:188–197

Bornet E, Flahault C (1888) Revision des Nostacacees heterocystees continues dans les principaux herbiers de France (quatrieme et dernier fragment). Ann Sci Nat Bot 7:177–262

Bravo I, Fraga S, Figueroa RI et al (2010) The intricacies of dinoflagellate pellicle cysts: the example of *Alexandrium minutum* cysts from a bloom-recurrent area (Bay of Baiona, NW Spain). Deep Sea Res II 57:166–174

Brosnahan ML, Farzan S, Keafer BA et al (2014) Complexities of bloom dynamics in the toxic dinoflagellate *Alexandrium fundyense* revealed through DNA measurements by imaging flow cytometry coupled with species-specific rRNA probes. Deep Sea Res II 103:185–198

Brosnahan ML, Velo-Suárez L, Ralston DK et al (2015) Rapid growth and concerted sexual transitions by a bloom of the harmful dinoflagellate *Alexandrium fundyense* (Dinophyceae). Limnol Oceanogr 60:2059–2078

Burford MA, Hamilton DP, Wood SA (2018) Emerging HAB research issues in freshwater environments. In: Glibert PM, Berdalet E, Burford M et al (eds) Global ecology and oceanography of harmful algal blooms. Springer, Cham, pp 381–402

Cayetano AC, Yñiguez AT, Villanoy CL et al (2013) Improving the validity and robustness of a harmful algal bloom model through genetic algorithm-based optimization. J Environ Sci Manage Special Issue 1:21–28

Chepurnov VA, Mann DG, Sabbe K et al (2005) Sexual reproduction, mating system chloroplast dynamics and abrupt cell size reduction in *Pseudo-nitzschia pungens* from the North Sea (Bacillariophyta). Eur J Phycol 40:379–395

Choi CJ, Brosnahan ML, Sehein TR, Anderson DM, Erdner DL (2017) Insights into the loss factors of phytoplankton blooms: The role of cell mortality in the decline of two inshore Alexandrium blooms. Limnol Oceanogr 62:1742–1753

Corrales RA, Crisostomo R (1996) Variation of *Pyrodinium* cyst density in Manila Bay, Philippines. In: Yasumoto T, Oshima Y, Fukuyo Y (eds) Harmful and toxic algal blooms. IOC of UNESCO, Paris, pp 181–184

Corrales RA, Reyes M, Martin M (1995) Notes on the encystment and excystment of *Pyrodinium bahamense* var. *compressum in vitro*. In: Lassus P, Arzul G, Erard-Le Denn E et al (eds) Harmful marine algal blooms. Lavoisier Science Publ., Paris, pp 573–578

Cox PA, Banack SA, Murch SJ et al (2005) Diverse taxa of cyanobacteria produce β-*N*-methylamino-L-alanine, a neurotoxic amino acid. Proc Natl Acad Sci USA 102:5074–5078

Crespo BG, Keafer BA, Ralston DK, Lind H, Dawson F, Anderson DM (2011) Dynamics of Alexandrium fundyense blooms and shellfish toxicity in the Nauset Marsh System of Cape Cod (Massachusetts, USA). Harmful Algae 12:26–38

Czerny J, Barcelos e Ramos J, Riebesell U (2009) Influence of elevated CO_2 concentrations on cell division and nitrogen fixation rates in the bloom-forming cyanobacterium *Nodularia spumigena*. Biogeosciences 6:1865–1875

D'Alelio D, Amato A, Luedeking A et al (2009) Sexual and vegetative phases in the planktonic diatom *Pseudo-nitzschia multistriata*. Harmful Algae 8:225–232

D'Alelio D, Ribera d'Alcalà M, Dubroca L et al (2010) The time for sex: a biennial life cycle in a marine planktonic diatom. Limnol Oceanogr 55:106–114

Doucette GJ, Cembella AD, Boyer GL (1989) Cyst formation in the red tide dinoflagellate *Alexandrium tamarense* (Dinophyceae): effects of iron stress. J Phycol 25:721–731

Eichner M, Rost B, Kranz SA (2014) Diversity of ocean acidification effects on marine N_2 fixers. J Exp Mar Biol Ecol 457:199–207

Fauchot J, Saucier FJ, Levasseur M et al (2008) Wind-driven river plume dynamics and toxic *Alexandrium tamarense* blooms in the St. Lawrence estuary (Canada): a modeling study. Harmful Algae 7:214–227

Fehling J, Davidson K, Bolch C et al (2006) Seasonality of *Pseudo-nitzschia* spp. (Bacillariophyceae) in western Scottish waters. Mar Ecol Prog Ser 323:91–105

Feifel KM, Fletcher SJ, Watson LR et al (2015) *Alexandrium* and *Scrippsiella* cyst viability and cytoplasmic fullness in a 60-cm sediment core from Sequim Bay, WA. Harmful Algae 47:56–65

Franks PSJ, Anderson DM (1992) Alongshore transport of a toxic phytoplankton bloom in a buoyancy current: *Alexandrium tamarense* in the Gulf of Maine. Mar Biol 112:153–164

Fryxell GA, Garza SA, Roelke DL (1991) Auxospore formation in an Antarctic clone of *Nitzschia subcurvata* Hasle. Diatom Res 6:235–245

Furio EF, Azanza RV, Fukuyo Y et al (2012) Review of geographical distribution of dinoflagellate cysts in Southeast Asian coasts. Coastal Mar Sci 35:20–33

Galat DL, Verdin JP, Sims LL (1990) Large-scale patterns of *Nodularia spumigena* blooms in Pyramid Lake, Nevada, determined from Landsat imagery: 1972-1986. Hydrobiologia 197:147–164

GEOHAB (2001) Global ecology and oceanography of harmful algal blooms, science plan. Glibert P, Pitcher G (eds) SCOR and IOC, Baltimore and Paris, 86 pp

Giddings SN, MacCready P, Hickey BM et al (2014) Hindcasts of potential harmful algal bloom transport pathways on the Pacific Northwest coast. J Geophys Res Oceans 119:2439–2461

Glibert PM, Pitcher GC, Bernard S et al (2018) Advancements and continuing challenges of emerging technologies and tools for detecting harmful algal blooms, their antecedent conditions and toxins, and applications in predictive models. In: Glibert PM, Berdalet E, Burford M et al (eds) Global ecology and oceanography of harmful algal blooms. Springer, Cham, pp 339–357

Hallegraeff GM (1998) Transport of toxic dinoflagellates via ships ballast water: bioeconomic risk assessment and efficacy of possible ballast water management strategies. Mar Ecol Prog Ser 168:297–309

Hallegraeff GM (2010) Ocean climate change, phytoplankton community responses, and harmful algal blooms: a formidable predictive challenge. J Phycol 46:220–235

Hallegraeff GM, Bolch CJ (1991) Transport of toxic dinoflagellate cysts via ships' ballast water. Mar Poll Bull 22:27–30

He R, McGillicuddy DJ Jr, Keafer BA et al (2008) Historic 2005 toxic bloom of *Alexandrium fundyense* in the western Gulf of Maine: 2. Coupled biophysical numerical modeling. J Geophys Res Oceans 113(C07040). https://doi.org/10.1029/2007JC004601

Hense I (2007) Regulative feedback mechanisms in cyanobacteria-driven systems: a model study. Mar Ecol Prog Ser 339:41–47

Hense I, Beckmann A (2006) Towards a model of cyanobacteria life cycle – effects of growing and resting stages on bloom formation of N_2-fixing species. Ecol Model 195:205–218

Hense I, Beckmann A (2015) A theoretical investigation of the diatom cell size reduction-restitution cycle. Ecol Model 317:66–82

Hense I, Burchard H (2010) Modelling cyanobacteria in shallow coastal seas. Ecol Model 221:238–244

Hense I, Meier HM, Sonntag S (2013) Projected climate change impact on Baltic Sea cyanobacteria: climate change impact on cyanobacteria. Clim Change 119:391–406

Holtermann KE, Bates SS, Trainer VL et al (2010) Mass sexual reproduction in the toxigenic diatoms *Pseudo-nitzschia australis* and *P. pungens* (Bacillariophyceae) on the Washington coast. J Phycol 46:41–52

Huber AL (1984) *Nodularia* (Cyanobacteriaceae) akinetes in the sediments of the peel-harvey estuary, Western Australia: potential inoculum source for *Nodularia* blooms. Appl Environ Microbiol 47(2):234–238

Ishikawa A, Hattori M, Imai I (2007) Development of the "plankton emergence trap/chamber (PET Chamber)", a new sampling device to collect in situ germinating cells from cysts of microalgae in surface sediments of coastal waters. Harmful Algae 6:301–307

John U, Litaker RW, Montresor M et al (2014) Formal revision of the *Alexandrium tamarense* species complex (Dinophyceae) taxonomy: the introduction of five species with emphasis on molecular-based (rDNA) classification. Protist 165(6):779–804

Kahru M, Elmgren R (2014) Multidecadal time series of satellite-detected accumulations of cyanobacteria in the Baltic Sea. Biogeosciences 11:3619–3633

Karlson AML, Nascimento FJA, Suikkanen S et al (2012) Benthic fauna affects recruitment from sediments of the harmful cyanobacterium *Nodularia spumigena*. Harmful Algae 20:126–131

Kooistra WHCF, Gersonde R, Medlin LK et al (2007) The origin and evolution of the diatoms: their adaptation to a planktonic existence. In: Falkowski PG, Knoll AH (eds) Evolution of primary producers in the sea. Elsevier Academic Press, Burlington, pp 207–250

Kruger T, Monch B, Oppenhauser S et al (2010) LC-MS/MS determination of the isomeric neurotoxins BMAA (β-N-methylamino-L-alanine) and DAB (2, 4-diaminobutyric acid) in cyanobacteria and seeds of *Cycas revoluta* and *Lathyrus latifolius*. Toxicon 55(2):547–557

Lane JQ, Raimondi PT, Kudela RM (2009) Development of a logistic regression model for the prediction of toxigenic *Pseudo-nitzschia* blooms in Monterey Bay, California. Mar Ecol Prog Ser 383:37–51

Lefebvre KA, Robertson A (2010) Domoic acid and human exposure risks: a review. Toxicon 56:218–230

Lefebvre KA, Bargu S, Kieckhefer T, Silver MW (2002) From sanddabs to blue whales: the pervasiveness of domoic acid. Toxicon 40:971–977

Lehtimaki J, Moisander P, Sivonen K et al (1997) Growth, nitrogen fixation, and nodularin production by two Baltic Sea cyanobacteria. Appl Environ Microbiol 63:1647–1656

Lelong A, Hégaret H, Soudant P et al (2012) *Pseudo-nitzschia* (Bacillariophyceae) species domoic acid and amnesic shellfish poisoning: revisiting previous paradigms. Phycologia 51:168–216

Li Y, He R, McGillicuddy DJ Jr et al (2009) Investigation of the 2006 *Alexandrium fundyense* bloom in the Gulf of Maine: In-situ observations and numerical modeling. Cont Shelf Res 29:2069–2082

Lilly EL, Halanych KM, Anderson DM (2007) Species boundaries and global biogeography of the *Alexandrium tamarense* complex (Dinophyceae). J Phycol 43:1329–1338

Lim HC, Teng ST, Leaw CP et al (2013) Three novel species in the *Pseudo-nitzschia pseudodelicatissima* complex: *P. batesiana* sp. nov., *P. lundholmiae* sp. nov., and *P. fukuyoi* sp. nov. (Bacillariophyceae) from the Strait of Malacca, Malaysia. J Phycol 49:902–916

Lundholm N, Clarke A, Ellegaard M (2010) A 100-year record of changing *Pseudo-nitzschia* species in a sill-fjord in Denmark related to nitrogen loading and temperature. Harmful Algae 9:449–457

Margulis L, Sagan D (1985) Origins of sex. Yale University Press, New Haven

Matrai P, Thompson B, Keller M (2005) Circannual excystment of resting cysts of *Alexandrium* spp. from eastern Gulf of Maine populations. Deep Sea Res II 52:2560–2568

McGillicuddy DJ Jr, Anderson DM, Lynch DR et al (2005) Mechanisms regulating large-scale seasonal fluctuations in *Alexandrium fundyense* populations in the Gulf of Maine: Results from a physical-biological model. Deep Sea Res Part II 52:2698–2714

McGillicuddy DJ Jr, Townsend DW, He R et al (2011) Suppression of the 2010 *Alexandrium fundyense* bloom by changes in physical, biological, and chemical properties of the Gulf of Maine. Limnol Oceanogr 56:2411–2426

McGillicuddy DJ Jr, Townsend DW, Keafer BA et al (2014) Georges Bank: a leaky incubator of *Alexandrium fundyense* blooms. Deep Sea Res Part II 103:163–173

McQuoid MR (2005) Influence of salinity on seasonal germination of resting stages and composition of microplankton on the Swedish west coast. Mar Ecol Prog Ser 289:151–163

Mertens KN, Wolny J, Carbonell-Moore C et al (2015) Taxonomic re-examination of the toxic armored dinoflagellate *Pyrodinium bahamense* Plate 1906: can morphology or LSU sequencing separate *P. bahamense* var. *compressum* from var. *bahamense*? Harmful Algae 41:1–24

Moeys S, Frenkel J, Lembke C et al (2016) A sex-inducing pheromone triggers cell cycle arrest and mate attraction in the diatom *Seminavis robusta*. Sci Rep 6:19252

Montresor M, Di Prisco C, Sarno D et al (2013) Diversity and germination patterns of diatom resting stages at a coastal Mediterranean site. Mar Ecol Prog Ser 484:79–95

Montresor M, Vitale L, D'Alelio D et al (2016) Sex in marine planktonic diatoms: insights and challenges. Perspect Phycol. https://doi.org/10.1127/pip/2016/0045

Morquecho L, Alonso-Rodriguez R, Martinez-Tecuapacho GA (2014) Cyst morphology, germination characteristics, and potential toxicity of *Pyrodinium bahamense* in the Gulf of California. Bot Mar 57:303–314

O'Neil JM, Davis TW, Burford MA et al (2012) The rise of harmful cyanobacteria blooms: the potential roles of eutrophication and climate change. Harmful Algae 14:313–334

Olenin S, Gollasch S, Jonusas S et al (2000) En-route investigations of plankton in ballast water on a ship's voyage from the Baltic Sea to the open Atlantic coast of Europe. Int Rev Hydrobiol 85:577–596

Onda DFL, Lluisma AO, Azanza RV (2014) Development, morphological characteristics and viability of temporary cysts of *Pyrodinium bahamense* var. *compressum* (Dinophyceae) in vitro. Eur J Phycol 49(3):265–275

Paerl HW, Huisman J (2009) Climate change: a catalyst for global expansion of harmful cyanobacterial blooms. Environ Microbiol Rep 1:27–37

Paerl HW, Hall NS, Calandrino ES (2011) Controlling harmful cyanobacterial blooms in a world experiencing anthropogenic and climatic-induced change. Sci Total Environ 409:1739–1745

Pfiester LA, Anderson DM (1987) Dinoflagellate life cycles and their environmental control. In: Taylor FJR (ed) The biology of dinoflagellates. Blackwell Scientific Publications, Ltd., Oxford, pp 611–648

Pilskaln CH, Hayashi K, Keafer BA et al (2014) Benthic nepheloid layers in the Gulf of Maine and *Alexandrium* cyst inventories. Deep Sea Res II 103:55–65

Pitcher GC, Figueiras FG, Kudela RM et al (2018) Key questions and recent research advances on harmful algal blooms in eastern boundary upwelling systems. In: Glibert PM, Berdalet E, Burford M et al (eds) Global ecology and oceanography of harmful algal blooms. Springer, Cham, pp 205–227

Plate L (1906) *Pyrodinium bahamense* n.g, n.sp. die. Leucht-Peridinee des "Feuersees" von Nassau, Bahamas. Arch Protistenkd 7:411–429

Quijano-Scheggia S, Garcés E, Andree K et al (2009) Homothallic auxosporulation in *Pseudo-nitzschia brasiliana* (Bacillariophyta). J Phycol 45:100–107

Raine R, Berdalet E, Yamazaki H et al (2018) Key questions and recent research advances on harmful algal blooms in stratified systems. In: Glibert PM, Berdalet E, Burford M et al (eds) Global ecology and oceanography of harmful algal blooms. Springer, Cham, pp 165–186
Ralston DK, Brosnhan ML, Fox SE et al (2015) Temperature and residence time controls on an estuarine harmful algal bloom: modeling hydrodynamics and *Alexandrium fundyense* in Nauset estuary. Estuar Coasts 38:2240–2258
Ralston DK, Keafer BA, Brosnahan ML et al (2014) Temperature dependence of an estuarine harmful algal bloom: resolving interannual variability in bloom dynamics using a degree-day approach. Limnol Oceanogr 59:1112–1126
Rathaille AN, Raine R (2011) Seasonality in the excystment of *Alexandrium minutum* and *Alexandrium tamarense* in Irish coastal waters. Harmful Algae 10(6):629–635
Rengefors K, Karlsson I, Hansson L-A (1998) Algal cyst dormancy: a temporal escape from herbivory. Proc R Soc Lond B 265:1353–1358
Rinehart KL, Harada K, Namikoshi M et al (1988) Nodularin, microcystin, and the configuration of Adda. J Am Chem Soc 110:8557–8558
Rines JEB, Donaghay PL, Dekshenieks MM et al (2002) Thin layers and camouflage: hidden *Pseudo-nitzschia* spp. (Bacillariophyceae) populations in a fjord in the San Juan Islands, Washington, USA. Mar Ecol Prog Ser 225:123–137
Roy S, Montresor M, Cembella A (2018) Key questions and recent research advances on harmful algal blooms in fjords and coastal embayments. In: Glibert PM, Berdalet E, Burford M et al (eds) Global ecology and oceanography of harmful algal blooms. Springer, Cham, pp 187–203
Ruggiero MV, Sarno D, Barra L et al (2015) Diversity and temporal pattern of *Pseudo-nitzschia* species (Bacillariophyceae) through the molecular lens. Harmful Algae 42:15–24
Sarno D, Zingone A, Montresor M (2010) A massive and simultaneous sex event of two *Pseudo-nitzschia* species. Deep Sea Res Part II 57:248–255
Scalco E, Amato A, Ferrante M et al (2015) The sexual phase of the diatom *Pseudo-nitzschia multistriata*: cytological and time-lapse cinematography characterization. Protoplasma:1–11. https://doi.org/10.1007/s00709-015-0891-5
Scalco E, Stec K, Iudicone D et al (2014) The dynamics of sexual phase in the marine diatom *Pseudo-nitzschia multistriata* (Bacillariophyceae). J Phycol 50:817–828
Sekula-Wood E, Schnetzer A, Benitez-Nelson CR et al (2009) Rapid downward transport of the neurotoxin domoic acid in coastal waters. Nat Geosci 2:272–275
Siringan FP, Azanza RV, Macalalad NJH et al (2008) Temporal changes in the cyst densities of *Pyrodinium bahamense* var. *compressum* and other dinoflagellates in Manila Bay, Philippines. Harmful Algae 7:523–531
Sison-Mangus MP, Jiang S, Tran KN et al (2013) Host-specific adaptation governs the interaction of the marine diatom, *Pseudo-nitzschia* and their microbiota. ISME J. https://doi.org/10.1038/ismej.2013.138
Smayda TJ (1997) Harmful algal blooms: their ecophysiology and general relevance to phytoplankton blooms in the sea. Limnol Oceanogr 42:1137–1153
Solow AR, Beet AR, Keafer BA et al (2014) Testing for simple structure in a spatial time series with an application to the distribution of *Alexandrium* resting cysts in the Gulf of Maine. Mar Ecol Prog Ser 501:291–296
Steidinger K, Tester LS, Taylor FJR (1980) A redescription of *Pyrodinium bahamense* var. *compressa* (Bohm) stat. nov. from Pacific red tides. Phycologia 19:329–337
Stock CA, McGillicuddy DJ Jr et al (2005) Evaluating hypotheses for the initiation and development of *Alexandrium fundyense* blooms in the western Gulf of Maine using a coupled physical-biological model. Deep Sea Res Part II 52:2715–2744
Stoyneva MP (2016) Allochtonous planktonic algae recorded during the last 25 years in Bulgaria and their possible dispersal agents. Hydrobiologia 764:53–64
Suikkanen S, Kaartokallio H, Halfors S et al (2010) Life cycle strategies of bloom-forming, filamentous cyanobacteria in the Baltic Sea. Deep Sea Res Part II 57(3):199–209
Sullivan JM, Swift E, Donaghay PL et al (2003) Small-scale turbulence affects the division rate and morphology of two red-tide dinoflagellates. Harmful Algae 2:183–199

Teng ST, Lim HC, Lim PT et al (2014) *Pseudo-nitzschia kodamae* sp. nov. (Bacillariophyceae), a toxigenic species from the Strait of Malacca, Malaysia. Harmful Algae 34:17–28

Timmerman AHV, McManus MA, Cheriton OM et al (2014) Hidden thin layers of toxic diatoms in a coastal bay. Deep Sea Res Part II 101:129–140

Toth GB, Noren N, Selander E et al (2004) Marine dinoflagellates show induced life-history shifts to escape parasite infection in response to water-borne signals. Proc R Soc B Biol Sci 271:733–738

Townsend DW, McGillicuddy DJ Jr, Thomaza MA et al (2014) Nutrients and water masses in the Gulf of Maine-Georges Bank region: variability and importance to blooms of the toxic dinoflagellate *Alexandrium fundyense*. Deep Sea Res Part II 103:238–263

Trainer VL, Bates SS, Lundholm N et al (2012) *Pseudo-nitzschia* physiological ecology, phylogeny, toxicity, monitoring and impacts on ecosystem health. Harmful Algae 14:271–300

Vahtera E, Crespo BG, McGillicuddy DJ Jr et al (2014) *Alexandrium fundyense* cyst viability and germling survival in light vs. dark at a constant low temperature. Deep Sea Res Part II 103:112–119

Velo-Suárez L, Brosnahan ML, Anderson DM et al (2013) A quantitative assessment of the role of the parasite *Amoebophrya* in the termination of *Alexandrium fundyense* blooms within a small coastal embayment. PLoS One 8(12):e81150. https://doi.org/10.1371/journal.pone.0081150

Villanoy CL, Azanza RV, Altemerano A et al (2006) Attempts to model the bloom of *Pyrodinium*, a tropical toxic dinoflagellate. Harmful Algae 5:156–183

Villanoy CL, Corrales RA, Jacinto GS et al (1996) Towards the development of a cyst-based model for *Pyrodinium* red tides in Manila Bay, Philippines. In: Yasumoto T, Oshima Y, Fukuyo Y (eds) Harmful and toxic algal blooms. IOC-UNESCO, Paris, pp 189–192

Wall D, Dale B (1969) The hystrichospaerid resting spore of the dinoflagellate *Pyrodinium bahamense* Plate 1906. J Phycol 5:140–149

Yñiguez AT, Cayetano A, Villanoy CL et al (2012) Investigating the roles of intrinsic and extrinsic factors in the blooms of *Pyrodinium bahamense* var. *compressum* using an individual-based model. Proc Environ Sci 13:1462–1476

Part IV
Harmful Algal Blooms in Specific Habitats and Biomes

Chapter 9
Key Questions and Recent Research Advances on Harmful Algal Blooms in Stratified Systems

Robin Raine, Elisa Berdalet, Hidekatsu Yamazaki, Ian Jenkinson, and Beatriz Reguera

9.1 Introduction

The GEOHAB Core Research Project (CRP) on HABs in Stratified Systems was a crosscutting project focused on the relevance of physical structure to fundamental life processes of harmful microorganisms such as growth (together with nutrient and light availability), reproduction, life cycle, and ecological interactions. Advances in this area have heavily depended on the development of innovative instruments to observe and adequately sample these environments, as well as on improvements in numerical modelling techniques (GEOHAB 2011, 2013; Berdalet et al. 2014).

The implementation of this CRP was marked by three main events. First, an Open Science Meeting (OSM) held in Paris, on 5–8 December 2005, launched the project. Meeting participants identified research priorities for the study of HABs on stratified systems and the appropriate approaches and advanced instrumentation that were available at that time (GEOHAB 2008). Next, in June 2009, participants in a GEOHAB modelling workshop held in Galway, Ireland, analysed advances in modelling tools to implement the CRP objectives (GEOHAB 2011), and a group of

R. Raine (✉)
Earth and Ocean Sciences, National University of Ireland, Galway, Ireland
e-mail: robin.raine@nuigalway.ie

E. Berdalet
Institute of Marine Sciences (CSIC), Barcelona, Catalonia, Spain

H. Yamazaki
Tokyo University of Marine Science and Technology, Tokyo, Japan

I. Jenkinson
ACRO, La Roche-Canillac, France and Institute of Oceanology, Qingdao, China

B. Reguera
Instituto Español de Oceanografía, Centro Oceanográfico de Vigo, Vigo, Spain

P.M. Glibert et al. (eds.), *Global Ecology and Oceanography of Harmful Algal Blooms*, Ecological Studies 232, https://doi.org/10.1007/978-3-319-70069-4_9

papers showing recent progress was published in a special issue of the *Journal of Marine Systems* (McGillicuddy 2010). At the sunset of the GEOHAB Programme, a workshop was held in August 2012 at the Monterey Bay Aquarium Research Institute (MBARI) to review the advances in understanding of HABs in stratified systems and to identify extant gaps in knowledge to orient future research towards improved modelling and prediction (GEOHAB 2013; Berdalet et al. 2014). A final product of the CRP was a special issue in *Deep-Sea Research Part II* (Raine et al. 2014) that included studies on HAB events in a variety of stratified systems addressed with experimental, technical, and modelling tools and covering different scales from the cellular, small-scale to the mesoscale. A recent review of the outcomes and challenges at the end of the GEOHAB Programme concerning HABs in Stratified Systems can also be found in Berdalet et al. (2017a).

Several key terms are fundamental to this CRP.

Stratification The term "stratification" means "layering" of the water column, and it can occur in a variety of ecosystems. Fjords, open and semi-enclosed coastal zones, shelf seas, and upwelling areas exhibit varying levels and types of stratification with impacts on key phytoplankton drivers such as nutrient and irradiance conditions. Stratified water columns have a vertical structure typically composed of a wind-mixed "surface mixed layer" (SML) and a tidally mixed bottom layer (BML), separated by a pycnocline. The persistence and strength of stratification often appears as one of the most important conditions for phytoplankton development, but the particular mechanisms favouring the selection of noxious taxa as different as, for example, *Karenia* and *Pseudo-nitzschia* in these environments are not known. In addition, advances in technology have shown that the SML is very often not a homogeneous layer of uniform density. In fact, the formation of small density discontinuities is found which supports the existence of subsurface layers of plankton. This is unlike the well-mixed SML applied by Sverdrup (1953) to predict the onset of the North Atlantic spring bloom who used a simple conservation mass model. In addition, the pycnocline of stratified coastal water is often a stepwise series of density discontinuities where accumulations of phytoplankton occur. These subsurface, often thin, layers of plankton are common in many different ecosystems and have relevance in the promotion of HABs (Gentien et al. 2005; GEOHAB 2008).

Thin Layers Subsurface thin layers (TLs) of harmful algae were an important research focus within the stratified systems CRP. Their length scales are in the order of kilometres in the horizontal and from several metres (<5 m) down to millimetres in the vertical dimension (McManus et al. 2003). They have distinct physical, chemical, and biological signatures from the surrounding water within the vertical structure of the SML of a stratified water column (Fig. 9.1) and can persist for periods of days or longer. The organisms in these layers, including many HAB species, seem to respond to a physicochemical structure that favours their proliferation within the layer (Klausmeier and Litchman 2001). Consequences of their existence include many challenges for their detection, monitoring, and observation of such species (Escalera et al. 2012).

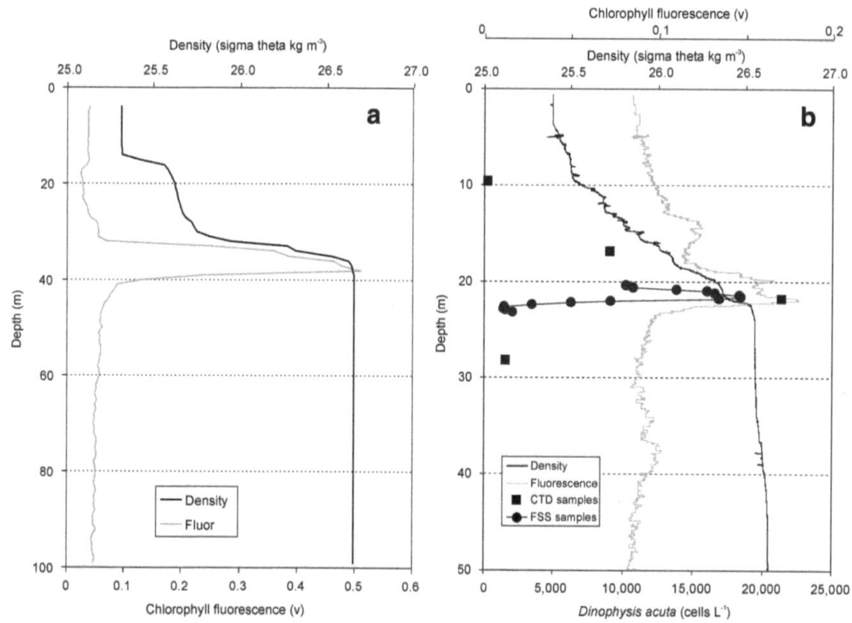

Fig. 9.1 (**a**, *left panel*) A typical profile of a subsurface thin layer of phytoplankton, as denoted by chlorophyll fluorescence located at the seasonal pycnocline in a stratified shelf sea. Note the number of vertical density discontinuities (at least three are visible) in the so-called surface mixed layer, in this instance extending down to approximately 37 m depth. (**b**, *right panel*) A thin layer of *Dinophysis acuta* sampled using the fine-scale sampler (FSS) off the south coast of Ireland in July 2014. A very high density of cells is visible at 21.5 m depth. Samples taken with a CTD rosette system just prior to the FSS deployment are also shown

Here we summarize progress in understanding the physical, chemical, biological, and ecological processes that interact over a large range of scales and govern the structure and dynamics of HABs in stratified systems, in particular in TLs, which has been made during the GEOHAB lifetime. The chapter is structured following the key questions that were addressed at the first OSM in Paris (GEOHAB 2008) to understand and predict HABs in stratified systems. We also identify scientific questions to be pursued within the next 10 years of research and that may benefit from further international cooperation.

9.2 Key Questions 1: What Are the Turbulence Length Scales Relevant to Harmful Phytoplankton and the Formation of Thin Layers? How Can These Be Measured?

The dynamics of planktonic communities are markedly driven by the physical properties of the aquatic ecosystems. Physical forcing shapes the structure of the pelagic environment and modulates, directly and indirectly, many biological processes (Mann and Lazier 1991; Kiørboe 1997). At the scale of a few to tens of metres advective and turbulent flows transport organisms around the water column determining, for instance, the inorganic nutrient and irradiance regime to the autotrophic organisms. Further, development and maintenance of plankton communities clearly depend on circulation patterns, which transport cells, influence sedimentation and resuspension, and determine accumulation. The familiar patchiness of passive materials in the ocean is the result of turbulent diffusion, a direct consequence of turbulent motion. In addition, at small cell scale, the transport of molecules in and out of the cells (e.g., Karp-Boss et al. 1996), the swimming behaviour of the organisms (e.g., Kamykowski et al. 1998 and references therein), or other physiological aspects (Estrada and Berdalet 1998) may also be directly affected by small-scale turbulence and shear.

Clearly, improvements in the characterization of turbulence, both at theoretical and natural levels and especially at the scales of biological relevance, have facilitated a better understanding of the physical–biological interactions that modulate phytoplankton dynamics, including that of HABs.

Turbulence is a random, three-dimensional motion, with the velocity and vorticity irregularly distributed in time and space. Turbulence is characterized by an energy transfer from large to small scales where dissipation of kinetic energy is taking place. Turbulent diffusion is the net advection due to correlation between the velocity and concentration fluctuations.

The complex nature of turbulent flow was successfully characterized by the inertial subrange theory of Kolmogorov (1941). Essentially, kinetic energy is imparted to the world ocean by forcing, such as heat exchanges, wind, and tides, at various spatiotemporal scales. The energy of large eddies is transmitted progressively to smaller and smaller scales, through the energy cascade, whose velocity power spectrum follows a $k^{-5/3}$ universal shape, being k the length scale of the various eddies, until the motion is resisted by the molecular viscosity of the water. The Kolmogorov theory predicts a universal spectral slope in the wave number range:

$$L_0^{-1} \ll k \ll \eta^{-1}$$

where L_0 is the spatial scale of external forcing and η the Kolmogorov length scale. Through the energy cascade, as long as the eddies are large, no energy is dissipated. Eventually, when the eddies reach a small enough size, the Kolmogorov length scale, molecular viscosity becomes important and acts to resist and damp the gradients in flow velocity, and the turbulent energy is finally converted to heat and dissipated. The amount of energy transferred down the cascade is equal to the energy dissipated by viscosity, which can be expressed by the rate of turbulent energy dissipation, epsilon, ε, [$L^2\,T^{-3}$], (W kg^{-1} or $m^2\,s^{-3}$). ε is the most important observable parameter describing the dynamics of turbulence. When turbulence is in an isotropic condition, acting in every direction equally, ε can be estimated from a cross-stream shear component, such as that:

$$\varepsilon = 7.5\nu\left\langle \left(\frac{\partial u}{\partial z}\right)^2 \right\rangle$$

where z is the vertical coordinate, u is a horizontal turbulence velocity component, and ν is the viscosity of the fluid. Accordingly, the Kolmogorov length scale is defined from a simple dimensional argument, determined from a combination of the kinetic energy dissipation rate and viscosity:

$$\eta = \left(\nu^3/\varepsilon\right)^{1/4}$$

Concerning TLs, observations indicate moderate turbulent levels associated with many layers (McManus et al. 2003). In consequence, the onset, maintenance, and decay of TLs would depend on the balance between the diffusive effects of turbulence and convergence mechanisms (Fig. 9.2). This has been described in the conceptual framework of Stacey et al. (2007). While turbulent (or molecular) diffusion acts to broaden layers, the combination of the different convergence mechanisms favours a persistent and stable layer structure. In this framework, layer thickness is characterized as a function of the local conditions of the diffusive and convergence parameters. Theoretical layer thickness values were compared with observations of TL thickness in Eastport, Washington, USA (Dekshenieks et al. 2001; Alldredge et al. 2002; Rines et al. 2002), concluding that either the

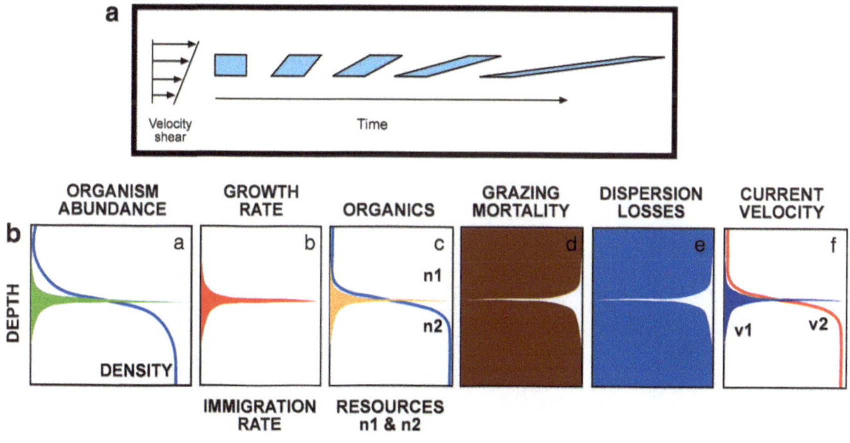

Fig. 9.2 (**a**, *upper panel*) Thin layer formation from velocity shear. Modified from Stacey et al. (2007). (**b**, *bottom panel*) General conceptual model of mechanisms controlling thin layers. Modified from Donaghay and Osborn (1997). Organism abundance increases in TLs are due to biological processes such as in situ growth rate, favoured by higher nutrient (inorganic or organic) resources, cell immigration, and dispersion losses, while mortality by grazing can be increased or not. Concurrent convergence physical dynamics will favour the formation of the TL structure as well

buoyancy or straining mechanism could contribute to the maintenance of the observed phytoplankton layers. Subsequent studies however noted that in environments affected by internal waves, their dynamics should also be taken into consideration in order to describe vertical oscillations of TLs, in addition to the aforementioned convergence and diffusion mechanisms (Cheriton et al. 2009).

In these pioneer studies, shear, stratification, and turbulent kinetic energy dissipation rate were estimated from water velocity measurements using a 1200 kHz acoustic Doppler current profiler (ADCP) attached to the side of the vessel. Subsequent progress in technology has allowed estimates of these parameters with shear probes mounted on free-fall microstructure turbulence profilers, such as the Self-Contained Autonomous Microstructure Profiler or SCAMP (Precision Measurement Engineering) that measures temperature, salinity, and pressure at 100 Hz, the microstructure profiler MSS (ISW Wassermesstechnik), the coastal vertical profiler MVP (Rockland Scientific), and the free-falling profiler TurboMAP and TurboMAP-G (JFE Advantech, Foloni-Neto et al. 2014; Fig. 9.3). Epsilon values estimated in natural conditions, in the context of stratification and TLs, were in the range of 10^{-7}–10^{-8} W Kg^{-1} or lower from use of a range of instrumentation. In Eastsound, the water column was moderately to weakly turbulent where TLs were observed. High ε values (10^{-7} m^2 s^{-3}) were measured in the upper part of the water column concurrently with the greatest shear and much lower in the pycnocline where maximum chlorophyll values were found (McManus et al. 2003). In a 24-h experiment in Lake Biwa, Japan using a TurboMAP-L deployed at 15-min intervals (Fig. 9.4), ε values were in the range of 10^{-8} W Kg^{-1} within a TL (at the local peak in chlorophyll) to 10^{-6} W kg^{-1} in the adjacent mixed layers (Yamazaki et al. 2010). Concurrent measurements of biological parameters at fine scale, including TLs, have also been made possible through the fine-scale sampler (FSS, Lunven et al. 2005) capable of taking 15 samples, 20 cm apart, over a vertical

Fig. 9.3 A quasi-horizontal turbulence profiler TurboMAP-Glider, courtesy of H. Yamazaki

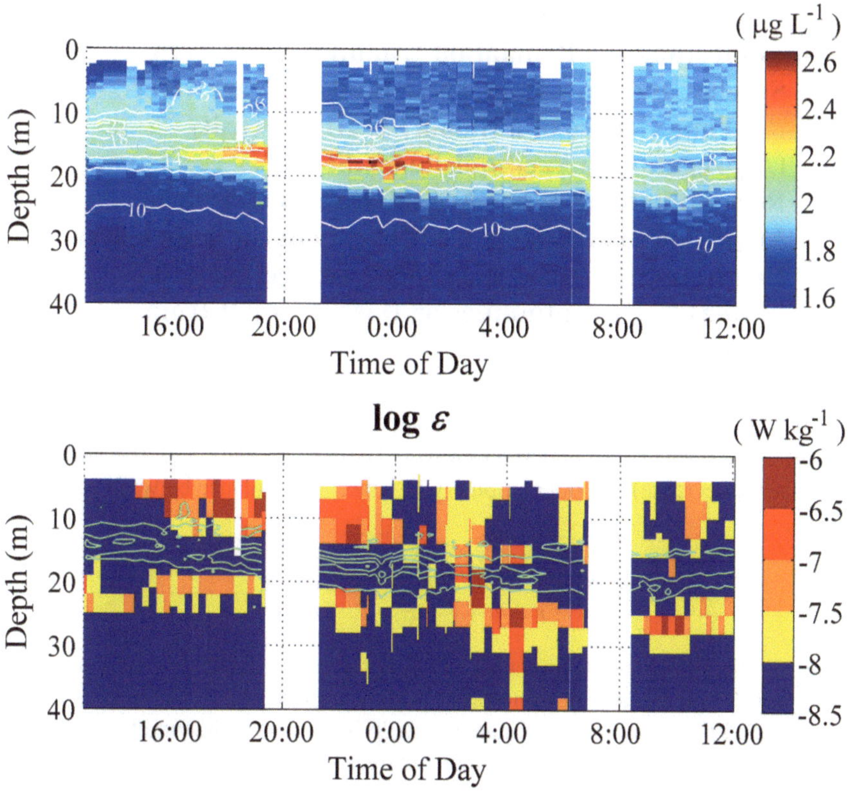

Fig. 9.4 24-h experiment using a TurboMAP-L deployed at 15-min intervals in order to study small-scale biophysical structures at a station in Lake Biwa (Japan) (Yamazaki et al. 2010). *Upper figure*: chlorophyll distribution (μg L^{-1}) and temperature structure (contoured in °C) measured over the survey period. The surface water temperature remained at ~27 °C during the experiment, although slight warming/cooling took place during day/night. The surface layer was well mixed down to approximately 12 m in depth (not shown). The temperature below the surface mixed layer dropped sharply to <10 °C at a depth <30 m. A pronounced chlorophyll peak layer of about 1 m thickness appeared in the strongly stratified water region of the water column where the temperature dropped from 18 to 14 °C. *Bottom figure*: kinetic energy dissipation rate (right scale, W kg^{-1}) depicted with the contours of chlorophyll concentration (1.6–2.6 μg L^{-1} range) corresponding to the figure above. The local peak in chlorophyll clearly appeared in the low dissipation rate region where ε did not exceed 10^{-8} W Kg^{-1}. In this situation, a swimming velocity scale of a cell can be as small as 10^{-8} m s^{-1} (0.01 μm s^{-1}) to maintain the layer thickness against turbulent diffusion. Swimming scales of phytoplankton can exceed 100 μm s^{-1} and thus the behavioural aspect becomes important for the observed layer thickness

distance of 3 m and the high-resolution IFREMER particle size analyser profiler (IPSAP; Lunven et al. 2012; Fig. 9.1) that combines particle size analysis with microphotographic video, tow-body systems such as the Scanfish (undulating CTD) and Acrobat (Sea Sciences), and autonomous underwater vehicles like the MBARI Dorado (e.g., Ryan et al. 2010).

9.3 Key Questions 2: What Are the Main Processes Controlling the Population Evolution of a Given Species, and How Does Their Ranking Vary Over the Time Course of a Bloom? Is High Density (Aggregation) a Prerequisite for Sexual Recombination? What Are the Gains of Vertical Migration? Is the High Cell Density in the Thin Layer Due to In Situ Division and Passive Accumulation in Density Discontinuities or Does It Result from Aggregation?

The best way to address this question is a population dynamics approach, applied to the different stages of a population growth. The net growth r, i.e., the numerical increase (or decrease) of a population P with time, is the balance between gains and losses:

$$dP/dt = rP$$
$$r = \mu + I - E - g - m$$

where μ is the intrinsic division rate, I and E represent physically driven imports and exports, g is grazing, and m is mortality (sedimentation, cell lysis, parasitism, and infections). The capability to estimate in situ division rates of the target species may be crucial to determine whether a sudden increase of a population has been due to advective processes, to high division rates, or to both. Challenges and opportunities to estimate cell abundance and in situ growth rates with high spatial resolution are reviewed in Berdalet et al. (2014). New possibilities are offered by underwater optical imaging systems (Jaffe et al. 2013), the in situ Video Fluorescence Analysis instrument, which can resolve fluorescent particles ranging from 6 μm to several millimetres in a 3500 μm deep measurement chamber (Lunven et al. 2012), or the LISST-HOLO (Sequoia Scientific), a holographic system that provides detailed images of 25–2500 micron plankton (Ryan et al. 2008).

These synoptic methods could be adapted or combined with traditional tools to provide species-specific estimations of the cell numbers of targeted HAB species. This is generally not a problem in high-biomass harmful events such as those caused by *Pseudo-nitzschia* spp., *Noctiluca* spp., and *Phaeocystis* spp. These techniques could well provide solutions to the elucidation of both cryptic blooms, a term adopted to denote the appearance of poisoning at higher trophic levels with no prior evidence of a HAB (McManus et al. 2008), as well as harmful organisms present at very low concentrations such as *Dinophysis* spp., *Alexandrium* spp., and *Gymnodinium catenatum*, which cause shellfish harvesting closures when present at 10^2–10^3 cells L^{-1}.

Gathering of in situ observations of physiological processes in phytoplankton is still well behind that of physical measurements. The need to move the microscope to the sea has encouraged development of both digital image recognition software and in situ imagery systems. The imaging flow cytometer FlowCytoBot has shown very encouraging results. For example, Campbell et al. (2010) show its use applied to a *Dinophysis* bloom in the Gulf of Mexico. Further developments in high-resolution imaging instruments such as the uncabled Imaging FlowCytoBot (IFCB, Brosnahan et al. 2015), now able to simultaneously image and profile, will further improve our observation capabilities [see also Chap. 18, Glibert et al. (2018)].

The questions formulated in this section will be addressed in turn along the main phases of a bloom, initiation, maintenance, and decline, with particular reference to TLs.

9.3.1 Bloom Initiation

Some harmful dinoflagellates are cyst-forming species, where excystment after a period of dormancy (overwinter) initiates blooms the following spring. In *Alexandrium*, seasonal studies have shown that high excystment rates are found in early spring indicating an environmental or biological control on this process. An endogenous clock or secondary dormancy patterns have been suggested as controls (Matrai et al. 2005; Ní Rathaille and Raine 2011), although the concept of degree-day is currently under investigation (Fischer 2017). The degree-day approach has also been used as an early warning indicator for *Alexandrium* blooms in localized areas in the Gulf of Maine (Ralston et al. 2014). High-resolution observations with the IFCB instrument revealed that division and life cycle transition rates in these blooms were much higher than those expected from previous laboratory experiments (Brosnahan et al. 2015).

Holoplanktonic species overwinter in low numbers (<100 cells L^{-1}) and sea-water warming and thermal stratification are crucial for blooms to develop. This appears to be the case for blooms of *Karenia mikimotoi* along the western European shelf. A number of low winter concentrations (5–10 cells L^{-1}) have been observed in the Bay of Brest, France, and it has been hypothesized that the population originates from southern Brittany (Gentien 1998). Blooms in Norwegian waters in 1988 and off the west of Scotland in June 2006 are both thought to have been seeded by blooms during the previous year, with the remnants of the original events overwintering on the shelf (Davidson et al. 2009). Individual-based modelling has also shown the necessity of an overwintering population on the shelf, coupled with advection to simulate a subsequent bloom of *K. mikimotoi* off Scotland in 2006.

In a somewhat similar fashion, advection of overwintered populations may also be the precursor to blooms of *Dinophysis acuta*. Summer stratification of shelf waters around Ireland promotes the formation of tidal fronts at boundaries of stratified and tidally mixed waters. These fronts are rich in resources, including the potential ciliate preys which trigger *D. acuta* division rate. Stratified water adjacent to these fronts have been identified by Raine et al. (2017) as the locations of large ($<10^4$ cells L^{-1}) *D. acuta* populations in the Celtic Sea in midsummer, which were not present elsewhere in the tidally mixed water. Only low numbers (<200 cells L^{-1}) were present in the extensive stratified area of the Celtic Sea at this time. The hypothesis is that low-density populations of *D. acuta* advect onto the shelf from the shelf break, or even further offshore, and once on the shelf are eventually transported into fertile regions where they can bloom.

Upwelling fronts are vertical fronts. Off northwestern Spain, moderate upwelling pulses sharpen surface discontinuities adjacent to thermally stratified water. This triggers the development of high-density populations of diatoms, including toxigenic *Pseudo-nitzschia* species, followed by those of *Dinophysis acuminata*. Upwelling circulation has been proposed to bring the inoculum populations of *Dinophysis* into the Galician Rías [Velo-Suárez et al. 2014; see also Chap. 11, Pitcher et al. (2018)].

In these examples, stratification is key to providing conditions for growth, whether this is from a light-nutrient-mixing basis or simply food for mixotrophs. Stratification is also key to transporting developed blooms into aquaculture sites where they can do harm. For example, fast but narrow baroclinic coastal currents result from density gradients caused by the increased effect of tidal mixing where stratified water comes towards the coastline (Hill et al. 2008). These are known to transport harmful populations (Farrell et al. 2012). Stratification is also a requirement for wind-forced exchanges bringing populations into the bays of southwest Ireland or the upwelling-downwelling pulses bringing populations into the Galician Rías.

9.3.2 Bloom Maintenance

Once the population builds up, a subsurface chlorophyll maximum may become a TL. The possible mechanisms contributing to it have been elegantly synthesized in the conceptual model by Donaghay and Osborn (1997) (Fig. 9.2). Biological processes favouring accumulation of cells in the TLs include in situ growth and immigration rates, increased nutrient availability, and decreased dispersion losses, while grazing could be either facilitated or suppressed in TLs (e.g., Greer et al. 2013). These biological processes would be modulated by physical interactions as described, for instance, by Stacey et al. (2007; see above Question 1).

Yamazaki et al. (2014) developed a numerical model to simulate TL formation due to a phototaxis-driven diel vertical migration (DVM). By embedding the model in a one-dimensional turbulence model, they showed that TL formation took place when wind speeds were below 5 m s^{-1}. Instrumental in the TL formation was dinoflagellate behaviour avoiding photoinhibition from excessive light. These results are consistent with the original formulation of Lasker (1978) that phytoplankton aggregation, and thus good feeding and successful recruitment of anchovy larvae, became possible if wind speeds remained below 4 m s^{-1} over 4 consecutive days.

Steinbuck et al. (2009) investigated the dynamics of a TL of *Akashiwo sanguinea* in Monterey Bay, California, situated in the thermocline of a 20.5 m water column, between a strongly turbulent SML and a weakly turbulent stratified interior. The differential mixing in combination with directed swimming established an asymmetric layer with a stronger chlorophyll *a* gradient at the base of the layer than at the top. These authors applied a Eulerian advection-diffusion model and a Lagrangian particle-tracking model to explain the mid-column convergence needed to maintain the TL and concluded that it was most likely swimming (by motile *A. sanguinea*) that balanced the diffusion by turbulent mixing on both sides of the layer and thereby maintained the observed chlorophyll *a* gradients, as opposed to sinking or growth processes. Ross and Sharples (2008) also examined the use of motility for creating and maintaining a layer of high cell density in a tidally energetic shelf sea environment and concluded that the ability to swim was essential in stratified environments to maintain high cell densities in the thermocline, providing motile cells with a direct competitive advantage.

Swimming behaviour would also explain the aggregation into high-density layers of *D. acuminata*, always associated with the daily subsurface thermocline during very hot days throughout a 17-d cruise in Ría de Pontevedra, Spain. The cell maxima was preceded by advection of cells from shelf waters during downwelling. In addition, in situ estimates of μ showed an actively dividing population (Velo-Suárez et al. 2008). During the same cruise, coupling between maximum values of shear and buoyancy frequency shaped a subsurface chlorophyll maximum (SCM) dominated by a mature population of *Pseudo-nitzschia*, mixed with *Chaetoceros socialis*, into a TL (Velo-Suárez et al. 2010). A later study carried out in the same location in May–June 2007 (Díaz et al. 2014) noted the relevance of considering the role of short-term variability, as the magnitude of the spring-neap tidal and circadian variability was observed to modulate the formation and dissipation of TL of *Pseudo-nitzschia* populations (Fig. 9.5). Interestingly, diatoms of the genus *Pseudo-nitzschia*, very common TL organisms in upwellling systems (Trainer et al. 2012), are not flagellated swimmers, but far from being nonmotile, they form long colonies showing a gliding movement when observed under the microscope. Probably this kind of motion confers them an adaptive advantage to aggregate in marked vertical density gradients in dynamic upwelling systems.

TLs of active swimmers do not always coincide with the maximum chlorophyll layer, the layer usually associated with the pycnocline region. During a summer cruise in southwestern Ireland, a 10×7 km patch of a *D. acuta* population showing a good division rate ($\mu = 0.55$ day^{-1}) formed a 5 m thick TL which did not coincide with the chlorophyll maximum. The layer moved with a similar speed and direction as the coastal flow, which ran parallel to the coast in the form of a coastal jet with speed of the order of 6.5–7 km day^{-1} (Farrell et al. 2012).

Moving to a new group of microalgae, a high density TL of a non-identified haptophyte developed under strong thermohaline stratification in the mouth of the Loire River (Bay of Biscay, France) during the exceptional hot summer of 2006. Measurements of the vertical distribution of horizontal shear showed that the population, located at ca. 20 m depth, was confined to a layer lying between two

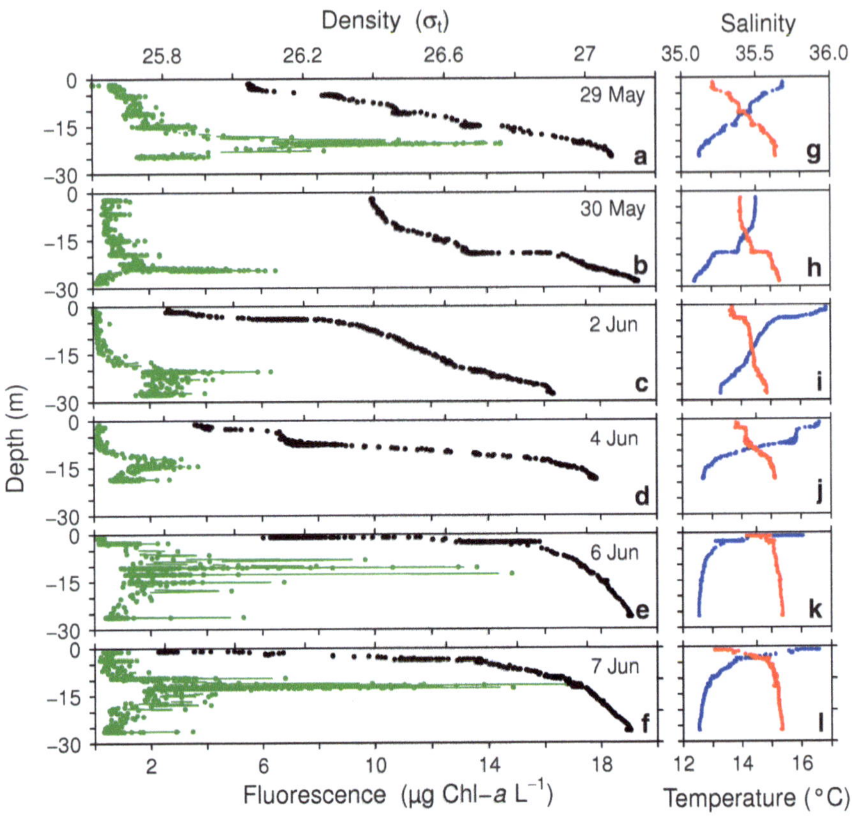

Fig. 9.5 Vertical profiles of fluorescence at a station in Ría de Pontevedra on 29th of May showing a TL of phytoplankton (14.5 µg *chl a* equiv. L^{-1}) (**a**), dominated by *Pseudo-nitzschia seriata* group (2×10^6 cells L^{-1}) between 18 and 19 m depth in the 27 s_t isopycnal under downwelling conditions (**g**). Southerlies caused a vertical displacement of a weakened (6.5 µg *chl a* equiv. L^{-1}) TL to 25 m depth on 30th of May (**b, h**) and its further erosion from 2nd to the 4th of June (**c, i** and **d, j**). A new TL was found in the upper water column under a new upwelling pulse during the second half of the cruise (17.2 µg *chl a* equiv. L^{-1} at 11–12 m) on 7th of June and again associated with the 27 s_t isopycnal (**e, k** and **f, l**) and dominated by the *Pseudo-nitzschia* spp. *Colour codes*: density (*black*), *chl a* fluorescence (*green*), temperature (*red*), and salinity (*blue*)

density discontinuities with high shear. Horizontally, the TL was well constrained by the strong estuarine front (Farrell et al. 2014).

All the above examples have shown the interaction of physical (upwelling, stratification, advection) and biological (swimming, high division rates) processes to form the TL. But there are also biological interactions that may contribute to the high densities reached (see Question 4). In general, we could expect "active" behaviour-driven TLs in the case of swimming dinoflagellates (phototactism, mating promoting aggregation, preying). Concerning TL of diatoms, they are usually located at the pycnocline, where neutral buoyancy is reached (see Question 1).

9.3.3 *Bloom Decline and TL Erosion*

Physical dispersion is a key factor contributing to the collapse of TLs (as, e.g., described by Stacey et al. 2007). In TLs of *Pseudo-nitzschia* in the Galician Rías, downwelling was the driving force eroding the TL of a senescent population (Velo-Suárez et al. 2010). These eroded populations, if not in their last physiological stages, can be reintroduced in the next upwelling pulse to form new TLs (Díaz et al. 2014). Natural life processes of plankton, such as cell encystment, and biological interactions, such as parasitism and bacterial lysis, may also have a relevant contribution to the bloom collapse. There is increasing information on these processes from laboratory experiments, but in situ observations are scarce due to the difficulty of quantifying live specimens in the water column. Indirect information has been obtained staining the target species with different fluorochromes (DAPI, SYBR Green) (González-Gil et al. 2010) or fluorescent probes to estimate percentages of viable cells and other physiological features, as well as parasites prevalence. Nevertheless, there are no demonstrated cases of bloom decay controlled by parasites.

9.4 Key Questions 3: How Can We Quantify Modifications in Turbulence by Phytoplankton Through Changes in the Viscosity of Its Physical Environment? What Are the Effects of a More Viscous and Localized Layer on the Turbulent Spectra? To What Extent Do Such Local Changes Contribute to the Maintenance of a Thin Layer? Can the Changes in Vertical Diffusivity Be Inferred?

These questions arise from the recognition of the importance of the "niche engineering" capacity of organisms in both terrestrial and aquatic environments. Phytoplankton can modify the environment due to their size and shape properties and colony formation capacity, interacting with buoyancy dynamics. Furthermore, by releasing dissolved organic matter (DOM) and producing particulate organic matter (POM) and colloids (exopolymeric substances, EPS), they can modify water viscosity and influence the rheological properties of the water. Recent developments in nano- and microfluidics, including "lab-on-a-chip," have shown that surfaces in fluids exert influence from nanometre to millimetre scale into the fluid (Rothstein 2010; Wong et al. 2011; Jenkinson 2014). This concerns pico- to microplankton, including the sculptured surfaces of diatoms, the microfibrils around pro- and eukaryotes, and the micrometre-scale feeding appendages of copepods and other zooplankton. Many dinoflagellates, a group that includes a relatively high number of HAB-producing species, usually produce high amounts of mucus during their blooms (e.g., Smayda 2002 and references therein). This could constitute a niche engineering strategy facilitating, among others, increased viscosity to damp small-scale turbulence, which has been shown to have certain direct damage or interference of physiological processes (e.g., Estrada and Berdalet 1998; Berdalet et al. 2011). Along with increased swimming speed capacity, mucus and toxin production could contribute to the blooming capacity of dinoflagellates overwhelming other species (Smayda 2002).

Useful summaries of the state of the art on the measurements of the rheological properties of seawater at that time have been provided in GEOHAB (2008, 2011). Briefly, viscosity, ν, of seawater and freshwater is composed of a Newtonian aquatic component ν_W due to water (and salts) plus a non-Newtonian organic component, ν_E , which is due mainly to EPS and produced mainly from phytoplankton and bacteria. The viscosity of the water, ν_W, is independent of shear rate and depends only on the temperature and salinity, and in most natural waters (temperatures from 0 to 30 °C, salinities from 0 to 40 psu) it varies by no more than a factor of 2 from 1 mPa s (Miyake and Koizumi 1948). On the other hand, ν_E varies with shear rate, it is mostly shear thinning, and it probably reflects the molecular structure of the phytoplankton-secreted EPS.

Measurements of rheology properties have been conducted on phytoplankton-poor Mediterranean seawater and on seawater from the North Sea during blooms of *Phaeocystis* and *Noctiluca* (Jenkinson and Biddanda 1995) and *Phaeocystis* blooms in Antarctica (Seuront et al. 2006, 2007; Kesaulya et al. 2008). In general, chlorophyll concentration was related to ν_E, with certain variations probably attributed to the physiological state of the colonies and the presence of heterotrophic bacteria (which would in turn affect the levels of EPS in the medium).

Based on published measurements of the rheological properties of *K. mikimotoi* cultures, a model has been published of how phytoplankton EPS may change the dynamics and the thickness of pycnoclines (Jenkinson and Sun 2011). The model predicts that the degree of modification will be extremely dependent on the relationship between excess viscosity and length scale, which will itself be a property of EPS lumpiness.

Using the zero-dimensional Kolmogorov model of turbulence, Jenkinson (1986) suggested how the biorheological modification of natural waters by phytoplankton might change the characteristics of turbulence, in particular increasing the length and time scales of Kolmogorov eddies for any given value of ε. Unfortunately, the experiments made so far were not conclusive, and the hypothesis that phytoplankton EPS would change characteristics in isotropic ocean turbulence remains open (Jenkinson and Sun 2010).

The Kolmogorov model will always be useful as a null hypothesis with which to compare the effects of non-Newtonian fluids, even if it should become violated. An obvious length scale to target for measurements of viscosity (and elasticity) would be the Kolmogorov length, L_k. From the viewpoint of TL dynamics (Jenkinson and Sun 2011), the vertical length scales of layer thickness and step thickness should be targeted. For processes taking place around organisms, such as diffusion in shear of nutrients, toxins, or dissolved allelopathic substances, encounter dynamics, and patch dispersal, Deborah numbers (Jenkinson and Wyatt 1992) would be useful in identifying the relevant length and time scales. For allelopathic action by harmful algae that use megadalton polymer complexes for cell-cell recognition and adhesion (Yamasaki et al. 2009; Blossom et al. 2012), the development, dynamics, and strength of adhesion are important, as they are in similar mechanisms in higher organisms. Atomic force microscopy (AFM), a tool widely used by micro- and nano-rheologists, will be likely useful to investigate cell-cell attraction, repulsion, and adhesion forces in flow fields, not only in such allelopathic action but also in processes such as flagellate grazing and mating (Schmid 1993), as well as in biofilm-like physical structuring of harmful algal and other microbial ecosystems in both the pelagos (Blossom et al. 2012) and the benthos (Berdalet et al. 2017b). AFM and hanging-drop electrochemistry are already being used (Svetlic et al. 2006) to study EPS mechanics and molecular dynamics.

Finally, in the context of HABs, it is important to note that some of the effects caused by increased viscosity due to phytoplankton EPS production occur at different microscales. Partial and/or total clogging of the fish gills (Jenkinson et al. 2007), mollusks, tunicates, sponges, polychaetes, etc. occur at 1 nm–1 mm scales. Rising organic matter and adsorption to the air-sea surface can result in reduction of air-sea gas exchange (Calleja et al. 2009) and wave and ripple damping (Carlson 1987), at scales of 10 μm–10 m. Flocculation into mucous aggregates, thus increasing sinking or rising speed and hence vertical organic flux (Mari et al. 2012), occurs at scales of 100 μm–1 mm and could also reinforce pycnoclines (Jenkinson and Sun 2011) at 10 cm–10 m. Trapping of toxins close to metabolically active surfaces, such as cell membranes and gills (Jenkinson et al. 2007), would take place at 10 nm–1 mm.

9.5 Key Questions 4: What Nutritional Opportunities Do Thin Layers Provide to Phytoplankton, Especially to the Species Selected Within Thin Layers? Is Growth Rate in the Thin Layer Higher Due to a Local Enhancement in Resources? Do the Accumulated Breakdown Products, Resulting from the Rain and Decomposition of Particles Sinking from the Overlying Water Column and Captured in the Thin Layer, Enhance or Impede Phytoplankton Growth, and Do They Select for the Species Present? What Is the Role of the Microbial Loop in Thin Layers?

The existence of fine-scale chemical gradients and chemically distinct TLs in stratified coastal environments was demonstrated by Hanson and Donaghay (1998) in a fjord in the US Pacific Northwest. The study also provided one of the first calculations of nutrient transfer across a TL boundary. Analogous studies were carried out by Lunven et al. (2005) in the Loire River plume (Bay of Biscay, France) showing nitrogen and silicate incorporation by phosphorus-limited diatoms within a TL. The research revealed periods of dissolved inorganic nutrient availability and limitation within the seasonal succession in the TLs. In the Gulf of Finland (Baltic Sea), subsurface TLs of the dinoflagellate *Heterocapsa triquetra* just below the strongest density gradient in the seasonal thermocline, coinciding with the nitracline depth, were reported by Kononen et al. (2003).

If TLs form in the upper part of the nutricline, the co-occurrence of favourable nutrient and light conditions will support growth (e.g., Cheriton et al. 2009). However, both phytoplankton and nutrient patterns follow intense spatial and temporal variability within the TL as has been revealed by the detailed study by Ryan et al. (2010) in Monterey Bay. In this case, the vertical migratory behaviour of the dinoflagellate was coupled to nitrate availability, which in turn was dependent on the complex mesoscale hydrodynamics in the area, including tidal mixing and wind regime. The phytoplankton abundances in the TL doubled during a 4-day period, in parallel with enhancement of stratification and internal tide-driven nutrient supply. Similar detailed studies conducted by Timmermann et al. (2014) suggested the potential phosphorus limitation of the toxigenic *Pseudo-nitzschia* accumulated in a subsurface TL in the area. In that particular situation, high domoic acid levels were also recorded in the TL. Altogether these examples illustrate how the dynamics of nutrients in TLs influence the physiological state of phytoplankton and, in the case of toxic species, modulate the harmful impacts.

TLs may provide advantages for mixotrophic phytoplankton by favouring high prey concentration (e.g., *Dinophysis* spp., Velo-Suárez et al. 2008). In addition, breakdown products from senescent cells can accumulate in the density gradient (pycnocline), thus resulting in increased bacterial production rates (McManus et al. 2003). Nevertheless, more investigation is required concerning the dynamics of the microbial loop in TLs. At this scale, mainly laboratory experiments and modelling efforts have documented the importance of small-scale turbulence and shear on contact rates (and hence coagulation and predation processes) between suspended plankters (Rothschild and Osborn 1988; Kiørboe 1997). In turn, phytoplankton in TLs also constitute preys for selective primary and/or secondary consumers (Greer et al. 2013).

9.6 Key Question 5: Are Allelopathy and "Chemical Warfare" at Work In Situ Within TLs?

Microalgal species, even with moderate division rates, may form blooms by reducing grazing mortality through killing potential protist predators or by secreting allelochemicals that inhibit their competitors' growth (Tillmann and John 2002; Tang and Gobler 2010). Results from use of solid phase adsorption toxin trackers

(SPATT; MacKenzie et al. 2004) have revealed that most shellfish toxin (DSP, PSP, ASP) producers release to the water a significant proportion of "dissolved toxins." At late stages of the population, it may surpass the level of intracellular toxins. If toxins are released in the water, either an allelopathic, sexual pheromone-like, or some kind of chemical signalling role would be expected (Wyatt and Jenkinson 1997).

It is important to bear in mind that in most laboratory experiments with toxin producers, cell densities between one and two orders of magnitude higher than those usually found in the field have been used. Harmful effects on other microalgae or microzooplanktonic organisms were density dependent and could not be reproduced with densities similar to those observed in the field. In some cases, the harmful effects were not due to the shellfish-poisoning toxins but to other non-identified compounds. This was the case with high-density cultures of *Dinophysis*, which appeared to immobilize their ciliate prey (*Mesodinium rubrum*) by secreting a toxic mucus trap. Comparisons with controls showed that the allelopathic substance was different from DSP toxins, and free PUFAs found in the mucus were considered as the potential killing compound (Mafra et al. 2016).

Flynn (2008) concluded that allelopathy does not aid the growth of a poor competitor for resources using models, which considered variable prey selection and stoichiometry. This would be the case unless some other factors initially elevate the biomass of the producer and hence increase the levels of allelopathic chemicals to exceed a critical threshold. In other words, allelopathy may contribute to the maintenance of a TL but not to its formation.

9.7 Conclusions and Next Steps

Key achievements on the understanding of HABs in stratified systems, in particular on processes at fine and small scales, resulted from international coordinated research conducted over the lifetime of GEOHAB. Some priorities still need to be addressed in the coming years, with the overall objective of developing better predictive tools and strategies to prevent and mitigate HAB impacts. International collaboration of research teams, and appropriate funds, will continue to facilitate progress. Some key issues for future research encompass the following topics:

- Stratification is a key factor modulating phytoplankton bloom dynamics, and likely, there are also common strategies involved in the proliferation of cells in TLs. However, the particular mechanisms leading to the selection of different noxious taxa such as *Karenia, Dinophysis*, or *Pseudo-nitzschia* are not known.
- Different convergent and diffusion mechanisms, with physical, biological, and chemical components, have been proposed to explain the formation, dissipation, and maintenance of thin layers. Still, most of our understanding remains at a theoretical level due to the difficulties involved in efficiently monitoring TL dynamics in real time.

- Technical advances offer good perspectives for future progress towards the observation of cells in situ. New mooring and profiling instruments and autonomous underwater vehicles (AUVs) are facilitating the comprehension of the HABs dynamics, while a reduction in price will facilitate their use on routine monitoring in HAB-threatened environments.
- It is particularly necessary to "move the microscope to the sea" not only to identify and visualize the microorganisms but to estimate the biological rates (growth, grazing, excystment, and encystment) in situ involved in their behaviour. These parameters are fundamental to describe the population dynamics of HAB species.
- Particularly challenging is the characterization of the chemical interactions among microorganisms, which can facilitate out-competition of HAB species over the others and escaping from grazers. The physiological role of the "known" toxins is not understood yet, while more chemical compounds may be involved in allelochemical responses observed mainly in laboratory experiments. Still, experimentation under more realistic, similar to natural conditions is required.
- Microorganisms are niche engineers. By excreting and utilizing DOM, they create microenvironmental habitats while modifying the physical structure at nanoscale. These processes are involved in TLs and other environments and have ecological and biogeochemical implications. Exploring the universe at this scale is one of the most important and challenging perspectives for future research.

Acknowledgments This chapter is a contribution of the GEOHAB Core Research Project on HABs in Stratified Systems. It is a tribute to our colleague and friend Patrick Gentien who passed away in 2010. Patrick inspired and fostered research of the fine- and small-scale processes involved in HAB dynamics.

References

Alldredge AL, Cowles TJ, MacIntyre S et al (2002) Occurrence and mechanism of formation of a dramatic thin layer of marine snow in a shallow Pacific fjord. Mar Ecol Prog Ser 233:1–12

Berdalet E, Llaveria G, Simó R (2011) Modulation of small-scale turbulence on methylsulfoniopropionate (DMSP) concentration in an *Alexandrium minutum* (Dinophyceae) culture: link with toxin production. Harmful Algae 10:88–95

Berdalet E, McManus MA, Ross ON et al (2014) Understanding harmful algae in stratified systems: review of progress and future directions. Deep Sea Res Part II 101:4–20

Berdalet E, Montresor M, Reguera B (2017a) Harmful algal blooms in fjords, coastal embayments, and stratified systems: recent progress and future research. Oceanography 30(1):46–57

Berdalet E, Tester PA, Chinain M (2017b) Harmful algal blooms in benthic systems: recent progress and future research. Oceanography 30(1):36–45

Blossom HE, Daugbjerg N, Hansen PJ (2012) Toxic mucus traps: a novel mechanism that mediates prey uptake in the mixotrophic dinoflagellae *Alexandrium pseudogonyaulax*. Harmful Algae 17:40–53

Brosnahan ML, Velo-Suárez L, Ralston DK (2015) Rapid growth and concerted sexual transitions by a bloom of the harmful dinoflagellate *Alexandrium fundyense* (Dinophyceae). Limnol Oceanogr 60:2059–2078
Calleja ML, Duarte CM, Prairie YT et al (2009) Evidence for surface organic matter modulation of air-sea CO_2 gas exchange. Biogeosciences 6:1105–1114
Campbell L, Olson RJ, Sosik HM et al (2010) First harmful *Dinophysis* (Dinophyceae, Dinophysiales) bloom in the U.S. revealed by automated imaging flow cytometry. J Phycol 46:66–75
Carlson DJ (1987) Viscosity of sea-surface slicks. Nature 329:823–825
Cheriton OM, McManus MA, Stacey MT et al (2009) Physical and biological controls on the maintenance and dissipation of a thin phytoplankton layer. Mar Ecol Prog Ser 378:55–69
Davidson K, Miller P, Wilding TA et al (2009) A large and prolonged bloom of *Karenia mikimotoi* in Scottish waters in 2006. Harmful Algae 8:349–361
Dekshenieks MM, Donaghay PL, Sullivan JM et al (2001) Temporal and spatial occurrence of thin phytoplankton layers in relation to physical processes. Mar Ecol Prog Ser 223:61–71
Díaz PA, Ruiz-Villarreal M, Velo-Suárez L et al (2014) Tidal and wind-event variability and the distribution of two groups of *Pseudo-nitzschia* species in an upwelling-influenced ría. Deep Sea Res Part II 101:163–179
Donaghay PL, Osborn TR (1997) Toward a theory of biological-physical control of harmful algal bloom dynamics and impacts. Limnol Oceanogr 42:1283–1296
Escalera L, Pazos Y, Doval MD et al (2012) A comparison of integrated and discrete depth sampling for monitoring toxic species of *Dinophysis*. Mar Pollut Bull 64:106–113
Estrada M, Berdalet E (1998) Effects of turbulence on phytoplankton. In: Anderson DM, Cembella AD, Halleagraaf GM (eds) Physiological ecology of harmful algal blooms, NATO ASI series G: ecological series, vol 41. Springer, Berlin, Heidelberg, New York, pp 601–618
Farrell H, Gentien P, Fernand L et al (2012) Scales characterising a high density thin layer of *Dinophysis acuta* Ehrenberg and its transport within a coastal jet. Harmful Algae 15:36–46
Farrell H, Gentien P, Fernand L et al (2014) Vertical and horizontal controls of a haptophyte thin layer in the Bay of Biscay, France. Deep Sea Res Part II 101:80–94
Fischer AD (2017) Alexandrium catenella cyst dynamics in a coastal embayment: temperature dependence of dormancy, germination, and bloom initiation. PhD Thesis, Massachusetts Institute of Technology, 165pp
Flynn KJ (2008) Attack is not the best form of defense: lessons from harmful algal bloom dynamics. Harmful Algae 8:129–139
Foloni-Neto H, Lueck R, Mabuchi Y et al (2014) A new quasi-horizontal glider to measure biophysical microstructure. J Atmos Oceanic Technol 31:2278–2293. https://doi.org/10.1175/JTECH-D-13-00240.1
Gentien P (1998) Bloom dynamics and ecophysiology of the *Gymnodinium mikimotoi* species complex. In: Anderson DM, Cembella AD, Hallegraeff GM (eds) Physiological ecology of harmful algal blooms Physiological ecology of harmful algal blooms, NATO ASI series G: ecological series, vol 41. Springer, Berlin, Heidelberg, New York, pp 155–173
Gentien P, Donaghay PL, Yamazaki H et al (2005) Harmful algal blooms in stratified environments. Oceanography 18:172–183
GEOHAB (2008) Global ecology and oceanography of harmful algal blooms, GEOHAB core research project: HABs in stratified systems. Gentien P, Reguera B, Yamazaki H et al (eds) SCOR and IOC, Paris, France, and Newark, Delaware, USA, 59 pp
GEOHAB (2011) GEOHAB modelling: linking observations to predictions, a workshop report. McGillicuddy DJ Jr, Glibert PM, Berdalet E et al (eds) IOC and SCOR, Paris and Newark, Delaware, 85 pp
GEOHAB (2013) Global ecology and oceanography of harmful algal blooms, GEOHAB core research project on HABs in stratified systems. Workshop on advances and challenges for understanding physical-biological interactions in HABs in stratified systems. McManus MA, Berdalet E, Ryan J et al (eds) IOC and SCOR, Paris and Newark, Delaware, U.S. 62 pp

Glibert PM, Pitcher GC, Bernard S et al (2018) Advancements and continuing challenges of emerging technologies and tools for detecting harmful algal blooms, their antecedent conditions and toxins, and applications in predictive models. In: Glibert PM, Berdalet E, Burford M et al (eds) Global ecology and oceanography of harmful algal blooms. Springer, Cham, pp 339–357

González-Gil S, Velo-Suárez L, Gentien P et al (2010) Phytoplankton assemblages and characterization of a *Dinophysis acuminata* population during an upwelling–downwelling cycle. Aquat Microb Ecol 58:273–286

Greer AT, Cowen RK, Guigand CM et al (2013) Relationships between phytoplankton thin layers and the fine-scale vertical distributions of two trophic levels of zooplankton. J Plankton Res 35:939–956

Hanson AK Jr, Donaghay PL (1998) Micro- to fine-scale chemical gradients and layers in stratified coastal waters. Oceanography 111:10–17

Hill AE, Brown J, Fernand L et al (2008) Thermohaline circulation of shallow tidal seas. Geophys Res Lett 35:L11605. https://doi.org/10.1029/2008GL033459

Jaffe JS, Franks PJS, Briseno C et al (2013) Advances in underwater fluorometry, from bulk fluorescence to planar laser imaging of individuals. In: Watson J, Zielinski O (eds) Subsea optics and imaging. Woodhead Publishing Limited, Cambridge. https://doi.org/10.1533/9780857093523.3.536

Jenkinson IR (1986) Oceanographic implications of non-newtonian properties found in phytoplankton cultures. Nature 323:435–437

Jenkinson IR (2014) Nano- and microfluidics, rheology, exopolymeric substances and fluid dynamics in calanoid copepods. In: Seuront L (ed) Copepods: diversity, habitat and behavior. Nova Science Publishers, New York, pp 181–214

Jenkinson IR, Biddanda BA (1995) Bulk-phase viscoelastic properties of seawater relationship with plankton components. J Plankton Res 17:2251–2274

Jenkinson IR, Claireaux G, Gentien P (2007) Biorheological properties of intertidal organic fluff on mud flats and its modification of gill ventilation in buried sole *Solea solea*. Mar Biol 150:471–485

Jenkinson IR, Sun J (2010) Rheological properties of natural waters with regard to plankton thin layers. J Mar Syst 83:287–297

Jenkinson IR, Sun J (2011) A model of pycnocline thickness modified by the rheological properties of phytoplankton exopolymeric substances. J Plankton Rese 33:373–383

Jenkinson IR, Wyatt T (1992) Selection and control of Deborah numbers in plankton ecology. J Plankton Res 14:1697–1721

Kamykowski D, Yamazaki H, Yamazaki AK et al (1998) A comparison of how different orientation behaviours influence dinoflagellate trajectories and photoresponses in turbulent water columns. In: Anderson DM, Cembella AD, Hallegraeff GM (eds) Physiological ecology of harmful algal blooms, NATO ASI series G: ecological series, vol 41. Springer, Berlin, Heidelberg, New York, pp 581–599

Karp-Boss L, Boss E, Jumars PA (1996) Nutrient fluxes to planktonic osmotrophs in the presence of fluid motion. Oceanogr Mar Biol Annu Rev 34:71–109

Kesaulya I, Leterme SC, Mitchell JG et al (2008) The impact of turbulence and phytoplankton dynamics on foam formation, seawater viscosity and chlorophyll concentration in the eastern English Channel. Oceanologia 50:167–182

Kiørboe T (1997) Small-scale turbulence, marine snow formation, and planktivorous feeding. In: Marrasé C, Saiz E, Redondo JM (eds) Lectures on plankton and turbulence. Sci Mar 61(Supl. 1):141–158

Klausmeier CA, Litchman E (2001) Algal games: the vertical distribution of phytoplankton in poorly mixed water columns. Limnol Oceanogr 46:1998–2007

Kolmogorov AN (1941) Dissipation of energy in isotropic turbulence. Dokl Akad Nauk SSSR 32:19–21

Kononen K, Huttunen M, Hällfors S et al (2003) Development of a deep chlorophyll maximum of *Heterocapsa triquetra* Ehrenb. at the entrance to the Gulf of Finland. Limnol Oceanogr 48:594–607

Lasker R (1978) Fishing for anchovies off California. Mar Pollut Bull 9:320–321

Lunven M, Guillaud JF, Youenou A et al (2005) Nutrient and phytoplankton distribution in the Loire River plume (Bay of Biscay, France) resolved by a new Fine Scale Sampler. Estuar Coast Shelf Sci 65:94–108

Lunven M, Landeira JM, Lehaître M et al (2012) *In situ* video and fluorescence analysis (VFA) of marine particles: applications to phytoplankton ecological studies. Limnol Oceanogr Methods 10:807–823

MacKenzie L, Beuzenberg V, Holland P et al (2004) Solid phase adsorption toxin tracking (SPATT): a new monitoring tool that simulates the biotoxin contamination of filter feeding bivalves. Toxicon 44:901–918

Mafra LL Jr, Nagai S, Uchida H et al (2016) Harmful effects of *Dinophysis* to the ciliate *Mesodinium rubrum*: implications for prey capture. Harmful Algae 59:82–90

Mann KH, Lazier JRN (1991) Dynamics of marine ecosystems. Biological-physical interactions in the oceans. Blackwell Scientific Publications, Boston

Mari X, Torréton J-P, Trinh CBT et al (2012) Aggregation dynamics along a salinity gradient in the Bach Dang estuary, North Vietnam. Estuar Coast Mar Sci 96:151–158

Matrai P, Thompson B, Keller M (2005) Circannual excystment patterns of *Alexandrium* spp. from eastern Gulf of Maine populations. Deep Sea Res Part II 52:2560–2568

McGillicuddy D Jr (2010) Models of harmful algal blooms: conceptual, empirical and numerical approaches. J Mar Syst 83:105–107

McManus MA, Alldredge AL, Barnard A et al (2003) Changes in characteristics, distribution and persistence of thin layers over a 48-hour period. Mar Ecol Prog Ser 261:1–19

McManus MA, Kudela RM, Silver MV et al (2008) Cryptic blooms: are thin layers the missing connection? Estuar Coasts 31:396–401

Miyake Y, Koizumi M (1948) The measurement of the viscosity coefficient of seawater. J Mar Res 7:63–66

Ní Rathaille A, Raine R (2011) Seasonality in the excystment of *Alexandrium minutum* and *Alexandrium tamarense* in Irish coastal waters. Harmful Algae 10(6):629–635

Pitcher GC, Figueiras FG, Kudela RM et al (2018) Key questions and recent research advances on harmful algal blooms in eastern boundary upwelling systems. In: Glibert PM, Berdalet E, Burford M et al (eds) Global ecology and oceanography of harmful algal blooms. Springer, Cham, pp 205–227

Raine R, Berdalet E, McManus MA et al (eds) (2014) Harmful algal blooms in stratified systems. Deep Sea Res Part II 101:254

Raine R, Cosgrove S, Fennell S et al (2017) Origins of *Dinophysis* blooms which impact Irish aquaculture. In: Proenca LAO, Hallegraeff G (eds) Proceedings of the 17th conference on harmful algae, Florianopolis, Brazil, October 2016 (in press)

Ralston DK, Keafer BA, Brosnahan ML et al (2014) Temperature dependence of an estuarine harmful algal bloom: resolving interannual variability in bloom dynamics using a degree day approach. Limnol Oceanogr 59:1112–1126

Rines JEB, Donaghay PL, Dekshenieks MM et al (2002) Thin layers and camouflage: hidden *Pseudo-nitzschia* populations in a fjord in the San Juan Islands, Washington, USA. Mar Ecol Prog Ser 225:123–137

Ross ON, Sharples J (2008) Swimming for survival: the role of phytoplankton motility in turbulent environments. J Mar Syst 70:248–262

Rothschild BJ, Osborn TR (1988) Small-scale turbulence and plankton contact rates. J Plankton Res 10:465–474

Rothstein JP (2010) Slip on superhydrophobic surfaces. Annu Rev Fluid Mech 42:89–209

Ryan JP, McManus MA, Paduan JD et al (2008) Phytoplankton thin layers caused by shear in frontal zones of a coastal upwelling system. Mar Ecol Prog Ser 354:21–34

Ryan JP, McManus MA, Sullivan JM (2010) Interacting physical, chemical and biological forcing of phytoplankton thin-layer variability in Monterey Bay, California. Cont Shelf Res 30:7–16
Schmid CE (1993) Cell-cell-recognition during fertilization in *Ectocarpus siliculosus* (Phaeophyceae). Hydrobiologia 260/261:437–443
Seuront L, Lacheze C, Doubell M et al (2007) The influence of *Phaeocystis globosa* on microscale spatial patterns of chlorophyll a and bulk-phase seawater viscosity. Biogeochemistry 83: 173–188
Seuront L, Vincent D, Mitchell JG (2006) Biologically induced modification of seawater viscosity in the Eastern English Channel during a *Phaeocystis globosa* spring bloom. J Mar Syst 61: 118–133
Smayda TJ (2002) Turbulence, watermass stratification and harmful algal blooms: an alternative view and frontal zones as "pelagic seed banks". Harmful Algae 1:95–102
Stacey MT, McManus MA, Steinbuck JV (2007) Convergences and divergences and thin layer formation and maintenance. Limnol Oceanogr 52:1523–1532
Steinbuck JV, Stacey MT, McManus MA et al (2009) Observations of turbulent mixing in a phytoplankton thin layer: implications for formation, maintenance, and breakdown. Limnol Oceanogr 544:1353–1368
Sverdrup HU (1953) On conditions for the vernal blooming of phytoplankton. J Conseil 18: 287–295
Svetlic V, Balnois E, Žutic V et al (2006) Electrochemical detection of gel microparticles in seawater. Croat Chem Acta 79:107–113
Tang YZ, Gobler CJ (2010) Allelopathic effects of *Cochlodinium polykrikoides* isolates and blooms from the estuaries of Long Island, New York, on co-occurring phytoplankton. Mar Ecol Prog Ser 406:19–31
Tillmann U, John U (2002) Toxic effects of *Alexandrium* spp. on heterotrophic dinoflagellates: an allelochemical defence mechanism independent of PSP toxin content. Mar Ecol Prog Ser 230:47–58
Timmermann A, Timmerman HV, McManus MA et al (2014) Hidden thin layers of toxic diatoms in a coastal bay. Deep Sea Res Part II 101:129–140
Trainer VL, Bates SS, Lundholm N et al (2012) *Pseudo-nitzschia* physiological ecology, phylogeny, toxicity, monitoring and impacts on ecosystem health. Harmful Algae 14:271–300
Velo-Suárez L, González-Gil S, Gentien P et al (2008) Thin layers of *Pseudo-nitzschia* spp. and the fate of *Dinophysis acuminata* during an upwelling-downwelling cycle in a Galician Ría. Limnol Oceanogr 53:1816–1834
Velo-Suárez L, Reguera B, González-Gil S et al (2010) Application of a 3D Lagrangian model to explain the decline of a *Dinophysis acuminata* bloom in the Bay of Biscay. J Mar Syst 83: 242–252
Velo-Suárez L, González-Gil S, Pazos Y et al (2014) The growth season of *Dinophysis acuminata* in an upwelling system embayment: a conceptual model based on in situ measurements. Deep Sea Res Part II 101:141–151
Wong T-S, Kang SH, Tang SKY et al (2011) Bioinspired self-repairing slippery surfaces with pressure-stable omniphobicity. Nature 477:443–447
Wyatt T, Jenkinson IR (1997) Notes on *Alexandrium* population dynamics. J Plankton Res 19: 551–575
Yamazaki H, Honma H, Nagai T et al (2010) Multilayer biological structure and mixing in the upper water column of Lake Biwa during summer 2008. Limnology 11:63–70
Yamazaki H, Locke C, Umlauf L et al (2014) A Lagrangian model for phototaxis-induced thin layer formation. Deep Sea Res Part II 101:193–206
Yamasaki Y, Shikata T, Nukata A et al (2009) Extracellular polysaccharideprotein complexes of a harmful alga mediate the allelopathic control it exerts within the phytoplankton community. ISME J 3:808–817

Chapter 10
Key Questions and Recent Research Advances on Harmful Algal Blooms in Fjords and Coastal Embayments

Suzanne Roy, Marina Montresor, and Allan Cembella

10.1 Introduction

Fjords and coastal embayments are coastal systems that are partially surrounded by land. As such, they tend to be heavily affected by terrigenous runoff, sediment transport and other land–sea interactions. When the international Programme on Global Ecology and Oceanography of Harmful Algal Blooms (GEOHAB) was initiated more than 10 years ago, it was thought that we should pay particular attention to these systems, because harmful algal blooms (HABs) often prevail and because some of their characteristics differ from those of other systems examined in the GEOHAB Programme. As an example, coastal embayments are often shallow and thus particularly affected by the link between the water column and bottom sediments. Fjords on the other hand are characterized by estuarine-type circulation patterns, which may provide good retention areas for benthic cysts and subsequent vegetative growth of some harmful species.

The Core Research project (CRP) on HABs in Fjords and Coastal Embayments was initiated with an Open Science Meeting in April 2004 in Viña del Mar, Chile. This meeting served to examine the various types of coastal systems where HABs were recurrent and to identify key features of these systems (Cembella et al. 2005; GEOHAB 2010). The other major goal of that meeting was to identify key questions for research that should promote advances in our understanding of the

S. Roy (✉)
Institut des Sciences de la Mer, Université du Québec à Rimouski, Rimouski, Canada
e-mail: Suzanne_Roy@uqar.ca

M. Montresor
Stazione Zoologica Anton Dohrn, Napoli, Italy

A. Cembella
Alfred-Wegener-Institut, Helmholtz-Zentrum für Polar- und Meeresforschung, Bremerhaven, Germany

P.M. Glibert et al. (eds.), *Global Ecology and Oceanography of Harmful Algal Blooms*, Ecological Studies 232, https://doi.org/10.1007/978-3-319-70069-4_10

dynamics of HABs in fjords and coastal embayments. Seven key questions were proposed, and progress towards them is summarized herein.

These questions were reviewed nearly a decade later, at a second Open Science Meeting held in Victoria, Canada, in May 2012. An additional set of five target areas of research was proposed to further our understanding of the dynamics of HABs (GEOHAB 2013); these target areas were focused on those that could be accomplished in a short period of time. In a few cases and in adherence with GEOHAB's strategy, comparative field studies were undertaken with a view towards modelling and prediction of bloom occurrence and dynamics in semi-enclosed systems. The following text describes the progress that was accomplished during this GEOHAB Programme and examines areas for future research that will help understand and improve prediction of HABs in fjords and coastal embayments. Information on habitat preferences, plankton assemblage behaviour and thereby environmental scenarios which eventually lead to bloom prediction, should continue to advance through comparative studies of algal assemblages in similar ecosystems. These case studies described here were endorsed and encouraged by GEOHAB and yielded concrete results and approaches for further hypothesis testing.

10.2 Key Question 1: Are There Definable Adaptive Strategies for HAB Species in Confined and Semi-confined Systems?

Similar to the conclusion reached by Smayda and Trainer (2010) in their study of dinoflagellate blooms in upwelling systems, no clear evidence was found of a HAB flora specific to fjords and coastal embayments. However, one difference with more open coastal systems is the generally greater reliance on bottom-dwelling resting stages in semi-confined systems. HAB flora can differ according to the physical robustness of the system: stronger coupling between the benthic nutrient fluxes and the plankton is observed in coastal lagoons due to the higher surface-volume ratios and long residence times. These conditions favour the development of picoplankton. In contrast, river-dominated estuaries are characterized by more dynamic conditions and seasonal nitrate inputs that favour diatom blooms (Glibert et al. 2010). The relative importance of the life cycle grows with the decrease in physical robustness of the coastal systems. Confined systems are more similar to lakes, where it is easier to follow the dynamics of planktonic populations. The modification of coastal ecosystems due to the increased number of tourist harbours or other coastal infrastructures can create new sheltered areas where HABs of cyst-forming species have been frequently reported (Garcés and Camp 2012).

With a growing understanding of species circumscription in the last decade, it is now recognized that species are organized into distinct populations characterized by significant intraspecific genetic diversity. Several populations of a harmful taxon may be present in the same environment, responding to different environmental factors and blooming at different times. A recent example of this is the extensive genetic diversity that was recorded among *Alexandrium fundyense* populations in a shallow coastal embayment along the coast of Cape Cod (Richlen et al. 2012). These populations were genetically distinct from those in coastal offshore waters, and they showed a surprisingly rapid temporal genetic differentiation, possibly attributable to a distinct timing in germination of the resting cyst. This is not the only case in which populations inhabiting embayments or fjords turned out to have a distinct genetic fingerprint. The bloom-forming diatom *Skeletonema marinoi* produces resting stages that help maintain genetically distinct populations in the fjord (Godhe and Härnström 2010) and form seed banks of resting cells stored alive in anoxic sediments for decades (Härnström et al. 2011). These environments with a limited exchange with open waters represent ideal environmental settings to study adaptation and identify selective forces at the base of population differentiation.

The autecological approach, in which one or a few cultured isolates are taken to be a representative of the response and diversity of natural populations, should be replaced by a greater effort to be inclusive of variable strains (Burkholder and Glibert 2009). Future HAB research needs to focus on in situ studies or to isolate numerous strains from the local environment in order to reflect the high degree of intraspecific diversity and the adaptation of the different strains to the particular environmental factors they face.

10.3 Key Question 2: What Is the Importance of Life History Transitions and Cyst Distribution in Bloom Initiation and Maintenance: Endogenous Seed Beds Versus Exogenous Introductions?

A large amount of work has been devoted to the life history of harmful species in the past decade, in part due to a number of large research programmes such as the GEOHAB-endorsed European SEED Programme (on Life Cycle Transformations among HAB Species, Garcés et al. 2010) and the American ECOHAB-Gulf of Maine and GOMTOX Programme (on the Dynamics of *A. fundyense* Distributions in the Gulf of Maine, Anderson et al. 2005, 2014). Both programmes highlighted two common features among harmful algal events: the presence of a sexual phase within the life cycle of *A. fundyense* and the production of benthic resting stages. Results from these programmes led to the recognition that a heteromorphic life stage represents an advantageous survival strategy because of the allocation into different ecological niches. Coupling information on cyst germination rates with an extensive mapping effort over the whole study area and with numerical modelling allowed the implementation of a conceptual model of the population dynamics of this dinoflagellate in the Gulf of Maine [McGillicuddy et al. (2005); see also Chap. 8, Azanza et al. (2018)]. Continued fieldwork provided clear evidence of the interannual variability of cyst beds in the region but also showed links between cyst distribution patterns and intensity/extension of blooms in the years immediately before and after cyst mapping surveys (McGillicuddy et al. 2014). Such links are, however, not clear in every case: Cosgrove et al. (2014) found no relationship between the intensity of blooms in Cork Harbour (Ireland) and cyst density in sediments determined during the previous winter. Resuspension was also required to prevent cyst stocks from becoming exhausted in the oxic layer of surface sediments. Cyst abundances in the nepheloid layer suggest that cyst germination might represent a significant inoculum for the seasonal bloom (Pilskaln et al. 2014). Vertical profiles of cyst distribution in the sediments are not homogeneous; abundances in the surface layer are influenced by the magnitude of cyst deposition and germination, but relatively stable deep cyst layers are produced by active bioturbation and may serve as long-term reservoirs (Shull et al. 2014). Challenging and still poorly explored is the assessment of germination success of cyst-forming dinoflagellates, which is a key parameter for population dynamic modelling.

Life cycle processes and their regulatory factors can differ widely, according to the species, and reversibility of some of the life cycle stages was shown to be possible (e.g., of the planozygote back into a vegetative cell instead of a cyst: Fig. 10.1, Figueroa et al. 2006; Bravo and Figueroa 2014). Some dinoflagellate species may produce asexual resting cysts as another survival strategy (Bravo et al. 2010). Molecular techniques have helped to identify specific life stages in the water column and in sediments (Erdner et al. 2010; Penna et al. 2010). These techniques also highlighted once more the importance of intraspecific diversity for many HAB species, leading to strain-specific population growth strategies (Tobin et al. 2013),

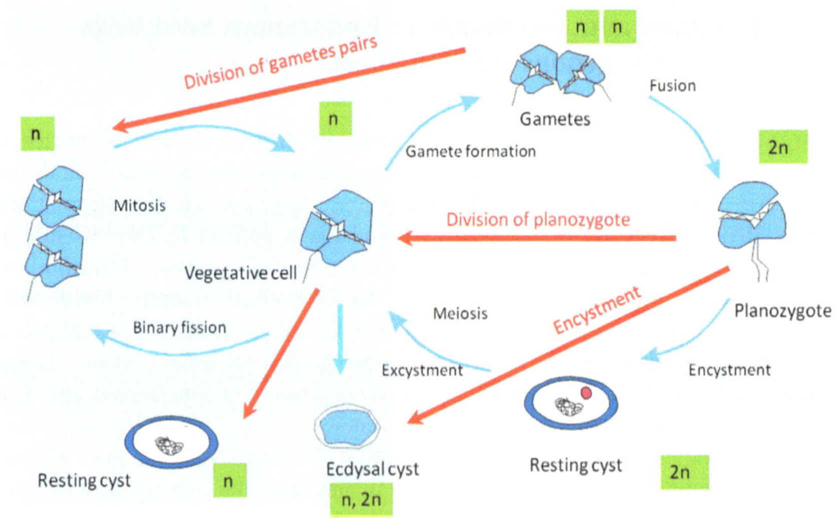

Fig. 10.1 Typical life cycle of dinoflagellates, with new features confirmed during the SEED project (from R. I. Figueroa, unpub., and reproduced from GEOHAB 2013)

behavioural traits (Harvey et al. 2015) and intra-regional variability (e.g. Varela et al. 2012). The toxic *A. fundyense* (ribotype Group I) and the non-toxic *A. tamarense* (ribotype Group III) are cryptic species (John et al. 2014) previously clustered within the *A. tamarense* species complex. The two species are capable of successful mating and production of cysts; however, they have extremely limited viability after germination (Brosnahan et al. 2010). This provides a further 'biological' support to the circumscription of the two species and also suggests the possibility to use the introduction of non-toxic strains as a mitigation strategy for bloom control.

There is increasing recognition that life cycle strategies including resting stages play an important role in population dynamics and in the maintenance of species diversity (e.g. Jones and Lennon 2010). Modelling of population dynamics thus needs to take into account the various life cycle stages and the dynamics of transition between them (Estrada et al. 2010; Warns et al. 2012). The population dynamics of harmful diatoms, such as *Pseudo-nitzschia* species, that do not include resting stages in their life cycle might be constrained by the occurrence of sexual reproduction that is responsible for the production of new large cells and thus on the maintenance of the population in a certain environment (D'Alelio et al. 2010). However, tools need to be developed to assist in situ identification of these life cycle stages, in order to increase the field data on these stages and to validate models that incorporate this information. A promising approach is the use of an Imaging FlowCytobot to follow the production of gametes and zygotes by *A. fundyense* and to estimate in situ cell division rates (Brosnahan et al. 2015).

Semi-enclosed systems such as fjords and coastal embayments are also ideal environments to study the impact of parasites on HAB populations, due to the fact that species are usually well retained within these environments. Several dinoflagellate species have been shown to be affected by parasites that can impair bloom development and enhance decline due to mass mortalities of host cells (e.g., Chambouvet et al., 2008). The parasite *Amoebophrya* has been shown to infect about 30% of natural populations of *A. fundyense* in a shallow bay; the peak of infection was concomitant to a massive sexual event and encystment of the host dinoflagellate, suggesting a possible escape strategy to the infection via the formation of gametes and production of resistant cysts (Velo-Suárez et al. 2013).

10.4 Key Question 3: How Do Physical Dispersion and Aggregation Processes Within a Semi-confined Basin Affect HAB Growth and Distribution?

Studies on physical dispersion and aggregation have been spurred partly by a growing interest in thin layers of plankton in the ocean [see also Chap. 9, Raine et al. (2018)], but physical factors that affect cell retention are also critical for semi-

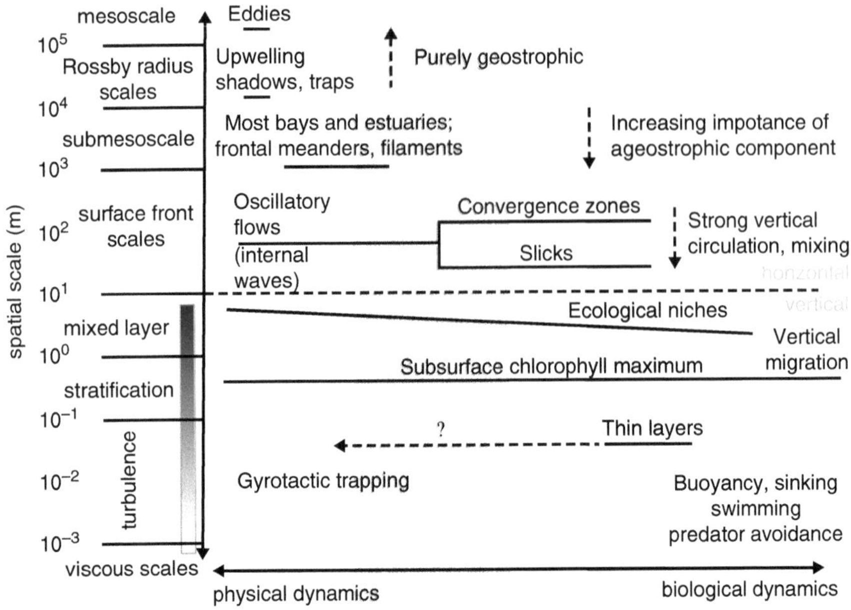

Fig. 10.2 Time and space scales of the physical and biological dynamics that control phytoplankton pattern and persistence in coastal and embayments settings (from A.J. Lucas, unpubl., and reproduced from GEOHAB 2013)

confined environments. Among these, stratification is a key factor (e.g., Díaz et al. 2011), but the transport mechanisms that bring harmful algae nearshore or inside bays, such as the wind-driven exchange in bays of southwestern Ireland (Raine et al. 2010) or changes in offshore upwelling associated with decadal climate variability (Sekula-Wood et al. 2011), are also potentially important. Lucas and Largier (2013) recently reviewed the range of physical factors that influence pattern and persistence of HABs in coastal environments. Although processes ranging over several scales can affect HABs in semi-confined environments, the importance of sub-mesoscale (<10 km) and high-frequency processes (Ryan et al. 2010), such as meanders developing on fronts, internal waves, wind-driven upwelling and turbulence, has been emphasized (Fig. 10.2). The influence may be direct (on the cells themselves) or indirect (on nutrient fluxes or light availability). A better understanding of these processes requires a greater availability of in situ physical and biological data, as well as higher resolution sampling (Dale and Murphy 2014). Such work should also help identify when and where biological factors, such as the swimming speed of some cells or life history stage transitions, represent an effective strategy, considering the local physical conditions (Persson et al. 2013). This information would be helpful for modellers.

10.5 Key Question 4: What Is the Relative Contribution of Nutrient Flux and Supply Ratios to HAB Dynamics in Eutrophic Versus Non-eutrophic Coastal Embayments?

Contrasting views have been published in recent literature on the role of nutrient loading and ratios in regulating HAB dynamics and global proliferation. Davidson et al. (2012, 2014) examined the evidence that nutrient ratios and forms influence the occurrence of HABs and concluded that it was difficult to establish a causal link between an increase in HABs and change in either nitrogen (N) or phosphorus (P) as the limiting nutrient. In contrast, several of the studies reviewed by Glibert and Burford (2017) point to a global expansion of HABs that is related to increasing nutrient loads and that the changes in nutrient ratios affect the type of plankton that can proliferate in a given environment [see also Chap. 12, Glibert et al. (2018a)]. Part of the explanation for this discrepancy may lie in the fact that the potential for regulation of processes at high nutrient concentrations (Glibert et al. 2013) has been disregarded in the studies by Davidson and colleagues. Nevertheless, there are many cases where inorganic macronutrients play an unknown, subsidiary or minor role in HAB bloom formation and successional processes in estuaries and coastal embayments. The relationship between HABs and nutrient loadings seems to be more consistent in some regions of the world, exemplified by a 100-year record of diatoms from sediment cores in a Danish fjord that showed a positive correlation between N-loading and the relative abundance of *Pseudo-nitzschia pungens* (Lundholm et al. 2010).

In addition to inorganic nutrients, organic nutrients have been recognized for several decades to play a critical role in phytoplankton nutrition (e.g., Berman and Bronk 2003). Organic nutrients can support the growth of a range of HAB species, including the brown tide *Aureococcus anophagefferens* (Berg et al. 1997). Gobler et al. (2011) used ecogenomics to identify the growth conditions of this fish-killing pelagophyte; high turbidity and high levels of dissolved organic matter were among the key factors that enabled this harmful species to outcompete co-occurring phytoplankton in estuaries of the eastern United States and South Africa. In contrast to diatom blooms in Thau Lagoon, France, that are due to conventional (nitrate, phosphate) nutrient pulses, local blooms of *A. catenella* are due to organic nutrient enrichment (Collos et al. 2014). Hence the organic forms of nutrients should be given closer attention, particularly for semi-confined coastal environments. Mixotrophy is another nutritional strategy that may be more widespread than previously thought (Flynn et al. 2013), and it seems to be particularly important for several harmful dinoflagellates, haptophytes and raphidophytes, especially when dissolved nutrients are insufficient or in imbalanced proportions [Glibert and Burford 2017; see also Chap. 7, Flynn et al. 2018]. These alternative metabolic strategies offer supplementary resources under both nutrient-poor and eutrophic conditions.

10.6 Key Question 5: What Is the Importance of Spatial Scale and Retention Time in the Expression and Effects of Allelochemicals/Toxins in Semi-confined Systems?

Semi-confined systems such as fjords and coastal embayments are ideal settings not only for the accumulation of cells as occurs during HABs but also as accumulation sites for secondary metabolites from these cells (e.g. allelochemicals, toxins). The physical constraints on advection and dispersion of blooms in semi-enclosed systems offer maximum potential for understanding expression and transduction of chemical signals in natural plankton assemblages at the micro- and mesoscale. Chemical ecological interactions among planktonic members of marine food webs affect and may even regulate 'top-down' processes, such as competition, predator–prey relationships and chemical communication within and among certain species (Fig. 10.3). Consideration of allelochemical interactions in the plankton (reviewed by Cembella 2003) has revealed that chemically mediated effects of phycotoxin-producing HAB taxa are complex, but the evidence does not support a general primary role for the known phycotoxins (with karlotoxins as a notable exception) as defensive compounds against protistan predators or competitors. A putative

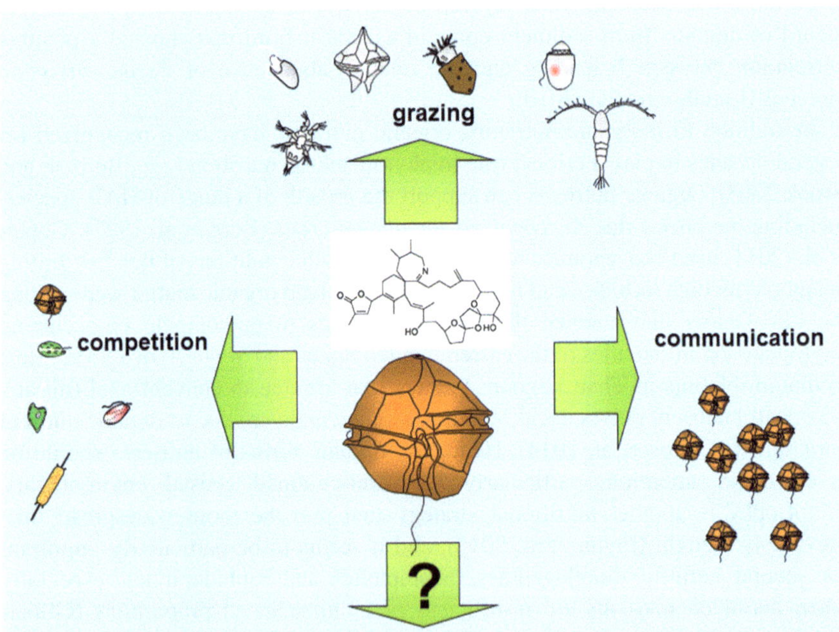

Fig. 10.3 Schematic of chemical ecological interactions mediated by toxins/allelochemicals and potentially affecting growth, bloom dynamics and life history transitions (from A.D. Cembella et al., unpub. and reproduced from GEOHAB 2013)

defence role for phycotoxins against most metazoan competitors (e.g. copepods) is equivocal but appears to be valid in a limited number of cases.

Allelochemical activity unrelated to the known phycotoxins may cause cell lysis and mortalities of many plankton components, particularly among species favouring coastal and brackish waters in embayments and lagoons. Allelochemicals are clearly involved in nutrient acquisition and feeding, as is well established for the mixotrophic prymnesiophyte *Prymnesium parvum*, which releases lytic compounds that immobilize or kill motile prey before ingestion (Skovgaard and Hansen 2003). Significantly, many HAB species known to produce allelochemicals against contemporaneous plankton (Cembella 2003) are also responsible for fish kills, particularly in aquaculture facilities in estuaries and coastal embayments, where blooms and their toxic metabolites, e.g., of the karlotoxin-producing dinoflagellate *Karlodinium veneficum*, are often most concentrated and persistent. A few recent studies have also demonstrated an elicited behavioural response of exposure to allelochemicals, even in the absence of ingestion by grazers, as mediated by waterborne chemical cues. In fact, exposure to copepod waterborne cues can result in a shift up in toxin production in toxigenic dinoflagellates, such as certain strains of *Alexandrium* species, as an apparent defensive response to potential predation (Wohlrab et al. 2010). Although in situ field studies on bidirectional allelochemical and toxin interactions remain rather limited in scope, they provide the basis for elucidating co-evolutionary mechanisms between HAB species and their predators and competitors in natural bloom populations.

10.7 Key Question 6: How Do Embayment Morphology, Bathymetry and Hydrodynamics Affect HAB Dynamics?

Various physical mechanisms influence phytoplankton biomass, productivity and community structure in coastal environments. Planktonic organisms, due to their small size, are strongly influenced by hydrodynamics, which in turn are affected by the bathymetry and morphology of semi-confined systems. Individually, these mechanisms rarely explain the changes in phytoplankton—multiple scale factors often need to be considered (Lucas et al. 2011). Some factors are often mentioned in association with HABs in semi-confined environments, such as freshwater inputs of nutrients and humic acids or wind-driven water exchange (e.g., Raine et al. 2010). A case study endorsed by GEOHAB examined the dynamics of putative HAB taxa and related phycotoxins in plankton from aquaculture sites in Alfacs and Fangar bays from the Ebro Delta, Spain, and sought to determine if agricultural freshwater inflow forms microniches which favour bloom development. Two major stratification patterns were evident during weekly cruises, with separation of the bloom period into two regimes of water stratification, which apparently complied with habitat preferences of harmful taxa (Busch et al. 2017). In this scenario, *Dinophysis* spp. was assigned to stratified areas and *Karlodinium* spp. to a lower degree of stratification. The diatom genus *Pseudo-nitzschia* was heavily represented in Fangar Bay regardless of regimes. The presence of *Karlodinium* spp. was associated to moderate salinity and increasing temperatures in both embayments. This study provides an example that nearby embayments can be affected by different harmful algal species or different bloom dynamics, according to differences in embayment morphology, bathymetry and hydrodynamics.

10.8 Key Question 7: Are the Effects of Human Activities (e.g. Aquaculture) and Global Climate Change on HAB Dynamics Magnified in Enclosed and Semi-enclosed Embayments?

The influence of aquaculture in general, and fish farming in particular, on HAB occurrence and dynamics is unclear, but substantial progress is being made in quantifying and modelling such impacts [e.g., Bouwman et al. 2011, 2013a, b; Burkholder and Shumway 2011; see also Chaps. 4, 12, Glibert et al. (2018a, b)]. While some think that the increase in organic waste material resulting from fish farming could have an effect on the development and the occurrence of HABs, especially in semi-confined environments, others contend that no causal link exists between the two. A particularly compelling example of a putative link between fish farms and HABs are the events in Norway, whereby a decrease in fish farms in the 1990s in western Norway was associated with a decrease in the occurrence of the fish-killing species *P. parvum*, which increased again only after fish farming was reintroduced in the area in 2005 (Johnsen et al. 2010). With the growing importance of aquaculture, as wild fish stocks are declining, there is an increasing need to understand both the effects of HABs on caged fish or farmed shellfish, as well as the influence of these human activities on phytoplankton communities, particularly HAB species (Davidson et al. 2012, 2014). A modelling exercise by Bouwman et al. (2011, 2013b) examined the different impacts of various aquaculture activities in terms of nitrogen and phosphorus discharge (global hindcasts and future projections). They pointed out that marine shellfish production is an important contributor to nutrient loading of coastal seas, particularly in Eastern Asia. This production causes changes in nutrient ratios and increases reduced and organic forms of nitrogen that can be used by some harmful algae. Limiting the impact of these discharges could be achieved with integrated aquaculture systems that include seaweed production, as they have a nutrient withdrawal role.

A few recent studies have examined how climate change might influence HABs in the oceans. Hallegraeff (2010) pointed out the multifactorial interactions associated with higher temperature, enhanced surface stratification, heavy precipitation and storm events which may cause changes in land runoff and nutrient availability, among other effects. Climate-relevant studies often take into account only a few factors, neglecting complex factor interactions. This issue and the complexity associated with physiological acclimation and genetic strain diversity, with studies often based on only a single or very few strains—the autecological fallacy—has led to contradictory species or strain-specific responses in some cases. It is thus difficult to predict the future of HABs in a changing climate, as pointed out in a recent review by Wells et al. [2015; see also Chap. 5, Wells and Karlson (2018)]. These authors highlight the lack of uniform experimental protocols, which limits the quantitative cross-investigation comparisons essential to the advancement of our knowledge. Some have attempted to model future HABs for specific regions undergoing major rapid changes, such as the Arctic: in their simulation analysis, Walsh et al. (2011) predicted increases in HABs (particularly toxigenic dinoflagellates producing saxitoxin analogues) for the region north of the Bering Strait, based on the relaxation of grazing pressure by herbivores due to overfishing as well as the development of human activities associated with ice retreat. A recent GEOHAB field programme in west Greenland and Iceland, ARCHEMHAB (Cembella et al. 2016), is examining effects on biogeography and diversity of HABs and their associated toxins which may be related to environmental regime shifts, particularly in Arctic fjords. The high species diversity in the microplankton fraction on the Greenland coast showed a strikingly high representation of cryptic dinoflagellate taxa. No definitive linkages to HAB-relevant regime shifts or environmental driving forces such as glacial melting on population dynamics were defined, but such multifaceted expeditions provide the basis for future scenarios and modelling change processes of HAB distribution in Arctic waters.

Climate change also influences ocean acidification due to rising CO_2 in the ocean. Ocean acidification may be relevant to the dependency of HAB toxicity on specific N sources (Fu et al. 2012). Low pH has been shown to inhibit nitrification and could ultimately shift the speciation of N away from nitrate and towards reduced forms such as ammonium and organic N. This could affect HAB toxicity since saxitoxin production, for example, may be enhanced when cultures are grown on ammonium as opposed to nitrate (e.g. Hamasaki et al. 2001).

The increasing spread of HAB blooms has been often related to ballast water discharges that might significantly promote the transport of alien species, including HAB taxa. This topic has been critically reviewed by Smayda (2007), who pointed out the need for detailed investigations to support human-mediated migration of species as well as the need for a better understanding of the ecological mechanisms that regulate colonization mechanisms and bloom development. Dispersal through ballast water or shellfish transplantation—often occurring in sheltered bays— has been supported only in some cases by detailed records of resting cysts in dated sediment cores coupled with genetic studies (e.g. *Gymnodinium catenatum*, Bolch and de Salas 2007). Mitigation measures, such as the exchange of ballast water from ports of origin with open ocean water, seem to be effective to prevent the transport of freshwater nuisance species. Such strategies, however, may be ineffective for coastal waters or even promote the spread of harmful marine species in some cases (Roy et al. 2012). Human-assisted and natural dispersions might be difficult to track, but confined habitats such as fjords and coastal bays are among the ideal sites to pursue these studies.

10.9 Future Research Priorities

Future research on HABs in fjords and coastal embayments should include at least two rather broad research themes. The first theme should address the influence of climate change, as this influence is expected to be more pronounced in coastal environments (particularly in polar latitudes). This should include factors such as temperature but also freshwater runoff, stratification, wind patterns, acidification, etc. The second research theme should deal with the influence of coastal aquaculture, particularly fish farming, on the development of HABs. We need to develop predictive capabilities regarding consequences to HABs in aquaculture-impacted environments. A production threshold may exist above which fish farming results in the development of blooms of ichthyotoxic algal species, which can in turn cause the collapse of this activity. This topic has attracted recent interest, due to the rapid development of aquaculture (e.g. the HABFISH project in Denmark: http://www.habfish.dk/).

A few other research topics have been identified which require immediate attention to improve our understanding of HABs in semi-confined environments and the development of better predictive tools.

- Resting stage formation, deposition and emergence fluxes should be determined in the field, with proven methods and identification of factors affecting the viability and abundance of these life history stages.
- Knowledge on small-scale processes of importance to HABs in coastal environments must be improved. Physical processes <10 km are particularly important for HABs in semi-confined environments, as well as their interactions with biological and chemical processes, such as life stage-dependent swimming behaviour, rheological properties, allelochemical interactions, etc.
- Species-specific databases containing information for HABs, particularly with respect to the ranges of phenotypic and genotypic variation, must be enhanced. This will require advances in development and application of molecular methods to determine cryptic and intraspecific diversity, identify toxic and non-toxic strains of the same species and determine the factors that affect strain-specific variations in toxicity.

Along with these priorities, there is an urgent need to develop new tools to identify and quantify the various life stages (gametes, zygotes, etc.) of HAB taxa in the field, including cell mortality, and to determine the nutritional mode of HA taxa (autotrophy, heterotrophy, mixotrophy). Eventual changes of these alternative nutritional modes may be associated with, for example, changes in nutrient loadings in coastal environments. The role of toxins and other allelochemicals in structuring the food web, in cell defence and in nutrient acquisition is another active field of research that is of special interest for fjords and coastal embayments because these secondary metabolites can be in greater concentrations in spatially constrained environments where intensive blooms can develop. Many of these research themes and requirements will be addressed within GlobalHAB, the new international Programme on HABs on a global scale.

Acknowledgements This is a contribution of the GEOHAB Core Research Project on HABs in Fjords and Coastal Embayments. Support for preparation of this manuscript is gratefully acknowledged within the PACES II Research Programme (Topic II Coast: WP3) of the Alfred-Wegener-Institut, Helmholtz Zentrum für Polar- und Meeresforschung under Earth and Environment, Helmholtz Gemeinschaft, Germany.

References

Anderson DM, McGillicuddy DJ, Degrasse SL et al (2014) Preface to: harmful algae in the Gulf of Maine: oceanography, population dynamics, and toxin transfer in the food web. Deep Sea Res II 103:1–5

Anderson DM, McGillicuddy D, Townsend D et al (2005) The ecology and oceanography of toxic *Alexandrium fundyense* blooms in the Gulf of Maine. Deep Sea Res II 52:2365–2876

Azanza RV, Brosnahan ML, Anderson DM et al (2018) The role of life cycle characteristics in harmful algal bloom dynamics. In: Glibert PM, Berdalet E, Burford M et al (eds) Global ecology and oceanography of harmful algal blooms. Springer, Cham, pp 133–161

Berg GM, Glibert PM, Lomas MW et al (1997) Organic nitrogen uptake and growth by the chrysophyte *Aureococcus anophagefferens* during a brown tide event. Mar Biol 129:377–387

Berman T, Bronk DA (2003) Dissolved organic nitrogen: a dynamic participant in aquatic ecosystems. Aquat Microb Ecol 31:279–305

Bolch CJS, De Salas MF (2007) A review of the molecular evidence for ballast water introduction of the toxic dinoflagellates *Gymnodinium catenatum* and the *Alexandrium "tamarensis* complex" to Australasia. Harmful Algae 6:465–485

Bouwman AF, Beusen AHW, Glibert PM et al (2013a) Mariculture: significant and expanding cause of coastal nutrient enrichment. Environ Res Let 8(044026):5. https://doi.org/10.1088/1748-9326/8/4/044026

Bouwman AF, Beusen AHW, Overbeek CC et al (2013b) Hindcasts and future projections of global inland and coastal nitrogen and phosphorus loads due to finfish aquaculture. Rev Fish Sci 21:112–156

Bouwman AF, Pawlowski M, Liu C et al (2011) Global hindcasts and future projections of coastal nitrogen and phosphorus loads due to shellfish and seaweed aquaculture. Rev Fish Sci 19: 331–357

Bravo I, Figueroa RI (2014) Towards an ecological understanding of dinoflagellate cyst functions. Microorganisms 2:11–32

Bravo I, Figueroa RI, Garcés E et al (2010) The intricacies of dinoflagellate pellicle cysts: the example of *Alexandrium minutum* cysts from a bloom-recurrent area (Bay of Baiona, NW Spain). Deep Sea Res II 57:166–174

Brosnahan ML, Kulis DM, Solow AR et al (2010) Outbreeding lethality between toxic Group I and nontoxic Group III *Alexandrium tamarense* spp. isolates: predominance of heterotypic encystment and implications for mating interactions and biogeography. Deep Sea Res II 57(3–4): 175–189

Brosnahan ML, Velo-Suárez L, Ralston DK et al (2015) Rapid growth and concerted sexual transitions by a bloom of the harmful dinoflagellate *Alexandrium fundyense* (Dinophyceae). Limnol Oceanogr 60:2059–2078
Burkholder JM, Glibert P (2009) The importance of intraspecific variability in harmful algae—preface to a collection of topical papers. Harmful Algae 8:744–745
Burkholder JM, Shumway SE (2011) Bivalve shellfish aquaculture and eutrophication. In: Shumway SE (ed) Shellfish and the environment. Wiley, New York, pp 155–215
Busch JA, Andree KA, Diogène J et al (2017) Spatio-temporal patchiness of harmful phytoplankton in comparative environmental regimes of two coastal embayments in the Ebro Delta (NW Mediterranean). Mar Ecol Prog Ser (in press)
Cembella AD (2003) Chemical ecology of eukaryotic microalgae in marine ecosystems. Phycologia 42:420–447
Cembella AD, Ibarra DA, Diogene J et al (2005) Harmful algal blooms and their assessment in fjords and coastal embayments. Oceanography 18(2):158–171
Cembella A, Zielinski O, Anderson D, Graeve M et al (2016) ARCHEMHAB: interactions and feedback mechanisms between hydrography, geochemical signatures and microbial ecology, with a focus on HAB species diversity, biogeography and dynamics. Cruise Report MSM21/3, DFG-Senatskommission für Ozeanographie, Bremen, Germany
Chambouvet A, Morin P, Marie D et al (2008) Control of toxic marine dinoflagellate blooms by serial parasitic killers. Science 322:1254–1257
Collos Y, Jauzein C, Ratmaya W et al (2014) Comparing diatom and *Alexandrium catenella/tamarense* blooms in Thau lagoon: importance of dissolved organic nitrogen in seasonally N-limited systems. Harmful Algae 37:84–91
Cosgrove S, Ní Rathaille A, Raine R (2014) The influence of bloom intensity on the encystment rate and persistence of *Alexandrium minutum* in Cork Harbor, Ireland. Harmful Algae 31: 114–124
D'Alelio D, Ribera D'Alcalà M, Dubroca L et al (2010) The time for sex: a biennial life cycle in a marine planktonic diatom. Limnol Oceanogr 55:106–114
Dale B, Murphy M (2014) A retrospective appraisal of the importance of high-resolution sampling for harmful algal blooms: lessons from long-term phytoplankton monitoring at Sherkin Island, S.W. Ireland. Harmful Algae 40:23–33
Davidson K, Gowen RJ, Harrison PJ et al (2014) Anthropogenic nutrients and harmful algae in coastal waters. J Environ Manag 146:206–216
Davidson K, Gowen RJ, Tett P et al (2012) Harmful algal blooms: How strong is the evidence that nutrient ratios and forms influence their occurrence? Estuar Coast Shelf Sci 115:399–413. https://doi.org/10.1016/j.ecss.2012.09.019
Díaz P, Molinet C, Caceres MA et al (2011) Seasonal and intratidal distribution of *Dinophysis* spp. in a Chilean fjord. Harmful Algae 10:155–164. https://doi.org/10.1016/j.hal.2010.09.001
Erdner DL, Percy L, Keafer B et al (2010) A quantitative real-time PCR assay for the identification and enumeration of *Alexandrium* cysts in marine sediments. Deep Sea Res II 57(3–4):279–287
Estrada M, Solé J, Anglès S et al (2010) The role of resting cysts in *Alexandrium minutum* population dynamics. Deep Sea Res II 57(3–4):308–321
Figueroa RI, Bravo I, Garcés E (2006) The multiple routes of sexuality in *Alexandrium taylori* (Dinophyceae) in culture. J Phycol 42:1028–1039
Flynn KJ, Mitra A, Glibert PM et al (2018) Mixotrophy in HABs: by whom, on whom, when, why, and what next. In: Glibert PM, Berdalet E, Burford M et al (eds) Global ecology and oceanography of harmful algal blooms. Springer, Cham, pp 113–132
Flynn KJ, Stoecker DK, Mitra A et al (2013) Misuse of the phytoplankton-zooplankton dichotomy: the need to assign organisms as mixotrophs within plankton functional types. J Plankton Res 35(1):3–11. https://doi.org/10.1093/plankt/fbs062
Fu FX, Tatters AO, Hutchins DA (2012) Global change and the future of harmful algal blooms in the ocean. Mar Ecol Prog Ser 470:207–233. https://doi.org/10.3354/meps10047

Garcés E, Camp J (2012) Habitat changes in the Mediterranean Sea and the consequences for harmful algal blooms formation. In: Stambler N (ed) Life in the Mediterranean Sea: a look at habitat changes. Nova Science Publishers Inc., pp 519–541

Garcés E, Montresor M, Lewis J et al (2010) Phytoplankton life cycles and their impacts on the ecology of harmful algal blooms. Deep Sea Res II 57:159–161

GEOHAB (2010) Global ecology and oceanography of harmful algal blooms, GEOHAB core research project: HABs in fjords and coastal embayments. Cembella A, Guzmán L, Roy S, et al (eds) IOC and SCOR, Paris, France, and Newark, Delaware, 57 pp

GEOHAB (2013) Global ecology and oceanography of harmful algal blooms, GEOHAB core research project: HABs in fjords and coastal embayments, second open science meeting. Roy S, Pospelova V, Montresor M et al (eds) IOC and SCOR, Paris, France and Newark, Delaware USA, 52 pp

Glibert PM, Al-Azri A, Allen JI et al (2018a) Key questions and recent research advances on harmful algal blooms in relation to nutrients and eutrophication. In: Glibert PM, Berdalet E, Burford M et al (eds) Global ecology and oceanography of harmful algal blooms. Springer, Cham, pp 229–259

Glibert PM, Beusen AHW, Harrison JA et al (2018b) Changing land, sea- and airscapes: sources of nutrient pollution affecting habitat suitability for harmful algae. In: Glibert PM, Berdalet E, Burford M et al (eds) Global ecology and oceanography of harmful algal blooms. Springer, Cham, pp 53–76

Glibert PM, Boyer J, Heil C et al (2010) Blooms in lagoons: different from those of river-dominated estuaries. In: Kennish M, Paerl H (eds) Coastal lagoons: critical habitats of environmental change. Taylor and Francis, pp 91–114

Glibert PM, Burford MA (2017) Globally changing nutrient loads and harmful algal blooms. Oceanography 30:44–55

Glibert PM, Kana TM, Brown K (2013) From limitation to excess: the consequences of substrate excess and stoichiometry for phytoplankton physiology, trophodynamics and biogeochemistry, and the implications for modeling. J Mar Syst 125:14–28

Gobler CJ, Berry DL, Dyhrman ST et al (2011) Niche of harmful algal *Aureococcus anophagefferens* revealed through ecogenomics. Proc Natl Acad Sci USA 108:4352–4357. https://doi.org/10.1073/pnas.1016106108

Godhe A, Härnström K (2010) Linking the planktonic and benthic habitat: genetic structure of the marine diatom *Skeletonema marinoi*. Mol Ecol 19:4478–4490

Hallegraeff GM (2010) Ocean climate change, phytoplankton community responses, and harmful algal blooms: a formidable predictive challenge. J Phycol 46:220–235. https://doi.org/10.1111/j.1529-8817.2010.00815.x

Hamasaki K, Horie M, Tokimitsu S et al (2001) Variability in toxicity of the dinoflagellate *Alexandrium tamarense* isolated from Hiroshima Bay, western Japan, as a reflection of changing environmental conditions. J Plankton Res 23:271–278. https://doi.org/10.1093/plankt/23.3.271

Härnström K, Ellegaard M, Andersen TJ et al (2011) Hundred years of genetic structure in a sediment revived diatom population. Proc Natl Acad Sci USA 108:4252–4257

Harvey EL, Menden-Deuer S, Rynearson TA (2015) Persistent intra-specific variation in genetic and behavioral traits in the raphidophyte, *Heterosigma akashiwo*. Front Microbiol 6:1277. https://doi.org/10.3389/fmicb.2015.01277

John U, Litaker W, Montresor M et al (2014) Formal revision of the *Alexandrium tamarense* species complex (Dinophyceae) taxonomy: the introduction of five species with emphasis on molecular-based (rDNA) classification. Protist 165:779–804

Johnsen TM, Eikrem W, Olseng CD et al (2010) *Prymnesium parvum*: the Norwegian experience. JAWRA 46(1):6–13. https://doi.org/10.1111/j.1752-1688.2009.00386.x

Jones SE, Lennon JT (2010) Dormancy contributes to the maintenance of microbial diversity. Proc Natl Acad Sci USA 107:5881–5886

Lucas AJ, Largier JL (2013) The influence of physical variability on HAB patterns and persistence in bays. In: GEOHAB, Roy S, Pospelova V, Montresor M, Cembella A (eds) Global ecology and oceanography of harmful algal blooms, GEOHAB Core Research Project: HABs in fjords and coastal embayments. Second open science meeting: progress in interpreting life history and growth dynamics of harmful algal blooms in fjords and coastal environments. Intergovernmental Oceanographic Commission and Scientific Committee on Oceanic Research, Paris and Newark, DE, pp 50–51 (abstract only)

Lucas AJ, Dupont CL, Tai V et al (2011) The green ribbon: multiscale physical control of phytoplankton productivity and community structure over a narrow continental shelf. Limnol Oceanogr 56:611–626. https://doi.org/10.4319/lo.2011.56.2.0611

Lundholm N, Clarke A, Ellegaard M (2010) A 100-year record of changing *Pseudo-nitzschia* species in a sill-fjord in Denmark related to nitrogen loading and temperature. Harmful Algae 9:449–457. https://doi.org/10.1016/j.hal.2010.03.001

McGillicuddy DJ Jr, Anderson DM, Lynch DR et al (2005) Mechanisms regulating large-scale seasonal fluctuations in *Alexandrium fundyense* populations in the Gulf of Maine: results from a physical-biological model. Deep Sea Res II 52:2698–2714

McGillicuddy DJ, Townsend DW, Keafer BA et al (2014) Georges Bank: a leaky incubator of *Alexandrium fundyense* blooms. Deep Sea Res II 103:163–173

Penna A, Battocchi C, Garcés E et al (2010) Detection of microalgal resting cysts in European coastal sediments using a PCR-based assay. Deep Sea Res II 57(3–4):288–300

Persson A, Smith BC, Wikfors GH et al (2013) Differences in swimming pattern between life cycle stages of the toxic dinoflagellate *Alexandrium fundyense*. Harmful Algae 21–22:36–43

Pilskaln CH, Hayashi K, Keafer BA et al (2014) Benthic nepheloid layers in the Gulf of Maine and *Alexandrium* cyst inventories. Deep Sea Res II 103:55–65

Raine R, Berdalet E, Yamazaki H et al (2018) Key questions and recent research advances on harmful algal blooms in stratified systems. In: Glibert PM, Berdalet E, Burford M et al (eds) Global ecology and oceanography of harmful algal blooms. Springer, Cham, pp 165–186

Raine R, McDermott G, Silke J et al (2010) A simple short range model for the prediction of harmful algal events in the bays of southwestern Ireland. J Mar Syst 83:150–157. https://doi.org/10.1016/j.jmarsys.2010.05.001

Richlen ML, Erdner DL, McCauley LAR et al (2012) Extensive genetic diversity and rapid population differentiation during blooms of *Alexandrium fundyense* (Dinophyceae) in an isolated salt pond on Cape Cod, MA, USA. Ecol Evol 2(10):2583–2594

Roy S, Parenteau M, Casas-Monroy O et al (2012) Coastal ship traffic: a significant introduction vector for potentially harmful dinoflagellates in eastern Canada. Can J Fish Aquat Sci 69: 627–644

Ryan JP, McManus MA, Sullivan JM (2010) Interacting physical, chemical and biological forcing of phytoplankton thin-layer variability in Monterey Bay, California. Cont Shelf Res 30:7–16. https://doi.org/10.1016/j.csr.2009.10.017

Sekula-Wood E, Benitez-Nelson C, Morton S et al (2011) *Pseudo-nitzschia* and domoic acid fluxes in Santa Barbara Basin (CA) from 1993 to 2008. Harmful Algae 10:567–575. https://doi.org/10.1016/j.hal.2011.04.009

Shull DH, Kremp A, Mayer LM (2014) Bioturbation, germination and deposition of *Alexandrium fundyense* cysts in the Gulf of Maine. Deep Sea Res II 103:66–78

Skovgaard A, Hansen PJ (2003) Food uptake in the harmful alga *Prymnesium parvum* mediated by excreted toxins. Limnol Oceanogr 48:1161–1166

Smayda TJ (2007) Reflections on the ballast water dispersal-harmful algal bloom paradigm. Harmful Algae 6:601–622

Smayda TJ, Trainer VL (2010) Dinoflagellate blooms in upwelling systems: seeding, variability, and contrasts with diatom bloom behaviour. Prog Oceanogr 85:92–107. https://doi.org/10.1016/j.pocean.2010.02.006

Tobin ED, Grünbaum D, Patterson J et al (2013) Behavioral and physiological changes during benthic-pelagic transition in the harmful alga, *Heterosigma akashiwo*: potential for rapid bloom formation. PLoS One 8(10):e76663. https://doi.org/10.1371/journal.pone.0076663

Varela D, Paredes J, Alves-de-Souza C et al (2012) Intraregional variation among *Alexandrium catenella* (Dinophyceae) strains from southern Chile: morphological, toxicological and genetic diversity. Harmful Algae 15:8–18. https://doi.org/10.1016/j.hal.2011.10.029

Velo-Suárez L, Brosnahan ML, Anderson DM et al (2013) A quantitative assessment of the role of the parasite *Amoebophrya* in the termination of *Alexandrium fundyense* blooms within a small coastal embayment. PLoS One 8(12):e81150. https://doi.org/10.1371/journal.pone.0081150

Walsh JJ, Dieterle DA, Chen FR et al (2011) Trophic cascades and future harmful algal blooms within ice-free Arctic Seas north of Bering Strait: a simulation analysis. Prog Oceanogr 91(3): 312–343. https://doi.org/10.1016/j.pocean.2011.02.001

Warns A, Hense I, Kremp A (2012) Modelling the life cycle of dinoflagellates: a case study with *Biecheleria baltica*. J Plankton Res 35(2):379–392. https://doi.org/10.1093/plankt/fbs095

Wells ML, Karlson B (2018) Harmful algal blooms in a changing ocean. In: Glibert PM, Berdalet E, Burford M et al (eds) Global ecology and oceanography of harmful algal blooms. Springer, Cham, pp 77–90

Wells ML, Trainer VL, Smayda TJ et al (2015) Harmful algal blooms and climate change: learning from the past and present to forecast the future. Harmful Algae 49:68–93

Wohlrab S, Iversen MH, John U (2010) A molecular and co-evolutionary context for grazer induced toxin production in *Alexandrium tamarense*. PLoS One 5(11):e15039. https://doi.org/10.1371/journal.pone.0015039

Chapter 11
Key Questions and Recent Research Advances on Harmful Algal Blooms in Eastern Boundary Upwelling Systems

Grant C. Pitcher, Francisco G. Figueiras, Raphael M. Kudela, Teresa Moita, Beatriz Reguera, and Manuel Ruiz-Villareal

11.1 Introduction

The scientific goal of the international Global Ecology and Oceanography of Harmful Algal Blooms (GEOHAB) Programme was to improve the prediction of harmful algal blooms (HABs; GEOHAB 2001). The approach of GEOHAB was to be comparative primarily through the establishment of Core Research Projects (CRP) in which international collaborative research would be conducted in ecosystems sharing common features [GEOHAB 2003; see also Chap. 3, Kudela et al. (2018)].

As one of the CRPs of GEOHAB, HABs in Upwelling Systems was initiated by means of an Open Science Meeting (OSM) in Lisbon, Portugal, in November 2003, which served to provide an overview of HABs in designated upwelling systems and to formulate key questions to be addressed as a requirement for advancing our understanding of the ecology and oceanography of HABs in these systems (GEOHAB 2005). A summary paper of the CRP—HABs in Upwelling Systems

G.C. Pitcher (✉)
Fisheries Research and Development, Cape Town, South Africa
e-mail: GrantP@daff.gov.za

F.G. Figueiras
Instituto de Investigaciones Marinas (CSIC), Vigo, Spain

R.M. Kudela
University of California Santa Cruz, Santa Cruz, CA, USA

T. Moita
Instituto Portugues do Mar e da Atmosfera (IPMA), Lisboa, Portugal

B. Reguera
Instituto Español de Oceanografía, Centro Oceanográfico de Vigo, Vigo, Spain

M. Ruiz-Villareal
Instituto Español de Oceanografía, Centro Oceanográfico de A Coruña, A Coruña, Spain

P.M. Glibert et al. (eds.), *Global Ecology and Oceanography of Harmful Algal Blooms*, Ecological Studies 232, https://doi.org/10.1007/978-3-319-70069-4_11

was published in *Oceanography* as part of a special issue on HABs as an initial step in the implementation of the project (Kudela et al. 2005). Here an invitation was extended to the HAB and oceanographic community to review the GEOHAB Science and Implementation Plans (GEOHAB 2001, 2003) and the CRP plan (GEOHAB 2005) and begin to frame future research objectives to address the goals of GEOHAB. In 2010 a comparative review of the ecology and oceanography of HABs in major eastern boundary upwelling systems was undertaken as part of the CRP with the publication of a special issue of *Progress in Oceanography*.

This chapter follows the conclusion of GEOHAB in 2015 and provides a summary of progress towards addressing the eight key questions identified at the OSM in 2003. It specifically addresses achievements in terms of the overall goal of GEOHAB of improved prediction of HABs supported by observation and modelling systems. We also highlight research activities that remain high priority and may benefit from further cooperative international research.

11.2 Key Question 1: Are There Definable Adaptive Strategies that Characterize HAB Species in Upwelling Systems?

Establishment of unique, species-specific characteristics and adaptations that contribute to the proliferation of HAB species in upwelling systems was expected to guide the construction and parameterization of predictive models. It was anticipated that the identification of the cellular properties that select for these species in upwelling systems would be assisted by a comparison of the species common to these systems. However, the list of harmful species in upwelling systems compiled by Trainer et al. (2010) caused Smayda (2010) to conclude, based on the cosmopolitan distribution of the HAB species common to upwelling systems, that there was "no HAB flora unique to upwelling systems".

The considerable diversity of the dinoflagellate species found in upwelling systems is considered to reflect the mosaic of multiple and shifting sub-habitats present within these systems. HABs in upwelling systems are most often attributed to dinoflagellates, and most of the distinct habitat types as defined by Smayda and Reynolds (2001), in which dinoflagellates may bloom, are encountered in upwelling systems, offering growth opportunities to a variety of dinoflagellate life forms. An example of the role of different mechanisms of bloom formation, based on different life strategies, in determining different spatial and seasonal patterns of occurrence, is provided by Bravo et al. (2010) in demonstrating the occupation of two very different sub-habitats within the Galician upwelling regime by two paralytic shellfish poisoning (PSP) toxin-producing dinoflagellates *Alexandrium minutum* and *Gymnodinium catenatum*. Whereas *A. minutum* thrives within the confines of small sheltered bays, e.g. Bahia de Baiona, and the recurrence of blooms is dependent on resting cysts, the bloom dynamics of *G. catenatum* are clearly linked to seasonal downwelling and the advection of *G. catenatum* into the rías from offshore, with little dependence on cyst germination.

Species-specific behaviour in the form of depth regulation and the development of thin layers (TLs) in upwelling systems have important consequences for the cross-shelf and alongshore transport of phytoplankton populations (Velo-Suárez et al. 2009; Moita et al. 2016). Off the northwest Iberian coast, *Dinophysis acuminata* and *Dinophysis acuta,* both known to cause diarrhetic shellfish poisoning (DSP), are found to concentrate in discrete layers with *D. acuminata* typically occurring closer to the surface, while *D. acuta* is usually found at the top of the pycnocline. Although vertical migration is considered an important mechanism in the establishment of these layers, observations of higher cell division within these layers are also likely to contribute to their formation (Velo-Suárez et al. 2009). High interannual variability in these *Dinophysis* blooms on the northwest Iberian coast has been attributed to strong variability in alongshore and cross-shelf currents associated with variable upwelling winds. However, in utilizing an individual-based model (IBM) coupled to a Regional Ocean Modelling System (ROMS) to investigate the role of these currents in the advection of *Dinophysis* blooms, the results were shown to be very sensitive to the vertical distribution of cells, thereby clearly demonstrating the need to consider species-specific behaviour in developing predictive capabilities.

The diversity of HAB species and the required explicit representation of the biology of these species for model development ensure that the need for species-specific information remains a priority (Pitcher 2012). However, this biological complexity needs to be matched with appropriate complexity in the representation of environmental and ecological parameters that characterize sub-habitats to further predictive capabilities.

11.3 Key Question 2: What Seeding Strategies Persist Within Upwelling Regions and Are They Consistent Among Regions?

Upwelling systems provide a unique selective pressure on phytoplankton population maintenance as they are inherently advective and need to counter washout. Blooms result from active cellular growth, physical accumulations of populations from far-field sources, local movements of the same population and a combination of these sources (Smayda and Trainer 2010). Nonetheless, the origin of the populations and the life history stages of the cells that inoculate blooms in upwelling systems are often unknown. Particularly during active upwelling when diatoms are in bloom, dinoflagellates may be found offshore, at fronts or in outer shelf currents where they may also accumulate vertically in subsurface layers. These refuge populations may serve as pelagic seed banks and provide sources of inocula for nearshore blooms during periods of upwelling relaxation (Smayda 2002).

Similar seeding mechanisms have been shown by Seegers et al. (2015) to exist for toxic *Pseudo-nitzschia* blooms responsible for amnesic shellfish poisoning (ASP) in Southern California except that the delivery of subsurface offshore populations to nearshore surface waters occurs during upwelling.

The role of life-cycle transitions in the seeding strategies of species-specific blooms has been considered to hold the key to the prediction of several HABs. In particular, bloom simulation through quantification of the benthic-pelagic coupling of life-cycle stages incorporating a benthic cyst has been considered likely to be predictive. However, progress in the parameterization of life-cycle transitions and their incorporation into species-specific numerical models has been limited in upwelling systems. Most dinoflagellates in upwelling systems are holoplanktonic in that they rely on vegetative cells to survive inhospitable growth conditions and to seed blooms (Smayda and Trainer 2010). This dominance by holoplanktonic dinoflagellates suggests that a cyst stage may provide little advantage for growth in upwelling systems.

Meroplanktonic or cyst-forming dinoflagellates are a minority in upwelling systems and tend also to be dominated by heterotrophic dinoflagellates (Pitcher and Joyce 2009). Nevertheless the prominence of the cyst-forming toxic species *Alexandrium catenella*, *Gymnodinium catenatum*, *Lingulodinium polyedrum* and *Protoceratium reticulatum* in upwelling systems suggests further exploration of the predictive value of a better understanding of the life-cycle stages of these species. Of interest in upwelling systems is the long-term irregularity of some dinoflagellate blooms and whether bloom failure may reflect failed seeding. As an example, observations of benthic resting cysts of *L. polyedrum* in the southern Benguela without evidence of pelagic blooms were made for many years prior to their appearance in 2014 (Pitcher and Joyce 2009), while in the Galician Rías Baixas *L. polyedrum,* known to cause red tides during the first half of the twentieth century, is now rare despite an abundance of cysts in the sediment (Trainer et al. 2010). These observations suggest that bloom behaviour cannot be inferred from the capacity of a species to produce resting cysts or from their presence or abundance in sedimentary deposits. Further complicating the role of dinoflagellate cysts are observations of varying dormancy periods. As an example the presence of a short dormancy period in *A. catenella* and *G. catenatum* suggests a rapid cycling between benthic and planktonic stages rather than an overwintering strategy (Joyce and Pitcher 2006; Amorim et al. 2014).

The role of species-specific life-cycle transitions in determining high variability at the decadal, annual and event scales of *Gymnodinium catenatum* blooms in the Iberian upwelling system has been investigated by Amorim et al. (2014). Observations of cyst distribution, physiology and viability were incorporated into particle-tracking experiments using a hydrodynamic model to reproduce shelf circulation within the study region. High intra- and interannual variability was evident from the model predictions of seeding with successful inoculation of coastal blooms most likely to occur from mid-shelf cyst beds during early summer. It may be that in upwelling systems, phytoplankton species are able to opportunistically exploit multiple seeding options conferred by their life cycle rather than depending on a single seeding mode in which case prediction is further complicated.

11.4 Key Question 3: How Do Small-Scale Physical Processes Affect HAB Growth and Dispersion in Upwelling Systems?

The importance of small-scale physical processes in effecting phytoplankton selection is typically observed in the vertical dimension. Cycles of mixing and stratification of the water column are known to be operative in regulating spatial and temporal differences in phytoplankton assemblages consistent with the conceptual model of Margalef (1978) in which functional groups of phytoplankton are variously adapted to an environment defined in terms of turbulence and nutrient availability. Recent progress in addressing the role of small-scale physical processes in HAB development has typically been achieved through technological advances in providing high-resolution observations in upwelling systems.

As an example, wirewalker profiling moorings have been used to demonstrate the influence of diurnal wind variability in forcing energetic inertial current oscillations instrumental in influencing phytoplankton distributions and productivity in the southern Benguela (Lucas et al. 2014). Strong shear caused by the vertical structure of the inertial oscillations driven by diurnal winds acts to weaken stratification, redistributing phytoplankton and mixing high nutrient waters into surface waters with consequent increases in phytoplankton biomass. This diurnal-inertial resonance-driven mechanism for nutrient flux into surface waters, embedded within the low-frequency advective vertical flux forced by Ekman dynamics, is considered to be of first order importance in the development of HABs during periods of stratification.

Monterey Bay has also been the site of rigorous investigation of vertical structure with reference to HABs. Intensive autonomous underwater vehicle (AUV) surveys have been undertaken to examine the spatio-temporal scales and processes of phytoplankton thin-layer development (Ryan et al. 2010). These studies have included establishment of the role played by species-specific phytoplankton properties in TL dynamics (Rines et al. 2010). As an example, layer formation by *Akashiwo sanguinea* was shown to be associated with strong migration and swarming behaviour. Migration was clearly responsive to nutrient levels with nighttime observations showing TLs aligned with a nitrate (NO_3^-) isopleth that diverged significantly from the isotherms owing to nutrient uptake during downward migration (Fig. 11.1). During these periods of stratification and thin-layer development, internal tidal pumping from Monterey Canyon onto the shelf is considered to support the growth of vertically migrating phytoplankton by enhancing nutrient availability at the thermocline. Laboratory and field data also demonstrate that both *A. sanguinea* and *Prorocentrum micans* will vertically migrate to

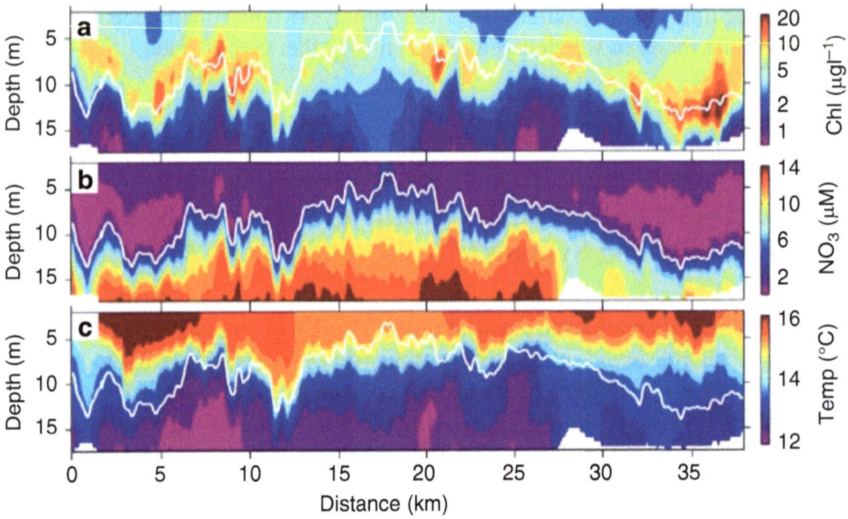

Fig. 11.1 An AUV survey in Monterey Bay showing vertical chlorophyll, NO_3^- and temperature distributions (from 20h05 on 28 August to 02h35 on 29 August 2005; white contour on all panels depicts the 3 μM NO_3^- isopleth). Chlorophyll distributions clearly aligned to the NO_3^- isopleth depicted diurnal vertical migration and nutrient uptake by the dinoflagellate *Akashiwo sanguinea*. Deepening of the nitracline in response to nighttime migration and the uptake of nutrients was evident in the nighttime divergence of the nitrate isopleth from the isotherms (Reproduced from Ryan et al. 2010 with permission of the publisher)

acquire both NO_3^- and iron, confirming the long-held belief that strong-swimming dinoflagellates are competitively favoured in regions with strong vertical gradients (Peacock and Kudela 2014).

On the Galician coast, differences in shear profiles have been shown to be important in the formation of TLs of toxin-producing species of *Pseudo-nitzschia* in the Ría de Pontevedra (Velo-Suárez et al. 2010). The coupling between maximum values of shear and buoyancy frequency was found to shape subsurface populations into TLs during upwelling events at which time the vertical variation of the horizontal inflow velocity was enhanced. Further to these studies, Díaz et al. (2014) demonstrated the importance of tidal forcing in modulating the distribution of *Pseudo-nitzschia* as determined by wind-driven circulation within the Ría. Circadian variability as regulated by tidal forcing was responsible for cell maxima appearing during low tides and maximum water column stratification.

Also of interest, in terms of small-scale physical processes, is the longshore variation of exposure of intertidal organisms to toxic *Pseudo-nitzschia* species as determined by surf zone hydrodynamics (Shanks et al. 2016). The ingress of *Pseudo-nitzschia* into surf zones and consequent contamination of filter feeders contrasted dramatically between reflective and dissipative surf zones. *Pseudo-nitzschia* and particulate domoic acid (DA) concentrations were shown to be markedly higher at dissipative beaches where rip currents were found to be conducive to the efficient exchange of surf zone water with offshore water.

Technological advances have been effective in achieving the high-resolution measurements required to demonstrate the further complexity created at the small scale by physically driven niche diversification and species-specific behaviour. However, the limited spatial resolution of many numerical models has prevented incorporation of this information into models for improved prediction.

11.5 Key Question 4: How Do Nutrient Supply Type and Ratios Determine HAB Population Dynamics in Upwelling Systems?

The important role of nutrients in the ordination of phytoplankton with environmental change is acknowledged by many authors (Margalef 1978; Wyatt 2014; Glibert 2016), and the clear linkage between nutrient availability and supply, and the physical dynamics of upwelling systems, indicates an obvious need to understand the nutrient ecophysiology of HAB species in attempting to predict HABs.

Predictable changes in the uptake of new and regenerated nitrogen (N), as depicted by changes in the f-ratio in association with an upwelling-downwelling cycle, are known to be accompanied by predictable changes in the phytoplankton community reflecting changing nutritional strategies. Recognition of these changes has led to attempts to establish the role of the nutrient regime in determining not only community succession but also the selection of harmful algal bloom species through measurements of N uptake rates, including nutrient uptake kinetics [see also Chap. 6, Glibert et al. (2018)]. As an example, Seeyave et al. (2009) compared nutrient uptake rates in the southern Benguela in assemblages dominated by *Pseudo-nitzschia*, *Alexandrium catenella* and *Dinophysis acuminata* (Fig. 11.2). Contrasting nutrient strategies were evident in that *Pseudo-nitzschia* spp. were found to display the highest maximum specific uptake rates for NO_3^- and NH_4^+, whereas *D. acuminata* displayed the highest affinity for NH_4^+ as shown by a higher α and the lowest V_{max} for NO_3^-. The study was able to conclude that the *Pseudo-nitzschia* spp. and *A. catenella* were velocity strategists adapted to utilizing high NO_3^- concentrations, whereas *D. acuminata* was shown to be an affinity strategist successful in N-depleted waters. Unfortunately, there have been very few attempts to determine species-specific nutrient requirements, and inconsistencies attributed to differences in nutrient history and/or interspecific differences are often observed. Moreover, the dynamics of nutrient kinetic relationships are rarely taken into consideration (e.g., Kana and Glibert 2016).

For upwelling regions there has been a general perception that anthropogenic nutrient inputs are small relative to that supplied by upwelling and are therefore unlikely to influence the development of HABs. However, recent studies on the Californian coast have shown that waste water effluent, riverine runoff and

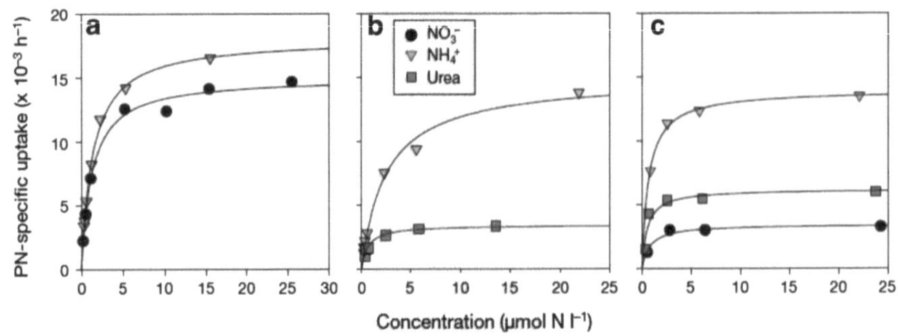

Fig. 11.2 Nutrient uptake versus concentration data fitted to the Michaelis–Menten equation using iterative, non-linear regression for phytoplankton assemblages in the southern Benguela dominated by (**a**) *Pseudo-nitzschia*, (**b**) *Alexandrium catenella* and (**c**) *Dinophysis acuminata* (Reproduced from Seeyave et al. 2009 with permission of the publisher)

submarine groundwater discharge and atmospheric deposition can provide a significant source of N for nearshore productivity (Howard et al. 2014; Lecher et al. 2015; Reifel et al. 2013). At the system or regional scale, these nutrient sources are smaller by at least an order of magnitude. However, at smaller spatial scales relevant to HAB development, natural and anthropogenic nutrient contributions are in some cases equivalent and are likely to stimulate bloom-forming phytoplankton.

11.6 Key Question 5: What Is the Role of Genetic Predisposition Versus Environmental Conditions in Toxin Production in Different Upwelling Systems Within a Given Genus or Species?

Although molecular techniques provide useful tools in determining ecophysiological responses of harmful algae in upwelling systems, progress in the evaluation of genotypic versus environmental variability in the toxigenicity of species is limited. Nevertheless genetic studies are increasingly demonstrating their value in the identification and delineation of harmful species, and molecular analysis may simply serve as a useful tool to distinguish morphologically similar species that may or may not produce toxins. As an example, Raho et al. (2008) undertook morphological, toxin and molecular analyses to discriminate *Dinophysis ovum* from the "*Dinophysis acuminata* complex" in the Galician Rías Baixas. This study showed that *D. ovum* which has in all likelihood often been misidentified as *D. acuminata* may be discriminated through analysis of mitochondrial genes. Such species-specific separation is fundamental to comparative studies and provided the first report of lipophilic shellfish toxins in *D. ovum*. The presence of only okadaic acid in both *D. ovum* and *D. acuminata* meant that these species were also indistinguishable on the basis of their toxin profile.

Molecular sequences from ribosomal DNA domains were used by Howard et al. (2009) to determine the existence of genetic differences among geographically distinct populations and between toxic and nontoxic isolates of dinoflagellate species known to produce yessotoxins (YTXs). Although the study firmly placed all known YTX-producing species within the order Gonyaulacales, the intraspecific diversity within the YTX-producing taxa was low, and it was generally difficult to distinguish genetically between isolates based on either toxicity or geographic origin. For example, there was no genetic divergence between cultures of *Protoceratium reticulatum* isolated from the California and Benguela current upwelling systems. The lack of diversity at this level was considered to indicate environmental control of toxicity.

The combination of molecular and morphological studies has been useful in demonstrating the presence of many cryptic and pseudo-cryptic species in various groups of algae. Their identity is important not only from a taxonomic point of view but also in the identification of different physiological traits including toxin production. Lundholm et al. (2012) used genes from the nucleus, mitochondria and plastids to reveal the presence of an unprecedented diversity of *Pseudo-nitzschia* off the coast of Washington State including a high degree of pseudo-cryptic diversity. The study also demonstrated considerable variation in toxicity among offspring and between parents and offspring providing further evidence of interclonal toxin variability in *Pseudo-nitzschia* species. Sison-Mangus et al. (2014) in turn showed varying DA production in several *Pseudo-nitzschia* species from the California current system to be linked to phylogenetically distinct bacterial associations. These molecular studies need to be extended to include assessment of the role of interacting environmental factors in determining the phenotypic plasticity of toxin production in *Pseudo-nitzschia* and other harmful species in upwelling systems.

11.7 Key Question 6: How Does Coastal Morphology and Bathymetry Affect HAB Dynamics in Upwelling Systems?

Although the timing of HABs in upwelling systems is typically controlled by wind stress fluctuations and buoyancy inputs at the event, seasonal and interannual scales, the spatial distribution of HABs within this temporal framework is controlled by mesoscale features that interrupt typical upwelling circulation patterns (Pitcher et al. 2010). Unlike offshore environments, the biophysical interaction in the coastal environment acquires a strong, spatially explicit component as a consequence of coastline configuration and orientation and bottom topography. These features determine the distribution of HABs through their influence on wind stress and water stratification and through their retentive circulation that minimizes dispersal by lateral advection. As an example, the identification of six HAB hotspots on the US West Coast is attributed to these coastline features and the elevated biomass with which they are associated (Frolov et al. 2013).

A spectrum of coastline configurations ranging from headlands, capes, peninsulas, rias, bays and estuaries have been assessed in terms of their vulnerability to HABs (Pitcher et al. 2010). Representing systems of increasing isolation from the open coast and consequent increasing retention times, these environments may offer the required balance between advective dispersion and enrichment, but also critical are water mass exchange processes with the adjacent coastal environments. Although the properties of stratification and retention are more pronounced the more isolated the bay from the open coast, limiting mechanisms of exchange with the coastal environment tend also to limit the introduction of coastal seed populations and the delivery of nutrients to these environments. Consequently high-biomass HABs have been shown to be more prominent in open bays with cyclonic retention. Examples include Monterey Bay, USA (Ryan et al. 2008); Paracas Bay, Peru (Kahru et al. 2008); Lisbon Bay, Portugal (Oliveira et al. 2009); and False Bay and St Helena Bay, South Africa (Pitcher et al. 2008).

11.8 Key Question 7: What Is the Relative Importance of Cross-Shelf and Along-Shore Advection for HABs in Different Upwelling Systems?

Cross-shelf and alongshore currents are prominent physical features of coastal upwelling systems, and their importance in the transport, accumulation and dispersion of HAB populations is now well recognized (Pitcher et al. 2010).

Cross-shelf currents, determined by fluctuations in wind direction associated with the upwelling-downwelling cycle, are known to be the cause of rapid changes in the phytoplankton at the event scale. In particular observations of the shoreward accumulation of dinoflagellate-dominated blooms when upwelling winds relax are common. As an example, Barton et al. (2016) have demonstrated the role of inward advection and subduction in the Ría de Vigo in concentrating phytoplankton in the middle and inner ría during periods of downwelling. Dinoflagellates capable of maintaining their position in the water column by swimming are particularly susceptible to these concentrating mechanisms. Velo-Suárez et al. (2014) have specifically demonstrated rapid increases of *D. acuminata* in the Galician Rías Baixas in response to wind relaxation and downwelling as depicted by Ekman transport estimates and water column temperature. Their conceptual model of the mechanisms by which *D. acuminata* blooms develop in a ría incorporates not only the physical transport of blooms but also their mixotrophic nutrition and behaviour. The biological processes include nutritional status as determined by availability of its ciliate prey *Mesodinium*, the possible role of parasitism and an overwintering stage.

There has been an increasing appreciation of the role of alongshore currents in dictating the distribution of HABs in upwelling systems. Poleward counter currents on the inner shelf which strengthen during the latter stages of the upwelling season have been shown to be particularly important in the transport of HABs. As an example, Escalera et al. (2010) have examined the role of the Portuguese Coastal Counter Current in the poleward transport of *Dinophysis acuta* and the occurrence of DSP in the Galician Rías Baixas in autumn. *Dinophysis acuta* was shown to develop off the Portuguese coast during summer where its decline towards the end of the upwelling season was associated with its appearance off the Galician coast which was attributed to transport by an inner shelf counter current. Maximum cell concentrations in the Galician Rías Baixas were found to correspond to periods when estimates of cell division were almost zero, thereby supporting the view that inshore poleward flow combined with cross-shelf advection was required to explain the abrupt cell increases in the rías at the end of the upwelling season.

The complexity of bloom transport pathways of *Pseudo-nitzschia* from development sites to razor clam beaches in the Pacific Northwest has been demonstrated by Hickey et al. (2013). A conceptual framework of transport mechanisms revealed the need to take into account seasonal changes in the prevailing currents in determining the source of toxic blooms (Fig. 11.3). During summer and autumn when upwelling favourable winds and equatorward flow predominate, the Juan de Fuca eddy serves as the source of *Pseudo-nitzschia*. At this time brief transitions to downwelling-favourable winds are required to move *Pseudo-nitzschia* blooms shoreward onto Washington coastal beaches. However, in early spring when poleward flow dominates, toxic *Pseudo-nitzschia* from Heceta Bank has been shown to be the likely source of toxin in razor clams on beaches in the Pacific Northwest. Within this region the buoyant Columbia River plume is also expected to influence transport pathways serving primarily as a barrier at some locations during summer and autumn and as a conduit during winter and spring.

Knowledge of these cross-shelf and alongshore transport mechanisms and pathways aid in demonstrating the potential use of circulation models driven by surface boundary forcing from weather forecasts in conjunction with real-time observations in the prediction of blooms.

11.9 Key Question 8: Are Climate Indicators Predictive of HAB Events in Upwelling Systems?

The growing need to establish the likely impact of climate change, and more specifically global warming, on the incidence of HABs provides a significant predictive challenge [see also Chap. 5, Wells and Karlson (2018)]. The increasing uncertainty in progressing from climate change to ocean response to biological impact casts considerable doubt on ecosystem scenario development. Nevertheless upwelling systems and the HABs common to these systems are likely to be sensitive

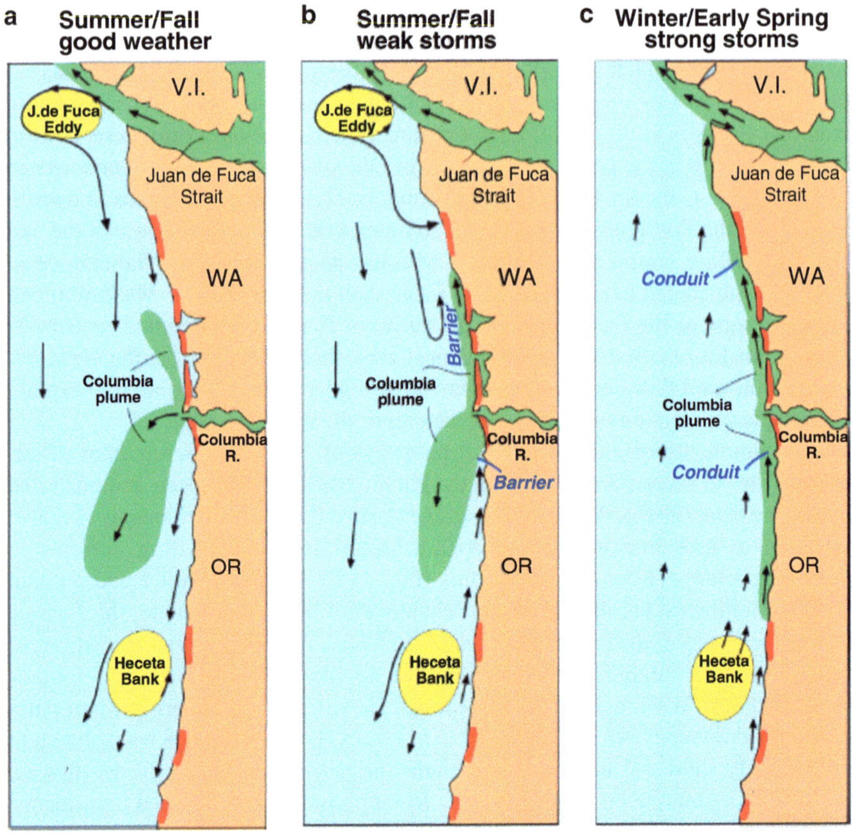

Fig. 11.3 A conceptual model of the transport of toxic *Pseudo-nitzschia* cells from the Juan de Fuca eddy and Heceta bank during summer and autumn under (**a**) prevailing upwelling favourable winds and (**b**) during weak reversals and (**c**) during late winter and early spring. *Arrows* depict surface currents, clamming beaches are marked in *red*, and freshwater plumes in *green* (Reproduced from Hickey et al. 2013 with permission of the publisher)

to climate change because of their strong dependency on environmental drivers. Consequently changing trends in HABs related to climate are emerging with increasing availability of long-term data sets.

Several studies have addressed long-term trends in the occurrence of HABs in the Iberian coastal system in response to reduced upwelling. Álvarez-Salgado et al. (2008) linked an increase in mussel farm harvesting closures in the Rías Baixas to an increase in HABs which was attributed to reduced upwelling and flushing of the Rías over a 40-year period. Subsequent work by Pérez et al. (2010) has described contrasting ecosystem responses on the outer and inner shelf to a weakening of upwelling off the Iberian coast, coincident with an increase in sea surface temperature. On the outer shelf the expected decrease in chlorophyll *a* was confirmed by remote sensing observations, but in the Rías chlorophyll *a* showed a positive trend.

Here weaker upwelling reduces flushing which favours stratification and local recycling of nutrients. The resulting positive trends in NO_2^-, NH_4^+ and PO_4^{3-} concentrations were linked to increases in dinoflagellate and *Pseudo-nitzschia* concentrations despite a decrease in the total abundance of diatoms.

As the primary cause of shellfish harvesting closures in Western Europe, including the Iberian coast, *Dinophysis* spp. are known to demonstrate considerable interannual variability in response to environmental conditions. Díaz et al. (2013, 2016) were able to describe the seasonal and interannual variability of *Dinophysis acuta* in the Galician Rías Baixas in relation to climate based on a descriptive analysis of a 30-year time series of this dinoflagellate. The study was also able to test a statistical model using climate indicators as predictors of these blooms. Three seasonal patterns in the occurrence of two *Dinophysis* species were identified under different climatic scenarios. Years with cool summers were characterized by the occurrence of only *D. acuminata*, which depicted a clear bimodal distribution, with maxima, in late spring and in late summer-early autumn. In years with moderate temperatures, short-lived blooms of *D. acuta* followed the *D. acuminata* blooms during the autumn transition. These blooms were dependent on the presence of good growth conditions off the Portuguese coast and their alongshore transport. In years with hot summers, *D. acuta* replaces *D. acuminata* in late summer. In these cases *D. acuta* shows a bimodal distribution, with a first maximum in late summer due to in situ growth and another maximum in the autumn transition linked to physical transport. Exceptional blooms during the late 1980s were characterized by the latter pattern associated with hotter than average summers and moderate upwelling. Prediction of these patterns of *D. acuta* was achieved through application of a generalized additive model incorporating an inoculum of *D. acuta*, a cumulative upwelling index, monthly average sea surface temperature and the tidal cycle and range (Díaz et al. 2016). The presence of an inoculum of *D. acuta* cells off the Portuguese coast is the most difficult factor to predict as these populations may undergo alongshore displacement as dictated by the strength of upwelling (Moita et al. 2016).

A shift towards a greater frequency and higher magnitude of *Pseudo-nitzschia* blooms has been reported for the California Current system (Sekula-Wood et al. 2011; Barron et al. 2013). A shift in 2000 to increasing *Pseudo-nitzschia* abundance and DA toxicity was reported by Sekula-Wood et al. (2011) for the Santa Barbara Basin following examination of a 15-year time series (1993–2008) of sediment trap samples. Attributed to increasing blooms of *P. australis*, the shift was considered to be related to climate variability associated with a change in phase of the North Pacific Gyre Oscillation and its influence on the upwelling of waters into the Santa Barbara Basin. The regional cooling of the Santa Barbara Basin in 1999 accompanied by a change to *Pseudo-nitzschia* species was also demonstrated through examination of a sediment core sample spanning the interval from 1748 to 2007 (Barron et al. 2013; Fig. 11.4).

McKibben et al. (2017) have subsequently provided evidence of climatic regulation of DA in shellfish over the past 20 years in the Northern California Current, with the timing of elevated DA coincident with warm phases of the Pacific Decadal

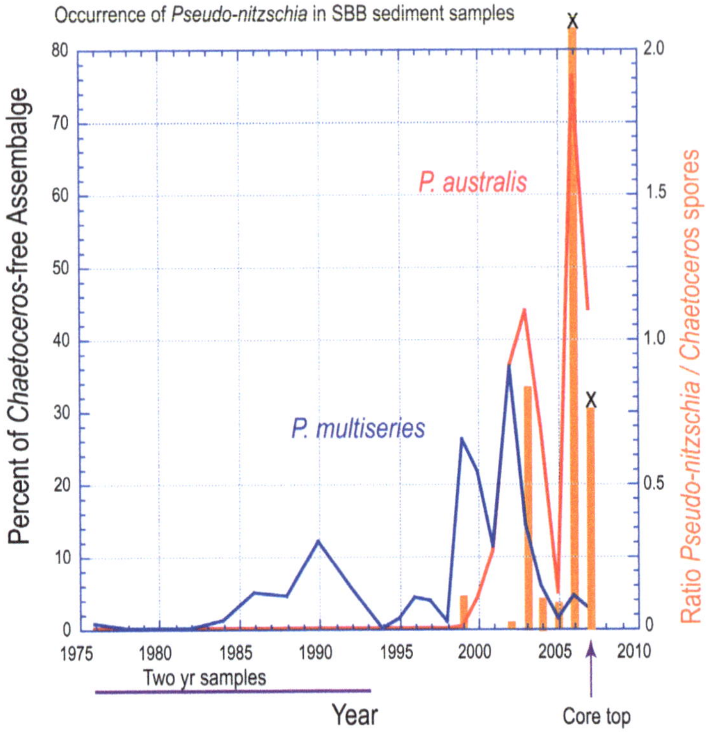

Fig. 11.4 A record *of Pseudo-nitzschia* species in a sediment core sample from the Santa Barbara basin expressed as relative abundance of *Chaetoceros*-free assemblages (*blue* and *red lines*) and as a ratio to *Chaetoceros* spores (*orange bars*). Years of recorded shellfish poisoning and marine mammal deaths are denoted by an X (Reproduced from Barron et al. 2013 with permission of the publisher)

Oscillation and the Oceanic Niño Index events. Ocean conditions in the northeast Pacific associated with warm phases of these indices, including changes in prevailing currents and advection of anomalously warm water onto the continental shelf, are considered to contribute to increases in this toxin. In these warm years, the spring biological transition date, indicative of the seasonal transition of the Northern California Current to southward flow, is later; alongshore currents tend to be weaker in association with warm Pacific Decadal Oscillation values; estuarine and offshore water temperatures are elevated; the copepod species richness anomaly is positive and there is a shift in copepod composition to species of a predominantly offshore or southern origin; and DA levels in the Oregon coast razor clams exceed closure values (McKibben et al. 2017). In 2015 extensive northeast Pacific warming was associated with an unprecedented, record-setting outbreak of DA poisoning along the North America West Coast attributed to *P. australis* (McCabe et al. 2016). This was the first *Pseudo-nitzschia* bloom to negatively impact both shellfish and fish harvest industries, as well as marine mammal health, along the entire US and

Canada West Coasts with prolonged closures of fisheries. If global warming causes these warm ocean regimes to become more persistent, as is hypothesized, DA events on the West Coast may increase in persistence and frequency (McKibben et al. 2017).

Apart from changes in the timing, location and magnitude of HABs, climate-related changes in the ocean are also likely to influence toxin production. Conditioning of an *Alexandrium catenella* culture isolated from California coastal waters to combinations of temperature, pCO_2 and phosphate concentrations has demonstrated important influences on growth, carbon fixation rates and total cellular toxicity (Tatters et al. 2013). Overall increased pCO_2 generally led to enhanced toxicity suggesting that acidification of the California Current system may exacerbate the toxic threat posed by this dinoflagellate, especially when combined with nutrient limitation. Similarly, Tatters et al. (2012) have demonstrated likely increases in cellular toxicity in the diatom *Pseudo-nitzschia fraudulenta* in the coastal California upwelling system in response to increasing anthropogenic acidification and episodic Si limitation.

Improved information on the linkages between HABs and climate will emerge only through the establishment and maintenance of long-term phytoplankton monitoring programmes adequately supported by environmental monitoring. At present our understanding and ability to predict how climate may select for HABs are severely limited by the scarcity of long-term records which are required for assessment of past climate variability on HABs and for the development and testing of predictive models.

11.10 HAB Prediction

Model development is fundamental to HAB prediction in upwelling systems but tends to lag our theoretical understanding [see also Chap. 19, Franks (2018)]. Although many HAB species are regular components of the seasonal succession of phytoplankton, few can be predicted with much success. Most HAB models to date have employed either conceptual or empirical approaches, and data limits have tended to restrict the development of new models. The mandala of Margalef (1978) offered one of the first models for the prediction of red tides in upwelling systems based on the interactive effects of habitat mixing and nutrient conditions. In its simplest form this model has been used in upwelling systems to temporally and spatially segregate diatom- and dinoflagellate-dominated communities based on their predictable response to different regimes of turbulence and nutrients (e.g. Pitcher and Nelson 2006). Several empirical or statistical models have been developed based on this ecological classification of diatoms and dinoflagellates.

Statistical models have been used with some success in predicting *Pseudo-nitzschia* abundance in the California and Iberian upwelling systems. These models are based on an ecological relationship between *Pseudo-nitzschia* species and potentially causative environmental variables. Anderson et al. (2009) used time-series data to define the ranges of environmental conditions associated with *Pseudo-nitzschia* bloom development in the Santa Barbara Channel. Based on these environmental conditions, statistical models incorporating satellite ocean colour and sea surface temperature were able to correctly estimate 98% of toxic bloom situations but with a significant rate of false-positive identification. Lane et al. (2009) similarly developed predictive logistic models of toxigenic *Pseudo-nitzschia* blooms in Monterey Bay. Models were developed for year-round and seasonal use. Chlorophyll *a* and $Si(OH)_4$ were predictors in all models, but period-specific inclusions of temperature, an upwelling index, river discharge and/or NO_3^- provided significant refinement to the models. For Lisbon Bay Palma et al. (2010) were also able to predict *Pseudo-nitzschia* blooms using statistical models in the form of regression analysis applied to the relationship of *Pseudo-nitzschia* blooms to sea surface temperature and an upwelling index. Models developed for three oceanographic seasons allowed forecasting of the presence of the genus in spring and summer, revealing a lag of 4–6 days between upwelling events and the appearance of *Pseudo-nitzschia*.

Statistical approaches have also been applied to long-term data sets from the Rías Baixas on the Iberian coast in the establishment of a correlation between wind patterns on the adjacent shelf, which control the flushing times of these large coastal inlets, and closures of mussel cultivation areas (Álvarez-Salgado et al. 2008). These relationships are established on the ecological basis of a dynamic segregation between diatoms and dinoflagellates determined by the residual circulation of the Rías Baixas. Dinoflagellates including those that produce lipophilic toxins have been favoured by a decrease in the extent and intensity of the upwelling season over the past 40 years and a consequent increase in the renewal time of the Rías (Álvarez-Salgado et al. 2008). Similar studies in the Rías de Ares-Betanzos (Rías Altas) have established correlations between mussel harvesting closures due to lipophilic toxins and both coastal winds and precipitation (Álvarez-Salgado et al. 2011). These studies showed that closures during summer can be predicted by the average continental runoff during the preceding spring, while closures during autumn can be predicted by average offshore Ekman transport during the preceding summer (Fig. 11.5). A rainy spring will therefore lead to extensive closures in summer and a windy summer to high closures in autumn. These relationships between meteorological conditions and harvesting closures allow predictions of shellfish toxicity based on predictions of climate in terms of precipitation and winds.

Hydrodynamic circulation models are now being used with some success in identifying and constraining the physical processes controlling bloom dynamics. These circulation models can be used to track bloom trajectories and are able to provide a useful tool for predicting the timing and spatial impact of blooms. Lagrangian particle tracking of HABs using numerical models of ocean circulation off the Iberian Peninsula has compared favourably with observations at coastal monitoring stations (Mateus et al. 2013; Pinto et al. 2016). The utilization of regional models in conjunction with information from in situ monitoring networks and remote sensing has been central to the development of HAB forecast systems on the Iberian coast as part of the ASIMUTH project (Applied Simulations and

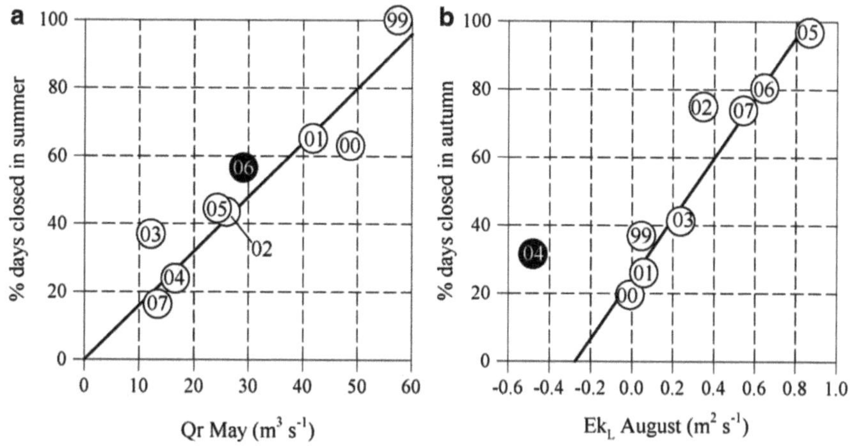

Fig. 11.5 Closure of mussel harvesting in Ría de Ares-Betanzos (Galician Rías Altas) due to lipophilic shellfish toxicity (**a**) during summer (June, July and August) as a function of continental runoff (Qr) during the preceding spring (May) and (**b**) during autumn (September, October and November) as a function of offshore Ekman transport (Ek_L) during the preceding summer (August). *Circled numbers* depict years: the average Qr of both April and May was used in 2006 (*black dot* in panel **a**), and the negative Ek_L of 2004 was not used in the regression (*black dot* in panel **b**) (Reproduced from Álvarez-Salgado et al. 2011 with permission of the publisher)

Integrated Modelling for the Understanding of Harmful Algal Blooms). Output from the Portuguese Coast Operational Model System (PCOM) and from the Regional Ocean Modelling System (ROMS) have been used in the development of forecast systems for the Iberian coast (Moita et al. 2016; Pinto et al. 2016; Silva et al. 2016) and Galician Rías Baixas (Ruiz-Villarreal et al. 2016), respectively. These models provide forecasts of oceanographic conditions important to HABs and are also coupled to Lagrangian particle-tracking models used to simulate alongshore transport of HABs and their movement between the Rías and shelf.

Numerical simulations employing the ROMS have also been used in predicting *Pseudo-nitzschia* blooms in the California Current system. The combination of satellite observations, statistical models and hydrodynamic models has been used to spatially map and predict *Pseudo-nitzschia* blooms and DA events in the Santa Barbara Channel (Anderson et al. 2011). For the Pacific Northwest Coast, a realistic numerical simulation of the region and particle-tracking experiments were used to investigate possible *Pseudo-nitzschia* transport pathways from previously identified hotspots of bloom formation, the Juan de Fuca eddy and Heceta Bank (Giddings et al. 2014). Overall these simulations confirmed the existence of northern and southern transport pathways and were also able to elucidate the influence of the Columbia River plume on potential pathways. Consequently hindcasts were able to capture most of the observed harmful *Pseudo-nitzschia* beach events, although numerous false positives were recorded. Predictability was considerably enhanced by the incorporation of phytoplankton biomass results from a coupled biogeochemical model that was shown to reduce the number of false positives significantly.

Model development has also been extended to provide predictions of the response of HABs to climate change. A heuristic model of toxic blooms of *Alexandrium* was developed for Puget Sound using long-term trends of PSP toxins in shellfish tissues (Moore et al. 2009). Examination of various time series of environmental conditions leading to toxic events allowed identification of a temporal window of opportunity for blooms which has been used to evaluate past and future bloom scenarios. Climate change projections from global climate models and regionally downscaled climate models for the Pacific Northwest were used to demonstrate a wider window of opportunity for *Alexandrium* blooms by the end of the twenty-first century (Moore et al. 2011). Compared to the present day, blooms are predicted to begin 2 months earlier and persist for 1 month longer, and these changes may be detectable within the next 30 years. These projections were further developed using a mechanistic modelling approach to evaluate the effects of various climate pathways on the proliferative phase of *Alexandrium* in Puget Sound (Moore et al. 2015). Experimentally derived growth responses to temperature and salinity combined with simulations of regional climate and hydrology have shown a 30-day-a-year increase in bloom favourable conditions by 2050.

11.11 Conclusions

System-based studies have been useful in advancing our knowledge of the ecology and oceanography of HABs, and the comparative approach, as endorsed by GEOHAB, is considered to have added value to regional or national science plans. Although comparison in many cases has been limited by the availability of data, GEOHAB has provided useful direction to HAB research in upwelling systems over the past decade, and the formulation of common research objectives and the development of common research approaches have contributed significantly to improving the availability of comparable data. The progress, as documented in this paper, in addressing the key questions formulated as part of the CRP—HABs in upwelling systems (GEOHAB 2005), now allows for greater focus in achieving the original goal of GEOHAB of improved prediction.

To further advance our predictive skills of HAB species, the development of biological models and their coupling to physical models will be required. The explicit representation of the biology of these species is an essential element to model development of their bloom dynamics. Even if the factors favouring or limiting the growth of a species can be identified, top-down processes also have a massive bearing on population numbers. Harmful species therefore need to be treated as components of communities in which there are complex interactions between the environment and the activities of a wide range of species, and this biological complexity needs to be matched with appropriate complexity in the representation of the environment.

The strong influence of physical forcing in upwelling systems has nevertheless allowed a great deal of understanding to be gained from physical models with passive tracers representing HABs. Physical transport models have provided a very useful first step in predicting the movement of blooms through advection. Models of coastal upwelling systems now incorporate advanced sub-grid-scale parameterizations for mixing, which allow good representation of several of the properties and features that characterize upwelling systems (GEOHAB 2011). The required representation of mesoscale activity represented by the weather and various coastline configurations and topographical structures can therefore be achieved, and equations describing the transport and fate of constituents in the water, such as HABs, can be coupled to the equations of hydrodynamic models. Near future advances in the prediction of HABs are most likely to be achieved through close co-ordination of observation and modelling efforts. With the advent of coastal observation systems, it is now possible to construct numerical simulations of ecosystem dynamics that assimilate data. Successful prediction of HABs as a possible outcome of this integration of real-time data into model systems as a component of operational forecasting of the ocean is most likely in upwelling systems owing to the strong influence of the physical environment on HAB development. These operational forecasts will contribute to our understanding and prediction of HABs and will also facilitate better management and mitigation of HABs in eastern boundary upwelling systems, as proposed in the Science and Implementation Plans of GlobalHAB, the successor to GEOHAB.

Acknowledgments This is a contribution of the GEOHAB Core Research Project on HABs in Upwelling Systems.

References

Álvarez-Salgado XA, Figueiras FG, Fernández-Reiriz MJ et al (2011) Control of lipophilic shellfish poisoning outbreaks by seasonal upwelling and continental runoff. Harmful Algae 10:121–129

Álvarez-Salgado XA, Labarta U, Fernández-Reiriz MJ et al (2008) Renewal time and the impact of harmful algal blooms on the extensive mussel raft culture of the Iberian coastal upwelling system (SW Europe). Harmful Algae 7:849–855

Amorim A, Nolasco R, Oliveira PB et al (2014) Seeding of *Gymnodinium catenatum* blooms in Iberian shelf waters. ICES CM 2014/H: 20

Anderson CR, Kudela RM, Benitez-Nelson C et al (2011) Detecting toxic diatom blooms from ocean colour and a regional ocean model. Geophys Res Lett 38:L04603. https://doi.org/10.1029/2010GL045858

Anderson CR, Siegel DA, Kudela RM et al (2009) Empirical models of toxigenic *Pseudo-nitzschia* blooms: potential use as a remote detection tool in the Santa Barbara Channel. Harmful Algae 8:478–492

Barron JA, Bukry D, Field DB et al (2013) Response of diatoms and silicoflagellates to climate change and warming in the California current during the past 250 years and the recent rise of the toxic diatom *Pseudo-nitzschia australis*. Quat Int 310:140–154

Barton ED, Torres R, Figueiras FG et al (2016) Surface water subduction during a downwelling event in a semienclosed bay. J Geophys Res Oceans 121:7088–7107

Bravo I, Fraga S, Isabel Figueroa R et al (2010) Bloom dynamics and life cycle strategies of two toxic dinoflagellates in a coastal upwelling system (NW Iberian Peninsula). Deep Sea Res II 57:222–234
Díaz PA, Reguera B, Ruiz-Villarreal M et al (2013) Climate variability and oceanographic settings associated with interannual variability in the initiation of *Dinophysis acuminata* blooms. Mar Drugs 11:2964–2981
Díaz PA, Ruiz-Villarreal M, Velo-Suárez L et al (2014) Tidal and wind-event variability and the distribution of two groups of *Pseudo-nitzschia* species in an upwelling-influenced Ría. Deep Sea Res II 101:163–179
Díaz PA, Ruiz-Villarreal M, Pazos Y et al (2016) Climate variability and *Dinophysis acuta* blooms in an upwelling system. Harmful Algae 53:145–159
Escalera L, Reguera B, Moita T et al (2010) Bloom dynamics of *Dinophysis acuta* in an upwelling system: in situ growth versus transport. Harmful Algae 9:312–322
Franks PJS (2018) Recent advances in modelling of harmful algal blooms. In: Glibert PM, Berdalet E, Burford MA et al (eds) Global ecology and oceanography of harmful algal blooms. Springer, Cham, pp 359–377
Frolov S, Kudela RM, Bellingham JG (2013) Monitoring harmful algal blooms in the era of diminishing resources: a case study of the U.S. West Coast. Harmful Algae 21–22:1–12
GEOHAB (2001) Global ecology and oceanography of harmful algal blooms, science plan. Glibert P, Pitcher G (eds) SCOR and IOC, Baltimore and Paris, 86 pp
GEOHAB (2003) Global ecology and oceanography of harmful algal blooms, implementation plan. Gentien P, Pitcher G, Cembella A et al (eds) SCOR and IOC, Baltimore and Paris, 36 pp
GEOHAB (2005) Oceanography of harmful algal blooms, GEOHAB core research project: HABs in upwelling systems. Pitcher G, Moita T, Trainer V et al (eds) Global ecology and IOC and SCOR, Paris and Baltimore, 88 pp
GEOHAB (2011) GEOHAB modelling: linking observations to predictions, a workshop report. McGillicuddy DJ Jr, Glibert PM, Berdalet E et al (eds) IOC and SCOR, Paris and Newark, Delaware, 85 pp
Giddings SN, MacCready P, Hickey BM et al (2014) Hindcasts of potential harmful algal bloom transport pathways on the Pacific Northwest coast. J Geophys Res Oceans 119:2439–2461
Glibert PM (2016) Margalef revisited: a new phytoplankton mandala incorporating twelve dimensions, including nutritional physiology. Harmful Algae 55:25–30
Glibert PM, Heil CA, Wilkerson F et al (2018) Nutrients and HABs: dynamic kinetics and flexible nutrition. In: Glibert PM, Berdalet E, Burford MA et al (eds) Global ecology and oceanography of harmful algal blooms. Springer, Cham, pp. 93–112
Hickey BM, Trainer VL, Kosro PM et al (2013) A springtime source of toxic *Pseudo-nitzschia* cells on razor clam beaches in the Pacific Northwest. Harmful Algae 25:1–14
Howard MDA, Smith GJ, Kudela RM (2009) Phylogenetic relationships of yessotoxin-producing dinoflagellates, based on the large subunit and internal transcribed spacer ribosomal DNA domains. Appl Environ Microbiol 75:54–63
Howard MDA, Sutula M, Caron DA et al (2014) Anthropogenic nutrient sources rival natural sources on small scales in the coastal waters of the Southern California Bight. Limnol Oceanogr 59:285–297
Joyce LB, Pitcher GC (2006) Cysts of *Alexandrium catenella* on the west coast of South Africa: distribution and characteristics of germination. Afr J Mar Sci 28:295–298
Kahru M, Mitchell BG, Diaz A et al (2008) MODIS detects a devastating algal bloom in Paracas Bay, Peru. EOS 85:465–472
Kana TM, Glibert PM (2016) On saturating response curves from the dual perspectives of photosynthesis and nitrogen acquisition. In: Glibert PM, Kana TM (eds) Aquatic microbial ecology and biogeochemistry: a dual perspective. Springer, Geneva, pp 93–104
Kudela R, Pitcher G, Probyn T et al (2005) Harmful algal blooms in coastal upwelling systems. Oceanography 18:184–197

Kudela RM, Raine R, Pitcher G et al (2018) Establishment, goals, and the legacy of the global ecology and oceanography of harmful algal blooms (GEOHAB) Program. In: Glibert PM, Berdalet E, Burford MA et al (eds) Global ecology and oceanography of harmful algal blooms. Springer, Cham, pp 27–49

Lane JQ, Raimondi PT, Kudela RM (2009) Development of a logistic regression model for the prediction of toxigenic *Pseudo-nitzschia* blooms in Monterey Bay, California. Mar Ecol Prog Ser 383:37–51

Lecher AL, Mackey K, Kudela R et al (2015) Nutrient loading through submarine groundwater discharge and phytoplankton growth in Monterey Bay, CA. Environ Sci Technol 49:6665–6673

Lucas AJ, Pitcher GC, Probyn TA et al (2014) The influence of diurnal winds on phytoplankton dynamics in a coastal upwelling system off southwestern Africa. Deep Sea Res II 101:50–62

Lundholm N, Bates SS, Baugh KA et al (2012) Cryptic and pseudo-cryptic diversity in diatoms – with descriptions of *Pseudo-nitzschia hasleana* sp. nov. and *P. fryxelliana* sp. nov. J Phycol 48:436–454

Margalef R (1978) Life-forms of phytoplankton as survival alternatives in an unstable environment. Oceanol Acta 1:493–509

Mateus M, Silva A, de Pablo H et al (2013) Using Lagrangian elements to simulate alongshore transport of harmful algal blooms. In: Mateus M, Neves R (eds) Ocean modelling for coastal management – case studies MOHID. IST Press, Lisbon, pp 235–247

McCabe RM, Hickey BM, Kudela RM et al (2016) An unprecedented coastwide toxic algal bloom linked to anomalous ocean conditions. Geophys Res Lett 43:10366–10376

McKibben SM, Peterson W, Wood AM et al (2017) Climatic regulation of the neurotoxin domoic acid. Proc Natl Acad Sci USA 114:239–244

Moita MT, Pazos Y, Rocha C et al (2016) Toward prediciting *Dinophysis* blooms off NW Iberia: a decade of events. Harmful Algae 53:17–32

Moore SK, Johnstone JA, Banas NS et al (2015) Present-day and future climate pathways affecting *Alexandrium* blooms in Puget Sound, WA, USA. Harmful Algae 48:1–11

Moore SK, Mantua NJ, Salathé EP (2011) Past trends and future scenarios for environmental conditions favoring the accumulation of paralytic shellfish toxins in Puget Sound shellfish. Harmful Algae 10:521–529

Moore SK, Mantua NJ, Trainer VL et al (2009) Recent trends in paralytic shellfish toxins in Puget Sound, relationships to climate, and capacity for prediction of toxic events. Harmful Algae 8:463–477

Oliveira PB, Nolasco R, Dubert J et al (2009) Surface temperature, chlorophyll and advection patterns during a summer upwelling event off central Portugal. Cont Shelf Res 29:759–774

Palma S, Mouriño H, Silva A et al (2010) Can *Pseudo-nitzschia* blooms be modeled by coastal upwelling in Lisbon Bay? Harmful Algae 9:294–303

Peacock MB, Kudela RM (2014) Evidence for active vertical migration by two dinoflagellates experiencing iron, nitrogen, and phosphorus limitation. Limnol Oceanogr 59:660–673

Pérez FF, Padín XA, Pazos Y et al (2010) Plankton response to weakening of the Iberian coastal upwelling. Glob Chang Biol 16:1258–1267

Pinto L, Mateus M, Sliva A (2016) Modeling the transport pathways of harmful algal blooms in the Iberian coast. Harmful Algae 53:8–16

Pitcher GC (2012) Harmful algae – the requirement for species-specific information. Harmful Algae 14:1–4

Pitcher GC, Bernard S, Ntuli J (2008) Contrasting bays and red tides in the southern Benguela upwelling system. Oceanography 21:82–91

Pitcher GC, Figueiras FG, Hickey BM et al (2010) The physical oceanography of upwelling systems and the development of harmful algal blooms. Prog Oceanogr 85:5–32

Pitcher GC, Joyce LB (2009) Dinoflagellate cyst production on the southern Namaqua shelf of the Benguela upwelling system. J Plankton Res 31:865–875

Pitcher GC, Neslon G (2006) Characteristics of the surface boundary layer important to the development of red tide on the southern Namaqua shelf of the Benguela upwelling system. Limnol Oceanogr 51:2660–2674

Raho N, Pizarro G, Escalera L et al (2008) Morphology, toxin composition and molecular analysis of *Dinophysis ovum* Schütt, a dinoflagellate of the "*Dinophysis acuminata* complex". Harmful Algae 7:839–848

Reifel KM, Corcoran AA, Cash C et al (2013) Effects of a surfacing effluent plume on a coastal phytoplankton community. Cont Shelf Res 60:38–50

Rines JEB, McFarland MN, Donaghay PL et al (2010) Thin layers and species-specific characterization of the phytoplankton community in Monterey Bay, California, USA. Cont Shelf Res 30:66–80

Ruiz-Villarreal M, García-García LM, Cobas M et al (2016) Modelling the hydrodynamic conditions associated with *Dinophysis* blooms in Galicia. Harmful Algae 53:40–52

Ryan JP, Gower JFR, King SA et al (2008) A coastal ocean extreme bloom incubator. Geophys Res Lett 35:L12602

Ryan JP, McManus MA, Sullivan JM (2010) Interacting physical, chemical and biological forcing of phytoplankton thin-layer variability in Monterey Bay, California. Cont Shelf Res 30:7–16

Seegers BN, Birch JM, Marin R III et al (2015) Subsurface seeding of surface harmful algal blooms observed through the integration of autonomous gliders, moored environmental sample processors, and satellite remote sensing in southern California. Limnol Oceanogr 60:754–764

Seeyave S, Probyn TA, Pitcher GC et al (2009) Nitrogen nutrition in assemblages dominated by *Pseudo-nitzschia* spp., *Alexandrium catenella* and *Dinophysis acuminata* off the west coast of South Africa. Mar Ecol Prog Ser 379:91–107

Sekula-Wood E, Benitez-Nelson C, Morton S et al (2011) *Pseudo-nitzschia* and domoic acid fluxes in Santa Barbara Basin (CA) from 1993 to 2008. Harmful Algae 10:567–575

Shanks AL, Morgon SG, MacMahan J et al (2016) Variation in the abundance of Pseudo-nitzschia and domoic acid with surf zone type. Harmful Algae 55:172–178

Silva A, Pinto L, Rodrigues SM et al (2016) A HAB warning system for shelfish in Portugal. Harmful Algae 53:33–39

Sison-Mangus MP, Jiang S, Tran KN et al (2014) Host-specific adaptation governs the interaction of the marine diatom, *Pseudo-nitzschia* and their microbiota. ISME J 8:63–76

Smayda TJ (2002) Turbulence, watermass stratification and harmful algal blooms: an alternative view and frontal zones as "pelagic seed banks". Harmful Algae 1:95–112

Smayda TJ (2010) Adaptations and selection of harmful and other dinoflagellate species in upwelling systems 1. Morphology and adaptive polymorphism. Prog Oceanogr 85:53–70

Smayda TJ, Reynolds CS (2001) Community assembly in marine phytoplankton: application of recent models to harmful dinoflagellate blooms. J Plankton Res 23:447–461

Smayda TJ, Trainer VL (2010) Dinoflagellate blooms in upwelling systems: seeding, variability, and contrasts with diatom bloom behaviour. Prog Oceanogr 85:92–107

Tatters AO, Flewelling LJ, Fu F et al (2013) High CO_2 promotes the production of paralytic shellfish poisoning toxins by *Alexandrium catenella* from southern California waters. Harmful Algae 30:37–43

Tatters AO, Fu F, Hutchins DA (2012) High CO_2 and silicate limitation synergistically increase the toxicity of *Pseudo-nitzschia fraudulenta*. PLoS One 7(2):e32116. https://doi.org/10.1371/journal.pone.0032116

Trainer VL, Pitcher GC, Reguera B et al (2010) The distribution and impacts of harmful algal bloom species in eastern boundary upwelling systems. Prog Oceanogr 85:33–52

Velo-Suárez L, Fernand L, Gentien P et al (2010) Hydrodynamic conditions associated with the formation, maintenance and dissipation of a phytoplankton thin layer in a coastal upwelling system. Cont Shelf Res 30:193–202

Velo-Suárez L, González-Gil S, Pazos Y et al (2014) The growth season of *Dinophysis acuminata* in an upwelling system embayment: a conceptual model based on in situ measurements. Deep Sea Res II 101:141–151

Velo-Suárez L, Reguera B, Garcés E et al (2009) Vertical distribution of division rates in coastal dinoflagellate *Dinophysis* spp. populations: implications for modeling. Mar Ecol Prog Ser 385:87–96

Wells ML, Karlson B (2018) Harmful algal blooms in a changing ocean. In: Glibert PM, Berdalet E, Burford MA et al (eds) Global ecology and oceanography of harmful algal blooms. Springer, Cham, pp 77–90

Wyatt T (2014) Margalef's mandala and phytoplankton bloom strategies. Deep Sea Res II 101: 32–49

Chapter 12
Key Questions and Recent Research Advances on Harmful Algal Blooms in Relation to Nutrients and Eutrophication

Patricia M. Glibert, Adnan Al-Azri, J. Icarus Allen, Alexander F. Bouwman, Arthur H.W. Beusen, Michele A. Burford, Paul J. Harrison, and Mingjiang Zhou

12.1 Introduction

Changing nutrient loads and increasing levels of atmospheric CO_2 are significant drivers of our "ever-changing world," and their adverse effects on aquatic bio-diversity and ecosystem health are well documented (e.g., Cloern 2001; Howarth et al. 2002; Heisler et al. 2008; Doney 2010; Glibert et al. 2014a; Glibert and

Paul J. Harrison was deceased at the time of publication.

P.M. Glibert (✉)
University of Maryland Center for Environmental Science, Horn Point Laboratory, Cambridge, MD, USA
e-mail: glibert@umces.edu

A. Al-Azri
College of Agricultural and Marine Sciences, Sultan Qaboos University, Muscat, Sultanate of Oman

J. Icarus Allen
Plymouth Marine Laboratory, Plymouth, UK

A.F. Bouwman • A.H.W. Beusen
Faculty of Geosciences, Department of Earth Sciences—Geochemistry, Utrecht University, Utrecht, The Netherlands

PBL Netherlands Environmental Assessment Agency, Bilthoven, The Netherlands

M.A. Burford
Australian Rivers Institute and School of Environment, Griffith University, Nathan, Australia

P.J. Harrison
Department of Earth and Ocean Sciences, University of British Columbia, Vancouver, BC, Canada

M. Zhou
Institute of Oceanology, Chinese Academy of Sciences, Qingdao, China

P.M. Glibert et al. (eds.), *Global Ecology and Oceanography of Harmful Algal Blooms*, Ecological Studies 232, https://doi.org/10.1007/978-3-319-70069-4_12

Burford 2017). Human activities, such as the use of synthetic fertilizers, energy production, and expansion of industrialized agriculture, aquaculture, and concentrated animal operations, have had tremendous impacts on the global cycling of nutrients and carbon (C) on land, in freshwaters, and in coastal waters (e.g., Smil 2001; Galloway et al. 2002; Doney 2010; Bouwman et al. 2011, 2013b). Coasts are "peopled seascapes," and, through their complex relationships with the oceans, people "pollute and build in the coastal zone, consume seafood and thereby support excessive fisheries, extract and burn fossil fuels and contribute to global change" (Shackeroff et al. 2009). An important consequence of such activities in aquatic systems is eutrophication. This results in harmful algal blooms (HABs) that can produce toxins, cause fish kills directly or through the formation of hypoxia/anoxia, alter trophic interactions and broadly affect food webs and biodiversity, and potentially impact human health through various routes of exposure (GEOHAB 2001). As recognized in the special issue of *Harmful Algae* on Eutrophication and HABs that was published in 2008, the complexity of the relationships between HABs and nutrients has been well recognized (Glibert et al. 2008a):

> Historically, our conceptual understanding of the relationship between nutrients and HABs was based upon the simplistic notion that more nutrients should yield higher algal biomass in a direct relationship. The frame of reference was restricted to noticeable blooms whose biomass was linearly related: if such high –biomass blooms did not occur, the HAB species was deemed unrelated to nutrient enrichment. Within the past decades, insights have advanced considerably. There is greater appreciation for interactive influences of the composition and relative proportion of availability of nutrient pools, the range of physiological responses by different phytoplankton species and functional groups, and the interactions of other dynamic factors such as physics and grazing in controlling responses to cultural eutrophication by HAB and other algal populations. It is now recognized that HAB species can be directly and/or indirectly stimulated by nutrient over enrichment, and that chronic, subtle effects can be equally important or even more important than the obvious, acute effects. And, it is recognized that nutrient enrichment interacts with other major drivers, such as climate change, in both direct and indirect ways.

These statements ring equally true today, and these complexities have been extensively described, and much more is now understood regarding the many interactions and factors related to nutrients and HABs (e.g., Glibert and Burford 2017).

The GEOHAB Core Research Project (CRP) on HABs and Eutrophication, launched in 2005, fostered two international Open Science Meetings (Baltimore, Maryland, USA, 2005, and Beijing, China, 2009), bringing researchers together with expertise ranging from algal physiology to global modelling of nutrient exports. The GEOHAB Report on HABs in Eutrophic Systems (GEOHAB 2005) laid out a number of key questions that required international, collaborative efforts. These key questions and their progress are summarized in this chapter, along with a summary of areas that need further attention, recognizing that progress has been enormous and only a few highlights can be given in this short chapter. The work of the CRP on HABs and Eutrophication was further augmented by the formation of a Scientific Committee on Oceanic Research (SCOR) Working Group on Land-Based Nutrients and HABs. Given the pace at which nutrient loads continue to pollute the global landscape and the global expansion of HABs, continued international collaborative efforts in understanding changing nutrients and their relationships with HABs are not only necessary, but urgent.

12.2 Key Question 1: Are There Clusters or Specific Types of HAB Species that Are Indicative of Global HAB Increases?

There are clear examples of HAB genera that have increased in concert with eutrophication, and several are highlighted here. Among the species for which there is considerable evidence of a direct eutrophication relationship is *Prorocentrum minimum* (Heil et al. 2005; Glibert et al. 2008b, 2012). These global relationships underscore what has been well recognized at the regional scale: nutrients and organic loads from land-based sources promote the growth of *P. minimum*. In the Baltic Sea, the Black Sea, the Chesapeake Bay, and the Neuse River Estuary of the Albemarle-Pamlico estuarine system in the southeastern USA, long-term records of phytoplankton assemblage composition provide strong support that this species has flourished as these systems became increasingly eutrophic (Marasović et al. 1990; Moncheva et al. 1995; Heil et al. 2005; Hajdu et al. 2005; Pertola et al. 2005; Tango et al. 2005). A recent assessment of *P. minimum* in the Baltic Sea has linked its expansion in time and space, and its ecosystem impacts with human activities and associated nutrient pollution (Olenina et al. 2010). There are numerous descriptions of this species proliferating in waters influenced by freshwater inputs, especially freshwaters laden with organic forms of nutrients (e.g., Silva 1985; Granéli et al. 1985, 1989; Stonik 1995). In South Australia, *P. minimum* blooms have been documented near sewage outfalls (Cannon 1990). Blooms in the Navichiste Lagoon Complex on the east side of the Gulf of California have been linked to high NO_3^- and NH_4^+ in offshore waters (Martinez-Lopez et al. 2008). Along the coast of Mexico, where *P. minimum* only became established about 30 years ago, most blooms are associated with nutrient-rich shrimp ponds or other aquaculture (Sierra-Beltran et al. 2005), and similarly in Queensland, Australia, such blooms have also been reported in intensive, eutrophic shrimp ponds (Burford and Pearson 1998).

The global expansion of *P. minimum* and its relationship to nutrients has been documented using spatially explicit global maps of nutrient pollution (see also Sect. 12.6). Global nutrient data, derived from the Global Nutrient Export from Watersheds (Global NEWS) effort (Seitzinger et al. 2005; Dumont et al. 2005; Harrison et al. 2005a, b), allowed identification of relationships between documented occurrences of *P. minimum* and nutrients as total nitrogen (TN) and phosphorus (TP) export (Fig. 12.1a, here, illustrating trends for N only), and by dominant form of N, as sewage, fertilizer, manure, or fixation of atmospheric sources (Fig. 12.1b), and the extent to which the export of dissolved organic N (DON)

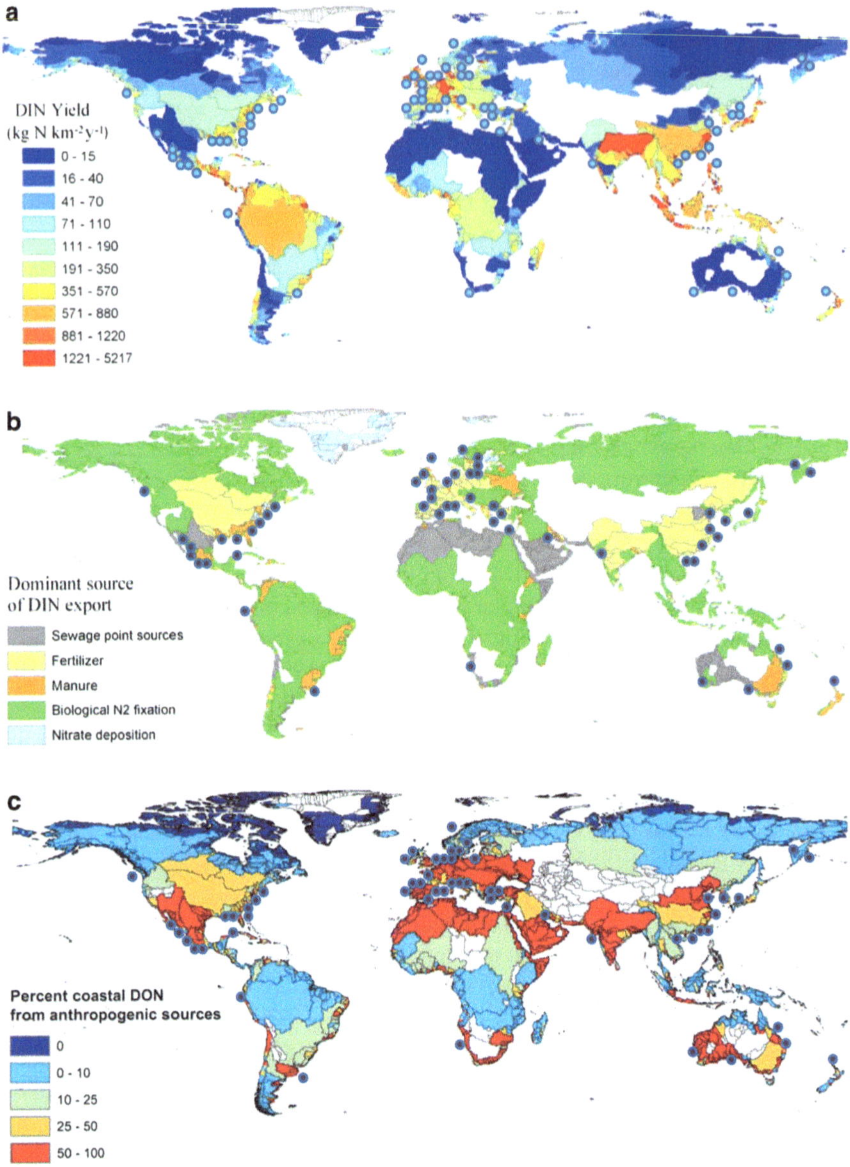

Fig. 12.1 (**a**) Global NEWS model of DIN yield by watershed (kg N km^{-2} year^{-1}) by exoreic basins (from Dumont et al. 2005); (**b**) dominant source of DIN export in exoreic basins as predicted by the Global NEWS models (from Dumont et al. 2005); and (**c**) percent anthropogenic contribution to DON export from exoreic basins as estimated from Global NEWS models (from Harrison et al. 2005a) and the global distribution of *P. minimum* indicated by *circles* (from Heil et al. 2005). Figures reproduced from Glibert et al. (2008b) with permission from the publisher

was attributable to anthropogenic sources [Fig. 12.1c; Glibert et al. 2008b; see also Chap. 4, Glibert et al. (2018)].

Among the most spectacular examples of the association between harmful planktonic *Prorocentrum* species and eutrophication is the recent development of massive blooms of the related species, *P. donghaiense*, along the coast of China, especially the East China Sea, in the region formed by the convergence of the Yangtze (Changjiang) River plume and the Taiwan Warm Current [Lu and Goebel 2001; Zhou et al. 2003, 2008; Lu et al. 2005; see also Chap. 15, Yu et al. (2018)]. Such blooms were not reported until 1995 (Lu et al. 2005), but massive blooms, up to 10,000 km^2 in areal extent, have occurred in late spring virtually every year since, and the bloom of 2005 was particularly well studied (Zhou et al. 2008; Li et al. 2009, 2010; Lu et al. 2011). The rapid rise in the export of N from the expansion of fertilizer use in the Changjiang River watershed has been strongly associated with the expansion of these blooms (Zhou et al. 2008; Li et al. 2009). This area is the most developed in China, with over 440 million people living in the watershed (Li et al. 2009). The annual dissolved inorganic N (DIN) load from coastal waters influenced by the Changjiang River has reached 1.4×10^6 t, which is higher than DIN loads from the Mississippi or Amazon rivers (Goolsby and Battaglin 2001; Duan et al. 2008; Li et al. 2009). Blooms form when the N-rich, but comparatively P-poor, Changjiang River plume mixes with upwelled water from the Taiwan Warm Current, which is comparatively P-replete (Tang et al. 2000; Fang 2004), forming a front where the N:P ratio is suitable for growth (Li et al. 2009). Blooms of *P. minimum* and *P. donghaiense* are also now a regular feature in the southern waters of Korea (Lee and Kim 2008; Lu et al. 2011). These blooms have been linked to fish and shrimp kills with considerable economic consequences (Lee and Kim 2008).

The dinoflagellate *Cochlodinium (Margalefidinium) polykrikoides* represents another clear example where there has been global expansion, and blooms appear to be related to nutrient pollution. This species is now widespread, from the western Pacific to the eastern Atlantic, the Indian Ocean, the Arabian Gulf, and the Mediterranean (Kudela et al. 2008b; Matsuoka et al. 2010; Kudela and Gobler 2012). As a fish killer, it has become particularly problematic in regions where aquaculture has expanded, such as Korea and Japan where it has caused considerable economic damage (GEOHAB 2010). In the USA, it has become a recurrent late-summer bloom former in Chesapeake Bay (e.g., Mulholland et al. 2009) and in other east coast lagoons when seasonal storms deliver a pulse of nutrients to the estuary. In the Arabian Gulf, a massive nearly 9-month-long *C. polykrikoides* bloom occurred in 2008–2009 and ultimately stretched from Oman to coastal Iran and caused fish kills as well as impacts to desalination plants throughout the region (Richlen et al. 2010; Al-Azri et al. 2014; Fig. 12.2). It is not clear what triggered this bloom in this particular year, nor why *Noctiluca scintillans*, the more frequent bloom former in the region over the past decade, was supplanted. A common feature of the antecedent nutrient conditions of this and other *C. polykrikoides* blooms is that phosphate (PO_4^{3-}) concentrations were on average 50–100% higher than in non-bloom sites, and waters were comparatively enriched in ammonium (NH_4^+) rather than nitrate (NO_3^-; Al-Azri et al. 2014).

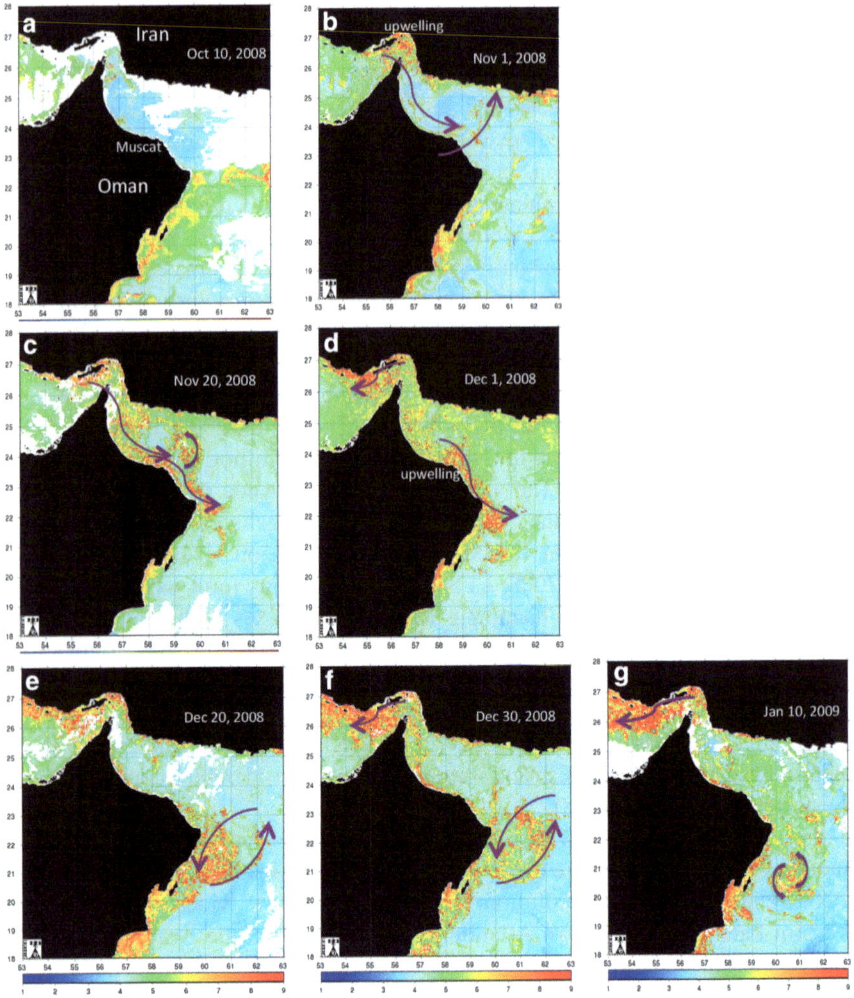

Fig. 12.2 **(a–g)** Satellite images showing the development of blooms in the Sea of Oman and Arabian Sea in 2008 and their dissipation in January 2009. The *arrows* superimposed on the images are intended to draw attention to the major mesoscale features and are not meant to imply measured flow. The chlorophyll scale on the images is intentionally reported dimensionless to underscore the relative patterns, not absolute biomass accumulations. Figure reproduced from Al-Azri et al. (2014) with permission from the publisher

Moreover, the mixotrophic properties of this species have long been known (Larsen and Sournia 1991; Jeong et al. 2004). Its growth rate has been shown to increase by almost a factor of two when growing mixotrophically, compared to phototrophically (Jeong et al. 2004). It is thus likely that *C. polykrikoides* derives a substantial fraction of its nutrition via mixotrophy during its blooms.

Clearly the evidence is broad and deep that HABs are expanding globally in concert with increasing nutrient loads. That certain taxa should be favoured under some nutrient enrichment conditions should be of no surprise, given the suite of adaptive strategies of many HABs related to nutrient availability, proportion, and form (e.g., Heisler et al. 2008; Glibert and Burkholder 2011; Glibert and Burford 2017, and references therein). The question of whether there are HAB species that track anthropogenic nutrient changes is clearly answered; what needs more thorough elucidation is the specific adaptations of these species that allow them to thrive under such conditions (e.g., Glibert and Burkholder 2006; Glibert and Burford 2017).

12.3 Key Question 2: To What Extent Do Residence Time and Other Physical Processes Impact the Relationship Between Nutrient Loading and HAB Proliferation?

There is considerable evidence that residence time and hydrodynamics affect the relationships between nutrients and HAB proliferation. These relationships are mainly mediated through the forms of nutrient that are either delivered primarily with freshwater flow or are promoted through water-column retention. A global review of estuarine coastal typology has highlighted the differences in nutrient retention between different types of coastal features and between different nutrients [N and P; Dürr et al. 2011; see also Chap. 4, Glibert et al. (2018)]. Not only are the differences in nutrient retention between different estuarine types substantial, but there is a general trend towards increased retention at higher latitudes (Laruelle 2009). Water-column retention is also being altered globally through the construction of dams and reservoirs as well as coastal alteration through construction of harbours and artificial canals for residential development.

One of the clearest examples of flow relationships with the propensity for HABs comes from the comparison of river-dominated estuaries versus coastal lagoons (Glibert et al. 2010). Coastal lagoons represent a class of estuary characterized as shallow, highly enclosed, and typically quiescent in terms of wind, current, and wave energy relative to their deeper, more dynamic counterparts, river-dominated estuaries (Madden et al. 2010). Coastal lagoons also tend to have high surface to volume ratios and can have high wetland area to water volume ratios as well (Bricker et al. 2007; Madden et al. 2010). Differences in the quality of the nutrient pool, the seasonal timing of the delivery and source of nutrients, as well as in the resident phytoplankton community can lead to fundamentally different types of algal blooms in coastal lagoons than in river-dominated estuaries. Lagoonal blooms are often dominated by picoplankton, such as *Aureococcus anophagefferens*, which has bloomed in the lagoons of Long Island, NY, and Maryland Coastal Bays, USA, and more recently the Bohai Sea [see Chap. 15, Yu et al. (2018)]. In fact, blooms of a related species, *Aureoumbra lagunensis*, which occurred in Laguna Madre, Texas,

USA, in the 1990s, lasted nearly 9 years (Buskey et al. 2001). Another example is *Synechococcus elongatus*, which has formed massive blooms in Florida Bay, USA, and these blooms may be also sustained for long periods of time. High-biomass picoplankton blooms in lagoons may reach near monospecific proportions. During blooms in Florida Bay in 2002 and 2005, for example, *Synechococcus* spp. comprised >99% of the phytoplankton assemblage and reached 10^8 cells L^{-1} (Glibert et al. 2004; Hitchcock et al. 2007). Similarly, in the Chincoteague Bay, USA, *A. anophagefferens* has been found to comprise 85–95% of the phytoplankton community during bloom periods (e.g., Wazniak and Glibert 2004). Sustained blooms of picoplankton such as these have been termed ecosystem disruptive algal blooms (EDABs; Sunda et al. 2006). Blooms in river-dominated estuaries, by contrast, tend to be highly seasonal and dominated by larger-sized phytoplankton (i.e., >10 μm), typically diatoms. Furthermore, blooms in lagoons tend to be supported by regenerated nutrients that may be lower in concentration but which are made available through higher regeneration rates from both water-column and benthic processes. Highest phytoplankton biomass in riverine systems generally occurs in the spring, following maximum runoff, mainly of NO_3^- (e.g., Pennock 1985; Malone et al. 1996). Only in summer do temperate river-dominated systems resemble those of coastal lagoons when temperature is warm, flow is reduced, and the spring diatom bloom begins to be succeeded by flagellates, cyanobacteria, and other picoplankton that thrive on the nutrients regenerated from the spring bloom (Malone et al. 1996). Episodic blooms of large dinoflagellates occur in both types of systems but are generally not long-lived.

Anthropogenic activities are also altering hydrology in many regions, which in turn leads to changes in nutrient export to aquatic systems. River discharge is being altered throughout the world due to dam construction and water abstraction. Dams of a capacity of more than 1 MW that have been constructed, or are under construction, number in the thousands (Zarfl et al. 2015), and extremely large dams (with capacity of >1 GW) are primarily located in Asia along the Changjiang River basin (e.g., the Three Gorges Dam) and along the Amazon basin. These dams have significant and complex effects on river flow and accordingly on environmental conditions conducive to HABs [see also Chap. 4, Glibert et al. (2018)]. In addition to the reductions in overall river flow that occurs from dam construction, large river systems can become fragmented, preventing free movement of organisms, and severe modification of river flow alters temperature regimes, dramatically reduces sediment transport (Vörösmarty et al. 2010; Lehner et al. 2011; Liermann et al. 2012), and also alters nutrient loads and proportions in downstream waters (Beusen et al. 2016). For example, retention of Si upstream, following construction of the Three Gorges Dam in China, has been considered to be an important factor leading to altered nutrient proportions favouring HABs in East China Sea (e.g., Zhang et al. 2015).

Many estuaries have controlled flow to meet management goals, but these controls have flow on effects on nutrients. Estuarine flow is also highly regulated in the San Francisco Bay Delta (Kimmerer 2002; Glibert et al. 2011), and strong, bivariate correlations have been reported between flow and organisms of all levels in the food web. In that estuary, sufficient flow is rigorously maintained through strictly established flow criteria in order to increase the low salinity habitat for endangered fish. Of particular significance is that over the past several decades (prior to the drought beginning in 2013), the phytoplankton community of the Bay Delta shifted from large diatoms, which were common up to the mid-1970s, to smaller flagellates and cyanobacteria since the mid-1980s (Lehman et al. 2005, 2008; Glibert et al. 2011), a consequence of managed flow, increased pollution by N, and altered grazing communities. Another example from the USA where flow has been substantially altered by human activities is that of the Everglades, Florida. In the 1940s, much of the natural flow was channelized and redirected. New efforts to restore the Everglades is resulting in higher flows of freshwater to the northern regions of Florida Bay, and this is having some effects on the phytoplankton assemblages there as well (e.g., Shangguan et al. 2017).

Important lessons emerge from these examples. Systems that naturally differ in their typology, and therefore in their nutrient retention, can create differing conditions for HABs, based on both total nutrient loads and the forms of nutrients that result. Altered flow, including the many mechanisms by which flow is being anthropogenically changed, also alters nutrients, in terms of total loads and proportions, and these changes can create "windows of opportunity" for HABs to develop or at least create conditions where phytoplankton assemblages change in composition.

12.4 Key Question 3: How Do Feedbacks and Interactions Between Nutrients and the Planktonic, Microbial Food Web Impact HABs and Their Detrimental Effects?

Grazer interactions and other feedbacks within the microbial food web are complex and clearly contribute to HABs. The clearest effect is when there are reductions in grazer abundance that might result from direct or indirect interactions with HABs. Local reductions in grazer abundances may also be in direct response to HABs (e.g., avoidance or mortality induced by the HABs, Granéli and Johansson 2003) or in response to the effects of past HAB events. The responses of zooplankton and other grazers of toxic algae are often species-specific in terms of behavioural responses and toxin susceptibility. Zooplankton impaired by ingesting harmful algae may also be more susceptible to predation and thus may become important vectors for transferring toxins in the pelagic food web. Alternatively, zooplankton faecal pellets may be important sources of toxin to benthic communities. Herbivorous fish can accumulate and transfer toxins and even cause mass mortalities of the marine birds that consume them. During food web transfers, toxins may be bioaccumulated, excreted, degraded, or structurally modified. In order to understand the effects of HAB species on the marine food web, both the direct as well as the indirect, more subtle pathways by which toxins are transferred and transformed, and the differential susceptibility of marine organisms to these toxins must be characterized. In some cases, HABs may insult, but not kill, other members of the food web, leaving them stressed and therefore more susceptible to other sources of mortality, such as infection by pathogens. Burkholder et al. (2018) have published a comprehensive analysis of food web effects of HABs.

Nutrients have important cascading effects through the food web. Phytoplankton vary in terms of lipid and protein content and also in terms of elemental stoichiometry. The net effect of the stoichiometric regulation of consumers, together with different nutrient requirements of phytoplankton, is that not only are grazers affected by food quality, they, in turn, affect food quality by altering the composition of nutrients available to them through nutrient recycling (Sterner and Elser 2002). Grazers release nutrients directly, but they also consume the primary consumers of nutrients, and other grazers, which are also nutrient regenerators (Glibert 1998). Whether grazers are highly constrained or more flexible in their stoichiometry has large consequences for nutrient regeneration, as those grazers that are highly constrained will, by necessity, regenerate the nutrients that are consumed but not needed to maintain their biomass. By regenerating the nutrients not needed, but retaining the nutrients that are needed, positive feedbacks are developed (Elser et al. 2000; Sterner and Elser 2002). For example, in principle, grazers with strict stoichiometry feeding on phytoplankton that are N-rich will excrete proportionately more N than those grazing on phytoplankton that are more balanced in their N:P or N:C ratio. By excreting more N, the condition of excess N is maintained for the phytoplankton (Sterner and Elser 2002). Such a condition creates wide variations in the stoichiometry of producers and consumers and presents further cellular challenges for producers to regulate both the limiting and saturating nutrient cellular levels. Preferential grazing will also affect the flow of regenerated nutrients. This can create the potential for shifts in dominant bloom species. This process of stoichiometric interaction is seen at all levels of the food web, leading to a complex interaction of stoichiometry and trophic cascades.

An important food web interaction at the microbial level, often involving HABs, is mixotrophy [see also Chap. 7, Flynn et al. (2018)]. Many different types of HAB species that thrive in eutrophying or eutrophic estuarine and coastal marine habitats, including cyanobacteria, dinoflagellates, and ochrophytes (diatoms, golden flagellates, brown tide algae, and raphidophycean flagellates), are mixotrophs (Burkholder et al. 2008). Most phytoflagellates (heterokontophytes, haptophytes, and cryptophytes) are capable of consuming bacteria and small eukaryotic cells (Unrein et al. 2007; Zubkov and Tarran 2008; Stukel et al. 2011). Dinoflagellates exhibit a wide range of capabilities, from those that express phagotrophy only upon extreme nutrient starvation to exclusive phagotrophic forms. Even the latter organisms may gain at least a transitory advantage from ingestion (Raven et al. 2009; Jeong et al. 2010; Hansen 2011) or other mixotrophs (Minnhagen et al. 2011). It is also apparent that the stoichiometry of predator and prey quality also has strong effects on the success of mixotrophs (Lundgren et al. 2016; Lin et al. 2017). The mixotrophs, *Prymnesium parvum* and *Karlodinium veneficum*, have both been shown to graze differentially on prey depending not only on their nutritional condition but also that of the prey, preferring prey that may help them to balance their nutritional requirements (i.e., the mixotroph in a high N condition will preferentially graze on prey in a comparatively high P condition and vice versa; Lundgren et al. 2016; Lin et al. 2017).

Importantly, it appears from several studies that mixotrophy also may permit growth to be sustained or even accelerated during periods of apparent water-column nutrient deficiency or imbalance. The benefits of mixotrophy for the dinoflagellates *C. polykrikoides* were noted above. The same growth advantage has been observed for *Karenia brevis*. Toxic *K. brevis* has been found to graze on the cyanobacterium *Synechococcus* sp., as well as on cryptophytes (Jeong et al. 2005; Adolf et al. 2008a; Glibert et al. 2009a). In laboratory experiments, it has been shown that *K. brevis* can graze from ~1 to 80 *Synechococcus* cells h^{-1} (Jeong et al. 2005; Glibert et al. 2009a), depending on the predator/prey ratio, with the growth rate of *K. brevis* increasing as the supply of *Synechococcus* increased. In contrast, *P. parvum* (as *Prymnesium patelliferum*) and *Chrysochromulina* spp. that ingested prey under light- and nutrient-sufficient conditions maintained similar growth rates with versus without phagotrophy (Larsen et al. 1993). It is clear that many mixotrophs increase their feeding when in a condition of nutrient imbalance, but strict nutrient limitation is not necessary. Moreover, there is much to be done to fully understand the relationships between mixotrophy and toxin production, which may be synergistic.

In all, factors such as the relative availability of the N (P and C) substrates, the nutritional status of the component organisms, and the number of trophic interactions, along with environmental parameters such as ambient light and temperature, all determine the extent of the dynamic metabolic balance in the use of one substrate vs another and, in turn, the relative success of (or not) of primary producers—including HABs—and their consumers. Once the balance is tipped, a new dynamic emerges. That nutrition plays a key role in this process should be obvious with our contemporary understanding of phytoplankton physiology and ecological stoichiometry; unravelling the multifaceted effects of nutrients, trophodynamics and HABs in an imbalanced and otherwise dynamic system will continue to present many challenges to be addressed in the future (Glibert 2017).

12.5 Key Question 4.0: Do Anthropogenic Alterations of the Food Web, Including Overfishing and Aquaculture Activities, Synergistically Interact with Nutrients to Favour HABs?

The interactions of trophodynamics and HABs at the lower end of the food web are also altered in complex ways when the upper levels of food webs are changed due to reductions in fish pressure, on the one hand, through overfishing or, on the other hand, through enhancement of fish (or shellfish) abundance through aquaculture. While these effects are many, the key aspect summarized here is how fish (or shellfish) abundance alters the availability and composition of nutrients that may, in turn, support HABs.

As outlined above, grazers alter the rate of nutrient regeneration and can alter the composition of that nutrient. The regeneration of nutrients from higher trophic levels becomes an increasingly important pathway when animals are at densities above that found in their natural habitat, as is the case with aquaculture. With regard to aquaculture impacts, all intensive culture systems alter the natural cycling of nutrients, with increased release of dissolved nutrients and solid excreta compared to natural environments, as well as the potential alterations of sediment biogeochemistry (Honkanen and Helminen 2000; Wu et al. 1994). Efficiency of nutrient retention by the reared organisms is on average 35% (Bouwman et al. 2013a; Fig. 12.3). In systems with exogenous feed addition, there is also breakdown of undigested feed. Estimates of dissolved and particulate N and P release, based on modelling that accounts for feed practices and conversion ratios by species (Bouwman et al. 2011, 2013b), show that dissolved nutrients are more N-rich, but particulate nutrients tend to be more P-rich in finfish systems and crustacean ponds than molluscan culture. Nutrient release products also vary considerably from the Redfield ratio (Redfield 1934). Further alteration of nutrient concentrations and stoichiometry results due to fluxes of N and P from sediments to water via remineralization of nutrients in excess feed, faeces or pseudofaeces (Burford and Longmore 2001).

The different release products, their forms, and proportions, in different aquaculture conditions, may contribute to which algal species proliferate due to the nutrient preferences of different algal groups (e.g., Heisler et al. 2008; Glibert et al. 2016). Toxic and fish-killing algae are commonly associated with finfish and molluscan production (Wu et al. 1994; Honkanen and Helminen 2000; GEOHAB 2010), while high-biomass bloom-forming algae are more commonly associated with pond production (Alonso-Rodríguez and Páez-Osuna 2003; Azanza et al. 2005). Of particular concern in both shellfish and finfish mariculture is the change in forms of N associated with excretion and microbial remineralization. Between 7 and 12% of the dissolved N waste of finfish consists of urea, the remainder most often is NH_4^+ (Kaushik and Cowey 1991), and 13% of dissolved N waste from shrimp excretion and leaching of feed waste in shrimp ponds has been found to be organic N (Burford and Williams 2001). As has been well described, there is mounting evidence that chemically reduced forms of N, including urea, differentially stimulate the growth of some types of HABs or may fuel the production of algae or bacteria on which mixotrophic HABs may feed (Glibert et al. 2006, 2016; Heisler et al. 2008; Kudela et al. 2008a; Flynn et al. 2013; Glibert and Burford 2017). The relationship between alteration in nutrient composition, not just total quantity, and the development of HABs is supported by examples from freshwater, estuarine, and marine waters globally (reviewed by Heisler et al. 2008; Anderson et al. 2002; Glibert and Burkholder 2006, 2017; Glibert et al. 2016). The fact that some HAB species are disproportionately more toxic under conditions of increasing

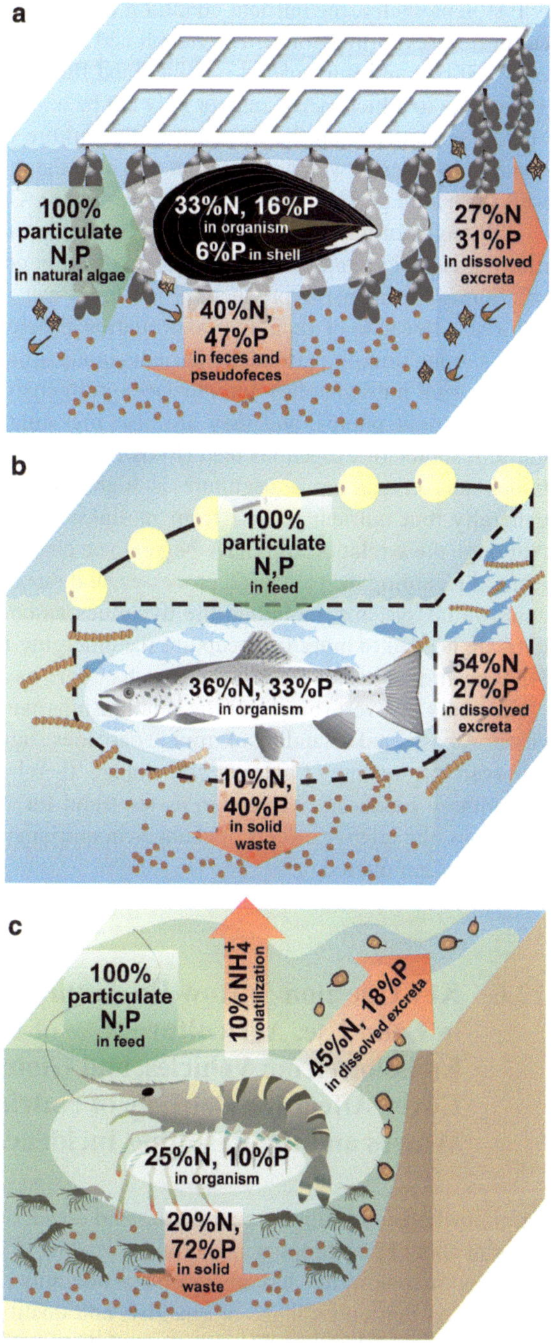

Fig. 12.3 Three types of mariculture and the associated N and P budgets in % of the intake (food or feed intake is 100%): (**a**) mollusc production in raft systems; (**b**) finfish production in cages; (**c**) crustacean production in ponds. The proportion of nutrients in dissolved versus solid form depends on the digestibility of the food or feed. The larger the digestible fraction of a nutrient in the various food or feed ingredients, the larger the fraction in dissolved form.
Figure reproduced from Bouwman et al. (2013a) under creative commons licence

N:P stoichiometry, the nutrient condition often occurring in finfish and crustacean culture, raises further concerns (Granéli and Flynn 2006; Sun et al. 2011; Fu et al. 2012; Glibert and Burford 2017; Glibert and Burkholder 2017; see also Sect. 12.7). The extent to which HABs may or may not be associated with specific aquaculture operations depends on the intensity of the culture operation and the overall husbandry practices as well as the extent to which the receiving waters are retentive or well flushed (e.g., Honkanen and Helminen 2000; Glibert et al. 2005).

Although normally considered beneficial environmentally because of their large filtering capacity, and because exogenous feed is not used when raised in culture conditions, in intensive production, molluscs can also become point sources of nutrients. Because of their low assimilation efficiency, molluscs act as nutrient pumps, transforming particulate nutrients in algal biomass to dissolved and particulate detrital nutrients; fish and crustacea in intensive culture conditions similarly act as nutrient pumps, and they are also ingesting and recycling the additional nutrient source of exogenous feed (Bouwman et al. 2011). Nutrient pollution from finfish and crustacean aquaculture is higher than that from molluscan culture generally, but nutrient release from molluscs nevertheless can be locally large. Although these relationships have begun to be characterized, there is much yet to be learned regarding the pathways, processes, and rates of nutrient regeneration within aquaculture systems, including the dynamics associated with the benthos. These data are needed to more fully capture the emerging links between aquaculture and HAB proliferations in regions where such culture has grown in volume or intensity.

In terms of overfishing and the relationships with HABs, much less is understood. A basic understanding of trophic cascades suggests that as top consumers are removed, the cycling of nutrient will change, thereby altering the flow of nutrients for primary producers. Moreover, as the trend for "fishing down the food chain" continues, the relative effects of changes in nutrient regeneration by top consumers will be further altered.

12.6 Key Question 5: How Do Anthropogenic Changes in Land Use, Agricultural Use of Fertilizer, NOx Emissions from Vehicles, and Global Changes in Land Cover Affect the Delivery of Nutrients to Coastal Waters and the Resulting Incidences of HABs?

Our knowledge of changes in land use, land cover, and the multiple changes being brought about by anthropogenic activities leading to altered delivery of nutrients to fresh and marine waters has advanced considerably [see also Chap. 4, Glibert et al. (2018)]. As nutrients change in quantity, and quality, the likelihood of HAB proliferations also changes. The success of HABs lies at the intersection of the physiological adaptations of the harmful algal species and/or strain (population), the environmental conditions, interactions with co-occurring organisms (both biogeochemically and trophodynamically), and physical dynamics that alter abiotic conditions and/or aggregate or disperse cells (or can alter abiotic conditions in a favourable or unfavourable manner), in turn promoting or inhibiting their growth (Glibert and Burford 2017).

Rapid progress is being made in spatially explicit modelling that is advancing our understanding of both regional and global variability in nutrients reaching the coastal zone. Among the various models being applied in regional to global estimates of nutrient export are the Spatially Referenced Regressions on Watersheds (SPARROW), Soil and Water Assessment Tool (SWAT), and Nutrient Export from Watersheds (NEWS) models. The NEWS system of models is unique in that it can be used to estimate magnitude and sources of multiple elements (N, P, C and Si) and their different forms (particulate, dissolved inorganic, and organic; Seitzinger et al. 2005). This latter system of models, based on data from more than 5000 exoreic basins, takes into account natural sources such as N_2 fixation and P weathering as well as anthropogenic sources (nonpoint inputs from fertilizer by crop type, N_2 fixation by crops, atmospheric N deposition, and manure by animal species and point sources from sewage, as estimated by human population and treatment level; Seitzinger et al. 2002, 2005, 2010; Dumont et al. 2005; Harrison et al. 2005a, b; Mayorga et al. 2010). It is these models that were applied in Key Question 1 when global patterns of *P. minimum* and nutrient export were compared. Another such approach is the Net Anthropogenic Nitrogen Input (NANI) model, first introduced by Howarth et al. (1996), that calculates N loads to watersheds as the sum of fertilizer use, N deposition, N_2 fixation, and net food and feed imports. The NANI models have been applied to many regions, most recently to estimate N fluxes from Indian watersheds (Swaney et al. 2014). The NANI models also illustrate the gaseous dry deposition of N that is much higher in urban areas and near highways due to local deposition of vehicle emissions (Howarth 2008).

Atmospheric deposition is also increasing due to many sources, among which is volatilization of NH_3 from agricultural lands and animal operations. An estimated 30% of the N deposited on these agricultural farmlands and intensive animal operations is volatilized to the atmosphere as NH_3, with most of this deposited near the site of emission (Howarth et al. 2002). Howarth (2008) reported, based on statistics that are now outdated but nonetheless illustrative (Holland et al. 1999), that on average worldwide deposition has probably doubled in the tropics and increased more than sixfold in the north temperate zone as a result of human activities. Collectively, the message from these models is the same: there is substantial export of both N and P to aquatic systems, but in general, there is a greater export of N than P (Sutton et al. 2013; Glibert et al. 2014a). Peñuelas et al. (2012) have estimated that when fertilizer use and atmospheric deposition are considered, the molar N:P ratio of that deposited on land is in the range of 44–47, while that of land plants is generally 22–30, but of even greater concern is that the N:P ratio of input to oceans is 114–370, which is 10–20 times the stoichiometric ratio of plankton characterized by Redfield (1934).

A recent model is the Integrated Model to Assess the Global Environment-Global Nutrient Model (IMAGE-GNM), which is a coupled hydrology-biogeochemistry model (Beusen et al. 2015, 2016) that describes delivery through aquifers, riparian zones, surface runoff, and atmospheric deposition as well as point sources to streams and rivers and subsequently in-stream retention processes. Beusen et al. (2016) used IMAGE-GNM to simulate N and P transport and retention by rivers for the twentieth century accounting for the construction of major dams and reservoirs and showed that N:P ratios in water draining to the world's oceans is rapidly increasing in many parts of the world. This is probably due to the legacy of nutrient management in agriculture during the 1970s and 1980s. For example, N concentrations in many rivers do not respond to increased agricultural N use efficiency, and as a result, European water quality is threatened by rapidly increasing N:P ratios (Bouwman et al. 2017). Latecomers to big agriculture like India and China will soon face their own nutrient legacies even as the current use of fertilizers decrease (Bouwman et al. 2017).

Examples from China again highlight the relationships between anthropogenic nutrient use and HAB proliferation. China's estuaries and coasts are among the most rapidly eutrophying in the world. With a population of more than 20 M people, Shanghai alone discharges more than 200 t of N daily to adjacent waters. Pollution by N in China from the heavy use of agricultural nutrient has led to the Changjiang River now being among the highest in the world in terms of annual N load. The spatial extent of the associated hypoxic region in the East China Sea now rivals that in the Gulf of Mexico receiving discharge of the Mississippi River (Chen et al. 2007; Zhu et al. 2011). At the mouth of the Changjiang River, molar NO_3^-:PO_4^{3-} values were 30–40 in the 1960s when HABs were comparatively rare in the East China Sea but rose to >250 by the late 1990s, numbering more than 100s. The scale of the HAB outbreaks has also increased, from 1000s of km^2 in 2000 to >15,000 km^2 by 2005 with many millions of dollars of lost production of fish due to hypoxia and direct toxic effects (Zhou et al. 2008; Li et al. 2009; Glibert et al. 2014a). Fertilizer N use in China has escalated from about ~0.5 million tonnes in the early 1960s to 42 million tonnes around 2010, with the fraction of urea increasing nearly fivefold over just the past two decades (Glibert et al. 2014a and references therein). More specifically, the use of N increased from 1980 to 2010 from ~0.5 to >1.2 tonnes km^{-2} $year^{-1}$ in the Changjiang River, from ~0.1 to ~0.2 tonnes km^{-2} $year^{-1}$ in the Yellow River, and from ~0.4 to >1.2 tonnes km^{-2} $year^{-1}$ in the Pearl River basins (Ti and Yan 2013). In parallel with these trends in nutrient loading, the number of HABs has increased in virtually all waters of China in the past three decades. In addition to these blooms, "green tides" have increased. These noxious macroalgal blooms (*Ulva prolifera*) received notoriety during the 2008 Qingdao Sailing Olympics, when the water was blanketed with thick green scums [Hu et al. 2010; see also Chap. 16, Liu and Zhou (2018)]. More recently "brown tides" have become recurrent in the Bohai Sea [Zhang et al. 2012; see also Chap. 15, Yu et al. (2018)].

Without question, much has been learned on a global scale regarding the export of nutrients, the stoichiometric balance of that export, and the complexity of factors that alter nutrients in total load and composition as they are transported from land to the sea or to freshwaters. Models of nutrient export have greatly advanced, but there remain many challenges. Data availability for model development and calibration for some parts of the world is very good, but for other regions, the data are poor. Improved spatial dimension of loading models is also needed. Dispersal of the nutrient and the interaction with the physical dynamics of the receiving water is still difficult to quantify. While river plume dynamics may be understood, plumes of nutrients from nonpoint sources are still largely difficult to resolve. Also, in spite of the range of examples provided here, there is much yet to be understood in terms of the role of different nutrients on toxin production. Coupling the physiological adaptations of the harmful algal species and/or strain (population), with the environmental conditions, and physical dynamics will remain a challenge.

12.7 Key Question 6: How Do the Stoichiometry and Quality of These Nutrient Sources Regulate the Biological Response, Including Toxins, Favouring HABs?

Considerable emerging evidence has shown that as nutrients change in total loads and in proportion, as already described, there are also effects on toxin production. Many cyanobacteria and marine dinoflagellate HABs have been shown to be more toxic when N is in stoichiometric excess over P, as is the case for many regions of the globe (reviewed by Glibert 2017). The evidence for toxin production increasing under non-stoichiometrically balanced conditions is considerable (e.g., Granéli and Flynn 2006; Hardison et al. 2012, 2013). Most notably, excess N over P availability has been related to the production of microcystin (MC), the toxin of the freshwater cyanobacteria *Microcystis* among other species, under controlled chemostat conditions and in natural populations (Van de Waal et al. 2009, 2010; Monchamp et al. 2014; Harris et al. 2016). As recently reviewed by Gobler et al. (2016), common cyanotoxins including MC, nodularins, cylindrospermopsins, and saxitoxins have amino acid precursors, either glutamine, arginine, or lutein, and they, in turn, depend on adequate N supply for their assimilation. In the dinoflagellate *Alexandrium tamarense*, saxitoxin production has been shown to increase by three- to fourfold under P deficiency (reviewed by Granéli and Flynn 2006). Additionally, a study of the dinoflagellate, *Karenia*, found that P-limited cells had a higher cellular toxin quota than P-replete cells (Hardison et al. 2013). Furthermore, under conditions of elevated N:P ratios, hemolytic activity per cell has been shown to increase by up to tenfold in the haptophytes *P. parvum* and *Chrysochromulina* (now *Prymnesium*) *polylepis* (Johansson and Granéli 1999). Similarly, at higher N:P ratios, neurotoxin production was shown to increase in the diatom *Pseudo-nitzschia* spp. (Bates et al. 1991; Pan et al. 1996; Sun et al. 2011). There are also examples of toxin production being stimulated at the opposite N:P spectrum. The freshwater cyanobacterium, *Cylindrospermopsis raciborskii*, was shown to increase the proportion of toxic strains when P was added to mesocosms (Burford et al. 2014), and the haptophyte *P. parvum* is also more toxic under low N:P (as well as high N:P) compared to balanced nutrient conditions (Granéli and Flynn 2006). Although not conclusively shown for any toxin, toxin production may be part of the complex suite of physiological processes involved in physiological "overflow metabolism" (Glibert et al. 2016; Glibert 2017).

12.8 Key Question 7: Do Climate Change and Climate Variability Have Impacts on Ecosystems that Augment the Impacts of Eutrophication and the Formation of HABs?

Nutrient changes and those of climate are complex on many levels, as has been highlighted in other chapters of this volume [see also Chap. 4, Glibert et al. (2018) and Chap. 5, Wells and Karlson (2018)] and elsewhere (e.g., Doney 2010; Fu et al. 2012; Wells et al. 2015). Climate controls many of the fundamental parameters regulating algal growth, including water temperature, nutrients, and light, and thus can be expected to influence changes in the species composition, trophic structure, and function of aquatic ecosystems.

Global temperatures are on the rise, a fact now well accepted to be related to anthropogenic activities (Intergovernmental Panel on Climate Change 2007, 2014). Average sea surface temperatures are expected to rise as much as 5 °C over the coming century, leading to a freshening of many oceanic regions due to ice melt and altered precipitation (e.g., Moore et al. 2008; Doney 2010; Fu et al. 2012, and references therein). These changes, in turn, will alter stratification, availability of nutrients and their forms and ratios, pCO_2, and light regimes among other factors (e.g., Boyd and Doney 2003), all of which control the extent to which HABs become established. Temperature affects growth rate, motility, germination, pigment content, enzyme reactions, photosynthesis, and various other processes, influencing the ability of cells to thrive in a particular area (e.g., Wells et al. 2015 and references therein). Increasing temperatures positively affect those taxa with higher temperature optima for growth and negatively influence those taxa that have lower temperature optima. Negative influences of increasing temperature can disproportionately affect diatoms, as they generally thrive in colder seasons, are more abundant in polar and temperate regions, and tend to have colder temperature optima than other microalgal groups (Harris 1986; Glibert et al. 2016, and references therein). Moreover, the uptake of NO_3^- and its reduction to NH_4^+ generally decreases at higher temperatures, especially above 15–18 °C (e.g., Lomas and Glibert 1999; Glibert et al. 2016), further suggesting that diatoms, which are more typically NO_3^- specialists, are negatively impacted as temperatures rise. In contrast, many cyanobacterial and dinoflagellates species, including HAB species, prefer warmer temperature conditions (e.g., Paerl and Huisman 2008; Paerl and Scott 2010). Temperatures also affect the community of organisms within which the harmful algal species live, including bacteria, viruses, competing phytoplankton taxa, and grazers (Wells et al. 2015 and references therein). Toxicity of many harmful algal species also increases with warming (Davis et al. 2009; Fu et al. 2012, and references therein). The combination of elevated pCO_2 together with nutrient limitation and altered nutrient ratios appears to be especially potent in affecting the toxicity of some harmful algal species. On the other hand, for some species, higher toxicity associated with warming may be associated with slower growth rates (e.g., Ogata et al. 1989; Lewis et al. 1993), but slower growth rates need not necessarily be detrimental to HAB formation if other factors, such as reduced grazing, come into play [see Chap. 7, Flynn et al. (2018)].

Examples can be found through the USA and European coasts, where long-term data are available, showing positive correlations between increasing mean water temperature and shifts in HAB species and the timing of their outbreaks. Warmer temperatures may contribute to range expansion of particular species. Higher temperatures promote increased water-column stability and increased thermal stratification that, in turn, favours known bloom-forming, toxigenic cyanobacterial species that control their vertical position through internal buoyancy regulation (e.g., Walsby 1975; Visser et al. 2016, and references therein). Under higher vertical mixing, diatoms are superior competitors; thus, as stratification increases, diatoms are more apt to sink out from the water column (Visser et al. 2016 and references therein). Moreover, at higher temperatures, water can become less viscous, buoyant cyanobacteria can float faster, and conversely sinking diatoms will sink faster (e.g., O'Neil et al. 2012 and references therein). With climate change, diatoms are seemingly negatively affected in several ways.

As mentioned above, warming trends are causing changes in the C cycle, resulting in acidification of the oceans and some estuaries. This contrasts with the high pH caused by high-biomass blooms that occur due to drawdown of CO_2. As acidification reduces pH, there is some evidence that some cyanobacteria can increase growth rates and therefore outcompete eukaryotic algae under such conditions (O'Neil et al. 2012 and references therein). Flynn et al. (2015), in a series of experiments coupled with modelling approaches, illustrated the complexity of effects of changes in pH coupled with eutrophication on phytoplankton species succession. Eutrophication increases the phytoplankton biomass that can be supported during a bloom, and the resultant uptake of dissolved inorganic C (DIC) during photosynthesis increases water-column pH. This increased pH can adversely affect plankton growth of some species, but not necessarily of others (e.g., some HABs). Using experimental analyses of the growth of three contrasting phytoplankton under different pH scenarios, coupled with mathematical models describing growth and death as functions of pH and nutrient status, Flynn et al. (2015) were able to show how different conditions of pH modify the scope for competitive interactions between phytoplankton species. Their work also suggested that, when coupled with effects on grazers, ocean acidification and eutrophication may increase the frequency of HABs, including blooms of mixotrophic species.

Climate change may further influence harmful algal species expansions due to altered precipitation patterns, including increases in droughts in some regions and increased frequency or intensity of storm events in other regions. Episodic storm events and climate variability affect the timing of freshwater flow, water residence times, the magnitude and timing of nutrient pulses, and resulting biotic responses (e.g., Miller et al. 2006; Burkholder et al. 2006; Heisler et al. 2008). As examples, within days after Hurricane Isabel in 2003 in the USA, a large phytoplankton bloom developed in Chesapeake Bay, linked to increased nutrient loads (Miller et al. 2005). Algal blooms and extensive hypoxia/anoxia occurred shortly after several hurricanes affected the lagoonal Neuse River Estuary in the 1990s (Burkholder et al. 2004, 2006). A bloom of the picocyanobacterium EDAB species *Synechococcus* in eastern Florida Bay, lasting more than 18 months, followed an injection of nutrients from the high freshwater discharge caused by Hurricanes Katrina, Rita, and Wilma in 2005 (Glibert et al. 2009b) and a similar response appears to have been initiated with the passage of Hurricane Irma in 2017.

Long-term changes in, or intensification of, climate forces such as monsoons or interannual oscillations, such as those related to the El Niño Southern Oscillation (ENSO), or longer-term cycles, such as North Atlantic Oscillation (NAO) and the Pacific Decadal Oscillation (PDO), can also alter conditions for HAB species. In the northern Iberian Peninsula, for example, the harmful dinoflagellate *Gymnodinium catenatum* was abundant during the mid-1980s when there was a transition from downwelling-favourable conditions to upwelling-favourable conditions, following a shift in the North Atlantic Oscillation (NAO) index (Alvarez-Salgado et al. 2003). In the late 2013, and continuing through much of 2016, anomalously warm water developed in the northeastern Pacific Ocean (e.g., Bond et al. 2015; Freeland and Whitney 2014), a feature associated with the unusually strong El Niño event and the Pacific Decadal Oscillation (PDO). The warm water moved over the continental margin, eventually extending from Southern California to Alaska by spring 2015. Coupled with seasonal upwelling, conditions were ideal for *Pseudo-nitzschia* that had sufficient nutrients in suitable forms, to proliferate, and suitable temperatures for rapid growth. Regulatory limits of domoic acid (DA) were exceeded along the entire coast for months, and toxin impacts were felt at many levels of the food chain, from razor clams and Dungeness crabs to sea lions and whales and porpoises (McCabe et al. 2016). This was the largest toxic *Pseudo-nitzschia* bloom on the West Coast thus far and portends of future outbreaks with conditions of increasing temperature and nutrient supply.

The interacting effects of pCO_2, temperature, and nutrient supply complicate the interpretation of effects of pCO_2 on HAB toxicity (e.g., Boyd and Hutchins 2012; Gobler et al. 2016). High CO_2 may also affect toxicity of HABs through various routes. An overall trend of increasing toxicity with increasing pCO_2 has been reported for numerous strains of cultured *Alexandrium ostenfeldii* (Kremp et al. 2012), but culturing approaches can affect the extent to which relationships were observed between elevated pCO_2 and toxicity in some harmful algal species (Van de Waal et al. 2011). The synthesis of at least some toxins is light dependent, as is the case for karlotoxin production by *K. veneficum* and saxitoxin production by *Alexandrium catenella* (Adolf et al. 2008b), suggesting that as photosynthesis is affected by changing pCO_2, toxin synthesis is also altered. In the toxigenic diatoms *Pseudo-nitzschia multiseries* and *P. fraudulenta*, concentrations of their toxin, DA, have increased at high CO_2/low pH in some studies (e.g., Sun et al. 2011; Tatters et al. 2012). This effect can be more pronounced when cells are nutrient-limited or when forms of N shift from oxidized to reduced forms (Glibert et al. 2016 and references therein).

The linkages between changes in nutrient supply to aquatic systems and climate are clearly complex. These complexities create even greater challenges for predicting how HABs may change. Using a global modelling approach, which coupled a climate change model in a coupled oceanographic-biogeochemical model, the potential effects of nutrient loading and climate change for end of the century conditions for two HAB genera, pelagic *Prorocentrum* and *Karenia*, were estimated for three regions of the globe, NW European Shelf-Baltic Sea system, NE Asia, and SE Asia (Glibert et al. 2014b). Projections for the different HABs applied a set of physiological "rules" based on suitable temperature, salinity, and nutrient conditions for growth. The nutrient conditions applied a nutrient ratio approach, rather than specific concentrations. Inasmuch as the goal was to identify the likelihood of conditions supportive of blooms, rather than their strength or magnitude, absolute concentrations were not necessary. Based on these models, an expansion in area and/or number of months annually conducive to development of these HABs along the NW European Shelf-Baltic Sea system and NE Asia was projected for both HAB genera, but no expansion (*Prorocentrum* spp.), or actual contraction in area and months conducive for blooms (*Karenia* spp.), was projected in the SE Asian domain (Glibert et al. 2014b). The implications of these projections, especially for Northern Europe, are shifts in vulnerability of coastal systems to HAB events, increased regional HAB impacts to aquaculture, increased risks to human health and ecosystems, and economic consequences of these events due to losses to fisheries and ecosystem services.

Clearly, current models provide some realizations of future states on a regional to global basis but are insufficient to address the response to HABs to the combined effects we are facing. Not only are advances needed in the coupling of land-use nutrient export models, but their coupling to oceanographic models and climate models is still in early stages of development. Considerable new data and computational capabilities will continue to be needed.

12.9 Conclusions

In sum, the evidence is unequivocal that, as concluded in the GEOHAB (2005) report on HABs in Eutrophic Systems and in the scientific consensus report by Heisler et al. (2008), HAB species can be directly and/or indirectly stimulated by nutrient over-enrichment, and nutrient enrichment interacts with other major drivers, such as hydrological changes, food web interactions, and climate change, in both direct and indirect ways as drivers of HABs. While over-enrichment of coastal waters by nutrients is a major pollution problem worldwide, it is not only total nutrient loads that promote HABs and alter microbial biodiversity. There are still many places and times where nutrient pollution continues to be minimal, and therefore nutrients and eutrophication are not a major contributor to HABs in those locations or time, but overwhelmingly we now recognize that, indeed, there are more blooms, more often, in new and different places, often lasting longer, with a range of toxicities, and many of these blooms are related to the global increase in nutrient pollution.

Many questions remain. Indeed, there are gaps to fill in all of the key questions listed here. Much needs to be done in parameterizing rates, characterizing traits, and how they are both externally driven and internally dynamically regulated. Many species are understudied. Work needs to advance in understanding the physiological responses to excess nutrient availability and relationships with toxicity, among other physiological processes. Improved understanding of when a nutrient becomes a "stress" and how these stresses are compounded by other stresses (such as light, temperature, pH) is sorely needed (Allen and Polimene 2011; Glibert et al. 2013; Glibert 2017).

A new emphasis on improved model formulations is needed, including efforts to incorporate dynamic balance models for physiology (Glibert et al. 2013). The plasticity of nutritional pathways, as well as the plasticity of food web interactions, including grazing, allelopathy, symbioses, and other interactions, creates immense challenges for model constructs. Variable stoichiometric parameterizations in models must also begin to recognize that physiological processes and organismal stoichiometry can and do vary even at growth-saturating substrate concentrations. Understanding and parameterizing these relationships and how they change under our new reality of altered environmental conditions (nutrient loads, forms, temperatures, pH) have implications for modelling current and projected changes in HABs as climate and nutrient change. Further advances in linking land-use nutrient models (by amount, form, and proportion), coastal typology, climate change, regional ocean models, and biogeochemical and physiological responses are needed. Through continued and new international and multidisciplinary collaborations, such as those that may be fostered through the GlobalHAB Programme, such advancements will be possible.

Acknowledgments This is a contribution of the GEOHAB Core Research project on HABs in Eutrophic Systems and of SCOR WG 132 on Land-Based Nutrient Pollution and Harmful Algal Blooms. It is contribution number 5407 from the University of Maryland Center for Environmental Science.

References

Adolf JE, Bachvaroff T, Place AR (2008a) Cryptophyte abundance drives blooms of mixotrophic harmful algae: a hypothesis based on *Karlodinium veneficum* as a model system. Harmful Algae 8:119–128

Adolf JE, Bachvaroff T, Place AR (2008b) Can cryptophytes trigger toxic *Karlodinium veneficum* blooms in eutrophic estuaries? Harmful Algae 81:19–128

Al-Azri A, Piontkovski SA, Al-Hashmi KA et al (2014) Mesoscale and nutrient conditions associated with the massive 2008 *Cochlodinium polykrikoides* bloom in the Sea of Oman/Arabian Gulf. Estuar Coasts 37:325–338
Allen JI, Polimene L (2011) Linking physiology to ecology: towards a new generation of plankton models. J Plankton Res 33:989–997. https://doi.org/10.1093/plankt/fbr032
Alonso-Rodríguez R, Páez-Osuna F (2003) Nutrients, phytoplankton and harmful algal blooms in shrimp ponds: a review with special reference to the situation in the Gulf of California. Aquaculture 219:317–336
Alvarez-Salgado XA, Figueiras FG, Perez FF et al (2003) The Portugal coastal counter current of NW Spain: new insights on its biogeochemical variability. Prog Oceanogr 56:281–321
Anderson DM, Glibert PM, Burkholder JM (2002) Harmful algal blooms and eutrophication: nutrient sources, composition, and consequences. Estuaries 25:562–584
Azanza RV, Fukuyo Y, Yap LG et al (2005) *Prorocentrum minimum* bloom and its possible link to a massive fish kill in Bolinae, Pangasian, northern Philippines. Harmful Algae 4:519–524
Bates SS, de Freitas ASW, Milley JE et al (1991) Controls on domoic acid production by the diatom *Nitzschia pungens* c.f. *multiseries* in culture: nutrients and irradiance. Can J Fish Aquat Sci 48:1136–1144
Beusen AHW, Bouwman AF, Van Beek LPH et al (2016) Global riverine N and P transport to ocean increased during the 20th century despite increased retention along the aquatic continuum. Biogeosciences 13:2441–2451. https://doi.org/10.5194/bg-13-2441-2016
Beusen AHW, Van Beek LPH, Bouwman AF et al (2015) Coupling global models for hydrology and nutrient loading to simulate nitrogen and phosphorus retention in surface water. Description of IMAGE-GNM and analysis of performance. Geosci Model Dev 8:4045–4067. https://doi.org/10.5194/gmd-4048-4045-2015
Bond NA, Cronin MF, Freeland H et al (2015) Causes and impacts of the 2014 warm anomaly in the NE Pacific. Geophys Res Lett 42:3414–3420
Bouwman AF, Beusen AHW, Glibert PM et al (2013a) Mariculture: significant and expanding cause of coastal nutrient enrichment. Environ Res Lett 8(044026):5. https://doi.org/10.1088/1748-9326/8/4/044026
Bouwman AF, Beusen AHW, Lassaletta L et al (2017) Lessons from temporal and spatial patterns in global use of N and P fertilizer on cropland. Sci Rep. https://doi.org/10.1038/srep40366
Bouwman AF, Beusen AHW, Overbeek CC et al (2013b) Hindcasts and future projections of global inland and coastal nitrogen and phosphorus loads due to finfish aquaculture. Rev Fish Sci 21:112–158
Bouwman AF, Pawlowski M, Liu C et al (2011) Global hindcasts and future projections of coastal nitrogen and phosphorus loads due to shellfish and seaweed aquaculture. Rev Fish Sci 19:331–357
Boyd PW, Doney SC (2003) The impact of climate change and feedback processes on the ocean carbon cycle. In: Fasham MJR (ed) Ocean biogeochemistry – the role of the ocean carbon cycle in global change. Springer, Berlin, pp 157–193
Boyd PW, Hutchins DA (2012) Understanding the responses of ocean biota to a complex matrix of cumulative anthropogenic change. Mar Ecol Prog Ser 470:125–135
Bricker SB, Longstaff B, Dennison W et al (2007) Effects of nutrient enrichment in the Nation's estuaries: a decade of change, NOAA coastal ocean program decision analysis series No. 26. National Center for Coastal Ocean Science, Silver Spring, MD, p 328
Burford MA, Davis TW, Orr PT et al (2014) Nutrient-related changes in the toxicity of field blooms of the cyanobacterium *Cylindrospermopsis raciborskii*. FEMS Microbiol Ecol 89:135–148
Burford MA, Longmore AR (2001) High ammonium production from sediments in hypereutrophic aquaculture ponds. Mar Ecol Prog Ser 224:187
Burford MA, Pearson D (1998) Effect of different nitrogen sources on phytoplankton composition in aquaculture ponds. Aquat Microb Ecol 15:277–284
Burford MA, Williams KC (2001) The fate of nitrogenous waste from shrimp feeding. Aquaculture 198:79–93

Burkholder JM, Dickey DA, Kinder C et al (2006) Comprehensive trend analysis of nutrients and related variables in a large eutrophic estuary: a decadal study of anthropogenic and climatic influences. Limnol Oceanogr 51:463–487

Burkholder JM, Eggleston D, Glasgow H et al (2004) Comparative impacts of two major hurricane seasons on the Neuse River and western Pamlico Sound ecosystems. Proc Natl Acad Sci USA 101:9291–9296

Burkholder JM, Glibert PM, Skelton H (2008) Mixotrophy, a major mode of nutrition for harmful algal species in eutrophic waters. Harmful Algae 8:77–93

Burkholder JM, Shumway SE, Glibert PM (2018) Food webs and ecosystem impacts of harmful algae. In: Shumway S, Burkholder JM, Morton SL (eds) Harmful algal blooms: a compendium desk reference. Wiley

Buskey EJ, Liu H, Collumb C et al (2001) The decline and recovery of a persistent Texas brown tide algal bloom in the Laguna Madre (Texas, USA). Estuaries 24:337–346

Cannon JA (1990) Development and dispersal of red tides in the Port River, South Australia. In: Granéli E, Sundstroem B, Edler L et al (eds) Toxic marine phytoplankton. Elsevier, New York, pp 110–115

Chen C-C, Gong G-C, Shiah F-K (2007) Hypoxia in the East China Sea: one of the largest coastal low oxygen areas in the world. Mar Environ Res 64:399–408

Cloern JE (2001) Our evolving conceptual model of the coastal eutrophication problem. Mar Ecol Prog Ser 210:223–253

Davis TL, Berry DL, Boyer GL et al (2009) The effects of temperature and nutrients on the growth and dynamics of toxic and non-toxic strains of *Microcystis* during cyanobacteria blooms. Harmful Algae 8:715–725

Doney SC (2010) The growing human footprint on coastal and open ocean biogeochemistry. Science 328:1512–1516

Duan SW, Liang T, Zhang S et al (2008) Seasonal changes in nitrogen and phosphorus transport in the lower Changjiang River before the construction of the Three Gorges Dam. Estuar Coast Shelf Sci 79:239–250

Dumont E, Harrison JA, Kroeze C et al (2005) Global distribution and sources of dissolved inorganic nitrogen export to the coastal zone: results from a spatially explicit, global model. Global Biogeochem Cycles 19:GB4S02

Dürr HH, Laruelle GG, van Kempen CM et al (2011) Worldwide typology of nearshore coastal systems: defining the estuarine filter of river inputs to the oceans. Estuar Coasts 34:441–458

Elser JJ, Sterner RW, Gorokhova E et al (2000) Biological stoichiometry from genes to ecosystems. Ecol Lett 3:540–550

Fang TH (2004) Phosphorus speciation and budget of the East China Sea. Cont Shelf Res 24: 1285–1299

Flynn KA, Clark DR, Mitra A et al (2015) Ocean acidification with (de)eutrophication will alter future phytoplankton growth and succession. Philos Trans R Soc B 282:20142604

Flynn KJ, Mitra A, Glibert PM et al (2018) Mixotrophy by HABs: by whom, on whom, when, why and what next. In: Glibert PM, Berdalet E, Burford M et al (eds) Global ecology and oceanography of harmful algal blooms. Springer, Cham, pp 113–132

Flynn KJ, Stoecker DK, Mitra A et al (2013) Misuse of the phytoplankton-zooplankton dichotomy: the need to assign organisms as mixotrophs within plankton functional types. J Plankton Res 35:3–11

Freeland H, Whitney F (2014) Unusual warming in the Gulf of Alaska. North Pac Mar Org (PICES) Press 22:51–52

Fu M, Wang Z, Pu X et al (2012) Changes in nutrient concentrations and N:P:Si ratios and their possible impacts on the Huanghai Sea ecosystem. Acta Oceanol Sinica 31:101–112

Galloway JN, Cowling EB, Seitzinger SP et al (2002) Reactive nitrogen: too much of a good thing? Ambio 31:60–63

GEOHAB (2001) Global ecology and oceanography of harmful algal blooms, science plan. Glibert P, Pitcher G (eds) SCOR and IOC, Baltimore and Paris, 86 pp

GEOHAB (2005) Global ecology and oceanography of harmful algal blooms: HABs in eutrophic systems. Glibert PM (ed) IOC and SCOR, Paris and Baltimore, 74 pp

GEOHAB (2010) Global ecology and oceanography of harmful algal blooms in Asia. Furuya K, Glibert PM, Zhou M et al (eds) IOC and SCOR, Paris and Newark, Delaware, 68 pp

Glibert PM (1998) Interactions of top-down and bottom-up control in planktonic nitrogen cycling. Hydrobiology 363:1–12

Glibert PM (2017) Eutrophication, harmful algae and biodiversity – challenging paradigms in a world of complex nutrient changes. Mar Pollut Bull 124:591–606. https://doi.org/10.1016/j.marpolbul.2017.04.027

Glibert PM, Allen JI, Artioli Y et al (2014b) Vulnerability of coastal ecosystems to changes in harmful algal bloom distribution in response to climate change: projections based on model analysis. Glob Chang Biol 20:3845–3858. https://doi.org/10.1111/gcb.12662

Glibert PM, Beusen AHW, Harrison JA et al (2018) Changing land-, sea- and airscapes: sources of nutrient pollution affecting habitat suitability for harmful algae. In: Glibert PM, Berdalet E, Burford M et al (eds) Global ecology and oceanography of harmful algal blooms. Springer, Cham, pp 53–76

Glibert PM, Boyer J, Heil C et al (2010) Blooms in lagoons: different from those of river-dominated estuaries. In: Kennish M, Paerl H (eds) Coastal lagoons: critical habitats of environmental change. Taylor and Francis, Boca Raton, FL, pp 91–114

Glibert PM, Burford MA (2017) Globally changing nutrient loads and harmful algal blooms: recent advances, new paradigms, and continuing challenges. Oceanography 30(1):58–69. https://doi.org/10.5670/oceanog.2017.110

Glibert PM, Burkholder JM (2006) The complex relationships between increasing fertilization of the earth, coastal eutrophication and proliferation of harmful algal blooms. In: Granéli E, Turner J (eds) Ecology of harmful algae. Springer, New York, NY, pp 341–354

Glibert PM, Burkholder JM (2011) Eutrophication and HABs: strategies for nutrient uptake and growth outside the Redfield comfort zone. Chin J Oceanol Limnol 29:724–738

Glibert PM, Burkholder JM (2017) Causes of harmful algal blooms. In: Shumway S, Burkholder JM, Morton SL (eds) Harmful algal blooms: a compendium desk reference. Wiley (in press)

Glibert PM, Burkholder JM, Granéli E et al (2008a) Advances and insights in the complex relationships between eutrophication and HABs: preface to the special issue. Harmful Algae 8:1–2

Glibert PM, Burkholder JM, Kana TM et al (2009a) Grazing by *Karenia brevis* on *Synechococcus* enhances their growth rate and may help to sustain blooms. Aquat Microb Ecol 55:17–30

Glibert PM, Burkholder JM, Kana TM (2012) Recent advances in understanding of relationships between nutrient availability, forms and stoichiometry and the biogeographical distribution, ecophysiology, and food web effects of pelagic and benthic *Prorocentrum* spp. Harmful Algae 14:231–259

Glibert PM, Fullerton D, Burkholder JM et al (2011) Ecological stoichiometry, biogeochemical cycling, invasive species and aquatic food webs: San Francisco Estuary and comparative systems. Rev Fish Sci 19:358–417

Glibert PM, Harrison JA, Heil CA et al (2006) Escalating worldwide use of urea – a global change contributing to coastal eutrophication. Biogeochemistry 77:441–463

Glibert PM, Heil CA, Hollander D et al (2004) Evidence for dissolved organic nitrogen and phosphorus uptake during a cyanobacterial bloom in Florida Bay. Mar Ecol Prog Ser 280:73–83

Glibert PM, Heil CA, Rudnick D et al (2009b) Florida Bay: status, trends, new blooms, recurrent problems. Contrib Mar Sci 38:5–17

Glibert PM, Kana TM, Brown K (2013) From limitation to excess: consequences of substrate excess and stoichiometry for phytoplankton physiology, trophodynamics and biogeochemistry, and implications for modelling. J Mar Syst 125:14–28. https://doi.org/10.1016/j.jmarsys.2012.10.004

Glibert PM, Maranger R, Sobota DJ et al (2014a) The Haber-Bosch–harmful algal bloom (HB-HAB) link. Environ Res Lett 9:105001 (13 p). https://doi.org/10.1088/1748-9326/9/10/105001

Glibert PM, Mayorga E, Seitzinger S (2008b) *Prorocentrum minimum* tracks anthropogenic nitrogen and phosphorus inputs on a global basis: application of spatially explicit nutrient export models. Harmful Algae 8:33–38
Glibert PM, Seitzinger S, Heil CA et al (2005) The role of eutrophication in the global proliferation of harmful algal blooms: new perspectives and new approaches. Oceanography 18(2):198–209
Glibert PM, Wilkerson FP, Dugdale RC et al (2016) Pluses and minuses of ammonium and nitrate uptake and assimilation by phytoplankton and implications for productivity and community composition, with emphasis on nitrogen-enriched conditions. Limnol Oceanogr 61:165–197
Gobler CJ, Burkholder JM, Davis TW et al (2016) The dual role of nitrogen supply in controlling the growth and toxicity of cyanobacterial blooms. Harmful Algae 54:87–97
Goolsby DA, Battaglin WA (2001) Long-term changes in concentrations and flux of nitrogen in the Mississippi River basin, USA. Hydrol Process 15:1209–1226
Granéli E, Edler L, Gedziorowska D et al (1985) Influence of humic and fulvic acids on *Prorocentrum minimum* (Pav.) Schiller. In: Anderson DM, White AW, Baden DG (eds) Toxic dinoflagellates. Elsevier, New York, NY, pp 201–206
Granéli E, Flynn K (2006) Chemical and physical factors influencing toxin content. In: Granéli E, Turner J (eds) The ecology of harmful algae. Springer, New York, NY, pp 229–241
Granéli E, Johansson N (2003) Increase in the production of allelopathic substances by *Prymnesium parvum* cells grown under N- or P-deficient conditions. Harmful Algae 2:135–145
Granéli E, Olsson P, Sundstrøm B et al (1989) In situ studies of the effects of humic acids on dinoflagellates and diatoms. In: Okaichi T, Anderson DM, Nomoto T (eds) Red tides: biology, environmental science and toxicology. Elsevier, New York, NY, pp 209–212
Hajdu S, Pertola S, Kuosa H (2005) *Prorocentrum minimum* (Dinophyceae) in the Baltic Sea: morphology, occurrence – a review. Harmful Algae 4:471–480
Hansen PJ (2011) The role of photosynthesis and food uptake for the growth of marine mixotrophic dinoflagellates. J Eukaryot Microbiol 58:203–214
Hardison DR, Sunda WG, Litaker RW et al (2012) Nitrogen limitation increases brevetoxins in *Karenia brevis* (Dinophyceae): implications for bloom toxicity. J Phycol 48:844–858
Hardison DR, Sunda WG, Shea D et al (2013) Increased toxicity of *Karenia brevis* during phosphate limited growth: ecological and evolutionary implications. PLoS One 8(3):e58545. https://doi.org/10.1371/journal.pone.0058545
Harris GP (1986) Phytoplankton ecology: structure, function and fluctuation. Chapman and Hall, London
Harris TD, Smith VH, Graham JL et al (2016) Combined effects of nitrogen to phosphorus and nitrate to ammonia ratios on cyanobacterial metabolite concentrations in eutrophic Midwestern USA reservoirs. Inland Wat 6:199–210
Harrison JA, Caraco NF, Seitzinger SP (2005a) Global patterns and sources of dissolved organic matter export to the coastal zone: results from a spatially explicit, global model. Global Biogeochem Cycles 19:GB4S04
Harrison JA, Seitzinger SP, Caraco N et al (2005b) Dissolved inorganic phosphorous export to the coastal zone: results from a new, spatially explicit, global model (NEWS-SRP). Global Biogeochem Cycles 19:GB4S03
Heil CA, Glibert PM, Fan C (2005) *Prorocentrum minimum* (Pavillard) Schiller – a review of a harmful algal bloom species of growing worldwide importance. Harmful Algae 4:449–470
Heisler J, Glibert PM, Burkholder JM et al (2008) Eutrophication and harmful algal blooms: a scientific consensus. Harmful Algae 8:3–13
Hitchcock G, Phlips E, Brand L et al (2007) Plankton blooms. In: Hunt JH, Nuttle W (eds) Florida Bay science program: a synthesis of research on Florida Bay. Fish and Wildlife Research Institute Technical Report TR-11. Florida Fish and Wildlife Research Institute, St Petersburg, FL, pp 77–91
Holland E, Dentener F, Braswell B et al (1999) Contemporary and preindustrial global reactive nitrogen budgets. Biogeochemistry 4:7–43

Honkanen T, Helminen H (2000) Impacts of fish farming on eutrophication: comparisons among different characteristics of ecosystem. Int Rev Hydrobiol 85:673–686

Howarth RW (2008) Coastal nitrogen pollution: a review of sources and trends globally and regionally. Harmful Algae 8:14–20

Howarth RW, Billen G, Swaney D et al (1996) Regional nitrogen budgets and riverine N & P fluxes for the drainages to the North Atlantic Ocean: natural and human in influences. Biogeochemistry 35:75–139

Howarth RW, Sharpley A, Walker D (2002) Sources of nutrient pollution to coastal waters in the United States: implications for achieving coastal water quality goals. Estuaries 25:656–676

Hu C, Li D, Chen C et al (2010) On the recurrent *Ulva prolifera* blooms in the Yellow Sea and East China Sea. Marine Science Faculty Publications Paper 58. http://scholarcommons.usf.edu/msc_facpub/58

Intergovernmental Panel on Climate Change (IPCC) Core Writing Team (2007) Climate change 2007: synthesis report. In: Pachauri RK, Reisinger A (eds) Contribution of working groups I, II and III to the fourth assessment report of the Intergovernmental Panel on Climate Change. IPCC, Geneva, p 104

Intergovernmental Panel on Climate Change (IPCC) core Writing Team (2014) Climate change 2014: synthesis report. In: Pachauri RK, Meyer LA (eds) Contribution of working groups I, II and III to the fifth assessment report of the Intergovernmental Panel on Climate Change. IPCC, Geneva

Jeong HJ, Park JY, Nho JH et al (2005) Feeding by red-tide dinoflagellates on the cyanobacterium Synechococcus. Aquat Microb Ecol 41:131–143

Jeong HJ, Yoo YD, Kim TH et al (2004) Mixotrophy in the phototrophic harmful alga *Cochlodinium polykrikoides* (Dinophyceae): prey species, the effects of prey concentration and grazing impact. J Eukaryot Microbiol 51:563–569

Jeong HJ, Yoo YD, Kim JS et al (2010) Growth, feeding and ecological roles of the mixotrophic and heterotrophic dinoflagellates in marine planktonic food webs. Ocean Sci J 45:65–91

Johansson N, Granéli E (1999) Influence of different nutrient conditions on cell density, chemical composition and toxicity of *Prymnesium parvum* (Haptophyta) in semi-continuous cultures. J Exp Mar Biol Ecol 239:243–258

Kaushik SJ, Cowey CB (1991) Dietary factors affecting nitrogen excretion by fish. In: Cowey CB, Cho CY (eds) Nutritional strategies and aquaculture waste. Proceedings of the 1st International symposium on nutritional strategies in management of aquaculture waste. University of Guelph, Ontario, Canada, pp 3–19

Kimmerer WJ (2002) Physical, biological and management responses to variable freshwater flow into the San Francisco Estuary. Estuaries 25:1275–1290

Kremp A, Godhe A, Egardt J et al (2012) Strain variability in the response of bloom-forming marine microalgae to changed climate conditions. Ecol Evol 2:1195–1207

Kudela RM, Gobler CJ (2012) Harmful dinoflagellate blooms caused by *Cochlodinium* sp.: global expansion and ecological strategies facilitating bloom formation. Harmful Algae 14:71–86

Kudela RM, Lane JQ, Cochlan WP (2008a) The potential role of anthropogenically derived nitrogen in the growth of harmful algae in California, USA. Harmful Algae 8:103–110

Kudela RM, Ryan JP, Blakeley MD et al (2008b) Linking the physiology and ecology of *Cochlodinium* to better understand harmful algal bloom events: a comparative approach. Harmful Algae 7:278–292

Larsen J, Eikrem W, Paasche E (1993) Growth and toxicity in *Prymnesium patelliferum* (Prymnesiophycae) isolated from Norwegian waters. Can J Bot 71:1357–1362

Larsen J, Sournia A (1991) Diversity of heterotrophic dinoflagellates. In: Patterson DJ, Larsen J (eds) The biology of free-living heterotrophic dinoflagellates. Clarendon Press, Oxford, pp 313–332

Laruelle GG (2009) Quantifying nutrient cycling and retention in coastal waster at the global scale. PhD dissertation, Geologica Ultraiectina. Mededelingen van de Faculteit Geowetenschappen Universiteit Utrecht No. 312, Utrecht University, Utrecht, p 226

Lee MO, Kim JK (2008) Characteristics of algal blooms in the southern coastal waters of Korea. Mar Environ Res 65:128–147

Lehman PW, Boyer G, Hall C et al (2005) Distribution and toxicity of a new colonial *Microcystis aeruginosa* bloom in the San Francisco Bay Estuary, California. Hydrobiologia 541:87–99

Lehman PW, Boyer G, Stachwell M et al (2008) The influence of environmental conditions on seasonal variation of *Microcystis* abundance and microcystins concentration in San Francisco Estuary. Hydrobiologia 600:187–204

Lehner B, Liermann CR, Revenga C et al (2011) High-resolution mapping of the world's reservoirs and dams for sustainable river-flow management. Front Ecol Environ 9:494–502

Lewis NI, Bates SS, McLachlan JL et al (1993) Temperature effects on growth, domoic acid production, and morphology of the diatom *Nitzschia-pungens f. multiseries*. In: Smayda TJ, Shimizu Y (eds) Toxic phytoplankton blooms in the sea. Elsevier Science Publications B V, Amsterdam, pp 601–606

Li J, Glibert PM, Zhou M et al (2009) Relationships between nitrogen and phosphorus forms and ratios and the development of dinoflagellate blooms in the East China Sea. Mar Ecol Prog Ser 383:11–26

Li J, Glibert PM, Zhou M (2010) Temporal and spatial variability in nitrogen uptake kinetics during harmful dinoflagellate blooms in the East China Sea. Harmful Algae 9:531–539

Liermann CR, Nilsson C, Robertson J et al (2012) Implications of dam obstruction for global freshwater fish diversity. Biosciences 62:539–548

Lin C-H, Accoroni S, Glibert PM (2017) Mixotrophy in the dinoflagellate *Karlodinium veneficum* under variable nitrogen:phosphorus stoichiometry: feeding response and effects on larvae of the eastern oyster (*Crassostrea virginica*). Aquat Microb Ecol 79:101–114. https://doi.org/10.3354/ameo01823.

Liu D, Zhou MJ (2018) Green tides of the Yellow Sea: massive free-floating blooms of *Ulva prolifera*. In: Glibert PM, Berdalet E, Burford M et al (eds) Global ecology and oceanography of harmful algal blooms. Springer, Cham, pp 317–326

Lomas MW, Glibert PM (1999) Temperature regulation of nitrate uptake: a novel hypothesis about nitrate uptake and reduction in cool-water diatoms. Limnol Oceanogr 44:556–572

Lu D, Goebel J (2001) Five red tide species in genus *Prorocentrum* including the description of *Prorocentrum donghaiense* Lu sp. nov. from the East China Sea. China Chin J Oceanol Limnol 19:337–344

Lu D, Goebel J, Qi Y et al (2005) Morphology and genetic study of *Prorocentrum donghaiense* Lu from the East China Sea, and comparison with some related *Prorocentrum* species. Harmful Algae 4:493–506

Lu D, Wang H, Huang H et al (2011) Morphological and genetic comparison of two strains of a *Prorocentrum* species isolated from Zhejiang coastal water of China and Masan Bay of Korea. Chin J Oceanol Limnol 29:832–839

Lundgren V, Glibert PM, Granéli E et al (2016) Metabolic and physiological changes in *Prymnesium parvum* when grown under, and grazing on, prey of variable nitrogen:phosphorus stoichiometry. Harmful Algae 55:1–12

Madden CJ, Smith R, Dettmann E et al (2010) Estuarine typology development and application. In: Glibert P, Madden C, Boynton W et al (eds) Estuarine nutrient criteria development: state of the science. EPA Office of Water. http://www2.epa.gov/sites/production/files/documents/nutrients-in-estuaries-november-2010.pdf

Malone TC, Conley DJ, Glibert PM et al (1996) Scales of nutrient limited phytoplankton productivity: the Chesapeake Bay example. Estuaries 19:371–385

Marasović I, Pucher-Petkovic T, Petrova-Karadjova V (1990) *Prorocentrum minimum* (Dinophyceae) in the Adriatic and Black Sea. J Mar Biol Assoc UK 70:473–476

Martinez-Lopez A, Escobedo-Urias DC, Ulloa-Perez AE et al (2008) Dynamics of a *Prorocentrum minimum* bloom along the northern coast of Sinaloa, Mexico. Cont Shelf Res 28:1693–1701

Matsuoka K, Takano Y, Kamrani E et al (2010) Study on *Cochlodinium polykrikoides* Margalef in the Oman Sea and the Persian Gulf from August 2008 to August 2009. Curr Dev Oceanogr 1:153–171

Mayorga E, Seitzinger SP, Harrison JA et al (2010) Global nutrient export from WaterSheds 2 (NEWS 2): model development and implementation. Environ Model Softw 25(7):837–853

McCabe RM, Hickey BM, Kudela RM et al (2016) An unprecedented coastwide toxic algal bloom linked to anomalous ocean conditions. Geophys Res Lett. https://doi.org/10.1002/2016GL070023

Miller, WD, Harding LW, Adolf JE (2005) The influence of Hurricane Isabel on Chesapeake Bay phytoplankton dynamics. In: Sellner KG (ed) Hurricane Isabel in perspective, Chesapeake Research Consortium Publication 05-160, Edgewater, Maryland, pp 155–160

Miller WD, Harding LW, Adolf JE (2006) Hurricane Isabel generated an unusual fall bloom in Chesapeake Bay. Geophys Res Lett 33:LO6612. https://doi.org/10.1029/2005GL025658

Minnhagen S, Kim M, Salomon P et al (2011) Active uptake of kleptoplastids by *Dinophysis caudata* from its ciliate prey *Myrionecta rubra*. Aquat Microb Ecol 62:99–108

Monchamp ME, Pick FR, Beisner BE et al (2014) Variation in microcystin concentration and composition in relation to cyanobacterial community structure. PLoS One 9(1):e85573. https://doi.org/10.1371/journal.pone0085573

Moncheva D, Petrova Kardjova V, Palasov A (1995) Harmful algal blooms along the Bulgarian sea coast and possible patterns of fish and zoobenthos mortalities. In: Lassus P, Arzul G, Erard-Le Denn E et al (eds) Harmful marine algal blooms. Lavoisier, Paris, pp 193–198

Moore SK, Trainer VL, Mantua NJ et al (2008) Impacts of climate variability and future change on harmful algal blooms and human health. Environ Health 7:S4. https://doi.org/10.1186/1476-069X-7-S2-S4

Mulholland MR, Morse RE, Boneillo GE et al (2009) Understanding causes and impacts of the dinoflagellate, *Cochlodinium polykrikoides*, blooms in the Chesapeake Bay. Estuaries 32:734–747

O'Neil JM, Davis TW, Burford MA et al (2012) The rise of harmful cyanobacteria blooms: the potential roles of eutrophication and climate change. Harmful Algae 14:313–334

Ogata T, Kodama M, Ishimaru T (1989) Effect of water temperature and light intensity on growth rate and toxin production of toxic dinoflagellates. In: Okaichi T, Anderson DM, Nemoto T (eds) Red tides, biology, environmental science and toxicology. Elsevier, New York, NY, pp 423–426

Olenina I, Wasmund N, Hajdu S et al (2010) Assessing impacts of invasive phytoplankton: the Baltic Sea case. Mar Pollut Bull 60:1691–1700

Paerl HW, Huisman J (2008) Blooms like it hot. Science 3320:57–58

Paerl HW, Scott JT (2010) Throwing fuel on the fire: synergistic effects of excessive nitrogen inputs and global warming on harmful algal blooms. Environ Sci Technol 44:7756–7758

Pan YL, Subba Rao DV, Mann KH (1996) Changes in domoic acid production and cellular chemical composition of the toxigenic diatom *Pseudo-nitzschia multiseries* under phosphate limitation. J Phycol 32:371–381

Pennock JR (1985) Chlorophyll distributions in the Delaware Estuary: regulation by light limitation. Estuar Coast Shelf Sci 21:711–725

Peñuelas J, Sardans J, Rivas-Ubach A et al (2012) The human-induced imbalance between C, N and P in Earth's life system. Glob Chang Biol 18:3–6

Pertola S, Kuosa H, Olsonen R (2005) Is the invasion of *Prorocentrum minimum* (Dinophyceae) related to the nitrogen enrichment of the Baltic Sea? Harmful Algae 4:481–492

Raven JA, Beardall J, Flynn KJ et al (2009) Phagotrophy in the origins of photosynthesis in eukaryotes and as a complementary mode of nutrition in phototrophs: relation to Darwin's insectivorous plants. J Exp Bot 60:3975–3987. https://doi.org/10.1093/jxb/erp282

Redfield AC (1934) On the proportions of organic derivatives in sea water and their relation to the composition of plankton. In: James Johnstone memorial volume. University of Liverpool Press, Liverpool, pp 176–192

Richlen ML, Morton SL, Jamali EA et al (2010) The catastrophic 2008-2009 red tide in the Arabian Gulf region, with observations on the identification and phylogeny of the fish-killing dinoflagellate *Cochlodinium polykrikoides*. Harmful Algae 9:163–172

Seitzinger SP, Harrison JA, Dumont E et al (2005) Sources and delivery of carbon, nitrogen, and phosphorus to the coastal zone: an overview of global nutrient export from Watersheds (NEWS) models and their application. Global Biogeochem Cycles 19:GB4S01. https://doi.org/10.1029/2005gb002606

Seitzinger SP, Kroeze C, Bouwman AF et al (2002) Global patterns of dissolved in organic and particulate nitrogen inputs to coastal systems: recent conditions and future projections. Estuaries 25:640–655

Seitzinger SP, Mayorga E, Bouwman AF et al (2010) Global river nutrient export: a scenario analysis of past and future trends. Global Biogeochem Cycles 24:GB0A08. https://doi.org/10.1029/2009GB003587

Shackeroff JM, Hazen EL, Crowder LB (2009) The oceans as peopled seascapes. In: McLeod K, Leslie H (eds) Ecosystem based management for the oceans. Island Press, Washington, DC, pp 33–54

Shangguan Y, Glibert PM, Alexander JA et al (2017) Nutrients and phytoplankton community composition in semi-enclosed lagoon systems in Florida Bay and their responses to changes in flow from Everglades restoration. Limnol Oceanogr 62:S327–S347. https://doi.org/10.1002/lno.10599

Sierra-Beltran AP, Cortes-Altamirano R, Cortes-Lara MC (2005) Occurrences of *Prorocentrum minimum* (Pavillard) in Mexico. Harmful Algae 4:507–518

Silva ES (1985) Ecological factors related to *Prorocentrum minimum* blooms in Obidos Lagoon (Portugal). In: Anderson DM, White A, Baden D (eds) Toxic dinoflagellates. Elsevier, New York, NY, pp 251–256

Smil V (2001) Enriching the Earth: Fritz Haber, Carl Bosch, and the transformation of world food. The MIT Press, Cambridge

Sterner RW, Elser JJ (2002) Ecological stoichiometry: the biology of elements from molecules to the biosphere. Princeton University Press, Princeton, NJ, p 439

Stonik IV (1995) A potentially toxic dinoflagellate, *Prorocentrum minimum*, in Amurskii Bay of the Sea of Japan. Russ J Mar Biol Assoc UK 20:314–320

Stukel MR, Landry MR, Selph KE (2011) Nanoplankton mixotrophy in the eastern equatorial Pacific. Deep Sea Res Part II 58:378–386

Sun J, Hutchins DA, Feng YY et al (2011) Effects of changing pCO_2 and phosphate availability on domoic acid production and physiology of the marine harmful bloom diatom *Pseudo-nitzschia multiseries*. Limnol Oceanogr 56:829–840

Sunda WG, Granéli E, Gobler CJ (2006) Positive feedback and the development and persistence of ecosystem disruptive algal blooms. J Phycol 42:963–974

Sutton MA, Bleeker A, Howard CM et al (2013) Our nutrient world: the challenge to produce more food and energy with less pollution. Centre for Ecology and Hydrology, Edinburgh

Swaney DP, Hong B, Paneer Selvam A et al (2014) Net anthropogenic nitrogen inputs and nitrogen fluxes from Indian watersheds: an initial assessment. J Mar Syst. https://doi.org/10.1016/j.marsys.2014.09.004

Tang TY, Tai JH, Yang YJ (2000) The flow pattern north of Taiwan and the migration of the Kuroshio. Cont Shelf Res 20:349–371

Tango PJ, Magnien R, Butler W et al (2005) Impacts and potential effects due to *Prorocentrum minimum* blooms in Chesapeake Bay. Harmful Algae 4:525–531

Tatters AO, Fu F-X, Hutchins (2012) High CO_2 and silicate limitation synergistically increase the toxicity of *Pseudo-nitzschia fraudulenta*. PLoS One 7(2):e32116. https://doi.org/10.1371/journal.pone.0032116

Ti C, Yan X (2013) Spatial and temporal variations of river nitrogen exports from major basins in China. Environ Sci Pollut Res 20:6509–6520

Unrein F, Massana R, Alonso-Saez L et al (2007) Significant year-round effect of small mixotrophic flagellates on bacterioplankton in an oligotrophic coastal system. Limnol Oceanogr 52:456–469
Van de Waal DB, Ferreruela G, Tonk L et al (2010) Pulsed nitrogen supply induces dynamic changes in the amino acid composition and microcystin production of the harmful cyanobacterium *Planktothrix agardhii*. FEMS Microbiol Ecol 74:430–438
Van de Waal DB, Verspagen JM, Finke JF et al (2011) Reversal in competitive dominance of a toxic versus non-toxic cyanobacterium in response to rising CO_2. ISME J 5:1438–1450
Van de Waal DB, Verspagen JMH, Lürling M et al (2009) The ecological stoichiometry of toxins produced by harmful cyanobacteria: an experimental test of the carbon-nutrient balance hypothesis. Ecol Lett 12:1326–1335
Visser PM, Verspagen JMH, Sandrini G et al (2016) How rising CO_2 and global warming may stimulate harmful cyanobacterial blooms. Harmful Algae 54:145–159
Vörösmarty CJ, McIntyre PB, Gessner MO et al (2010) Global threats to human water security and river biodiversity. Nature 467:555–561
Walsby T (1975) Gas vesicles. Annu Rev Plant Physiol 26:427–439
Wazniak CE, Glibert PM (2004) Potential impacts of brown tide, *Aureococcus anophagefferens*, on juvenile hard clams, *Mercenaria mercenaria*, in the Coastal Bays of Maryland, USA. Harmful Algae 3:321–329
Wells ML, Karlson B (2018) Harmful algal blooms in a changing ocean. In: Glibert PM, Berdalet E, Burford M et al (eds) Global ecology and oceanography of harmful algal blooms. Springer, Cham, pp 77–90
Wells ML, Trainer VL, Smayda TJ et al (2015) Harmful algal blooms and climate change: learning from the past and present to forecast the future. Harmful Algae 49:68–93
Wu RSS, Lam KS, Mackay DW et al (1994) Impact of marine fish farming on water quality and bottom sediment: a case study in the sub-tropical environment. Mar Environ Res 38:115–145
Yu R-C, Lü S-H, Liang Y-B (2018) Harmful algal blooms in the coastal waters of China. In: Glibert PM, Berdalet E, Burford M et al (eds) Global ecology and oceanography of harmful algal blooms. Springer, Cham, pp 309–316
Zarfl C, Lumsdon AE, Berlekamp J et al (2015) A global boom in hydropower dam construction. Aquat Sci 77:161–170
Zhang QC, Qiu LM, Yu RC et al (2012) Emergence of brown tides caused by *Aureococcus anophagefferens* Hargraves et Sieburth in China. Harmful Algae 19:117–124
Zhang J, Wu Y, Zhang YY (2015) Plant nutrients and trace elements from the Changjiang watersheds and East China Sea Shelf. In: Zhang J (ed) Ecological continuum from the Changjiang (Yangtze River) watersheds to the East China Sea continental margin. Springer International Publishing, Switzerland, pp 93–118
Zhou M, Yan T, Zou J (2003) Preliminary analysis of the characteristics of red tide areas in Changjiang River estuary and its adjacent sea. Chin J Appl Ecol 14:1031–1038
Zhou MJ, Shen ZL, RC Y (2008) Responses of a coastal phytoplankton community to increased nutrient input from the Changjiang (Yangtze) River. Cont Shelf Res 28:1483–1489
Zhu Z-Y, Zhang J, Wu Y et al (2011) Hypoxia off the Changjiang (Yangtze River) estuary: oxygen depletion and organic matter decomposition. Mar Chem 125:108–116
Zubkov MV, Tarran GA (2008) High bacterivory by the smallest phytoplankton in the North Atlantic Ocean. Nature 455:224–226

Chapter 13
Key Questions and Recent Research Advances on Harmful Algal Blooms in Benthic Systems

Elisa Berdalet and Patricia A. Tester

13.1 Introduction

Endemic in tropical areas, ciguatera fish poisoning (CFP), caused by the toxins produced by benthic dinoflagellates mostly of the genus *Gambierdiscus*, has significant long-term health effects. In recent years, CFP and new species of *Gambierdiscus* from subtropical waters and the Mediterranean Sea are being reported (Fraga and Rodríguez 2014; Fraga et al. 2011, 2016; Nishimura et al. 2014). Similarly, *Ostreopsis*, another benthic dinoflagellate genus traditionally found associated with *Gambierdiscus* in tropical areas, seems to be expanding to relatively colder waters and temperate latitudes (Rhodes 2011). In the tropics, *Ostreopsis siamensis* was first related to certain fish poisonings (palytoxicosis; e.g., Randall 2005). In the last two decades, frequent and intense outbreaks of *Ostreopsis* cf. *ovata* and *O. siamensis* have been documented in the Mediterranean and New Zealand waters, associated with faunal damage and human respiratory problems. Fortunately, seafood poisonings have not been reported in these areas. The scenario of more frequent BHAB events with range extensions to higher latitudes is especially troubling because the knowledge of BHAB species is nascent and, even in developed nations, the regulatory and monitoring infrastructure is not adequate to protect public health and marine resources (GEOHAB 2012).

The GEOHAB Core Research Project on Harmful Algal Blooms in Benthic Systems, BHAB CRP (GEOHAB 2012), started with an international Open Science Meeting (OSM) in Honolulu (Hawai'i, USA) in June 2010. In that meeting, scientists established the main priorities and summarized their research ideas in

E. Berdalet (✉)
Institute of Marine Sciences (CSIC), Barcelona, Catalonia, Spain
e-mail: berdalet@icm.csic.es

P.A. Tester
Ocean Tester LLC, Beaufort, NC, USA

P.M. Glibert et al. (eds.), *Global Ecology and Oceanography of Harmful Algal Blooms*, Ecological Studies 232, https://doi.org/10.1007/978-3-319-70069-4_13

the GEOHAB (2012) Report. That report included a summary of the current status of taxonomy, biogeography, genetic diversity, ecology, and physiology of the BHAB organisms and noted the lack of standardized sampling methods. The report also defined a consensus for the implementation of the BHAB CRP and stimulated a resurgence of interest in BHABs. The objectives and main questions were structured following the key research elements of the GEOHAB Science and Implementation Plans (GEOHAB 2001, 2003, respectively). Furthermore, the BHAB CRP included unique objectives compared with other GEOHAB CRPs, directly related to toxins and impacts on human health and ecosystems. Here, we synthesize the many advances and future needs for progress on BHAB research following the objectives addressed in the GEOHAB (2012) report and mainly focused on *Gambierdiscus* and *Ostreopsis*, the most important genera involved on BHAB events. However, we recognize that other BHAB taxa deserve further investigation.

13.2 Key Question 1: What Is the Biogeography and Biodiversity of BHABs and the Relationships Among Distributions of BHAB Species, Biodiversity, and Environmental Changes?

When dealing with toxic species, the taxonomy and genetic variability of target organisms are essential to determine their toxicity, population dynamics, and biogeography. Great efforts have been made to develop molecular methods to verify morphological identifications for *Gambierdiscus* and *Ostreopsis* species (Table 13.1). Among the expected achievements, the availability of PCR and qPCR assays for field samples has been, and will continue to be, extremely helpful for species verification and timely monitoring during *Gambierdiscus* and CFP events and *Ostreopsis* outbreaks.

13.2.1 Gambierdiscus

The history of the taxonomy of *Gambierdiscus* is well detailed in GEOHAB (2012). Here we summarize (Table 13.1) the recent changes that warrant mention. Since 2011, six new species of *Gambierdiscus* (*G. excentricus*, Fraga et al. 2011; *G. silvae*, Fraga and Rodríguez 2014; *G. scabrosus*, Nishimura et al. 2014; *G. balechii*, Fraga et al. 2016; *G. cheloniae*, Smith et al. 2016; and *G. lapillus*, Kretzschmar et al. 2016) have been described from opposite sides of the globe (Canary Islands, Honshu–Japan, the Celebes and Cook Islands, and the Great Barrier Reef, Australia); two *Gambierdiscus* species have been moved to a new genus, *Fukuyoa*, and an additional new *Fukuyoa* species was described (Gómez et al. 2015). Both the discovery of new species and new insights on the biogeography of *Gambierdiscus* have been due, in

Table 13.1 Species of the genera *Gambierdiscus* and *Fukuyoa* with indication of the location of the first descriptions and selected references

Species	Origin	References	qPCR
Atlantic Basin (Caribbean, Gulf of Mexico, and Eastern and Western Atlantic)			
F. paulensis (F. Gómez, D. Qiu, R.M. Lopes et S. Lin)	Ubatuba, Southern Brazil	Gómez et al. (2015)	
G. excentricus (S. Fraga)	Canary Islands	Fraga et al. (2011); Nascimento et al. (2012)	
G. silvae (S. Fraga et F. Rodríguez)	Canary Islands	Fraga and Rodríguez (2014)	
Gambierdiscus ribotype 1 = *G. silvae*	Caribbean	Litaker et al. (2009); Fraga and Rodríguez (2014)	
Gambierdiscus ribotype 2	Caribbean	Litaker et al. (2009); Ennaffah and Chaira (2015)	[1]
Pacific Basin and Indian Ocean			
G. australes (M.A. Faust et M. Chinain)	Australes Archipelago	Chinain et al. (1999); Litaker et al. (2009)	[2], [3]
G. balechii (S. Fraga, F. Rodríguez et I. Bravo)	Celebes Sea	Fraga et al. (2016)	
G. cheloniae (K.F. Smith, L. Rhodes et S.A. Murray)	Cook Islands	Smith et al. (2016)	
G. lapillus (A. L. Kretzschmar, M. Hoppenrath et S.A. Murray)	Great Barrier Reef	Kretzschmar et al. (2016)	
G. pacificus (M. Chinain et M.A. Faust)	Tuamotu Archipelago	Chinain et al. (1999); Litaker et al. (2009)	[3]
G. polynesiensis (M. Chinain M. Faust)	Australes and Tuamotu Archipelago	Chinain et al. (1999); Litaker et al. (2009)	[3]
G. scabrosus (T. Nishimura, S. Sato et M. Adachi)	Kashiwajima Island, Japan	Nishimura et al. (2014)	[2]
Gambierdiscus sp. ribotype 1 = *G. scabrosus*	Japan	Nishimura et al. (2013)	[2]
Gambierdiscus toxicus (R. Adachi et Y. Fukuyo)	Gambier Islands, French Polynesia	Adachi and Fukuyo (1979); Litaker et al. (2009)	[3]
Gambierdiscus sp. ribotypes 2 and 3	Japan	Nishimura et al. (2013); Kuno et al. (2010)	[2]
Gambierdiscus sp. ribotypes 4, 5, and 6	Republic of Kiribati	Xu et al. (2014)	
Mediterranean			
G. carolinianus (R.W. Litaker, M.W. Vandersea, M.A. Faust, S.R. Kibler, W.C. Holland et P.A. Tester)	Continental shelf, North Carolina, USA	Holland et al. (2013)	

(continued)

Table 13.1 (continued)

Species	Origin	References	qPCR
Co-occurring in the Atlantic, Pacific, and Indian Oceans			
G. belizeanus (M.A. Faust)	Belize, Central America, Caribbean Sea	Turquet et al. (2001); Litaker et al. (2009); Leaw et al. (2011)	[1]
G. caribaeus (M.W. Vandersea, R.W. Litaker, M.A. Faust, S.R. Kibler, W.C. Holland et P.A. Tester)	Belize, Central America, Caribbean Sea; Continental Shelf, North Carolina, USA	Litaker et al. (2009); Holland et al. (2013)	[1]
G. carolinianus (R.W. Litaker, M.W. Vandersea, M.A. Faust, S.R. Kibler, W.C. Holland et P.A. Tester)	Continental Shelf, North Carolina; Hawaii	Litaker et al. (2009); Lewis et al. (2016)	[1]
G. carpenteri (R.W. Litaker, M.A. Faust, W.C. Holland, M.W. Vandersea et P.A. Tester)	Belize, Caribbean Sea	Litaker et al. (2009); Holland et al. (2013); Kohli et al. (2014)	[1]
F. ruetzleri (M.A. Faust, R.W. Litaker, M.W. Vandersea, S.R. Kibler, W.C. Holland et P.A. Tester) F. Gómez, D.X. Qiu, R.M. Lopes et S. Lin[a]	Belize, Caribbean Sea	Litaker et al. (2009); Nishimura et al. (2013)	[1]
F. yasumotoi (M.J. Holmes) F. Gómez, D.X. Qiu, R.M. Lopes et S. Lin[b]	Singapore Island off Pulau Hantu	Holmes (1998); Litaker et al. (2009); Rhodes et al. (2014b)	

The species are presented according to their known geographic distributions. The availability of qPRC assays for some species is indicated in the column on the right; the specific qPCR methods are [1] Vandersea et al. (2012) and [2] Nishimura et al. (2016), while [3] indicates that the methods are in preparation by T. Darius (Institut Louis Malardé, French Polynesia) and will be available soon. [a]*Gambierdiscus ruetzleri* and [b]*Gambierdiscus yasumotoi* moved to new genus *Fukuyoa* (Gómez et al. 2015)

large measure, to the availability of sequence data in GenBank and PCR assays for members of this genus.

Reports of *Gambierdiscus* species in new locations, other than endemic areas in the tropical Pacific and Caribbean, and range extensions within oceanic basins are numerous (Table 13.1). Also, it is likely that many more *Gambierdiscus* species will be found as sampling efforts continue. Overall, six *Gambierdiscus* (and *Fukuyoa*) species are cosmopolitan and occur in both the Atlantic and Pacific-Indian basins: *F. ruetzleri*, *F. yasumotoi*, *G. belizeanus*, *G. caribaeus*, *G. carpenteri*, and *G. carolinianus* (Fig. 13.1a–c). Other *Gambierdiscus* are only known from one basin, in the Atlantic or the Pacific (Table 13.1).

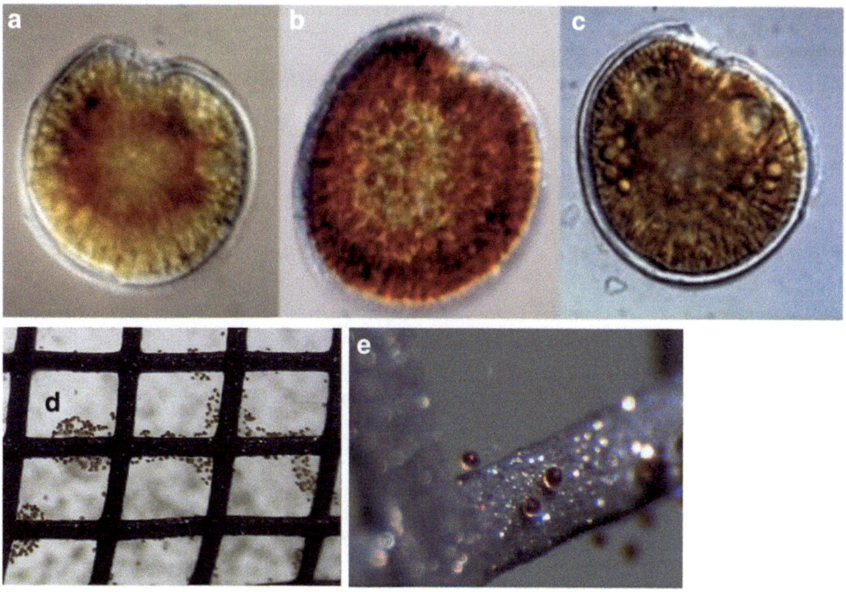

Fig. 13.1 Different *Gambierdiscus* species: *G. caribaeus* (**a**), *G. carolinianus* (**b**), and *G. carpenteri* (**c**); micrographs by M. Vandersea (NOAA, US). The artificial substrate, a window screen, deployed in situ (**d**) and micrograph of the cells attached to the mesh (**e**); photos by S. Kibler (NOAA, US)

13.2.2 Ostreopsis

The *Ostreopsis* genus was first described in the Gulf of Siam (Thailand) in 1901, with *O. siamensis* as type species. As noted in GEOHAB (2012), the high variability of the morphological features and lack of molecular characterizations of the holotypes hamper the conclusive discrimination among *Ostreopsis* species. Ongoing studies based on ITS-5.8S and LSU rDNA sequences (Penna et al. 2012, 2014) are making progress in establishing phylogenetic and phylogeographic links among the different isolates and described species, as well as in identifying cryptic species. An updated list of the *Ostreopsis* species is provided in Table 13.2.

New strains isolated from the coasts of the Atlantic Iberian Peninsula (David et al. 2013), Réunion Island (Indian Ocean) (Carnicer et al. 2015), and Cyprus (Tartaglione et al. 2016; Accoroni et al. 2016) revealed new morphotypes and phylogenetic trees that may result in new species. *Ostreopsis*. cf. *ovata* (Fig. 13.2a) is the most commonly reported species in the last two decades in the northern and southern coasts of the Mediterranean Sea, detected in French and Spanish Atlantic coastal waters as well as in the Pacific coasts of Japan and Russia (e.g., GEOHAB 2012; Selina et al. 2014; Vila et al. 2016 and references therein). Heightened concerns have increased the efforts to efficiently detect its presence. Currently qPCR assays are available for *O*. cf. *ovata* (Casabianca et al. 2013).

Table 13.2 Species of the genera *Ostreopsis* with selected references

Species	Geographic origin	Selected references
O. belizeanus (M.A. Faust)	Belize, Central America, Caribbean Sea	Faust (1999)
O. caribbeanus (M.A. Faust)	La Parguera, Puerto Rico, Caribbean Sea	Faust (1999)
O. heptagona (D.R. Norris)	Florida Keys, Gulf of Mexico	Norris et al. (1985)
O. labens (M.A. Faust and S.L. Morton)	Belize, Central America, Caribbean Sea	Faust and Morton (1995)
O. lenticularis (Y. Fukuyo)	Gambier Islands and Society Islands, French Polynesia, South Pacific Ocean	Fukuyo (1981); Norris et al. (1985); Larsen and Nguyen (2004)
O. marinus (M.A. Faust)	Belize, Central America, Caribbean Sea	Faust (1999)
O. mascaranensis (J.P. Quod)	Reunion Island, West Indian Ocean	Quod (1994)
O. ovata (Y. Fukuyo)	French Polynesia, New Caledonia and Ryukyu Islands, South and West Pacific Ocean	Fukuyo (1981); Penna et al. (2012); Sato et al. (2011)
O. cf. *ovata*	Atlantic/Mediterranean/Pacific (Japan Sea and Mediterranean, Ionian, and Tyrrhenian Seas) Atlantic/Indian/Pacific clade (including isolates from Belize) Pacific clade (Vietnam)	Penna et al. (2012)
O. cf. *ovata*, potential new species	Atlantic coast of the Iberian Peninsula strains	David et al. (2013)
O. siamensis (J. Schmidt)	Trat Province, Gulf of Thailand	Schmidt (1901); Fukuyo (1981)
O. cf. *siamensis*	Atlantic/Mediterranean clade	Penna et al. (2012); Sato et al. (2011)
O. cf. *siamensis*, potential new species	Atlantic coast of the Iberian Peninsula strains	David et al. (2013)
Ostreopsis spp., potential new species	Réunion Island coast	Carnicer et al. (2015)
Ostreopsis fattorussoi	Cyprus Island coast	Accoroni et al. (2016)

Updated from GEOHAB (2012)

13.3 Key Question 2: What Are the Relationships Between Eutrophication and Nutrient Transformation Pathways and BHAB Population Dynamics?

There is a need to clarify the possible link between eutrophication and nutrient dynamics in general and BHABs. The knowledge gap of the inorganic and organic nutrient requirements of *Gambierdiscus* and *Ostreopsis* species and their potential

Fig. 13.2 Different aspects of an *Ostreopsis* cf. *ovata* bloom in Sant Andreu de Llavaneres (NW Mediterranean). Micrographs (**a**) of natural samples; *Ostreopsis* cf. *ovata* and some benthic diatoms (*Coscinodiscus* sp.). *Ostreopsis* cells attach to the benthic macroalgae and macrofauna (**b** and **c**). Dense aggregates are released and float in the water column (**c**) and at the surface (**d**). Images by E. Berdalet and M. Vila, ICM-CSIC

mixotrophic capacity is large, and modulation of toxin production by nutrients is speculative.

13.3.1 Gambierdiscus

Blooms of *Gambierdiscus* do not usually occur in eutrophic environments, although nutrients could favour the growth of the macroalgae that provide the optimal substrate for these epiphytic species (Lapointe et al. 2010). GEOHAB (2012) summarized the uncertainties concerning the *Gambierdiscus* nutritional requirements from the early laboratory studies. Overall, species use a variety of inorganic and organic phosphorus (P) and nitrogen (N) forms including phosphate, nitrate, ammonium, glycerophosphate, urea, amino acids, and even putrescine. However, the responses were species-specific suggesting the dominant nutrient source could be a selective factor of *Gambierdiscus* species. Some studies (Lartigue et al. 2009), although not conclusive, have addressed the possible role of macroalgal-released substances as sources of micronutrients or chelating agents and the potential mixotrophic capacity of *Gambierdiscus* (Faust 1998).

13.3.2 Ostreopsis

As noted in GEOHAB (2012), the available data in the Mediterranean and the Pacific do not draw a clear causal relationship between (inorganic and organic) nutrient supply and *Ostreopsis* sp. blooms. For instance, *Ostreopsis* blooms in New Zealand occurred in open coastal waters with low nutrient levels (Shears and Ross 2009), and the direct link with nutrients in the NW Mediterranean could not be established (Cohu et al. 2013). Accoroni et al. (2015) suggested a conceptual model of the onset of the bloom in the Conero Riviera (Northern Adriatic Sea) that depended on an increased supply of PO_4 (and low N:P ratios) concurrently with temperatures above 25°C and calm waters (Fig. 13.3). Thereafter, the maintenance of the bloom would occur through other adaptive mechanisms (e.g., allelopathy and mixotrophy). Nutrient limitation would be involved in the decline as suggested by the double-walled cysts found in sediments in the Adriatic, which are also formed in N-depleted cultures (Accoroni et al. 2014). Similar to *Gambierdiscus*, the potential mixotrophic capacity of *Ostreopsis* (Pinna et al. 2015) needs to be determined while an indirect role of nutrients on the macrophyte substrate cannot be ruled out.

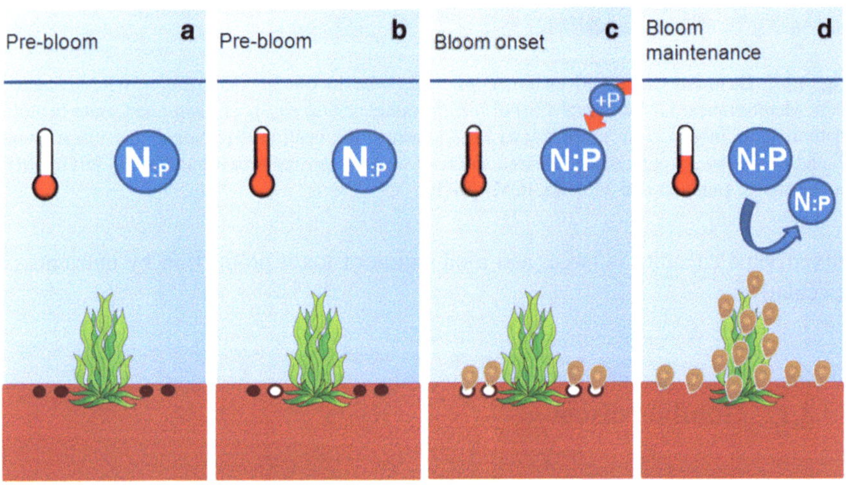

Fig. 13.3 Diagram corresponding to the conceptual model on the annual *Ostreopsis* cf. *ovata* blooms in the northern Adriatic Sea proposed by Accoroni et al. (2015). The model integrates the synergic effects of hydrodynamics, temperature, and the N:P ratio of water column nutrients. Panel **a** represents the typical pre-bloom condition, showing low temperature and very high N:P ratio. Panel **b** represents the pre-bloom condition just before the bloom onset: although the temperature (>25.8°C) is optimal for cyst germination, the high N:P ratio does not allow the start of the bloom. Bloom onset (panel **c**) occurs at T >25.8°C and at N:P ratio (around 24) balanced by an injection of P-rich water, allowing both cyst germination and cell proliferation. Panel **d** shows the optimal environmental conditions for the bloom maintenance: temperature is not limiting, as bloom peak occurs at T around 20.8°C, and the slight increase of N:P ratio accompanied by a general decrease of N and P could be attributable to *O.* cf. *ovata* growth. Reproduced from Accoroni et al. (2015) with permission of the publisher

Nutrient availability could influence toxin levels, but the available data are scarce (but see Chap. 12, Glibert et al. 2018), on the progress in unravelling the relationships between nutrients and toxins in pelagic HABs. Most studies have found increased toxicity towards the stationary phase, when limitation occurs (e.g., Guerrini et al. 2010; Nascimento et al. 2012). However, in at least one study, in the late stationary phase, N- and P-limited cultures had a 50% lower toxin content than in the control (initially nutrient replete) cultures (Vanucci et al. 2012). Pezzolesi et al. 2016 suggested that in the environment toxin production is sustained by a low and constant nutrient supply, able to maintain appropriate cellular C:N (>12) or C:P (>170) ratios for long periods.

13.4 Key Question 3: Are There Particular Characteristics and Adaptations of BHAB Species That Determine When and Where They Occur and Produce Harmful Effects?

BHABs occur in relatively shallow waters (Mangialajo et al. 2011; Xu et al. 2014), attached to different substrates (mainly macrophytes, but also rocks, sand, or corals), and are exposed to high light and temperature in variable environments (Figs. 13.1d and 13.2b–d). There is a need to understand the physiological adaptations, likely species-specific, to the variations in the physicochemical environmental conditions at small, local scales. Research should also consider biological interactions, including the effects of toxins on marine animals at lethal and sublethal levels, allelochemistry, competition and/or succession with other microalgae and bacteria, chemical interactions with the substrate, and the role of the mucus (Fig. 13.2b, c). Other factors include parasite and viral infections, induction of resting stages, or endogenous clocks versus external triggers that favour cell proliferation.

13.4.1 Gambierdiscus

Bravo et al. (2014) examined the variation in nuclear shapes, cell sizes, and division processes in a single strain of *Gambierdiscus* sp., established from a sample obtained from the Celebes Sea (SW Pacific Ocean). For the first time, they observed meiosis and gametogenesis and the occurrence of division processes in both the sexual and asexual cycle, not just during the asexual phase as traditionally understood. The authors proposed a life cycle including a sexual (haploid) and asexual (diploid) phases. Gaps in knowledge still remain regarding when mitosis and meiosis occur.

The physiological studies conducted by Kibler et al. (2012) with seven Caribbean species (*G. australes*, *G. belizeanus*, *G. caribaeus*, *G. carolinianus*, *G. ruetzleri*, ribotype 2, and *G. pacificus*) indicated that optimal growth of *Gambierdiscus* would occur at temperatures between 25 and 29°C and salinities between 20 and 25, with growth rates up to 0.3 divisions day^{-1}. *Gambierdiscus* appears to have a strong resistance to the potentially damaging levels of PAR and UV radiation allowing its growth at surface, highly illuminated layers, although it has a capacity, like most dinoflagellates, to grow at lower light intensities (Kibler et al. 2012; GEOHAB 2012). New studies on the temperature and salinity tolerance (Yoshimatsu et al. 2014; Tawong et al. 2016; Xu et al. 2016) of *G. scabrosus* and ribotypes 3, 4, and 5 corroborate the general trends described by Kibler et al. (2012). However, the ranges appear to be wider: optimal and suboptimal growth occurs in the range of 21.0 and 32.5°C and salinities of 25 and 45, with exceptional growth rates up to 0.48 divisions day^{-1} under certain experimental conditions. For all strains tested growth was higher at 110–400 μmol photons m^{-2} s^{-1} than at 55 μmol photons m^{-2} s^{-1}. It was recognized that *G. caribaeus*, *G. carpenteri*, and *G. pacificus* were more tolerant to environmental conditions than were the other species tested (Xu et al. 2016). The value of these tedious and exacting physiological experiments goes beyond the ability to culture *Gambierdiscus* species for experimental work. Physiological data are used to inform models of habitat suitability or forecast which species may be favoured or disadvantaged by climate changes (Gingold et al. 2014; Kibler et al. 2015; see Sect. 13.6 and Wells et al. 2015).

13.4.2 Ostreopsis

Most studies on adaptive strategies of this BHAB have been conducted on *O.* cf. *ovata* that thrives in the Mediterranean coasts. Morphological differentiation of sexual and asexual reproduction, temporary, and resting cysts has been observed both in laboratory and field samples (Bravo et al. 2012; Accoroni et al. 2014). Unveiling the possible role of the different morphotypes, varying in shape and size, displayed by *Ostreopsis* spp. cells (e.g., "small cells," "dark cells," double-walled cysts), can shed light on the cell and life history, as well as other less well-understood physiological aspects. Identifying the latency phase of the life cycle and factors that favour the formation and germination of the different kinds of cysts sporadically observed in shallow epiphytic samples, both from bloom and non-bloom seasons (Bravo et al. 2012), should also help to understand the blooms recurrence.

New data from field and laboratory studies are not conclusive and cannot currently support the hypothesized expansion of the *Ostreopsis* spp. blooms driven by global warming. Optimal culture conditions appear to be between 22 and 30°C, especially around 25°C (Accoroni and Totti 2016 and references therein), but experiments do not show increased growth rate with increased temperature. In the Northern Mediterranean, the temperature control of the *Ostreopsis* spp. blooms is not clear either (Mangialajo et al. 2011). In general, the temperature window of 23–27.5°C would favour maximal abundances, although the blooms will also depend on local wind and hydrodynamic conditions (Cohu et al. 2013). The higher cell abundances mostly occur in midsummer (end of July) in the Northwestern basin and in late summer-early fall in the Adriatic. In the Atlantic coast of the Iberian Peninsula, absolute temperature cannot alone explain the *Ostreopsis* distribution (David et al. 2012). However, warming effects may also be indirect, negatively affecting competitor or consumer species, and these relationships are not currently known.

The response to salinity, as to temperature, is also species- and strain-specific. In temperate coastal areas, cell abundances are high at 30–39 salinity range and very low in river plume-affected areas (salinity <25) (Pezzolesi et al. 2012; Accoroni and Totti 2016 and references therein). However, because low salinities are associated with land runoff and potential nutrient supply, discrimination of the individual role of these two factors is difficult. Neither can the role of light intensity in *Ostreopsis* bloom dynamics be easily determined (Scalco et al. 2012). Recent experiments are not conclusive and often contradictory as to their preferred depth, methods of avoiding photodamage, and preference for calm environments.

13.5 Key Question 4: Are There Mechanisms Underlying BHAB Population and Community Dynamics Across Ecosystem Types That Are Revealed Through Comparative Studies?

Similar to the other GEOHAB CRPs, the comparative research approach (GEOHAB 2001) should aid in understanding the occurrence of BHAB species in natural ecosystems with different characteristics. For example, comparisons between and among populations experiencing different carbonate content or pH content (as in underwater volcanic habitats; Di Cioccio et al. 2014), mangrove habitats compared to seagrass- or macroalgae-dominated communities, limestone vs. volcanic islands, and pebbles vs. sandy substrates can be insightful, as can comparisons of systems with different fish sizes or different toxin levels. In the Caribbean, CFP is generally associated with large fish, while in the Pacific small fish are known to cause CFP (Lehane and Lewis 2000). Other environmental factors include the role of coastal oceanography processes (accumulation and transport) and covariation of depth with light quality and intensity (Cohu et al. 2013; Xu et al. 2016). Comparing the similarities and differences among the communities found in microhabitats (Caribbean, Pacific, Mediterranean) can also help resource managers and health officials with strategies for monitoring and management. Finally, the accumulated and successful knowledge on the toxic aerosol-producing planktonic species, *Karenia brevis*, which at high cell abundances can be fatal to marine fauna, can help inform research and prevention approaches appropriate for the *Ostreopsis* blooms in the affected habitats (Fleming et al. 2011).

13.6 Key Question 5: What New Observation and Modelling Approaches Are Available to Help in the Detection and Prediction of BHABs?

This objective included the need to establish effective detection technologies and monitoring tools and strategies, accompanied by suitable infrastructure in the affected areas. The necessity to generate data to develop predictive models of environmental conditions that may be conducive to bloom development is also recognized.

The apparent increased occurrence of BHABs in both temperate and tropical regions (Rhodes 2011) could be due to the anthropogenic forcing linked to the alteration of coastal ecosystems due to excessive human use and global warming. To ascertain this trend, improved detection and observation monitoring capacities are required. A major impediment identified by GEOHAB (2012) was the ability to sample BHAB cells quantitatively so cell abundance could be normalized and compared across studies. The use of artificial substrate (Fig. 13.1e), rather than direct collection of macrophytes, corals, or other benthic substrates, was suggested as a method to standardize sampling of BHAB species (Tester et al. 2014) and it has proven successful in a number of studies (Tan et al. 2013). Jauzein et al. (2016) validated the screen loading time (24 h) and optimized the artificial substrate method of Tester et al. (2014) finding it suitable for sampling *Ostreopsis* as well as *Gambierdiscus*. In the Mediterranean, new collection and counting methods, including the artificial substrate, were developed and tested through standardization exercises among different laboratories to facilitate the establishment of a common data basis for BHABs in that area (M.C. Chiantore, pers. comm.).

Progress on modelling is occurring, but it is limited by the lack of ecophysiological parameters and long-term data series of BHAB species abundances. New ecological and physiological data should allow development of models on BHAB dynamics (Accoroni et al. 2015) and forecast future scenarios for species selection (Gingold et al. 2014; Kibler et al. 2015).

13.7 BHAB Toxins

Although research on toxins was explicitly not included among the scientific elements and objectives in the initial GEOHAB Programme Science Plan (GEOHAB 2001), as the needs for unravelling ecological and oceanographic processes were emphasized, toxins were incorporated in the BHAB CRP (GEOHAB 2012). Research on ciguatoxins (CTXs) and maitotoxins (MTXs) from *Gambierdiscus* and palytoxins (PLTXs) and analogues from *Ostreopsis* have benefitted from major advances in new-generation innovations in analytical technology. The accurate characterization of toxin profiles is extremely helpful to clarify the interspecific and intraspecific toxicity of BHAB species and isolates.

Table 13.3 Ciguatoxin and maitotoxin activity of *Gambierdiscus* and *Fukuyoa* species by analytical method and location

Species	CTX activity	MTX activity	Method	Reference
G. australes				
French Polynesia RAV-92	Detect		RBA	Chinain et al. (2010a)
French Polynesia MUR-6	Detect		RBA	Chinain et al. (2010a)
French Polynesia MUR-14	Detect		RBA	Chinain et al. (2010a)
French Polynesia TB-1	Detect		RBA	Chinain et al. (2010a)
French Polynesia RAI-5	NA/ND		RBA	Chinain et al. (2010a)
French Polynesia MG-4	ND		RBA	Chinain et al. (2010a)
Cook Islands CAWD216	ND	ND	LC-MS	Rhodes et al. (2010, 2014a)
Cook Islands CAWD149	ND	Detect	LC-MS	Rhodes et al. (2010, 2014a)
Hawaii W B Gam 3		Detect	Ca^{2+} flux assay	Lewis et al. (2016)
Hawaii CAWD185	ND	Detect	LC-MS	Rhodes et al. (2014a)
Japan S080911_1	Detect	Detect	MBA	Nishimura et al. (2013)
G. balechii				
Celebes Sea	Detect		MBA	Fraga et al. (2016)
G. belizeanus				
St. Barthelemy STB-1	Detect		RBA	Chinain et al. (2010a)
Caribbean (averaged)		Detect	Hemolytic assay	Holland et al. (2013)
St. Barts CCMP 399	Detect	Detect	Ca^{2+} flux assay	Lewis et al. (2016)
G. caribaeus				
Belize Gam 19	ND	ND	Ca^{2+} flux assay	Lewis et al. (2016)
Hawaii Pat HI Jar 2 Gam2		Detect	Ca^{2+} flux assay	Lewis et al. (2016)
Thai strain	Detect		MBA	Tawong et al. (2016)
G. carolinianus				
Caribbean (averaged)		Detect	Hemolytic assay	Holland et al. (2013)
Belize Dive 1 Gam 1	Detect	Detect	Ca^{2+} flux assay	Lewis et al. (2016)
North Carolina Kenny 6	ND	Detect	Ca^{2+} flux assay	Lewis et al. (2016)
Hawaii Pat HI Jar3 Gam 9	ND	Detect	Ca^{2+} flux assay	Lewis et al. (2016)
G. carpenteri				
Australia (NSW)		MTX-like	MBA	Kohli et al. (2014)
Belize GT4	ND	Detect	Ca^{2+} flux assay	Lewis et al. (2016)
Pat HI Jar7 Gam 11	ND	Detect	Ca^{2+} flux assay	Lewis et al. (2016)
G. cheloniae				
Cook Islands CAWD 232	ND	Detect	MBA	Smith et al. (2016)
Cook Islands CAWD 236	ND	Detect	MBA	Smith et al. (2016)
G. excentricus				
Canary Islands VGO790	Detect	Detect	N2A	Fraga et al. (2011)
Canary Islands VGO790	Detect	Detect	N2A	Fraga et al. (2011)
Canary Islands VGO790	Detect	Detect	N2A	Fraga et al. (2011)

(continued)

Table 13.3 (continued)

Species	CTX activity	MTX activity	Method	Reference
G. lapillus				
Great Barrier Reef	Detect	Detect	MBA, LC-MS	Kretzschmar et al. (2016)
G. pacificus				
French Polynesia HO91	ND		RBA	Chinain et al. (2010a)
French Polynesia MUR4	ND		RBA	Chinain et al. (2010a)
French Polynesia ET-2	ND		RBA	Chinain et al. (2010a)
Cook Islands CAWD213	ND	ND	LC-MS	Rhodes et al. (2010, 2014a)
G. polynesiensis				
French Polynesia TB92	Detect		RBA	Chinain et al. (2010a)
French Polynesia RAI-1	Detect		RBA	Chinain et al. (2010a)
French Polynesia RG-92	Detect		RBA	Chinain et al. (2010a)
TB-92	Detect	Detect	MBA	Chinain et al. (2010a)
French Polynesia MG-7	Detect		MBA	Chinain et al. (2010a)
Cook Islands CAWD212	Detect	ND	LC-MS	Rhodes et al. (2010, 2014a)
G. silvae				
Canary Islands				Fraga and Rodríguez (2014)
G. scabrosus				
Japan	Detect	Detect	MBA	Nishimura et al. (2013)
G. ribotype 2				
Caribbean (averaged)		Detect	Hemolytic assay	Holland et al. (2013)
St. Maartens Gam 6	Detect	Detect	Ca^{2+} flux assay	Lewis et al. (2016)
Puerto Rico mixed PR Gam 3	Detect	Detect	Ca^{2+} flux assay	Lewis et al. (2016)
G. toxicus				
Australian and Polynesian strains	Detect	Detect	MBA	Holmes et al. (1990)
Tahiti FP GTT91	ND		RBA	Chinain et al. (2010a)
Tahiti FP HIT-0	ND		RBA	Chinain et al. (2010a)
Tahiti FP HIT-25	Detect		RBA	Chinain et al. (2010a)
Tahiti FP PAP1	Detect		RBA	Chinain et al. (2010a)
La Reunion FP REN-1	ND		RBA	Chinain et al. (2010a)
F. ruetzleri				
Belize *Gam*1	Detect	Detect	Ca^{2+} Flux Assay	Lewis et al. (2016)
F. cf. *yasumotoi*				
New Zealand CAWD210, 211	ND	ND	LC-MS	Rhodes et al. (2010, 2014a)
Gambierdiscus ribotypes				
Gam ribotype 1 (Caribbean)	Detect	Detect	N2A	Holland et al. (2013)

(continued)

Table 13.3 (continued)

Species	CTX activity	MTX activity	Method	Reference
Gam ribotype 1 KW070922_1 Japan	Detect	Detect	MBA	Nishimura et al. (2013)
Gam ribotype 2 M080829_2 Japan	ND	ND	MBA	Nishimura et al. (2013)
Gam ribotype 3 WI11G Japan	Detect	Detect	MBA	Nishimura et al. (2013)
Gambierdiscus ribotype 4 Republic af Kiribati	Detect		MBA	Xu et al. (2014)
Gambierdiscus ribotype 5 Republic af Kiribati	Detect		MBA	Xu et al. (2014)
Gambierdiscus ribotype 6 Republic af Kiribati	Detect		MBA	Xu et al. (2014)

CTX, ciguatoxin(s); MTX, maitotoxin(s); RBA, receptor binding assay; LC-MS, liquid chromatography-mass spectrometry; N2A, mouse neuroblastoma cells; MBA, (whole) mouse bioassay. The term "Caribbean averaged" indicates that data from multiple isolates from different locations having the same toxicity patterns were combined. ND, not detected; NA, non-acclimated strain

Other relevant questions include the modulation of toxin production by environmental factors and the clarification of the trophic transfer processes involved in toxin bioaccumulation and which may ultimately affect human health.

13.7.1 Gambierdiscus

Cell culture techniques have improved in tandem with functional assays (mouse bioassay, neuroblastoma cell line (N2A) and receptor binding assays (RBA)) and increased instrument sensitivity. Thus, the validation of these functional assays by LC-MS is providing preliminary data about species-specific *Gambierdiscus* toxicity. A summary of the data including the lipid-soluble CTX and water-soluble MTX fractions obtained from *Gambierdiscus* extracts, subsequent to the taxonomic revision by Litaker et al. (2009), shows progress (Table 13.3). Note, however, that the MTX toxic potential is not yet clear (Chinain et al. 2010a; Holland et al. 2013). Overall, concern for toxin transfer into the marine food web should be heightened where *G. polynesiensis* or *G. scabrosus* in the Pacific or *G. excentricus* and *G. silvae* in the Caribbean/Atlantic are detected.

Several new studies have documented CTX in the food web components of coral reef ecosystems, invertebrates, and lionfish. Silva et al. (2015) found significant CTX in two species of echinoderms among the macrofauna tested from the northwestern Moroccan coast. In French Polynesia, some tourists recently suffered severe CFP-like symptoms after they consumed gastropods (*Tectus niloticus*) collected in local waters (Gatti et al. 2015). In the coral reef ecosystem in the Republic of Kiribati, Mak et al. (2013) detected Pacific ciguatoxin (P-CTX) in the 54%, 72%, and 76% of the tested herbivorous, omnivorous, and carnivorous fishes, respectively, but the toxin did not seem to depend on the predators' fat content. Studies applying $\delta^{15}N$ approaches have revealed interesting potentialities to determine CTX biomagnification in the food webs. Controlled feeding studies on *G. polynesiensis* cells by mullet (*Mugil cephalus*) indicated rapid absorption (within 3 h) as well as elimination (within 24 h) of ingested toxins (Ledreux et al. 2014). This study also suggested a time-dependent transformation of oxopene CTXs may be necessary for the bioconcentration in higher trophic levels.

CTX in Caribbean lionfish was the focus of one basin wide (12 locations in the Caribbean and Gulf of Mexico, Hardison et al. 2016a) and two regional "hot spot" studies in the US Virgin Islands (Robertson et al. 2014) and in the French Antilles (including a location not known for CFP) (Soliño et al. 2015). Resource managers there are trying to mitigate the effects of a voracious, invasive species by establishing a fishery for lionfish. An imperative question is whether lionfish are more toxic or any more frequently toxic than other reef fish that are fished commercially or eaten by recreational or subsistence fishers. Toxicity in lionfish (detection of Caribbean CTX above 0.1 ppb, the US Food and Drug Administration's -USFDA- guidance level of this toxin) was patchy and generally higher in known 'hot spots' or where higher incidences of CFP are reported.

An update on methodologies for CTX determination was provided by Caillaud et al. (2010) to help address the concerns about the first cases of CFP in Europe (Boada et al. 2010). Later on, Lewis and Vetter (2015) and Lewis et al. (2016) developed a new, rapid extraction procedure that facilitates the simultaneous detection of CTXs and MTXs from cultures of *Gambierdiscus* by LC/MS and a functional bioassay and that could potentially be applied to field samples collected during blooms. More recently, a fluorescence-based receptor binding assay ($RBA_{(F)}$) was developed by Hardison et al. (2016b) to provide an alternative method for screening fish samples for CTXs in facilities that are not certified to use isotopes.

13.7.2 Ostreopsis

Advances in the characterization of the toxin profile of the *Ostreopsis* genus have been made on *O.* cf. *ovata* and *O. siamensis* strains that proliferate in temperate waters. Interestingly, *O.* cf. *siamensis* from the Mediterranean and the Atlantic do not produce either ostreocins (OSTRs) or ovatoxins (OVTXs) and contain very weak levels of palytoxin (Ciminiello et al. 2013) in contrast to the ones isolated in Japanese waters (Ukena et al. 2001).

In natural samples of *O.* cf. *ovata* from the Mediterranean, OVTX-a often represents more than 90% of the total toxin intracellular content (Accoroni et al. 2011; Séchet et al. 2012) while PLTX contributes less than 10%. In cultures, toxin content increased from the exponential to the stationary phases (Nascimento et al. 2012; Pezzolesi et al. 2012; Vanucci et al. 2012), but the values were relatively lower under nutrient-deficient than replete conditions (see Sect. 13.3.2). Overall, the reported ranges for OVTX-a and PLTX in *O.* cf. *ovata* cells are, respectively,

10–75 pg $cell^{-1}$ and 1.5–2.5 pg $cell^{-1}$. The proportion between these toxins was maintained in the entire cells and the mucus (Giussani et al. 2015) and in macrofauna (sea urchins, mussels) collected in the French, Italian, and Greek Mediterranean coasts (Biré et al. 2013; Brissard et al. 2014; Ciminiello et al. 2015). In these samples, total toxin levels were above the threshold values (30 μg PLTX equivalent/kg shellfish flesh) established by the European Food Safety Authority (EFSA 2009). More information on the occurrence of PLTX-like compounds on seafood can be found in the revision by Aligizaki et al. (2011). Until now, no food-borne intoxications in humans related to PLTX and analogues have been reported in Europe, in contrast to the dramatic, although rare, cases in tropical areas (Randall 2005).

The newest analytical procedures on liquid chromatography-high-resolution multiple-stage mass spectrometry (LC-HRMS/MS) have identified novel molecules such as OVTX-g and isobaric PLTX in strains isolated from the south of Catalonia (NW Mediterranean) (García-Altares et al. 2015) and four new OVTXs (i, j_1, j_2, and k) in the *Ostreopsis fattorussoi* from Cyprus Island (Tartaglione et al. 2016), which were at very low levels.

Because OVTXs and OSTRs have not been isolated yet, functional assays are used to characterize their toxic effects. Bioassays have revealed differential sensitivity of model organisms (*Artemia salina* nauplii, *Aurelia aurita* ephyrae) exposed to intact *Ostreopsis* cells, with higher mortality especially during the late stationary phase of the dinoflagellate culture (Pezzolesi et al. 2012; Giussani et al. 2016), and the mucilage could play an active toxic role. However, in the case of the benthic marine organism mortalities (Shears and Ross 2009; review by Accoroni and Totti 2016), it is difficult to discriminate the direct toxic impact from the indirect effects (e.g., oxygen depletion, increased temperature) on the reproductive capacity in sea urchins (Migliaccio et al. 2016) or inflammation of internal organs in mussels (Carella et al. 2015) co-occurring with high biomass *Ostreopsis* blooms. Further, a major challenge is the characterization of the potential toxic effect of the variety of OVTXs and OSTRs in humans.

Finally, there is no reliable method for rapid screening of these toxins in seafood. Most existing toxicological data are derived from in vivo mouse assay and are related to acute effects of pure PLTX. Recently, Alloisio et al. (2016) explored, for the first time, the possibilities of in vitro neuronal networks coupled to a microelectrode array (MEA)-based system, as sensitive biosensors for the evaluation of *Ostreopsis* toxicity on mammalian cells.

13.8 Impacts on Human Health

As with toxins, public health was outside the mission of GEOHAB (2001), but the BHAB CRP was born with an explicit commitment to collaborate with the medical authorities to address, together, the challenge of preventing seafood intoxication associated to CFP. For human populations in the tropics and those in emergent areas like the Mediterranean, the alert posed by the *Ostreopsis* blooms has identified two common needs: better estimates of the incidences on human health and the coordination between the medical and the environmental monitoring.

13.8.1 *Health Disorders Associated with* Gambierdiscus *Outbreaks*

With an estimated 50,000 cases a year worldwide, CFP is the most common nonbacterial seafood illness associated with eating fish (Dickey and Plakas 2000; Friedman et al. 2017). CFP intoxication most often occurs in tropical and subtropical regions but can occur anywhere given the wide distribution of fish harvested from the affected regions and tourism. Typical CFP intoxication can cause gastrointestinal, cardiovascular (bradycardia with hypotension), and neurological (paresthesia) symptoms that can last from weeks to months, including the diagnostic hot-cold temperature reversal (dysesthesia) (Bagnis et al. 1979). CTXs bind voltage-gated sodium channels causing an influx of Na^+ into the cell, disrupting cellular functions including signal transmission in nerves.

Underreporting and misdiagnosis are common and limit the capacity to prevent and palliate CFP impacts on human health. Recent multidisciplinary research, integrating epidemiology, and *Gambierdiscus* ecology and toxin studies (Chinain et al. 2010b; Morin et al. 2016; Radke et al. 2013; Skinner et al. 2011; Tester et al. 2010; see also http://www.ilm.pf) are offering new insights to the problems in the Caribbean and the Pacific territories. In the USA, outbreaks of CFP and related information are reported by public health officials voluntarily to the National Outbreak Reporting System (NORS, http://www.cdc.gov/nors/index.html) or the One Health Harmful Algal Bloom System (OHHABS, http://www.cdc.gov/habs/ohhabs.html), at the Centers for Disease Control and Prevention (CDC) surveillance system.

13.8.2 *Health Disorders Associated with* Ostreopsis *Outbreaks*

PLTX-like toxins are powerful vasoconstrictors in mammals with a main mode of action affecting the sodium/potassium pump (Rhodes et al. 2002; Tubaro et al. 2011). In the tropics, *Ostreopsis* spp. were initially linked to seafood intoxications, although more studies are required (Randall 2005 and references therein). In cases of extreme exposure, PLTX causes vomiting, laboured breathing, kidney problems, and, in some cases, death. Evaluation of human health impacts related to *Ostreopsis* is especially challenging in tropical areas, where the PLTX group co-occurs with other seafood biotoxins (e.g., ciguatoxins, saxitoxins, and tetrodotoxin). The lack of surveillance and reporting of health-related disorders and the prevalence of the toxins in reef species that are not commercially harvested make this task more difficult.

Over the past two decades, extensive blooms of *Ostreopsis* cf. *ovata* and *O.* cf. *siamensis* have occurred in the Mediterranean (e.g., Algeria, France, Italy, Spain, Tunisia). During these blooms, beach users have experienced upper respiratory disorders (rhinorrhea), eye and nose irritation, fever, and general malaise and sought medical care in hospital emergency departments and primary healthcare centres (see Tubaro et al. 2011 and Vila et al. 2016 for information on the reported cases in the Mediterranean since 1998). A similar situation could also be occurring in certain beaches in Brazil (Proença et al. 2010). The health effects may be caused by exposures to aerosols containing biotoxins or allergenic substances produced by *Ostreopsis* cell fragments or accompanying microbiota (Casabianca et al. 2013; Ciminiello et al. 2014). Further cutaneous irritations have been attributed (but not yet proven) to the direct dermal contact with water containing *Ostreopsis* spp. cells (Tubaro et al. 2011). At present, simultaneous epidemiology and ecology approaches are underway to improve the understanding of *Ostreopsis* bloom dynamics and the possible links to human health effects. Although wind and water motion may favour aerosolization and transport, the physiological state of the microalga seems to play the key role in the production of certain irritating compounds (Vila et al. 2016). Given that PLTX-like compounds have been only rarely detected in the aerosol (Ciminiello et al. 2014), the link of these compounds with the respiratory symptoms is still unclear.

13.9 Summary and Recommendations

The extraordinary research efforts conducted in the last few years are bringing useful fundamental and applied knowledge that will help to characterize BHABs events, optimize the potential to monitor environmental conditions and BHAB events, and prevent or palliate their noxious impacts on human health and the environment. Examples of this progress include productive new ways to sample *Gambierdiscus* and *Ostreopsis* cells, new toxin extraction techniques as well as a modification of the RBA for CTX, and sophisticated characterization of the PLTX group of toxins. Importantly, joint international collaboration has been fundamental for these advances. The topics, key questions, and objectives identified by the international community in GEOHAB (2012) are still valid guides to future research on BHABs. Apart from continuing the advance on the four GEOHAB programme elements described in the previous sections, urgent aspects to be investigated involve:

- The implication of other benthic taxa (specially benthic cyanobacteria; Kerbrat et al. 2011) in BHAB events
- The inter- and intraspecific differences in toxicity of BHAB-causing organisms

- The role of environmental factors (e.g., nutrient availability, water motion, and temperature) in determining cell toxicity of BHAB species
- The impact of climate change (global warming and ocean acidification) and anthropogenically driven changes in the ecosystem structure (jellyfish outbreaks, fishing pressure, coral reef destruction), in the biogeographic distribution of the BHAB species, and in the frequency and intensity of their harmful events
- On the practical side, there is a need to have cost-effective methods for screening fish for CTX (and other toxins), pure, less expensive standards for the major toxins, and a multidisciplinary and international collaborative approach integrating medical professionals, end users, stakeholders, and policy makers to evaluate management solutions based on the scientific advances.

GlobalHAB, the new international collaboration Programme following GEOHAB, is assuming these objectives for the coming decade.

Acknowledgments This is a contribution of the GEOHAB Core Research Project on HABs in Benthic Systems.

References

Accoroni S, Glibert PM, Pichierri S et al (2015) A conceptual model of annual *Ostreopsis* cf. *ovata* blooms in the northern Adriatic Sea based on the synergic effects of hydrodynamics, temperature, and the N:P ratio of water column nutrients. Harmful Algae 45:14–25

Accoroni S, Romagnoli T, Colombo F et al (2011) *Ostreopsis* cf. *ovata* bloom in the northern Adriatic Sea during summer 2009: ecology, molecular characterization and toxin profile. Mar Poll Bull 62:2512–2519

Accoroni S, Romagnoli T, Penna A et al (2016) *Ostreopsis fattorussoi* sp. nov. (Dinophyceae), a new benthic toxic *Ostreopsis* species from the eastern Mediterranean Sea. J Phycol 52: 1064–1084

Accoroni S, Romagnoli T, Pichierri S et al (2014) New insights on the life cycle stages of the toxic benthic dinoflagellate *Ostreopsis* cf. *ovata*. Harmful Algae 34:7–16

Accoroni S, Totti C (2016) The toxic benthic dinoflagellates of the genus *Ostreopsis* in temperate areas: a review. Adv Oceanogr Limnol 7:1–15

Adachi R, Fukuyo Y (1979) The thecal structure of a marine toxic dinoflagellate *Gambierdiscus toxicus* gen. et sp. nov. collected in a ciguatera-endemic area. Bull Jpn Soc Sci Fish 45:67–71

Aligizaki K, Panagiota K, Milandri A et al (2011) Occurrence of palytoxin-group toxins in seafood and future strategies to complement the present state of the art. Toxicon 57:390–399

Alloisio S, Giussani V, Nobile M et al (2016) MEA platform as a sensitive tool to detect and evaluate *Ostreopsis* cf. *ovata* toxicity. Harmful Algae 55:230–237

Bagnis R, Kuberski T, Laugier S (1979) Clinical observations on 3,009 cases of ciguatera (fish poisoning) in the South Pacific. Am J Trop Med Hyg 28:1067–1073

Biré R, Trotereau S, Lemée R et al (2013) Occurrence of palytoxins in marine organisms from different trophic levels of the French Mediterranean coast harvested in 2009. Harmful Algae 28:10–22

Boada LD, Zumbado M, Luzardo OP et al (2010) Ciguatera fish poisoning on the West Africa Coast: an emerging risk in the Canary Islands (Spain). Toxicon 56:1516–1519

Bravo I, Figueroa RI, Fraga S (2014) Cellular and nuclear morphological variability within a single species of the toxigenic dinoflagellate genus *Gambierdiscus*: relationship to life-cycle processes. Harmful Algae 40:1–8

Bravo I, Vila M, Casabianca S et al (2012) Life cycle stages of the benthic palytoxin-producing dinoflagellate *Ostreopsis* cf. *ovata* (Dinophyceae). Harmful Algae 18:24–34

Brissard C, Herrenknecht C, Séchet V et al (2014) Complex toxin profile of French Mediterranean *Ostreopsis* cf. *ovata* strains, seafood accumulation and ovatoxins prepurification. Mar Drugs 12:2851–2876

Caillaud A, De La Iglesia P, Darius HT et al (2010) Update on methodologies available for ciguatoxin determination: perspectives to confront the onset of ciguatera fish poisoning in Europe. Mar Drugs 8:1838–1907

Carella F, Sardo A, Mangoni O et al (2015) Quantitative histopathology of the Mediterranean mussel (*Mytilus galloprovincialis* L.) exposed to the harmful dinoflagellate *Ostreopsis* cf. *ovata*. J Invert Pathol 127:130–140

Carnicer O, Tunin-Ley A, Andrée KB et al (2015) Contribution to the genus *Ostreopsis* in the Reunion Island (Indian Ocean): molecular, morphologic and toxicity characterization. Cryptogam Algol 36:101–119

Casabianca SA, Casabianca P, Riobó JM et al (2013) Quantification of the toxic dinoflagellate *Ostreopsis* spp. by qPCR assay in marine aerosol. Environ Sci Technol 47:3788–3795

Chinain M, Darius HT, Ung A (2010a) Growth and toxin production in the ciguatera-causing dinoflagellate *Gambierdiscus polynesiensis* (Dinophyceae) in culture. Toxicon 56:739–750

Chinain M, Darius HT, Ung A (2010b) Ciguatera risk management in French Polynesia: the case study of Raivavae Island (Australes Archipelago). Toxicon 56:674–690

Chinain M, Faust MA, Pauillac S (1999) Morphology and molecular analyses of three toxic species of *Gambierdiscus* (Dinophyceae): *G. pacificus*, sp. nov., *G. australes*, sp. nov., and *G. polynesiensis*, sp. nov. J Phycol 35:1282–1296

Ciminiello P, Dell'Aversano C, Dello Iacovo E et al (2013) Investigation of toxin profile of Mediterranean and Atlantic strains of *Ostreopsis* cf. *siamensis* (Dinophyceae) by liquid chromatography-high resolution mass spectrometry. Harmful Algae 23:19–27

Ciminiello P, Dell'Aversano C, Dello Lacovo E et al (2014) First finding of *Ostreopsis* cf. *ovata* toxins in marine aerosols. Environ Sci Technol 48:3532–3540

Ciminiello P, Dell'Aversano C, Dello Lacovo E et al (2015) Liquid chromatography–high-resolution mass spectrometry for palytoxins in mussels. Anal Bioanal Chem 407:1463–1473

Cohu S, Mangialajo L, Thibaut T et al (2013) Proliferation of the toxic dinoflagellate *Ostreopsis* cf. *ovata* in relation to depth, biotic substrate and environmental factors in the North West Mediterranean Sea. Harmful Algae 24:32–44

David H, Ganzedo U, Laza-Martínez A et al (2012) Relationships between the presence of (Dinophyceae) in the Atlantic coast of the Iberian Peninsula and sea-surface temperature. Cryptogam Algol 33:199–207

David H, Laza-Martínez A, Miguel I et al (2013) *Ostreopsis* cf. *siamensis* and *Ostreopsis* cf. *ovata* from the Atlantic Iberian Peninsula: morphological and phylogenetic characterization. Harmful Algae 30:44–55

Di Cioccio D, Buia MC, Zingone A (2014) Ocean acidification will not deliver us from *Ostreopsis*. In: Kim HG, Reguera B, Hallegraeff G et al (eds) Harmful Algae 2012. Proceedings of the 15th international conference on Harmful Algae. International Society for the Study of Harmful Algae, Changwon, Korea, pp 85–88

Dickey RW, Plakas SM (2000) Ciguatera: a public health perspective. Toxicon 56:123–136

EFSA (2009) Marine biotoxins in shellfish—palytoxin group. Opinion of the scientific panel on contaminants in the food chain. Eur Food Saf Auth J 7:1393–1430

Ennaffah B, Chaira K (2015) First report of *Gambierdiscus* in Moroccan Atlantic waters. Harmful Algae News 50:20

Faust MA (1998) Mixotrophy in tropical benthic dinoflagellates. In: Reguera B, Blanco J, Fernandez ML et al (eds) Harmful algae. Xunta de Galicia and IOC Commission of UNESCO, Paris, pp 390–393
Faust MA (1999) Three new *Ostreopsis* species (Dinophyceae): *O. marinus* sp. nov., *O. belizeanus* sp., and *O. caribbeanus* sp. nov. Phycologia 38:92–99
Faust MA, Morton SL (1995) Morphology and ecology of the marine dinoflagellate *Ostreopsis labens* sp. nov. (Dinophyceae). J Phycol 31:456–463
Fleming LE, Kirkpatrick B, Backer LC et al (2011) Review of Florida Red Tide and Human Health Effects. Harmful Algae 20:224–233
Fraga S, Rodríguez F (2014) Genus *Gambierdiscus* in the Canary Islands (NE Atlantic Ocean) with description of *Gambierdiscus silvae* sp. nov., a new potentially toxic epiphytic benthic dinoflagellate. Protist 165(6):839–853
Fraga S, Rodríguez F, Caillaud A et al (2011) *Gambierdiscus excentricus* sp. nov. (Dinophyceae), a benthic toxic dinoflagellate from the Canary Islands (NE Atlantic Ocean). Harmful Algae 11:10–22
Fraga S, Rodríguez F, Riobo P et al (2016) *Gambierdiscus balechii* sp. nov. (Dinophyceae), a new benthic toxic dinoflagellate from the Celebes Sea (SW Pacific Ocean). Harmful Algae 58:93–105
Friedman MA, Fernandez M, Backer L et al (2017) An updated review of ciguatera fish poisoning: clinical, epidemiological, environmental, and public health management. Mar Drugs 15:72. https://doi.org/10.3390/md15030072
Fukuyo Y (1981) Taxonomical study on benthic dinoflagellates collected in coral reefs. Bull Jpn Soc Sci Fish 47:967–978
García-Altares M, Tartaglione L, Dell'Aversano C et al (2015) The novel ovatoxin-g and isobaric palytoxin (so far referred to as putative palytoxin) from *Ostreopsis* cf. *ovata* (NW Mediterranean Sea): structural insights by LC-high resolution MS. Anal Bioanal Chem 407:1191–1204
Gatti CM, Darius HT, Chinain M et al (2015) First report of a mass-poisoning outbreak following the consumption of *Tectus niloticus* (gastropod) in French Polynesia: a novel pathway of ciguatera shellfish poisoning? Harmful Algae News 50:19–20
GEOHAB (2001) Global ecology and oceanography of harmful algal blooms, science plan. Glibert P, Pitcher G (eds), SCOR and IOC, Baltimore and Paris, 86 pp
GEOHAB (2003) Global ecology and oceanography of harmful algal blooms, implementation plan. Gentien P, Pitcher G, Cembella A et al. (eds), SCOR and IOC, Baltimore and Paris, 36 pp
GEOHAB (2012) Global ecology and oceanography of harmful algal blooms, GEOHAB core research project: HABs in benthic systems. Berdalet E, Tester P, Zingone A (eds), IOC and SCOR, Paris and Newark, 64 pp
Gingold DB, Strickland MJ, Hess JJ (2014) Ciguatera fish poisoning and climate change: analysis of National Poison Center data in the United States, 2001-2011. Environ Health Perspect 122:580–586
Giussani V, Sbrana F, Asnaghi V et al (2015) Active role of the mucilage in the toxicity mechanism of the harmful benthic dinoflagellate *Ostreopsis* cf. *ovata*. Harmful Algae 44:46–53
Giussani V, Costa E, Pecorino DD et al (2016) Effects of the harmful dinoflagellate *Ostreopsis* cf. *ovata* on different life cycle stages of the common moon jellyfish *Aurelia* sp. Harmful Algae 57:49–58
Glibert PM, Al-Azri A, Allen JI et al (2018) Key questions and recent research advances on harmful algal blooms in relation to nutrients and eutrophication. In: Glibert PM, Berdalet E, Burford M et al (eds) Global ecology and oceanography of harmful algal blooms, Springer, Cham, pp 229–259
Gómez F, Qui D, Lopes RM et al (2015) *Fukuyoa paulensis* gen. et sp. nov., a new genus for the globular species of the dinoflagellate *Gambierdiscus* (Dinophyceae). PLoS One 10(4): e0119676. https://doi.org/10.1371/journal.pone.0119676
Guerrini F, Pezzolesi L, Feller A et al (2010) Comparative growth and toxin profile of cultured *Ostreopsis ovata* from the Tyrrhenian and Adriatic Seas. Toxicon 55:211–220

Hardison DR, Holland WC, Darius HT et al (2016a) Ciguatoxin concentrations in invasive lionfish from the Caribbean regions: implications for fishery development and management. PLoS One 11(4):e0153348. https://doi.org/10.1371/journal.pone.0153348

Hardison DR, Holland WC, McCall JR et al (2016b) Fluorescent receptor binding assay for detecting ciguatoxin in fish. PLoS One. https://doi.org/10.1371/journal.pone.0153348

Holland WC, Litaker WR, Tomas CR et al (2013) Differences in the toxicity of six *Gambierdiscus* (Dinophyceae) species measured using an in vitro human erythrocyte lysis assay. Toxicon 65: 15–33

Holmes MJ (1998) *Gambierdiscus yasumotoi* sp. nov. (Dinophyceae), a toxic benthic dinoflagellate from southeastern Asia. J Phycol 34:661–668

Holmes MJ, Lewis RJ, Gillespie NC (1990) Toxicity of Australian and French Polynesia strains of *Gambierdiscus toxicus* (Dinophyceae) grown in culture: characterization of a new type of maitotoxin. Toxicon 28:1159–1172

Jauzein C, Fricke A, Mangialajo L et al (2016) Sampling of *Ostreopsis* cf. *ovat*a using artificial substrates: optimization of methods for the monitoring of benthic harmful algal blooms. Mar Pollut Bull. https://doi.org/10.1016/j.marpolbul.2016.03.047

Kerbrat AS, Amzil Z, Pawlowiez R et al (2011) First evidence of palytoxin and 42-hydroxy-palytoxin in the marine cyanobacterium *Trichodesmium*. Mar Drugs 9:543–560

Kibler S, Litaker RW, Holland WC et al (2012) Growth of eight *Gambierdiscus* (Dinophyceae) species: effects of temperature, salinity and irradiance. Harmful Algae 19:1–14

Kibler SR, Tester PA, Kunkel KE (2015) Effects of ocean warming on growth and distribution of dinoflagellates associated with ciguatera fish poisoning in the Caribbean. Ecol Model 316:194–210

Kohli GS, Murray SA, Neilan B et al (2014) High abundance of the potentially maitotoxic dinoflagellate *Gambierdiscus carpenteri* in temperate waters of New South Wales, Australia. Harmful Algae 39:134–145

Kretzschmar ALV, Harwood DT, Hoppenrath M et al (2016) Characterisation of *Gambierdiscus lapillus* sp. nov. (Gonyaulacales, Dinophyceae): a toxic new dinoflagellate from the Great Barrier Reef (Australia). J Phycol. https://doi.org/10.1111/jpy.12496-16-084

Kuno S, Kamikawa R, Yoshimatsu S et al (2010) Genetic diversity of *Gambierdiscus* spp. (Gonyaulacales, Dinophyceae) in Japanese coastal areas. Phycol Res 58:44–52

Lapointe BE, Langton R, Bedford BJ (2010) Land-based nutrient enrichment of the Buccoo Reef Complex and fringing coral reefs of Tobago, West Indies. Mar Pollut Bull 60:334–343

Larsen J, Nguyen NL (2004) Potentially toxic microalgae of Vietnamese waters. Opera Bot 140:5–216

Lartigue J, Jester ELE, Dickey RW et al (2009) Nitrogen source effects on the growth and toxicity of two strains of the ciguatera-causing dinoflagellate *Gambierdiscus toxicus*. Harmful Algae 8:781–791

Leaw CP, Lim PT, Tan TH et al (2011) First report of the benthic dinoflagellate, *Gambierdiscus belizeanus* (Gonyaulacales: Dinophyceae) for the east coast of Sabah, Malaysian Borneo. Phycol Res 59:143–146

Ledreux A, Brand H, Chinain M et al (2014) Dynamics of ciguatoxins from *Gambierdiscus polynesiensis* in the benthic herbivore *Mugil cephalus*: trophic transfer implications. Harmful Algae 39:165–174

Lehane L, Lewis RJ (2000) Ciguatera: recent advances but the risk remains. Int J Food Microbiol 61:91–125

Lewis R, Inserra M, Holland CW et al (2016) Rapid extraction and identification of maitotoxin and ciguatoxin-like toxins from Caribbean and Pacific *Gambierdiscus* using a new functional bioassay. PLoS One. https://doi.org/10.1371/journal.pone.0160006

Lewis R, Vetter I (2015) Ciguatoxin and ciguatera. Mar Freshwater Tox. https://doi.org/10.1007/978-94-007-6605-1_13-1

Litaker RW, Vandersea MW, Faust MA et al (2009) Taxonomy of *Gambierdiscus* including four new species, *Gambierdiscus caribaeus*, *Gambierdiscus carolinianus*, *Gambierdiscus carpenteri* and *Gambierdiscus ruetzleri* (Gonyaulacales, Dinophyceae). Phycologia 48:344–390

Mak YL, Wai T-C, Murphy MB et al (2013) Pacific ciguatoxins in food web components of coral reef systems in the Republic of Kiribati. Environ Sci Technol 47:14070–14079

Mangialajo L, Ganzin N, Accoroni S et al (2011) Trends in *Ostreopsis* proliferation along the Northern Mediterranean coasts. Toxicon 57:408–420

Migliaccio O, Castellano I, Di Cioccio D et al (2016) Subtle reproductive impairment through nitric oxide-mediated mechanisms in sea urchins from an area affected by harmful algal blooms. Nat Sci Rep 6:article 26086. https://doi.org/10.1038/srep26086

Morin E, Gatti C, Bambridge T et al (2016) Ciguatera fish poisoning: incidence, health costs and risk perception on Moorea Island (society archipelago, French Polynesia). Harmful Algae 60: 1–10

Nascimento SM, Corrêa EV, Menezes M (2012) Growth and toxin profile of *Ostreopsis* cf. *ovata* (Dinophyta) from Rio de Janeiro, Brazil. Harmful Algae 13:1–9

Nishimura T, Hariganeya H, Tawong W et al (2016) Quantitative PCR assay for detection and enumeration of ciguatera-causing dinoflagellate *Gambierdiscus* spp. (Gonyaulacales) in coastal areas of Japan. Harmful Algae 52:11–22

Nishimura T, Sato S, Tawong W et al (2013) Genetic diversity and distribution of the ciguatera-causing dinoflagellate *Gambierdiscus* spp. (Dinophyceae) in coastal areas of Japan. PLoS One 8(4):e60882. https://doi.org/10.1371/journal.pone.0060882

Nishimura T, Sato S, Tawong W et al (2014) Morphology of *Gambierdiscus scabrosus* sp. nov. (Gonyaulacales): a new epiphytic toxic dinoflagellate from coastal areas of Japan. J Phycol 50:506–514

Norris DR, Bomber JW, Balech E (1985) Benthic dinoflagellates associated with ciguatera from the Florida Keys. I. *Ostreopsis heptagona* sp. nov. In: Anderson DM, White AW, Baden DG (eds) Toxic dinoflagellates. Elsevier, New York, pp 39–44

Penna A, Battocchi C, Capellacci S et al (2014) Mitochondrial, but not rDNA, genes fail to discriminate dinoflagellate species in the genus *Ostreopsis*. Harmful Algae 40:40–50

Penna A, Fraga S, Battocchi C et al (2012) Genetic diversity of the genus *Ostreopsis* Schmidt: phylogeographical considerations and molecular methodology applications for field detection in the Mediterranean Sea. Cryptogam Algol 33:153–163

Pezzolesi L, Guerrini F, Ciminiello P et al (2012) Influence of temperature and salinity on *Ostreopsis* cf. *ovata* growth and evaluation of toxin content through HR LC-MS and biological assays. Water Res 46:82–92

Pezzolesi L, Vanucci S, Dell'Aversano C et al (2016) Effects of N and P availability on carbon allocation in the toxic dinoflagellate *Ostreopsis* cf. *ovata*. Harmful Algae 55:2012–2212

Pinna A, Pezzolesi L, Pistocchi R et al (2015) Modelling the stoichiometric regulation of C-rich toxins in marine dinoflagellates. PLoS One 10:e0139046. https://doi.org/10.1371/journal.pone.0139046

Proença LAO, Boemer GL, Dias JP et al (2010) Can the cases of airborne intoxication of beach users in south coast of Bahia (16°24′S–39°02′W) be related to microalgae? GEOHAB Open Science Meeting: HABs in Benthic Systems, Honolulu, Hawai'i, USA, 21–23 June 2010 (abstract only)

Quod JP (1994) *Ostreopsis mascarenensis* sp. nov. (Dinophyceae), dinoflagellé toxique associé a la ciguatera dans l'Océan Indien. Cryptogam Algol 15:243–251

Radke EG, Grattan LM, Smith CRL et al (2013) Ciguatera incidence in the U.S. Virgin Islands has not increased over a 30-year time period despite rising seawater temperatures. Am J Trop Med Hyg 88:908–913

Randall JE (2005) Review of clupeotoxism, an often fatal illness from the consumption of clupeoid fishes. Pac Sci 59:73–77

Rhodes L (2011) World-wide occurrence of the toxic dinoflagellate genus *Ostreopsis* Schmidt. Toxicon 57:400–407

Rhodes L, Harwood T, Smith K et al (2014a) Production of ciguatoxin and maitotoxin by strains of *Gambierdiscus australes, G. pacificus* and *G. polynesiensis* (Dinophyceae) isolated from Rarotonga, Cook Islands. Harmful Algae 39:185–190

Rhodes L, Papiol GG, Smith K et al (2014b) *Gambierdiscus* cf *yasumotoi* (Dinophyceae) isolated from New Zealand's sub-tropical northern coastal waters. NZ J Mar Freshwat Res 48:303–310

Rhodes L, Smith KF, Munday R et al (2010) Toxic dinoflagellates (Dinophyceae) from Rarotonga, Cook Islands. Toxicon 56:751–758

Rhodes L, Towers N, Briggs L et al (2002) Uptake of palytoxin-like compounds by shellfish fed *Ostreopsis siamensis* (Dinophyceae). NZ J Mar Freshwat Res 36:631–636

Robertson A, Garcia AC, Quintana HAF et al (2014) A potential human health threat for ciguatera fish poisoning in tropical waters. Mar Drugs 12:88–97

Sato S, Nishimura T, Uehara K et al (2011) Phylogeography of *Ostreopsis* along West Pacific Coast, with Special Reference to a Novel Clade from Japan. PLoS One 6(12):1–14. e27983. https://doi.org/10.1371/journal.pone.0027983

Scalco E, Brunet C, Marino F et al (2012) Growth and toxicity responses of Mediterranean *Ostreopsis* cf. *ovata* to seasonal irradiance and temperature conditions. Harmful Algae 17: 25–34

Schmidt J (1901) Peridiniales. Preliminary report of the botanical results of the Danish expedition to Siam (1899-1900). Botanische Pidsslerift, pp 212–221

Séchet V, Sibat M, Chomérat N et al (2012) *Ostreopsis* cf. *ovata* in the French Mediterranean coast: molecular characterisation and toxin profile. Cryptogam Algol 33:89–98

Selina MS, Morozova DI, Vyshkvartsev DI et al (2014) Seasonal dynamics and spatial distribution of epiphytic dinoflagellates in Peter the Great Bay (Sea of Japan) with special emphasis on *Ostreopsis* species. Harmful Algae 32:1–10

Shears NT, Ross PM (2009) Blooms of benthic dinoflagellates of the genus *Ostreopsis*: an increasing and ecologically important phenomenon on temperate reefs in New Zealand and worldwide. Harmful Algae 8:916–925

Silva M, Rodríguez I, Barreiro A et al (2015) First report of ciguatoxins in two starfish species: *Ophidiaster ophidianus* and *Marthasterias glacialis*. Toxins 7:3740–3757

Skinner MP, Brewer TD, Jonstone et al. (2011) Ciguatera fish poisoning in the Pacific Islands (1998-2008). PLoS Negl Trop Dis 5(12):e1416. https://doi.org/10.1371/journal.pntd.0001416

Smith KF, Rhodes L, Verma A et al (2016) A new *Gambierdiscus* species (Dinophyceae) from Rarotonga Cook Islands: *Gambierdiscus cheloniae* sp. nov. Harmful Algae 60:45–56

Soliño L, Widgy S, Pautonnier A et al (2015) Prevalence of ciguatoxins in lionfish (*Pterois spp.*) from Guadeloupe, Saint Martin, and Saint Barthélmy Islands (Caribbean). Toxicon 102:62–68

Tan T-H, Lim P-T, Mujahid A et al (2013) Benthic harmful dinoflagellate assemblages in a fringing reef of Sampadi Island, Sarawak, Malaysia. Mar Res Indones 38:11–21

Tartaglione L, Mazzeo A, Dell'Aversano C (2016) Chemical, molecular, and eco-toxicological investigation of *Ostreopsis* sp. from Cyprus Island: structural insights into four new ovatoxins by LC-HRMS/MS. Anal Bioanal Chem 408:915–932

Tawong W, Yoshimatsu T, Yamaguchi H et al (2016) Temperature and salinity effects and toxicity of *Gambierdiscus caribaeus* (Dinophyceae) from Thailand. Phycologia 55:274–278

Tester PA, Feldman RL, Nau AW et al (2010) Ciguatera fish poisoning and sea surface temperatures in the Caribbean Sea and the West Indies. Toxicon 56:698–710

Tester PA, Kibler SA, Holland WC et al (2014) Sampling harmful benthic dinoflagellates: comparison of artificial and natural substrate methods. Harmful Algae 39:8–25

Tubaro A, Durando P, Del Favero G et al (2011) Case definitions for human poisonings postulated to palytoxins exposure. Toxicon 57:478–495

Turquet J, Quod J-P, Ten-Hage L et al. (2001) Example of a *Gambierdiscus toxicus* flare-up following the 1998 coral bleaching event in Mayotte Island (Comoros, southwest Indian Ocean). In: Hallegraeff GM, Blackburn SI, Bolch CJ et al (eds), Harmful Algal Blooms 2000, Proceedings of the ninth international conference on harmful algal blooms, UNESCO, France, pp 50–55

Ukena T, Satake M, Usami M et al (2001) Structure elucidation of ostreocin D, a palytoxin analog isolated from the dinoflagellate *Ostreopsis siamensis*. Biosci Biotechnol Biochem 65: 2585–2588

Vanucci S, Pezzolesi L, Pistocchi R et al (2012) Nitrogen and phosphorus limitation effects on cell growth, biovolume, and toxin production in *Ostreopsis* cf. *ovata*. Harmful Algae 15:78–90

Vila M, Abós-Herràndiz R, Isern-Fontanet J et al (2016) Establishing the link between *Ostreopsis* cf. *ovata* blooms and human health impacts using ecology and epidemiology. Sci Mar 80(S1): 107–115

Wells ML, Trainer VL, Smayda TJ et al (2015) Harmful algal blooms and climate change: learning from the past and present to forecast the future. Harmful Algae 49:68–93

Xu Y, Richlen ML, Liefer J et al (2016) Influence of environmental variables on *Gambierdiscus* spp. (Dinophyceae) growth and distribution. PLoS One 11(4):e0153197. https://doi.org/10.1371/journal.pone.0153197

Xu Y, Richlen ML, Morton SL et al (2014) Distribution, abundance and diversity of *Gambierdiscus* spp. from a ciguatera-endemic area in Marakei, Republic of Kiribati. Harmful Algae 34: 56–68

Yoshimatsu T, Yamaguchi H, Iwamoto H et al (2014) Effects of temperature, salinity and their interaction on growth of Japanese *Gambierdiscus* spp. (Dinophyceae). Harmful Algae 35(2014):29–37

Part V
Spotlight on Harmful Algal Blooms in Asia

Chapter 14
Overview of Harmful Algal Blooms in Asia

Ken Furuya, Mitsunori Iwataki, Po Teen Lim, Songhui Lu, Chui-Pin Leaw, Rhodora V. Azanza, Hak-Gyoon Kim, and Yasuwo Fukuyo

14.1 Introduction

Harmful algal blooms (HABs) caused by various types of microalgae pose serious threats to coastal ecosystems, fisheries, and public health in the Asian region (GEOHAB 2010). Along with economic development and increases in the human population, records of HAB events have increased. Although HAB events themselves are a natural phenomenon caused by endemic microalgal species adapted to coastal environments, the increase of reports of red tides and range expansion of some species are the result of increased human activities, e.g., coastal development and transportation of organisms. This chapter reviews the current knowledge of harmful algal blooms (HABs) in Asian waters on the basis of existing knowledge, as summarized in GEOHAB (2010) and Fukuyo et al. (2011). Active research on HABs accelerated in the 2000s, with much of this being undertaken under the umbrella of GEOHAB Asia. The GEOHAB Asia Programme was

K. Furuya (✉) • Y. Fukuyo
Graduate School of Agricultural and Life Sciences, The University of Tokyo, Tokyo, Japan
e-mail: furuya@fs.a.u-tokyo.ac.jp

M. Iwataki
Asian Natural Environmental Science Center, The University of Tokyo, Tokyo, Japan

P.T. Lim • C.-P. Leaw
Institute of Ocean and Earth Sciences, University of Malaya, Bachok, Malaysia

S. Lu
Research Center for Harmful Algae and Marine Biology, Jinan University, Guangzhou, China

R.V. Azanza
Marine Science Institute, University of the Philippines, Quezon City, Philippines

H.-G. Kim
Department of Oceanography, Pukyong National University, Busan, Republic of Korea

P.M. Glibert et al. (eds.), *Global Ecology and Oceanography of Harmful Algal Blooms*, Ecological Studies 232, https://doi.org/10.1007/978-3-319-70069-4_14

initiated with two Open Science meetings in Asia, the first in Japan in 2007 and the second in Vietnam in 2008.

14.2 Previous and Well-Documented HAB Issues in Asia

14.2.1 Harmful Red Tides in Asia

Red tides in Asia have been attributed to many different species, and their outbreaks have had a range of effects. Red tide outbreaks in Asia, particularly those caused by *Noctiluca scintillans* and *Trichodesmium* spp., have long been known to occur in East Asia and Southeast Asia. Massive fish mortality due to such outbreaks was first reported in the 1980s (e.g., White et al. 1984), when fish farming and other aquaculture activities became more common in coastal areas, yet scientific studies remained scarce even through the 1990s.

One issue involving the entire region is outbreaks of the harmful red tide dinoflagellate *Cochlodinium* (*Margalefidinium*) *polykrikoides*. Once observed only in some parts of East Asia, especially in Japan and Korea, *C. polykrikoides* outbreaks have now been observed over a wide area, from tropical Southeast Asia to areas of the Persian Gulf (Azanza et al. 2008; Richlen et al. 2010). Its distribution has apparently expanded, putatively due to artificial vectors such as ballast water or the transportation of aquaculture products, although this claim still lacks evidence. However, *C. polykrikoides* may have existed across the vast waters of Asia already, unbeknownst to humans, as outbreaks are observed almost every year in East Asia, and their genotype is clearly differentiated from those distributed in North and Central America, based on rDNA sequencing (Iwataki et al. 2008). In addition, molecular studies on the genus *Cochlodinium* found that different, putatively nonindigenous genotypes have recently been distributed throughout Asia, e.g., along the coast of Sabah, Malaysia (Iwataki et al. 2008). More detailed research on their ecology, including their distribution, expansion, and mechanisms of settlement, is eagerly awaited.

Aside from members of *Cochlodinium*, one well-known high-biomass-causing species in Southeast Asia is the so-called green dinoflagellate, *Noctiluca*. Cells of the species *N. scintillans* possess a swarmer of the symbiotic green alga *Pedinomonas noctilucae* (pedinophyte) inside the vacuole. This organism has long been known to form massive red tides, but its outbreaks had failed to draw much interest because these events did not cause any massive fish mortalities despite a decrease in catch volume due to fish escaping from the affected area. However, green *Noctiluca* outbreaks appear to have grown in both duration and geographical extent with the eutrophication of coastal areas (Harrison et al. 2011), and their occurrence over a wide range in the Arabian Sea is starting to be discussed in relation to global environmental change [Gomes et al. 2014; see also Chap. 17, Goes et al. (2018)].

Another harmful red tide species found in Asia is an armoured dinoflagellate, *Heterocapsa circularisquama*, which forms red tides, killing bivalves such as the cultured oyster and pearl oyster (Horiguchi 1995). This species was first detected in

Uranouchi Bay, Japan, in 1988, and it subsequently expanded its distribution along the coasts of western Japan (Matsuyama 2012). In Asia Pacific, this species has so far been reported only in Japan and Hong Kong (Iwataki et al. 2002).

Other harmful red tide microalgae responsible for large fish kills in Asia are *Karenia mikimotoi* and *Chattonella antiqua/marina*. The unarmoured dinoflagellate *K. mikimotoi*, originally described from Japan, is one of the most harmful red tide species in temperate and tropical Asia. Occurrences of this species have been recorded in China, Japan, Korea, Hong Kong, and Singapore (Okaichi 2004; Qi et al. 2004; Leong et al. 2015). The raphidophyte *Chattonella antiqua/marina* has formed massive red tides associated with major fisheries impacts in Japan. It has been reported in both temperate and tropical Asia.

There are occasional reports of other red tide species causing harm such as the mass killing of cultured fishes. These include the dinoflagellates *Akashiwo sanguinea*, *Gonyaulax polygramma*, *Prorocentrum* spp. (especially *P. dentatum/donghaiense* and *P. minimum*), *Heterocapsa triquetra*, and *Scrippsiella trochoidea* and the diatoms *Pseudo-nitzschia* spp. and *Chaetoceros* spp. In 2002, a fish kill event in Bolinao, northwestern Philippines, was attributed to massive blooms of *P. minimum*, which resulted in heavy economic losses (Azanza et al. 2004; Fig. 14.1). However, the issue is far from resolved; little research has been done

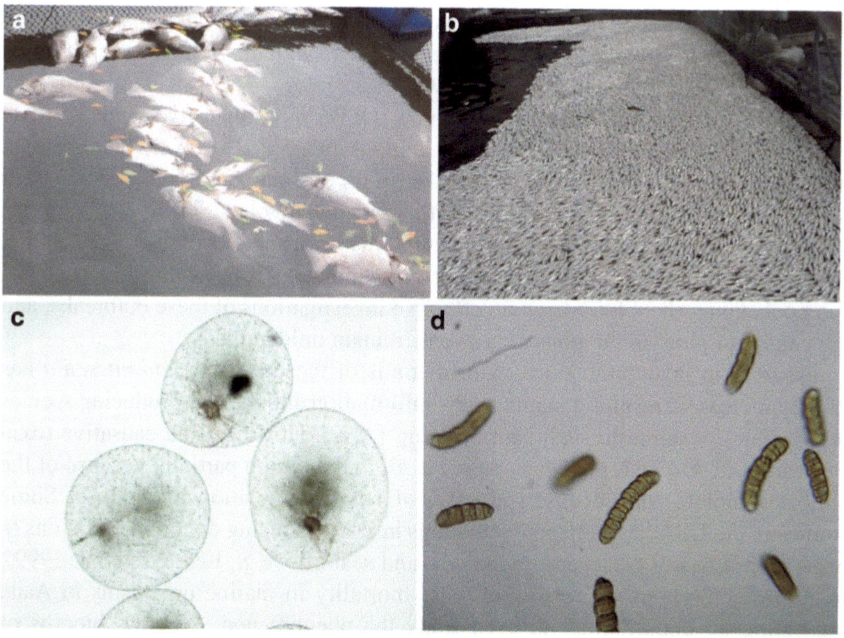

Fig. 14.1 Recent fish mass mortalities occurred in Asia. Red tides caused by *Karlodinium australe* in Johor Strait, Malaysia (**a**), and by *Prorocentrum minimum* in Bolinao, Philippines (**b**). Cells of green *Noctiluca* (**c**) and *Cochlodinium polykrikoides* (**d**). Photos courtesy of Wilson Mendoza (**b**), Takuo Omura (**c**), and Kazuya Takahashi (**d**)

on their population dynamics. The area has subsequently experienced blooms of several HAB species, and ongoing research on these species began in 2012.

14.2.2 Shellfish and Fish Poisoning in Asia

In addition to species described above that have long caused large-scale blooms, there are also numerous reports of toxic species that have occurred throughout Asia, typically affecting shellfish. Paralytic shellfish poisoning (PSP) occurred locally in the 1970s in locations such as Papua New Guinea and Sabah in Malaysia, but spread to the central-northern Philippines and eastern Indonesia in the 1980s (White et al. 1984). As knowledge of cases of shellfish poisoning has spread in newspapers and other media, reports of simultaneous poisoning incidents came from Sumatra in Indonesia and both sides of the Malay Peninsula. Several people seemed to have suffered shellfish poisoning at the same time, although the outbreak date and responsible species were unknown. However, the introduction of systems for monitoring concentrations of toxins in cultured shellfish in large marine areas in the Philippines and Sabah has sharply decreased the incidence of poisoning cases since the 1990s. In total, more than 2000 cases of PSP have been reported for Southeast Asia as a whole, approximately 10% of which were fatalities. The responsible dinoflagellate species was primarily *Pyrodinium bahamense*, but cases involving *Alexandrium tamiyavanichii*, *A. minutum*, and *Gymnodinium catenatum* were also found (Lim et al. 2012).

In terms of diarrheal shellfish poisoning (DSP), more than five kinds of toxic species have been observed in tropical regions in Asia. However, there remain many unanswered questions, including the origin of the toxins, as there have been almost no studies of the ecology of the species involved or how the toxins accumulate in shellfish. Because the symptoms of suspected cases of poisoning, such as diarrhoea and vomiting, cannot be distinguished from those of bacterial food poisoning, there have been no effective investigations of these outbreaks, and the causes of most of the poisoning events remain unidentified.

Progress in taxonomic research of diatoms of the genus *Pseudo-nitzschia* has rapidly increased, resulting in much new information about toxin-producing species responsible for amnesic shellfish poisoning (ASP). However, the causative toxin (domoic acid, DA) has not been studied at all in the region partially because of the almost complete lack of observed cases of DA accumulation in shellfish. Some studies in the USA have linked fluctuations in DA-producing diatom populations to mass mortalities of cetaceans, pinnipeds, and seabirds (e.g., Lefebvre et al. 1999). In contrast, there are no reports of mass mortality in marine organisms in Asia, presenting no opportunities to even study the phenomenon. Whereas blooms of *Pseudo-nitzschia* associated with ASP have never been reported in the Asian region, seasonal blooms of *Pseudo-nitzschia* species had been reported in the Sea of Japan and Russia (Stonik et al. 2011), with *P. calliantha*, *P. multiseries*, *P. multistriata*, and *P. pungens* blooms observed during the summer and autumn in Peter the Great Bay. Studies on DA in bivalves conducted in Vietnam confirmed that bivalves in the areas were contaminated by DA (Dao et al. 2009). However, no evidence showed that *Pseudo-nitzschia* was the main source of DA contamination in the shellfish. Later, two species, *P.* cf. *caciantha* and *P. fukuyoi*, from the area were proven toxic in culture studies (Dao et al. 2014, 2015).

Southeast Asia has begun to frequently report incidents of ciguatera fish poisoning (CFP), which is caused by benthic dinoflagellate species. However, diagnoses especially in remote areas have been inadequate, i.e., immediately labelling cases of fish poisoning as ciguatera, and none of the studies have precisely confirmed the causative toxin. A study in the Philippines, however, was able to confirm the presence of ciguatoxin in the blood samples of affected individuals using electrospray ionization mass spectrometry and included the epidemiology of 22 affected patients that showed medical indications of CFP (Mendoza et al. 2013). Nonetheless, there is no doubt that CFP outbreaks are occurring over wider areas, as demonstrated by observations of CFP due to consumption of the fish species, spotted knifejaw in Japan, a country where the disease had never been reported before. In Southeast Asia as well, the number of poisoning incidents and causative fish species has increased. Expanding the scope of investigations of causative species has become an urgent issue, with the objective of ensuring the safety of natural fish resources in coastal regions within Asia.

14.2.3 Toxic Species in the East China Sea

Large-scale algal blooms have been reported and especially well studied along the coasts of China [see also Chap. 15, Yu et al. (2018) and Chap. 16, Liu and Zhou (2018)]. *Prorocentrum donghaiense*, *Phaeocystis globosa*, as well as macroalgal green tide by *Ulva (=Enteromorpha) prolifera* are the most common causative species (GEOHAB 2010). Among them, *P. donghaiense* is the most important in the East China Sea (Lu et al. 2005), which has formed extensive harmful blooms every year since 1998. Normally the bloom starts in May and ends in late June. The scale of bloom area has been estimated as thousands of square kilometres every year since 1998. Although it is not a toxin producer, the extensive and prolonged *P. donghaiense* blooms alter the ecosystem by reducing phytoplankton biodiversity and zooplankton abundance (Lin et al. 2014). Additionally, *P. donghaiense* blooms change the zooplankton community from copepods to small jellyfish during the bloom. The blooms also have detrimental consequences for the recruitment of copepod *Calanus sinicus*, an important and dominant zooplankton species in the East China Sea waters (Lin et al. 2014). Ultimately, these negative impacts of the blooms at such large spatial and temporal scales can result in a major threat to fishery resources, an unbalanced marine ecosystem, and degradation of marine ecosystems (GEOHAB 2010).

Many investigations in China indicate that phycotoxin contamination of shellfish is a potential threat to the safety of seafood. Phycotoxins detected from the shellfish samples include both hydrophilic toxins, like paralytic shellfish toxins (PSTs), and lipophilic toxins, like diarrhetic shellfish toxins (DSTs). In the coastal waters of China, most of the PSTs in shellfish are associated with the toxic species in the genus *Alexandrium* (Yu et al. 2014). There has been an increased understanding of the diversity of *Alexandrium* spp. present in the coastal waters of China over the last several decades. The recorded species in the genus *Alexandrium* include *A. tamarense* species complex (two morphological species *A. tamarense* and *A. catenella*), *A. minutum*, *A. affine*, *A. ostenfeldii*, *A. andersonii*, *A. fraterculum*, *A. leei*, *A. pseudogonyaulax*, *A. tamutum*, etc. (Gu et al. 2013b; Yu et al. 2014). Among these species, *A. tamarense* species complex, *A. minutum*, and *A. ostenfeldii* are capable of producing PSTs. Among them, the *A. tamarense* species complex is the major one responsible for PSTs (Gao et al. 2015). Besides toxic species in the genus *Alexandrium*, another species of PST-producing dinoflagellates, *G. catenatum*, has been isolated from the Yellow Sea of China (Gu et al. 2013b). The strains of *G. catenatum* can produce 11-hydroxysulfate benzoate analogs and N-sulfocarmaboyl analogs of PSTs. The significance of *G. catenatum* in PST contamination of seafood is not fully understood.

Lipophilic toxins often detected in the contaminated seafood in China include okadaic acid (OA), azaspiracid (AZA), pectenotoxin (PTX), yessotoxins (YTX), gymnodimine (GYM), and spirolide (SPX) (Wu et al. 2015). Knowledge on the toxic algae associated with lipophilic toxins has continually increased in the last decade. Strains of *Azadinium poporum* producing AZA-2 were isolated from the China seas (Gu et al. 2013a), and YTX-producing strains of *Protoceratium reticulatum* were also established. The toxin profiles of *Dinophysis caudata*, *D. acuminata*, and *D. rotundata* were elucidated recently based on the LC-MS analysis of handpicked cells (Luo et al. 2014; Li et al. 2015). Based on the analysis, *D. caudata* was primarily identified as the toxic algae that led to a poisoning incident due to the consumption of cultivated mussels (*Mytilus galloprovincialis*) in May 2011.

14.3 HAB Genera and Events Recently Documented in Asia

Although harmful algae, at least the causative agents of fish kills and shellfish/fish poisoning, have previously been identified in Asia (GEOHAB 2010; Fukuyo et al. 2011), blooms of unrecorded harmful red tide species in Asia have caused mass mortalities of cultured fish. The planktonic dinoflagellate *Karlodinium australe* in Johor Strait in Malaysia (Lim et al. 2014) and *Cochlodinium polykrikoides* in Lampung Bay, Indonesia (Thoha et al. 2015), are such cases. Since previous distributions of these harmful species were not fully recognized, recent anthropogenic introduction of these dinoflagellates is difficult to demonstrate, but new evidence is accumulating. For several HAB species, their presence, various outbreaks, and their effects have been well characterized. These species are summarized in this section.

14.3.1 Karlodinium australe

Since the establishment of blooms of the unarmoured dinoflagellate genera, *Karenia*, *Karlodinium,* and *Takayama* (Daugbjerg et al. 2000; de Salas et al. 2003), more than 20 species of these genera have been described; some are reported as fish-killing red tides, and others are recognized as potentially toxic. In Asia, *K. mikimotoi* of the genus *Karenia*, *Karlodinium micrum*, and *K. veneficum* have mainly been reported. *K. veneficum* and *K. australe* were recently reported to cause massive blooms that were attributed to fish kill events. *Karlodinium veneficum* was observed in samples collected from the Johor Strait (Lim et al. 2014; Leong et al. 2015; Teng et al. 2016). Blooms of *K. australe* were found along the Johor Strait, in Southern Peninsular Malaysia (Fig. 14.2), as well as in Aman Island, Malacca Strait.

The first massive fish kills in Asia associated with *K. australe* occurred in 2014 in Malaysia (Lim et al. 2014). Recurrence of *K. australe* blooms was observed in the following year with even greater mariculture losses (Teng et al. 2016). During

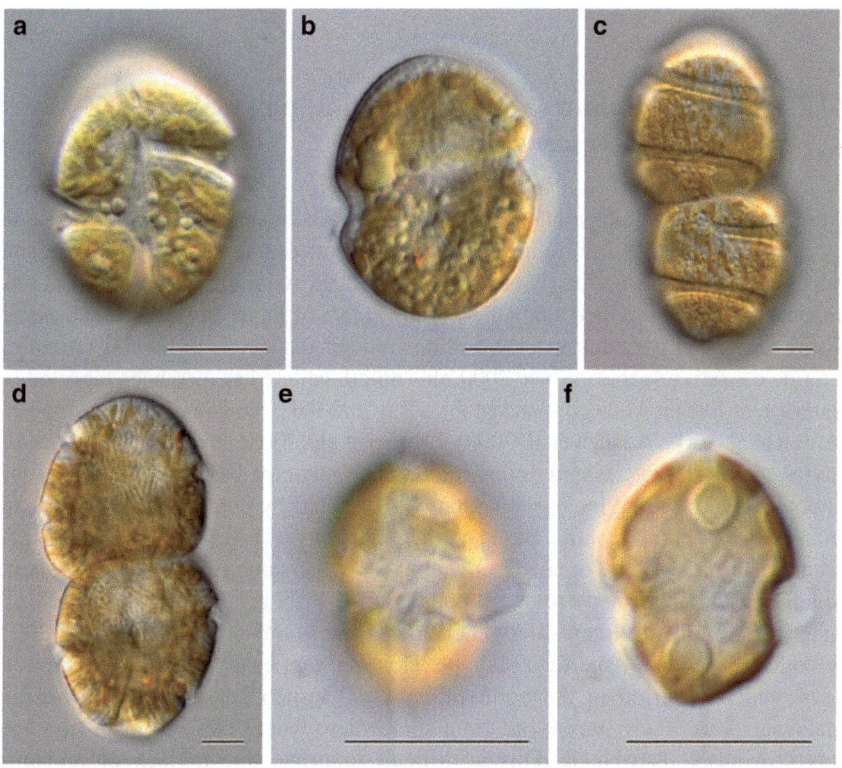

Fig. 14.2 Cells of *Karlodinium australe* (**a**, **b**), *Cochlodinium polykrikoides* (**c**, **d**), *Azadinium poporum* (**e**, **f**). *Scale bars* = 10 μm. Photos courtesy of Kazuya Takahashi

the blooms of *K. australe* in Johor Strait in both February 2014 and March 2015, water discolouration was observed during neap tides. The highest cell densities recorded during the incident were 2.34×10^6 cells L^{-1} in the 2014 bloom and even higher densities during the recurrence in 2015, 8×10^6 cells L^{-1} (Lim et al. 2014; Teng et al. 2016). The blooms not only caused massive fish mortality of farmed fish, but wild fish and crustaceans were also affected.

Fish affected by *Karlodinium* show various symptoms including suffocation, weakness, whitish spots on the scales, and cloudy eyes. The ichthyotoxin from *K. veneficum*, karlotoxin (KmTxs), was discovered after the fish kill events in the USA (Kempton et al. 2002). Karlotoxin was found to cause damage to fish gill epithelial tissues (Deeds and Place 2006; Place et al. 2012). Fish necropsies from the Johor Strait events and preliminary fish experiments showed that the fish died in a similar manner to those reported for *Karlodinium* toxins (Teng et al. 2016). Continued analytical characterization of the toxin associated with the fish kill events in Johor Strait will provide further insight into the fish kill mechanism. The bloom of *K. australe* in Malaysia is envisaged to be related to dredging work conducted in the area that promotes the resuspension of sediment. Release of nutrient-rich water into the water column tends to favour initiation and development of blooms.

14.3.2 Cochlodinium (Margalefidinium) polykrikoides *and* C. fulvescens

The unarmoured dinoflagellate, *C. polykrikoides,* is known as a harmful red tide species related to mass mortalities of fish mainly in coastal areas of East Asia and North and Central America (Iwataki et al. 2008). In Asia, many cases of fisheries impacts have been reported along the western coasts of Japan and southern coasts of Korea associated with this species, and it is recognized as one of the most harmful red tide causative species. Since 2000, this dinoflagellate has also been reported in waters near Sabah, Malaysia, and Manila Bay and Palawan Island, the Philippines (Anton et al. 2008; Azanza et al. 2008). Iwataki et al. (2008) analysed the molecular phylogeny of the species and differentiated their genotype based on LSU rDNA as ribotypes, whereas the local population was detected by comparison of sampling localities of *C. polykrikoides*. Based on the phylogeny of LSU rDNA, red tides that occurred in Japan and Korea were recognized as the same ribotype (East Asian), which was distinguished from the population distributed in Central and North America (American). Interestingly, cells occurring in Sabah, Malaysia, after 2000 belonged to the latter American ribotype. This fact implies that *C. polykrikoides* red tides, which occurred in the northern coast of Borneo Island including Sabah, Malaysia, were likely introduced from somewhere where the American ribotype was previously distributed. This is difficult to prove due to lack of previous information in Sabah coastal waters. Subsequently, *C. polykrikoides* of the American ribotype was also reported in the Arabian Gulf region, UAE, and Oman, during 2008–2009 (Richlen et al. 2010).

In October 2012, a large red tide of *C. polykrikoides* was first found in Lampung Bay, Indonesia, associated with mass mortalities of cultured fish (Thoha et al. 2015; Fig. 14.2). Red tides of *C. polykrikoides* have continuously occurred in Lampung Bay at least through 2015, and species composition of phytoplankton, especially for harmful species, has been monitored by local researchers in the last two decades. The species had not been detected until the first occurrence in 2012 (Muawanah, pers. comm.). Considering the continuous occurrence and detection of immotile cells, possibly from *C. polykrikoides* in Lampung Bay, this species has already adapted to the environment and settled in this area. The environmental conditions suitable for settlement of *C. polykrikoides* should be investigated to consider future distribution expansion in tropical Asian regions.

A species related to this harmful alga, *Cochlodinium fulvescens,* has recently been found and described from western Kyushu, Japan (Iwataki et al. 2007). Fortunately, fish kills have not been reported associated with *C. fulvescens* in the Asian coasts despite their occurrence in Indonesia, Japan, and Vietnam. This species is a planktonic unarmoured dinoflagellate that forms cell chains, therefore superficially resembling *C. polykrikoides* under light microscope. However, it possesses a shallow sulcus at the middle of two cingula in the dorsal view and granular chloroplasts by which the species can be distinguished from *C. polykrikoides* (Iwataki et al. 2007). An apical groove of this species was recently revealed by SEM, which resembles an eyespot pore and a connection to a sulcal extension that differs in the constricted middle part of the groove from *C. polykrikoides* (Iwataki et al. 2015).

14.3.3 Pseudo-nitzschia *spp.*

Species responsible for ASP, *Pseudo-nitzschia* spp., are common in Asian coasts, and species identifications have recently been confirmed based on molecular analysis and ultrastructural observation. Among all 46 described species of *Pseudo-nitzschia,* 17 species are known to produce DA (Trainer et al. 2012; Dao et al. 2014, 2015; Teng et al. 2014), and of those 46, 36 species of *Pseudo-nitzschia* have been recorded from China, Japan, Korea, Malaysia, the Philippines, Russia, Thailand, and Vietnam (Lim et al. 2012, 2014; Teng et al. 2014; Dao et al. 2015). *Pseudo-nitzschia sinica* was described from Dapeng Bay, China (Qi et al. 1994). It consists of unique frustule characters with a higher range of cell widths (3.7–4.6 μm) and a higher number of 10 μm striae (15–17) compared to other species in the *P. pseudodelicatissima* complex sensu lato. *Pseudo-nitzschia micropora* was the next species to be described from Asia, first discovered from Phuket Island, Thailand (Prisholm et al. 2002). The species *P. multiseries* was the first toxic species to have been proven to occur in the Sea of Japan, East Asia (Kotaki et al. 1999). After a decade, the number of toxic species increased to four with three new species: *P. kodamae*, *P.* cf. *caciantha,* and *P. fukuyoi* (Dao et al. 2014, 2015; Teng et al. 2014). In recent years, 15 potentially toxic species have been reported in Asia (Dao et al. 2014; Lim et al. 2013; Trainer et al. 2012; Teng et al. 2014).

Although at present large-scale blooms of *Pseudo-nitzschia* associated with ASP have never been reported in the Asian region. Seasonal blooms of *Pseudo-nitzschia* have been reported in the Sea of Japan, with *P. calliantha*, *P. multiseries*, *P. multistriata*, and *P. pungens* blooms observed during the summer and autumn in Peter the Great Bay, Russia (Stonik et al. 2011). Studies on bivalves conducted in Vietnam confirmed that they were contaminated by DA (Dao et al. 2009). However, there is no evidence that *Pseudo-nitzschia* was the main source of DA contamination in the shellfish. Later, two species, *P.* cf. *caciantha* and *P. fukuyoi*, from the area were proven toxic by culture studies (Dao et al. 2014, 2015).

In recent years, the diversity of *Pseudo-nitzschia* spp. has been extensively examined, especially isolates from the coast of Malaysia (Fig. 14.3). Through the results of ultrastructural observation and molecular comparison, five species have

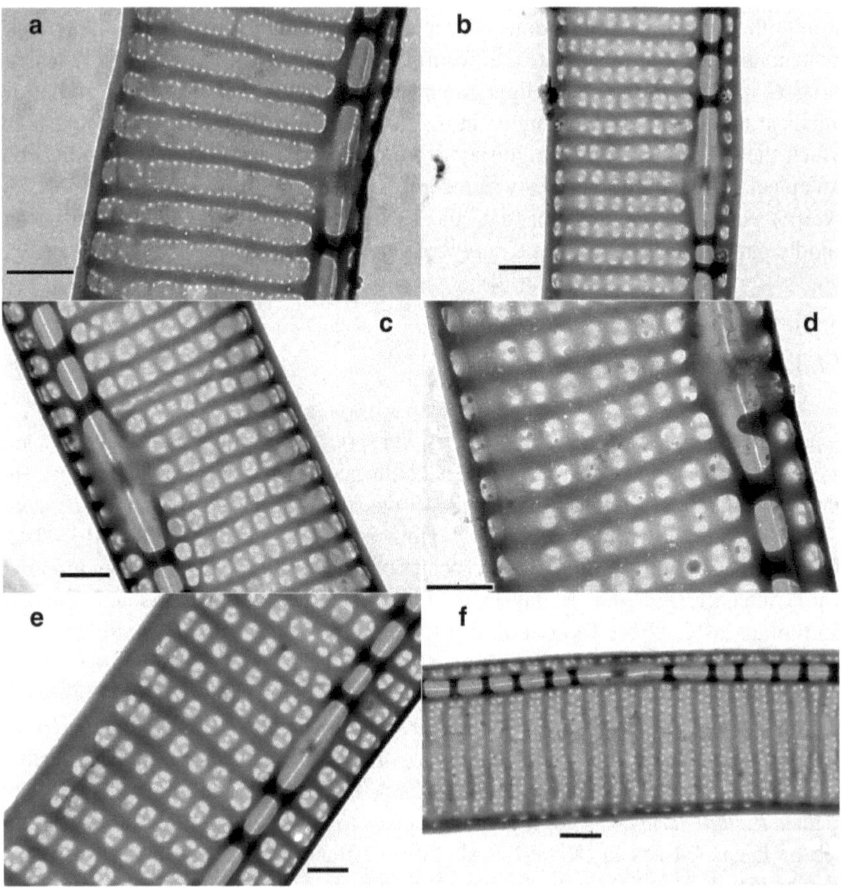

Fig. 14.3 *Pseudo-nitzschia* species recently described in Malaysian waters. *P. circumpora* (**a**), *P. batesiana* (**b**), *P. fukuyoi* (**c**), *P. lundholmiae* (**d**), *P. kodamae* (**e**), and *P. sabit* (**f**). *Scale bars* = 0.5 μm

been newly described and two more undescribed morphotypes were detected. *Pseudo-nitzschia circumpora* was described from Kuching, Borneo (Lim et al. 2012). It consists of a unique perforation in the poroids as compared to that in other *Pseudo-nitzschia* species. Genetically it is closer to *P. caciantha*. The presence of toxins in *P. circumpora* has yet to be confirmed. Lim et al. (2013) described three other species, *P. batesiana*, *P. fukuyoi*, and *P. lundholmiae* from Malacca Strait. Cells of these species have lower ranges of 10 μm striae (28–34) and a lower range of divided sector density (1–3 sectors) than species of the *P. pseudodelicatissima* complex sensu lato. *P. fukuyoi* and *P. lundholmiae* were closely related to *P. pseudodelicatissima* and *P. cuspidata*, whereas *P. batesiana* is the sister species to *P. circumpora*. Strains of *P. fukuyoi* from Malaysia were found to be nontoxic, whereas strains from Vietnam were toxic (Dao et al. 2015). The toxin content of *P. batesiana* and *P. lundholmiae* remains to be tested. Teng et al. (2014) reported a toxic species, *P. kodamae*, from Port Dickson, Malacca Strait. The species was morphologically similar to *P. caciantha* but can be distinguished by the characteristic of lower densities of striae in 10 μm (22–29). It is phylogenetically related to *P. hasleana*, and this species was highly toxic (Teng et al. 2014). *Pseudo-nitzschia sabit* was also described from Port Dickson, Malacca Strait (Teng et al. 2015). The species has a distinctive falcate cell valve, from which they form sicklelike colonies. It morphologically and genetically resembles *P. decipiens*. Strains of the species have tested nontoxic (Teng et al. 2015). These sequential taxonomic studies on *Pseudo-nitzschia* are revealing the species richness of *Pseudo-nitzschia* in Asian waters.

A comprehensive taxonomic database of the valid species of *Pseudo-nitzschia* and an interactive key for species identification have been developed (Teng et al. 2014; http://dmitriev.speciesfile.org/key.asp?key=Bacillariales&lng=En&i=1&keyN=2) (Fig. 14.4). This interactive key is designed to assist researchers in identifying *Pseudo-nitzschia* species based on the morphometric and ultrastructural data of all published species. It also compiles the distribution of all species of *Pseudo-nitzschia* based on more than 2000 published studies on *Pseudo-nitzschia* species.

14.3.4 Aureococcus anophagefferens

Brown tides are ecosystem-disruptive algal blooms (EDABs) in shallow estuaries, which have negative impacts on wild and cultured shellfish and result in loss of seagrass beds (Gobler and Sunda 2012). The negative effects were originally reported from the US east coast in 1985 and from Texas in 1990. The causative species of the former was subsequently described as *A. anophagefferens* and the latter as *Aureoumbra lagunensis*, both of which are tiny planktonic pelagophytes, approximately 2 μm in diameter.

In China, *A. anophagefferens* was recently identified as the major causative species of the blooms occurring in the coastal waters of Qinhuangdao in the Bohai Sea based on the results of pigment analysis and partial rDNA gene sequencing

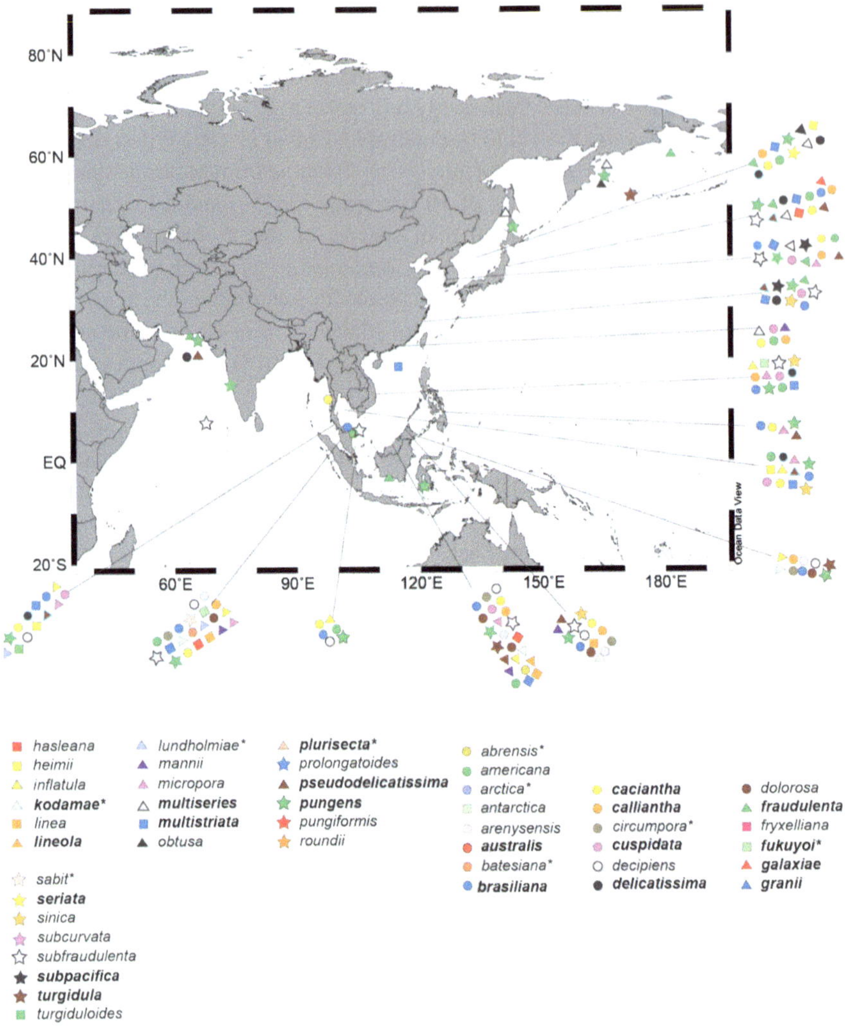

Fig. 14.4 Distribution of *Pseudo-nitzschia* species in Asia

(Kong et al. 2012; Zhang et al. 2012). Since the first record in 2009, brown tides in the Bohai Sea have generally started in late May and are sustained until the middle of July [see also Chap. 15, Yu et al. (2018)]. During brown tides, seawater is coloured dark brown, and the maximum cell abundance can reach 10^9 cells L^{-1} (Zhang et al. 2012). The coastal waters of Qinhuangdao, China, are an important mariculture zone for scallops *Argopecten irradians*. The occurrence of brown tides in this region has led to tremendous economic loss due to the inhibitory effects on the growth of scallops. The direct economic loss of the scallop mariculture industry in 2010 alone was estimated to be 0.2 billion RMB (Bulletin of Chinese Marine

Disasters 2010). China is the third country experiencing brown tides of *A. anophagefferens* in the world, besides the USA and South Africa (Gobler et al. 2005; Probyn et al. 2010; Zhang et al. 2012). From 2009 to 2015, brown tides have occurred consecutively in this region with varying scales and intensity.

14.3.5 Azadinium *spp.*

The thecate planktonic dinoflagellate genus *Azadinium* was first detected in the North Sea, with the species *A. spinosum* found to be the producer of azaspiracid (AZA), a group of lipophilic toxins (Tillmann et al. 2009). After the discovery of the causative species of azaspiracid shellfish poisoning (AZP), a total of ten species have been reported for the genus. Distribution of *Azadinium* species has also been reported at the species level. In East Asia, *A. poporum* and related species have been reported in Chinese and Korean waters (Fig. 14.2). *Azadinium dalianense* has also been described from cultures obtained from sediment samples of the Yellow Sea (Luo et al. 2013). A culture isolated from a sediment sample from Shiwha Bay, Korea, was identified as *Azadinium* cf. *poporum*, and cultures from sediments in the Bohai Sea were likewise identified as *A. poporum* (Potvin et al. 2012; Gu et al. 2013a). The AZA toxins have been detected, but the presence of *Azadinium* cells has not reported from Japan.

Chinese strains of *Azadinium poporum* produce diverse AZA profiles. The cell quota of AZA can reach a maximum of 23 fg cell^{-1} (Gu et al. 2013a). This species has been recorded in the South China Sea, the East China Sea, as well as the Bohai Sea (Gu et al. 2013a; Lu et al. 2014).

Another species, *A. dalianense*, was first described from the Yellow Sea of China (Luo et al. 2013). It possesses a plate formula of Po, cp, X, 4′, 3a, 6″, 6C, 5S, 6‴, 2⁗. *A. dalianense* shares identical hypothecal, cingular, and sulcal plates with the other *Azadinium* species, but it is unique in having only three apical and two anterior intercalary plates. Phylogenetic analyses revealed that *A. dalianense* was nested within *Azadinium* and formed a strongly supported clade with *A. poporum*. Liquid chromatography-mass spectrometry analyses have not detected any known AZAs (Luo et al. 2013).

14.3.6 Vulcanodinium rugosum

The dinoflagellate, *Vulcanodinium rugosum*, has been cultured from individual noncalcareous cysts isolated from surface sediment samples collected in Okinawa, Japan, and Guangxi, China (Smith et al. 2011; Zeng et al. 2012). Japanese *V. rugosum* strains produce pinnatoxin G (Smith et al. 2011), whereas Chinese strains produce pinnatoxin H (Selwood et al. 2014). Japanese *V. rugosum* strains are genetically close to those of Hawaii, but Chinese strains are closer to those of

New Zealand. Pinnatoxins were responsible for shellfish poisoning in Guangdong in 1990 (Zheng et al. 1990).

14.3.7 Alexandrium *spp.*

In Southeast Asia, a high diversity of *Alexandrium* has been recorded through the years (Fukuyo et al. 2011). In Vietnam, at least 15 *Alexandrium* species have been documented (Larsen and Nguyen 2004). Malaysia has reported eight species, with *A. taylori*, *A. peruvianum,* and *A. tamutum* as the most recently identified (Lim et al. 2005; Kon et al. 2013). For the Philippines, a total of eight *Alexandrium* species has been recorded as well [Azanza unpubl.; see also Chap. 8, Azanza et al. (2018)].

Taxonomic characterization of representatives from the genus *Alexandrium* has become more significant as PSP cases in the Philippines have been implicated with rising numbers of toxic *Alexandrium* blooms. At least five *Alexandrium* species were previously reported in the Philippines, i.e., *A. affine*, *A. fraterculus*, *A. leei*, *A. tamiyavanichii*, and *A. minutum* (Baula et al. 2011). Several additional *Alexandrium* species have been more recently observed, namely, *A. leei*, *A. pseudogonyaulax*, and *A. minutum*, the last two being new reports in Manila Bay (Azanza unpubl.).

The overall distribution of *Alexandrium* in the Philippines, an archipelagic country with several embayments, is still relatively unknown except in Manila Bay and Bolinao-Anda, Pangasinan. The species *A. minutum* has been considered as the main culprit of PSP cases reported in 2011 and 2013 in the Philippines. Those cases occurred even with the issuance of shellfish bans in five coastal areas of Bataan, as monitored by the Bureau of Fisheries and Aquatic Resources (BFAR). A maximum cell density of 6.4×10^5 cells L^{-1} was measured with a toxin level of 439 μg STX eq per 100 mL in water samples, based on HPLC analysis. The species *A. pseudogonyaulax* and *A. leei* have also been observed at very low concentrations (2–10 cells L^{-1}), but *A. minutum* has been found at very high cell densities during sampling in November 2013. Bolinao-Anda, Pangasinan, located in the northwestern Philippines, is another coastal area where reports of PSP cases have been associated with toxic *Alexandrium* blooms (Azanza and Benico 2013). Occurrence of *A. minutum* in this area has been confirmed (Baula et al. 2011; Azanza and Benico 2013).

14.4 Additional Newly Described or Documented HAB Species from Asia

In recent years, in addition to the species and events described above, several new HAB-causing microalgae have been described from Asian waters. These species are described in this section. Species belonging to the genera *Cochlodinium*

(Margalefidinium), *Pseudo-nitzschia*, and *Azadinium* are documented in the previous section.

14.4.1 Takayama xiamenensis

The dinoflagellate, *Takayama xiamenensis*, was recently described from Xiamen Harbor, East China Sea (Gu et al. 2013a, b). The vegetative cells are characterized by a deep sigmoid apical groove and a sulcus extending into the epicone as a distinctive fingerlike protrusion. Epifluorescence microscopy revealed a large, cup-shaped nucleus, filling most of the epicone. The chloroplasts are enveloped by two membranes, and some chloroplasts possessed large, bulging pyrenoids (Gu et al. 2013a). *T. xiamenensis* has been reported to bloom in Singaporean coastal waters (as *T. acrotrocha*; Tang et al. 2012).

14.4.2 Coolia malayensis

The genus *Coolia*, a fish-poisoning benthic dinoflagellate, is commonly found in diverse coastal ecosystems, e.g., coral reefs, seaweed beds, rocky shores, etc. Currently, there are seven species in the genus, with the harmful species being *Coolia monotis*. Studies on *Coolia* were intensified after 2000, mainly focused on the discovery of new species, i.e., *Coolia tropicalis*, *Coolia canariensis*, *Coolia areolata*, *C. malayensis*, *Coolia palmyrensis,* and *Coolia santacroce* (Karafas et al. 2015). Toxicity in one of the species (later identified as *C. tropicalis*, Mohammad-Noor et al. 2013) was confirmed with the presence of YTX-like compounds, cooliatoxins.

Coolia malayensis, first discovered from seaweed samples from the Malacca Strait, Malaysia, is widely distributed in the Pacific and Atlantic Oceans (Leaw et al. 2016). This species is generally distributed within the Pacific region, whereas its close sister species, *C. monotis*, is distributed mainly in the Atlantic and Mediterranean regions (Leaw et al. 2016). These two species can be distinguished both morphologically and with rDNA markers (Leaw et al. 2016; Rhodes et al. 2014). Its distribution across Asia is yet to be determined.

14.4.3 Gambierdiscus scabrosus

The dinoflagellate, *Gambierdiscus toxicus*, the type species of the genus, was first described from Gambier Islands, French Polynesia (Adachi and Fukuyo 1979). Species in the genus generally possess lenticular shape and are anterior-posteriorly compressed (Litaker et al. 2009). *Gambierdiscus scabrosus* was described from the southern part of Japan (Nishimura et al. 2014). It was previously grouped in *Gambierdiscus* sp. type 1. Morphology of *G. scabrosus* resembles *G. belizeanus* with a narrow 2'''' and areolated surface. The distinguishable features of *G. scabrosus* are the presence of the asymmetric shaped 3'' and the rectangular shaped 2' (Nishimura et al. 2014). The distribution of this species is also yet to be described for Asia.

14.4.4 Alexandrium pohangense

Despite the extensive taxonomic studies of the dinoflagellate, *Alexandrium*, from temperate to tropical Asia, a new phototrophic dinoflagellate, *Alexandrium pohangense*, was recently discovered and described from the coastal area of Pohang, South Korea, where it was named after the location (Lim et al. 2015). This species can be distinguished from species in the sister clade, *Alexandrium margalefii*, by the wide rectangular first apical plate that does not touch the second apical plate. Being a newly described species, the general distribution of this species in Asia is generally unknown, and it is unclear whether it is limited to temperate Asian regions.

14.5 Conclusions

The GEOHAB Programme has made a significant contribution to develop our understanding of HABs in Asian waters. Both types of HAB events, those caused by the toxin producers and the high-biomass formers, occur in the region. The frequency and severity of HAB events are increasing, and furthermore, their distribution within the region appears to be expanding. While research achievements of GEOHAB are major in terms of science, it should be emphasized that GEOHAB, in particular, GEOHAB Asia, has developed networks among scientific communities in the region enhancing international collaboration on HAB research. The serious impacts of HABs on both coastal ecosystems and human activities in Asia including aquaculture, public health, and coastal management have been well recognized, and various efforts towards countermeasures and mitigations against the negative impacts are being undertaken at regional, national, and local levels. Our knowledge acquired through the GEOHAB Programme and the Asian network are expected to enhance activities to tackle HAB issues in the region together with other intergovernmental activities including IOC-WESTPAC, NOWPAP, and PICES.

Acknowledgments We thank Wilson Mendoza, Takuo Omura, and Kazuya Takahashi for providing their original photos. This study was financially supported by a Grant-in-Aid of MEXT, Japan (24121001).

References

Adachi R, Fukuyo Y (1979) The thecal structure of a marine toxic dinoflagellate *Gambierdiscus toxicus* gen. et sp. nov. collected in a ciguatera-endemic area. Bull Jpn Soc Sci Fish 45:67–71

Anton A, Teoh PL, Mohd-Shaleh SR et al (2008) First occurrence of *Cochlodinium* blooms in Sabah, Malaysia. Harmful Algae 7:331–336

Azanza RV, Benico G (2013) Toxic *Alexandrium* blooms in fish farming sites in Bolinao, Pangasinan. J Environ Sci Manag Special Issue 1:44–49

Azanza RV, Brosnahan ML, Anderson DM et al (2018) The role of life cycle characteristics in harmful algal bloom dynamics. In: Glibert PM, Berdalet E, Burford M et al (eds) Global ecology and oceanography of harmful algal blooms. Springer, Cham, pp 133–161

Azanza RV, David LT, Borja RT et al (2008) An extensive *Cochlodinium* bloom along the western coast of Palawan, Philippines. Harmful Algae 7:324–330

Azanza RV, Fukuyo Y, Yap-Dejeto LG et al (2004) *Prorocentrum minimum* bloom and its possible link to a massive fish kill in Bolinao, Pangasinan, Northern Philippines. Harmful Algae 4:519–524

Baula IU, Azanza RV, Fukuyo Y et al (2011) Dinoflagellate cyst composition, abundance and horizontal distribution in Bolinao, Pangasinan, Northern Philippines. Harmful Algae 11:33–44

Bulletin of Chinese Marine Disasters, SOA (2010) http://www.soa.gov.cn/zwgk/hygb/zghyzhgb/2010nzghyzhgb/201212/t20121207_21453.html (in Chinese)

Dao VH, Lim PT, Ky PX et al (2014) Diatom *Pseudo-nitzschia* cf. *caciantha* (Bacillariophyceae), the most likely source of domoic acid contamination in the thorny oyster *Spondylus versicolor* Schreibers 1793 in Nha Phu Bay, Khanh Hoa Province, Vietnam. Asian Fish Sci 27:16–29

Dao VH, Phan BV, Teng ST et al (2015) *Pseudo-nitzschia fukuyoi* (Bacillariophyceae), a domoic acid-producing species from Nha Phu Bay, Khanh Hoa Province, Vietnam. Fish Sci 81:533–539

Dao VH, Takata Y, Omura T et al (2009) Seasonal variation of domoic acid in a bivalve *Spondylus versicolor* in association with that in plankton samples in Nha Phu Bay, Khanh Hoa, Vietnam. Fish Sci 75:507–512

Daugbjerg N, Hansen G, Larsen J et al (2000) Phylogeny of some of the major genera of dinoflagellates based on ultrastructure and partial LSU rDNA sequence data, including the erection of three new genera of unarmoured dinoflagellates. Phycologia 39:302–317

de Salas MF, Bolch CJS, Botes L et al (2003) *Takayama* gen. nov. (Gymnodiniales, Dinophyceae), a new genus of unarmored dinoflagellates with sigmoid apical grooves, including the description of two new species. J Phycol 39:1233–1246

Deeds JR, Place AR (2006) Sterol specific membrane interactions with the toxins from *Karlodinium micrum* (Dinophyceae) – a strategy for self-protection? Afr J Mar Sci 28:421–427

Fukuyo Y, Kodama M, Omura T et al (2011) Ecology and oceanography of harmful marine microalgae. In: Nishida S, Fortes MD, Miyazaki N (eds) Coastal marine science in Southeast Asia. TerraPub, Tokyo, pp 23–48

Gao Y, Yu R, Murray SA et al (2015) High specificity of a quantitative PCR assay targeting a saxitoxin gene for monitoring toxic algae associated with paralytic shellfish toxins in the yellow sea. Appl Environ Microbiol 81:6973–6981

GEOHAB (2010) GEOHAB Asia, global ecology and oceanography of harmful algal blooms in Asia: a regional comparative programme. Furuya K, Glibert PM, Zhou M et al (eds) IOC and SCOR, Baltimore and Paris, 69 pp

Gobler CJ, Lonsdale DJ, Boyer GL (2005) A synthesis and review of causes and impact of harmful brown tide blooms caused by the alga, *Aureococcus anophagefferens*. Estuaries 28:726–749

Gobler CJ, Sunda WG (2012) Ecosystem disruptive algal blooms of the brown tide species, *Aureococcus anophagefferens* and *Aureoumbra lagunensis*. Harmful Algae 14:36–45

Goes JI, Gomes HR, Al-Hashimi K et al (2018) Ecological drivers of green *Noctiluca* blooms in two monsoonally driven ecosystems. In: Glibert PM, Berdalet E, Burford M et al (eds) Global ecology and oceanography of harmful algal blooms. Springer, Cham, pp 327–336

Gomes HDR, Goes JI, Matondkar SGP et al (2014) Massive outbreaks of *Noctiluca scintillans* blooms in the Arabian Sea due to spread of hypoxia. Nat Commun 5:4862. https://doi.org/10.1038/ncomms5862
Gu H, Luo Z, Krock B et al (2013a) Morphology, phylogeny and azaspiracid profile of *Azadinium poporum* (Dinophyceae) from the China Sea. Harmful Algae 21–22:64–75
Gu H, Zeng N, Liu T et al (2013b) Morphology, toxicity, and phylogeny of *Alexandrium*, (dinophyceae) species along the coast of china. Harmful Algae 27:68–81
Harrison PJ, Furuya K, Glibert PM et al (2011) Geographical distribution of red and green *Noctiluca scintillans*. Chin J Oceanol Limnol 29:807–831
Horiguchi T (1995) *Heterocapsa circularisquama* sp. nov. (Peridiniales Dinophyceae): a new marine dinoflagellate causing mass mortality of bivalves in Japan. Phycol Res 43:129–136
Iwataki M, Kawami H, Mizushima K et al (2008) Phylogenetic relationship in the harmful dinoflagellate *Cochlodinium polykrikoides* (Gymnodiniales, Dinophyceae) inferred from LSU rDNA sequences. Harmful Algae 7:271–277
Iwataki M, Kawami H, Matsuoka K (2007) *Cochlodinium fulvescens* sp. nov. (Gymnodiniales, Dinophyceae), a new chain-forming unarmored dinoflagellate from Asian coasts. Phycol Res 55:231–239
Iwataki M, Takayama H, Takahashi K et al (2015) Taxonomy and distribution of the unarmored dinoflagellates *Cochlodinium polykrikoides* and *C. fulvescens*. In: Ohtsuka S, Suzaki T, Horiguchi T et al (eds) Marine protists: diversity and dynamics. Springer, Berlin, pp 551–565
Iwataki M, Wong MW, Fukuyo Y (2002) New record of *Heterocapsa circularisquama* (Dinophyceae) from Hong Kong. Fish Sci 68:1161–1163
Karafas S, York R, Tomas C (2015) Morphological and genetic analysis of the *Coolia monotis* species complex with the introduction of two new species, *Coolia santacroce* sp. nov. and *Coolia palmyrensis* sp. nov. (Dinophyceae). Harmful Algae 46:18–33
Kempton JW, Lewitus AJ, Deeds JR et al (2002) Toxicity of *Karlodinium micrum* (Dinophyceae) associated with a fish kill in a South Carolina brackish retention pond. Harmful Algae 1:233–241
Kon NF, Hii KS, Tan TH et al (2013) First record of marine dinoflagellate, *Alexandrium tamutum* (Dinophyceae) from Malaysia. Malays J Sci 32:81–88
Kong F, Rencheng YU, Zhang Q et al (2012) Pigment characterization for the 2011 bloom in Qinhuangdao implicated "brown tide" events in china. Chin J Oceanol Limnol 30:361–370
Kotaki Y, Koike K, Sato S et al (1999) Confirmation of domoic acid production of *Pseudo-nitzschia multiseries* isolated from Ofunato Bay, Japan. Toxicon 37:677–682
Larsen J, Nguyen NL (2004) Potentially toxic microalgae of Vietnamese waters. Opera Bot 14:5–216
Leaw CP, Tan TH, Lim HC et al (2016) New scenario for speciation in the benthic dinoflagellate genus *Coolia* (Dinophyceae). Harmful Algae 55:137–149
Lefebvre KA, Powell CL, Busman M et al (1999) Detection of domoic acid in northern anchovies and California sea lions associated with an unusual mortality event. Nat Toxins 7:85–92
Leong SCY, Lim LP, Chew SM et al (2015) Three new records of dinoflagellates in Singapore's coastal waters, with observations on environmental conditions associated with microalgal growth in the Johor Straits. Raffles Bull Zool 31:24–36
Li A, Sun G, Qiu J et al (2015) Lipophilic shellfish toxins in *Dinophysis caudata* picked cells and in shellfish from the east china sea. Environ Sci Pollut Res 22:3116–3126
Lim AS, Jeong HJ, Kim JH et al (2015) Description of the new phototrophic dinoflagellate *Alexandrium pohangense* sp. nov. from Korean coastal waters. Harmful Algae 46:49–61
Lim HC, Leaw CP, Tan TH et al (2014) A bloom of *Karlodinium australe* (Gymnodiniales, Dinophyceae) associated with mass mortality of cage-cultured fishes in West Johor Strait, Malaysia. Harmful Algae 40:51–62
Lim HC, Teng ST, Leaw CP et al (2013) Three novel species in the *Pseudo-nitzschia pseudodelicatissima* complex: *P. batesiana* sp. nov., *P. lundholmiae* sp. nov. and *P. fukuyoi* sp. nov. (Bacillariophyceae), from the Strait of Malacca, Malaysian. J Phycol 49:902–916

Lim PT, Usup G, Leaw CP et al (2005) First report of *Alexandrium taylori* and *Alexandrium peruvianum* (Dinophyceae) in Malaysia waters. Harmful Algae 4:391–400

Lim PT, Usup G, Leaw CP (2012) Harmful algal blooms in Malaysian waters. Sains Malays 41:1509–1515

Lin JN, Yan T, Zhang QC et al (2014) In situ detrimental impacts of *Prorocentrum donghaiense* blooms on zooplankton in the East China Sea. Mar Pollut Bull 88:302–310

Litaker RW, Vandersea MW, Faust MA et al (2009) Taxonomy of *Gambierdiscus* including four new species, *Gambierdiscus caribaeus*, *Gambierdiscus carolinianus*, *Gambierdiscus carpenteri* and *Gambierdiscus ruetzleri* (Gonyaulacales, Dinophyceae). Phycologia 48:344–390

Liu D, Zhou MJ (2018) Green tides of the Yellow Sea: massive free-floating blooms of *Ulva prolifera*. In: Glibert PM, Berdalet E, Burford M et al (eds) Global ecology and oceanography of harmful algal blooms. Springer, Cham, pp 317–326

Lu D, Goebel J, Qi Y et al (2005) Morphological and genetic study of *Prorocentrum donghaiense* Lu from the East China Sea, and comparison with some related *Prorocentrum* species. Harmful Algae 4:493–505

Lu D, Qi Y, Gu H et al (2014) Causative species of harmful algal blooms in chinese coastal waters. Biocontrol Sci Technol 145:145–168

Luo X, Yu R, Zhou M (2014) Analysis of toxins in cells of *Dinophysis acuminata* collected from the coastal waters of Qingdao with a LC-MS method. Mar Environ Sci 33:781–787

Luo Z, Gu H, Krock B et al (2013) *Azadinium dalianense*, a new dinoflagellate species from the Yellow Sea, China. Phycologia 52:625–636

Matsuyama Y (2012) Impacts of the harmful dinoflagellate *Heterocapsa circularisquama* bloom on shellfish aquaculture in Japan and some experimental studies on invertebrates. Harmful Algae 14:144–155

Mendoza C, Rabanes AC, Jimenez EC et al (2013) Detection of ciguatera fish poisoning in the Philippines. J Environ Sci Manag 1:50–55

Mohammad-Noor N, Moestrup Ø, Lundholm N et al (2013) Autecology and phylogeny of *Coolia tropicalis* and *Coolia malayensis* (Dinophyceae), with emphasis on taxonomy of *C. tropicalis* based on light microscopy, scanning electron microscopy and LSU rDNA. J Phycol 49:536–545

Nishimura T, Sato S, Tawong W et al (2014) Morphology of *Gambierdiscus scabrosus* sp. nov. (Gonyaulacales): a new epiphytic toxic dinoflagellate from coastal areas of Japan. J Phycol 50:506–514

Okaichi T (2004) Red-tide phenomena. In: Okaichi T (ed) Red Tides. Springer, pp 7–60

Place AR, Bowers HA, Bachvaroff TR et al (2012) *Karlodinium veneficum* – the little dinoflagellate with a big bite. Harmful Algae 14:179–195

Potvin É, Jeong HJ, Kang NS et al (2012) First report of the photosynthetic dinoflagellate genus *Azadinium* in the Pacific Ocean: morphology and molecular characterization of *Azadinium* cf. *poporum*. J Eukaryotic Microbiol 59:145–146

Prisholm K, Moestrup O, Lundholm N (2002) Taxonomic notes on the marine diatom genus *Pseudo-nitzschia* in the Andaman Sea near the island of Phuket, Thailand, with a a description of *Pseusdo-nitzschia micropora* sp. nov. Diatom Res 17:153–175

Probyn TA, Bernard S, Pitcher GC et al (2010) Ecophysiological studies on *Aureococcus anophagefferens*, blooms in Saldanha Bay, South Africa. Harmful Algae 9:123–133

Qi Y, Wang Y, Zheng L (1994) The taxonomy and bloom ecology of Pseudo-nitzschia on the coasts of China. In Nontji A, Soemdihardjo S, Iladude AG et al (eds) Proceedings of IOC-WESTPAC third international scientific symposium. Indonesian Institute of Science, Bali, pp 88–95

Qi YZ, Chen JF, Wang ZH et al (2004) Some observations on harmful algal bloom (HAB) events along the coast of Guangdong, southern China in 1998. Hydrobiologia 512:209–214

Rhodes L, Smith K, Gimenez Papiol G et al (2014) Epiphytic dinoflagellates in sub-tropical New Zealand, in particular the genus *Coolia* Meunier. Harmful Algae 34:36–41

Richlen ML, Morton SL, Jamali EA et al (2010) The catastrophic 2008–2009 red tide in the Arabian Gulf region, with observations on the identification and phylogeny of the fish-killing dinoflagellate *Cochlodinium polykrikoides*. Harmful Algae 9:163–172
Selwood AI, Wilkins AL, Munday R et al (2014) Pinnatoxin H: a new pinnatoxin analogue from a South China Sea *Vulcanodinium rugosum* isolate. Tetrahedron Lett 55:5508–5510
Smith KF, Rhodes LL, Suda S et al (2011) A dinoflagellate producer of pinnatoxin G, isolated from sub-tropical Japanese waters. Harmful Algae 10:702–705
Stonik IV, Orlova TY, Lundholm N (2011) Diversity of *Pseudo-nitzschia* H. Peragallo from the western North Pacific. Diatom Res 26:121–134
Tang YZ, Kong L, Morse RE et al (2012) Report of a bloom-forming dinoflagellate *Takayama acrotrocha* from tropical coastal waters of Singapore. J Phycol 48:455–466
Teng ST, Leaw CP, Lau WLS et al (2016) Recurrence of the harmful dinoflagellate *Karlodinium australe* along the Johor Strait. Harmful Algal News 52:5
Teng ST, Lim HC, Lim PT et al (2014) *Pseudo-nitzschia kodamae* sp. nov. (Bacillariophyceae), a toxigenic species from the Strait of Malacca, Malaysia. Harmful Algae 34:17–28
Teng ST, Lim PT, Lim HC et al (2015) A non-toxigenic but morphologically and phylogenetically distinct new species of *Pseudo-nitzschia, P. Sabit* sp. nov. (Bacillariophyceae). J Phycol 51:706–725
Thoha H, Rachman A, Sidabutar A et al (2015) First record of the harmful dinoflagellate *Cochlodinium polykrikoides* in Lampung Bay, Indonesia. Harmful Algae News 50:14–15
Tillmann U, Elbrächter M, Krock B et al (2009) *Azadinium spinosum* gen. et sp. nov. (Dinophyceae) identified as a primary producer of azaspiracid toxins. Eur J Phycol 44:63–79
Trainer VL, Bates SS, Lundholm N et al (2012) *Pseudo-nitzschia* physiological ecology, phylogeny, toxicity, monitoring and impacts on ecosystem health. Harmful Algae 14:271–300
White AW, Anraku M, Hooi K-K (1984) Toxic red tides and shellfish toxicity in Southeast Asia. In: Proceedings of a consultative meeting held in Singapore, 11–14 September 1984, Southeast Asian Fisheries Development Center and International Development Research Centre, p 133
Wu H, Yao J, Guo M et al (2015) Distribution of marine lipophilic toxins in shellfish products collected from the chinese market. Mar Drugs 13:4281–4295
Yu R-C, Gao Y, Luo X et al (2014) Toxic algae and phycotoxins in the coastal waters of China. In: Sun S, Adrianov AV, Lutaenko KA et al (eds) Marine biodiversity and ecosystem dynamics of the Northwest Pacific Ocean. Science Press, Beijing, pp 92–107
Yu R-C, Lü S-H, Liang Y-B (2018) Harmful algal blooms in the coastal waters of China. In: Glibert PM, Berdalet E, Burford M et al (eds) Global ecology and oceanography of harmful algal blooms. Springer, Cham, pp 309–316
Zeng N, Gu H, Smith KF et al (2012) The first report of *Vulcanodinium rugosum* (Dinophyceae) from the South China Sea with a focus on the life cycle. N Z J Mar Freshw Res 46:511–521
Zhang Q-C, Qiu L-M, R-C Y et al (2012) Emergence of brown tides caused by *Aureococcus anophagefferens* Hargraves et Sieburth in China. Harmful Algae 19:117–124
Zheng S, Huang F, Chen S et al (1990) The isolation and bioactivities of pinnatoxin. Chin J Mar Drugs 9:33–35

Chapter 15
Harmful Algal Blooms in the Coastal Waters of China

Ren-Cheng Yu, Song-Hui Lü, and Yu-Bo Liang

15.1 Introduction: Current Situation of HABs in the Coastal Waters of China

Harmful algal blooms are a natural ecological phenomenon caused by fast reproduction or accumulation of algae, which may have disastrous economic or ecological consequences, such as mass mortality of marine animals, seafood contamination, alteration of marine food webs, or deterioration of tourism industries. Typical HAB events include not only high-biomass blooms ("red tides" or "brown tides") formed by microalgae but also "green tides" caused by some macroalgal species [see also Chap. 16, Liu and Zhou (2018)]. In China, harmful algal blooms (HABs) are probably the most severe marine ecological problem. Under the support of Ministry of Science and Technology (MoST), National Natural Science Foundation of China (NSFC), Chinese Academy of Sciences (CAS), and State Ocean Administration (SOA), conducted under the umbrella of the Chinese Ecology and Oceanography of HABs (CEOHAB) Programme, a component of GEOHAB Asia [see also Chap. 14, Furuya et al. (2018)], many studies have been carried out to investigate the

R.-C. Yu (✉)
CAS Key Laboratory of Marine Ecology and Environmental Sciences, Institute of Oceanology, Chinese Academy of Sciences, Qingdao, China

University of Chinese Academy of Sciences, Beijing, China

Laboratory for Marine Ecology and Environmental Science, Qingdao National Laboratory for Marine Science and Technology, Qingdao, China
e-mail: rcyu@qdio.ac.cn

S.-H. Lü
Jinan University, Guangzhou, China

Y.-B. Liang
National Marine Environmental Monitoring Center, Dalian, China

P.M. Glibert et al. (eds.), *Global Ecology and Oceanography of Harmful Algal Blooms*, Ecological Studies 232, https://doi.org/10.1007/978-3-319-70069-4_15

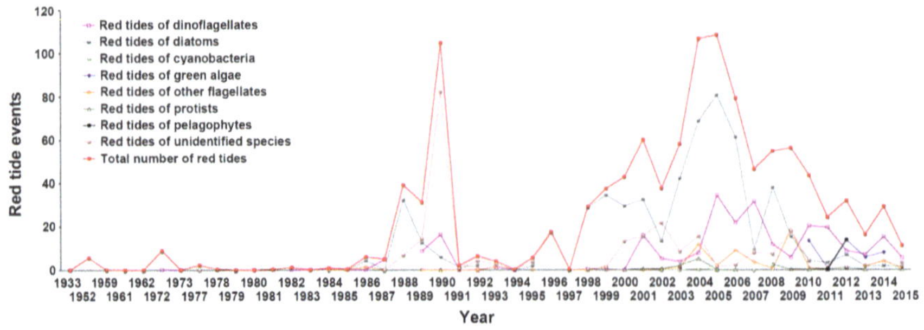

Fig. 15.1 Different red tide events recorded in the coastal waters of China (data from National Marine Environmental Monitoring Center, State Oceanic Administration)

mechanisms and impacts of HABs and to develop monitoring methods, control measures, and prevention strategies against HABs [see also Chap. 21, Yu et al. (2018)].

An increasing number of HAB events have been recorded in the coastal waters of China during the last two decades (Fig. 15.1). Before the 1990s, the recorded number of HABs (red tides with negative impacts) was quite limited. Most of the red tides were caused by nontoxic diatoms and affected a relatively small area of the Chinese marine waters. Since the year 2000, the recorded number of HAB events every year has reached 50–80. Many HABs have lasted for more than a month and have covered large areas up to thousands of square kilometres. In addition, the causative species of HABs also exhibited significant changes. The HABs are generally found in the coastal waters of the Bohai Sea, the East China Sea, and the South China Sea, and each sea area has its representative HAB problems. Green tides caused by green alga *Ulva* (*Enteromorpha*) *prolifera* have mainly taken place in the southern Yellow Sea and are described in more detail in Liu and Zhou (2018, Chap. 16).

15.1.1 Red Tides of Dinoflagellates in the East China Sea

In the East China Sea, the area adjacent to the Changjiang River estuary is the most notable region for red tide outbreaks (Zhou et al. 2001). Intensive red tides of dinoflagellates were first observed in this region at the beginning of the twenty-first century. Since then the red tides have become a recurrent phenomenon. The red tides of dinoflagellates, which can last from early May to the middle June each year, affect an area up to 10,000 km^2. The major causative species of the red tides, *Prorocentrum donghaiense* and *Karenia mikimotoi*, are harmful both to the aquaculture industry and to marine ecosystems. For instance, a red tide of *K. mikimotoi* in 2005 led to mass mortality of cultured fish in the sea area around Nanji Island, and the economic loss was estimated to be 30 million RMB (nearly US$4,500,000). In 2012, another red tide of *K. mikimotoi* destroyed the abalone aquaculture industry along the coast of Fujian Province, and the economic loss was around 2 billion RMB (nearly US$300,000,000).

The red tides of dinoflagellates in the East China Sea have been intensively studied during the last 15 years. The distribution pattern and dynamics of *P. donghaiense* red tides have been described, and the over-enrichment of nitrate from the Changjiang River has been shown to be a major reason for these dinoflagellate red tides (Zhou and Zhu 2005; Zhou et al. 2008). The ability of *P. donghaiense* to use organic forms of phosphorus (P) has been found to be an important adaptive strategy, allowing it to compete with rapid-growing diatoms under P-limiting conditions (Ou et al. 2008). Based on the role of nutrients in the formation of dinoflagellate red tides, nitrogen (N) reduction has been suggested as an important strategy to control the blooms in the long term (Zhou et al. 2008). Some questions, however, remain and will need further study in the future.

Although dinoflagellate red tides are generally dominated by *P. donghaiense*, in some years, the dominant species changes into fish-killing *K. mikimotoi*. The mechanism for this change in dominant species is still poorly understood. Furthermore, while the dinoflagellate red tides in the East China Sea are affected by nitrate transported from the Changjiang River, they are also affected by phosphate from the Taiwan warm current through upwelling [see also Chap. 12, Glibert et al. (2018a)]. The role of the Taiwan warm current and its variability in the formation of large-scale dinoflagellate blooms still need to be further examined to have a better understanding on the mechanisms causing dinoflagellate red tides.

15.1.2 Red Tides of **Phaeocystis globosa** *in the South China Sea*

The coastal waters along the South China Sea are also a notable region for red tides in China. Before the late 1990s, most of the red tides recorded in the South China Sea were confined in a relatively small region in bays such as Dapeng Bay, Daya Bay, and Shenzhen Bay. Large-scale red tides were seldom reported until the late 1990s, when an extensive red tide occurred in Zhelin Bay, Quangdong Province (Chen et al. 1999). This red tide caused mass mortality of cultured fish, and the economic loss was estimated to be 70 million RMB (about $10,000,000). This species can form giant gelatinous colonies in the coastal waters of China. The size of the colonies, with hundreds of cells embedded in a polysaccharide matrix (mucus), varies from several millimetres to over 2–3 cm (Shen et al. 2000). Since then, red tides of *P. globosa* have become a recurrent phenomenon, with significant interannual variation in scale and intensity. From the year 2010, intensive red tides of *P. globosa* started to appear in Beibu Gulf, an isolated area in the northwestern part of the South China Sea previously considered free of HABs. Red tides of *P. globosa* occurred in northern Beibu Gulf in 2014 and 2015 and posed potential threats to the operation of cooling systems of a nuclear power plant. Up to now, research on red tides of *Phaeocystis* sp. has been generally limited to specific biological studies, and knowledge on the ecology and oceanography of *P. globosa* in this region is still quite limited.

15.1.3 Brown Tides in the Bohai Sea and Yellow Sea

The Bohai Sea, a semi-enclosed inland sea, is another notable region for red tides in China. In the 1980s and 1990s, intensive red tides of dinoflagellates *Noctiluca scintillans*, *Prorocentrum* spp. and *Ceratium furca* were reported (Song et al. 2016). In 2009, a unique HAB started to appear in the coastal waters of Qinhuangdao, with significant negative impacts on the aquaculture industry of scallops, *Argopecten irradians*. Based on the pigment composition of phytoplankton samples collected during the bloom and the results of molecular clone library analysis, the major bloom-forming species was determined to be the pelagophyte, *Aureococcus anophagefferens* (Kong et al. 2012a; Zhang et al. 2012), the causative species of brown tides along the east coast of the USA. Henceforth, blooms of this species have become a recurrent phenomenon in this region. In 2010, the bloom affected an area over 3000 km^2 and led to economic loss of 200 million RMB (about US $30,000,000). This was the first brown tide event recorded in the coastal waters of the Pacific Ocean. The presence of *A. anophagefferens* has since been documented in other studies (Zhen et al. 2016; Xu et al. 2017). Besides the coastal waters of Qinhuangdao, a similar "brown tide" event was recorded in the Sanggou Bay of the northern Yellow Sea in 2011 (Kong et al. 2012b).

Brown tides have been considered to be ecosystem disruptive algal blooms (EDABs, e.g., Sunda et al. 2006). The emergence of brown tides in the east coast of the US led to reduction in shellfish resources and degeneration of seagrass beds. The Bohai Sea and the Yellow Sea are important spawning and nursing grounds and host many shellfish aquaculture areas. The consequences of the brown tides in the Bohai Sea and Yellow Sea attracted much attention. However, the reason why brown tides suddenly appeared in the Bohai Sea is still not known.

15.1.4 Toxic Algal Blooms in the Coastal Waters of China

Toxic algal blooms, as well as their associated seafood contamination issues, are another big concern in China. Blooms of toxic dinoflagellate, *Alexandrium* spp. associated with paralytic shellfish poisoning (PSP) have been reported in the northern Yellow Sea, the Changjiang River estuary and its adjacent waters, and some bays in the South China Sea (Gao et al. 2015). Blooms of toxic dinoflagellate, *Dinophysis* spp. have been reported in the Bohai Sea, the East China Sea, and the South China Sea. Recently, many toxic dinoflagellate species were identified, such as *Alexandrium fundyense*, *A. pacificum*, *Gymnodinium catenatum*, *Dinophysis acuminata, and D. fortii*, and toxin production characteristics have been studied, based on cyst germination experiment or single-cell analysis (Gu et al. 2013a, b; Li et al. 2015).

Many phycotoxins, including saxitoxin (STX) and its analogues, okadaic acid (OA), dinophysis toxins (DTXs), pectenotoxins (PTXs), yessotoxins (YTXs), azaspiracids (AZA), gymnodimines (GYM), and spirolides (SPX), have been detected in shellfish samples collected along the coast of China (Li et al. 2017; Liu et al. 2017). Some poisoning events have also been reported. In 2008, seven people were poisoned and one died in Jiangsu Province, due to the consumption of contaminated clams by paralytic shellfish toxins (PST). In May 2011, more than 200 people were poisoned in Zhejiang Province and Fujian Province by eating mussels with high levels of OA and DTXs (Li et al. 2012). In April 2016, another poisoning incident in Qinhuangdao was reported, due to the consumption of PST-contaminated mussels.

15.2 Evolution of HABs in the Coastal Waters of China

From the 1970s to now, HABs in China have undergone significant changes. First, the recorded bloom-forming species have been becoming more and more diversified. There have been blooms of diatoms, dinoflagellates, haptophytes, cyanobacteria, cryptophytes, raphidophytes, and pelagophytes (Fig. 15.1). Some bloom-forming species, such as *A. anophagefferens*, were never reported before recent years. Second, the emerging blooms have shifted from diatoms to flagellates that tend to be "harmful". Toxic or harmful algae have occupied an increasing proportion in the bloom-forming species over the last decade (Fig. 15.2). The dinoflagellate red tides in the East China Sea, brown tides in the Bohai Sea, and *Phaeocystis* red tides in the South China Sea all have had major impacts on the aquaculture industry and on the health of natural ecosystems. *Prorocentrum donghaiense*, although nontoxic, has been shown to exhibit strong inhibitory effects on reproduction of the keystone zooplankton species, *Calanus sinicus* in the East China Sea (Lin et al. 2014). Third, the cell size of bloom-forming species on average exhibits a trend of getting smaller. Cell size of some bloom-forming species, such as *Phaeocystis* spp. and *A. anophagefferens*, is only 2–8 μm. This makes it difficult to carry out routine monitoring and identification of harmful algal blooms based on traditional microscopic approaches.

HAB development in China is driven largely by land-based pollution and climate change. Many studies have suggested that the occurrence of HABs is closely related with coastal eutrophication caused by nutrient pollution (Zhou et al. 2008; Li et al. 2014). With rapid urbanization of coastal zones and increasing fertilizer use in large river basins, huge amounts of N and P are discharged into the sea in different forms, aggravating the coastal eutrophication problem [see also Chaps. 4, 12, Glibert et al. (2018a, b)]. For example, the Changjiang River estuary and its adjacent waters are characterized by high concentration of NO_3- and high N/P and N/Si ratios. In the coastal waters of Bohai Sea, however, the organic forms

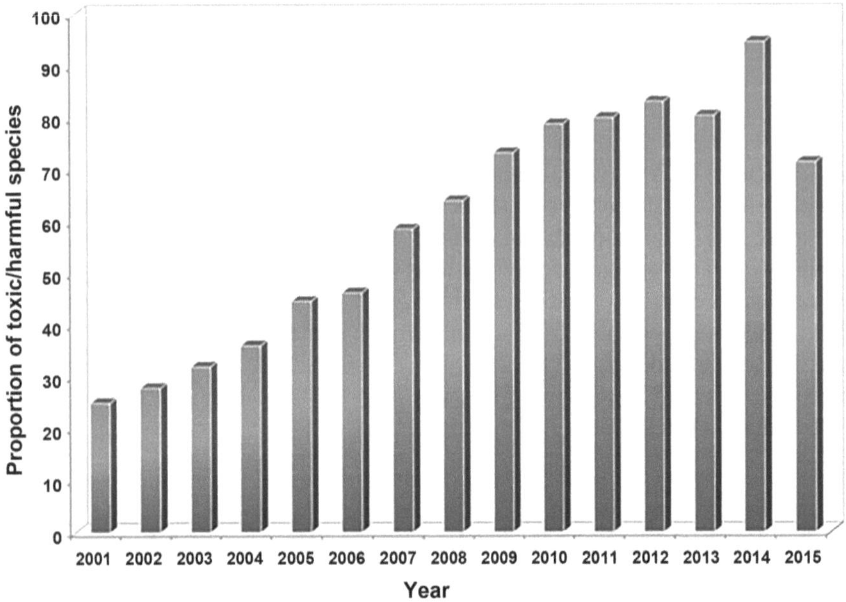

Fig. 15.2 Proportion of toxic/harmful species in recorded bloom-forming microalgal species in the coastal waters of China

of N are more prominent. This will lead to the proliferation of different groups of microalgae with distinct adaptive strategies. Climate change is another important factor likely affecting HABs. The rise in temperatures in the coastal waters of China is significant. Strong regional climate events, such as El Niño and La Niña, may alter oceanic currents and precipitation on land, affecting the occurrence of HABs [see also Chap. 5, Wells and Karlson (2018)]. Unfortunately, knowledge of this aspect is still very limited.

15.3 Strategies to Prevent HABs in the Coastal Waters of China

To support sound measures for managing HABs and to develop effective prevention strategies against HABs, there is an increasing need to carry out fundamental studies on the underlying mechanisms of HAB events, including the diversity of bloom-forming microalgal species, the adaptive features and competitive strategies of bloom-forming species, the major factors driving the evolution of HABs, and the relationship between HABs and the health of marine ecosystems. Furthermore, continued research on the economic and ecological consequences of HABs is urgent.

In addition, techniques for observing and predicting HABs should be emphasized. The diversification of bloom-forming species makes it a considerable challenge to detect bloom formers at the early stages. Therefore, it is necessary to develop sensitive and specific in situ monitoring facilities based on molecular biology, microscopic examination, and analytical chemistry. The development of numerical models for specific HABs will also increase the ability of early warning and prediction.

Finally, there is an urgent need for effective control technologies and prevention strategies for HABs. At present, modified clay has been successfully applied to control some HAB events in China [see Chap. 21, Yu et al. (2018)]. However, considering the intensive scale of most HABs occurred in China, measures to control land-based pollution and coastal eutrophication are more important to prevent HABs in the long run.

References

Chen JF, Xu N, Jiang TJ et al (1999) A report of *Phaeocystis globosa* bloom in coastal water of southeast China. J Jinan Univ 20(3):124–129 (in Chinese, with English abstract)

Furuya K, Iwataki M, Lim PT et al (2018) Overview of harmful algal blooms in Asia. In: Glibert PM, Berdalet E, Burford M et al (eds) Global ecology and oceanography of harmful algal blooms. Springer, Cham, pp 289–308

Gao Y, Yu RC, Zhang QC et al (2015) Distribution of *Alexandrium fundyense* and *A. pacificum* (Dinophyceae) in the Yellow Sea and Bohai Sea. Mar Pollut Bull 96:210–219

Glibert PM, Al-Azri A, Allen JI et al (2018a) Key questions and recent research advances on harmful algal blooms in relation to nutrients and eutrophication. In: Glibert PM, Berdalet E, Burford M et al (eds) Global ecology and oceanography of harmful algal blooms. Springer, Cham, pp 229–259

Glibert PM, Beusen AHW, Harrison JA et al (2018b) Changing land, sea- and airscapes: sources of nutrient pollution affecting habitat suitability for harmful algae. In: Glibert PM, Berdalet E, Burford M et al (eds) Global ecology and oceanography of harmful algal blooms. Springer, Cham, pp 53–76

Gu HF, Liu TT, Vale P et al (2013a) Morphology, phylogeny and toxin profiles of *Gymnodinium inusitatum* sp. nov., *Gymnodinium catenatum* and *Gymnodinium microreticulatum* (Dinophyceae) from the Yellow Sea, China. Harmful Algae 28:97–107

Gu HF, Zeng N, Liu TT et al (2013b) Morphology, toxicity, and phylogeny of *Alexandrium* (Dinophyceae) species along the coast of China. Harmful Algae 27:68–81

Kong FZ, Yu RC, Zhang QC et al (2012a) Pigment characterization for the 2011 bloom in Qinhuangdao implicated "brown tide" events in China. Chin J Limnol Oceanogr 30(3):361–370

Kong FZ, Yu RC, Zhang QC et al (2012b) Primary analyses on the causative species of a bloom in the Sanggou Bay. Mar Environ Sci 31(6):824–829 (in Chinese, with English abstract)

Li AF, Ma JJ, Cao JJ et al (2012) Toxins in mussels (*Mytilus galloprovincialis*) associated with diarrhetic shellfish poisoning episodes in China. Toxicon 60:420–425

Li AF, Sun G, Qiu JB et al (2015) Lipophilic shellfish toxins in *Dinophysis caudata* picked cells and in shellfish from the East China Sea. Environ Sci Pollut Res 22:3116–3126

Li HM, Tang HJ, Shi XY et al (2014) Increased nutrient loads from the Changjiang (Yangtze) River have led to increased harmful algal blooms. Harmful Algae 39:92–101

Li MH, Sun G, Qiu JB et al (2017) Occurrence and variation of lipophilic shellfish toxins in phytoplankton, shellfish and seawater samples from the aquaculture zone in the Yellow Sea, China. Toxicon 127:1–10

Lin JN, Yan T, Zhang QC et al (2014) *In situ* detrimental impacts of *Prorocentrum donghaiense* blooms on zooplankton in the East China Sea. Mar Pollut Bull 88:302–310

Liu D, Zhou M (2018) Green tides of the Yellow Sea: massive free-floating blooms of *Ulva prolifera*. In: Glibert PM, Berdalet E, Burford M et al (eds) Global ecology and oceanography of harmful algal blooms. Springer, Cham, pp 317–326

Liu Y, Yu RC, Kong FZ et al (2017) Lipophilic marine toxins discovered in the Bohai Sea using high performance liquid chromatography coupled with tandem mass spectrometry. Chemosphere 183:380–388

Ou LJ, Wang D, Huang BQ et al (2008) Comparative study of phosphorus strategies of three typical harmful algae in Chinese coastal waters. J Plankton Res 30(9):1007–1017 (in Chinese, with English abstract)

Shen PP, Wang Y, Qi YZ et al (2000) Growth characteristics and life cycle of *Phaeocystis globosa* Scherffel. Acta Hydrobiol Sin 24(6):635–643 (in Chinese, with English abstract)

Song NQ, Wang N, Lu Y et al (2016) Temporal and spatial characteristics of harmful algal blooms in the Bohai Sea during 1952–2014. Cont Shelf Res 122(1):77–84

Sunda WG, Graneli E, Gobler CJ (2006) Positive feedback and the development and persistence of ecosystem disruptive algal blooms. J Phycol 42(5):963–974

Wells ML, Karlson B (2018) Harmful algal blooms in a changing ocean. In: Glibert PM, Berdalet E, Burford M et al (eds) Global ecology and oceanography of harmful algal blooms. Springer, Cham, pp 77–90

Xu X, Yu ZM, Cheng FJ et al (2017) Molecular diversity and ecological characteristics of the eukaryotic phytoplankton community in the coastal waters of the Bohai Sea, China. Harmful Algae 61:13–22

Yu Z, Song X, Cao X, Liu Y (2018) Mitigation and control of harmful algal blooms. In: Glibert PM, Berdalet E, Burford M et al (eds) Global ecology and oceanography of harmful algal blooms. Springer, Cham, pp 403–423

Zhang QC, Qiu LM, Yu RC et al (2012) Emergence of brown tides caused by *Aureococcus anophagefferens* Hargraves et Sieburth in China. Harmful Algae 19:117–124

Zhen Y, Qiao L, Gu B et al (2016) Characteristics of eukaryotic microalgal community and its abiotic influencing factors during brown tide blooms near Qinhuangdao, China. Harmful Algae 57:1–12

Zhou MJ, Shen ZL, Yu RC (2008) Responses of a coastal phytoplankton community to increased nutrient input from the Changjiang (Yangtze) River. Cont Shelf Res 28:1483–1489

Zhou MJ, Zhu MY (2005) Progress of the project "Ecology and oceanography of harmful algal blooms in China". Adv Earth Sci 21(7):673–679 (in Chinese, with English abstract)

Zhou MJ, Zhu MY, Zhang J (2001) Status of harmful algal blooms and related research activities in China. Chin Bull Life Sci 13(2):54–59 (in Chinese, with English abstract)

Chapter 16
Green Tides of the Yellow Sea: Massive Free-Floating Blooms of *Ulva prolifera*

Dongyan Liu and Mingjiang Zhou

16.1 Introduction

Marine and estuarine ecosystems are undergoing various environmental pressures ranging from anthropogenic nutrients to climate change, which have resulted in a series of negative consequences, including biodiversity loss, ecosystem function deterioration, species outbreaks, and many other changes (Duarte 2009). Macroalgal blooms, an expanding worldwide phenomenon, are a form of HABs that frequently occurs along many coasts and estuaries. The macroalgal blooms formed by green algae [e.g., *Ulva* (*Enteromorpha*), *Chaetomorpha*, and *Cladophora*] are usually referred as "green tides" (Fletcher 1996). Most studies have demonstrated that increased nutrient loads, most often derived from land and human activities, are the dominant causes for these macroalgal blooms (Lapointe 1997; Valiela et al. 1997). In oligotrophic coral reefs and rocky shores, reduced grazing pressure related to overfishing has also been identified as being involved in a cascading effect controlling algal biomass and buffering the impact of nutrient enrichment (Burkepile and Hay 2006). Macroalgal blooms can produce a number of undesirable effects on marine and estuarine ecosystems, e.g., reducing benthic diversity and abundance, replacing seagrass meadows and algal beds, impacting the health of coral reefs, killing commercial shellfish and fish, releasing hydrogen sulphide (H_2S) and ammonia (NH_3) with harmful effects on fauna and humans during decomposition, and

D. Liu (✉)
State Key Laboratory of Estuarine and Coastal Research, East China Normal University, Shanghai, China
e-mail: dyliu@sklec.ecnu.edu.cn

M. Zhou
Key laboratory of Marine Ecology and Environmental Sciences, Institute of Oceanology, Chinese Academy of Sciences, Qingdao, China

P.M. Glibert et al. (eds.), *Global Ecology and Oceanography of Harmful Algal Blooms*, Ecological Studies 232, https://doi.org/10.1007/978-3-319-70069-4_16

stimulating phytoplankton blooms after decomposition (Valiela et al. 1992, 1997; Raffaelli et al. 1998; Nelson et al. 2008). Recent studies revealed that *Ulva* is an efficient carrier of toxic organic pollutants and heavy metals, and a bloom can greatly increase the risk of mid-trophic-level consumers and pose a health threat to humans (Cheney et al. 2014).

Generally, macroalgal blooms are confined within estuaries or coastal bays, but in recent years, massive free-floating macroalgal blooms have appeared, including two extraordinary cases, the "green tide" caused by *Ulva prolifera* in the Yellow Sea, China, since 2007 (Liu et al. 2013; Zhou et al. 2015), and the "golden tide" caused by *Sargassum* spp. in the Atlantic Ocean, USA (Gower and King 2011). Such blooms can cover thousands of square kilometres, have trans-regional or transnational impacts, require expenditure of billions of dollars for cleanup and emergency responses, and create large financial losses in aquaculture and tourism industries (Liu et al. 2013; Smetacek and Zingone 2013). Therefore, it is important to understand much more about this novel phenomenon, e.g., the mechanisms stimulating these massive macroalgal blooms, the species involved, and their ecological consequences. In this chapter, we summarize current knowledge on the massive free-floating macroalgal blooms in the Yellow Sea, China, and present the scientific questions for future research related to the ecological and economic impacts, aiming to provide sufficient information for understanding this important phenomenon.

16.2 Green Tides in the Yellow Sea

In summer of 2007, China Ocean News gave the first report on the free-floating green tide in the Yellow Sea (China Ocean News 2007, http://epaper.oceanol.com/zghyb/20070720/index.htm>). The bloom had been detected using satellite images and had reached a sea coverage of approximately 82 km^2 (Keesing et al. 2011). That report did not arouse much attention until late June 2008 when millions of tons of algal biomass blocked the waters and shores that were being used for the Olympic sailing events in Qingdao (Fig. 16.1). Eventually, over 16,000 people and 600 boats were involved in the algal cleanup to guarantee that the Olympic Games events could proceed and the safety of coastal activities. This involved removal of more than 1 million tons of green algae from the coast (Zhou et al. 2015). Costs for the cleanup and the emergency response were estimated at about 200 million RMB (approximately equal to US$30 million), and the economic losses for marine aquaculture industries and tourism were more than 500 million RMB (approximately equal to US$71 million) (Ye et al. 2011). In the following 8 successive years, green tides with a magnitude of at least a million tons of biomass and a coverage of thousand square kilometres have reoccurred every summer in the Yellow Sea (Table 16.1), and it is hard to see this situation changing in the near future. Since 2008, numerous reports and scientific articles related to the onset of green tides in the Yellow Sea have been published discussing the causative species, environmental mechanisms, and proposed mitigating policies. Here, we summarize the known information on this unique phenomenon.

Fig. 16.1 Green tides in Qingdao coast (from Zhou et al. 2015)

Table 16.1 Records of green tides in the Yellow Sea during the summer of 2008–2015

Year	Maximum distribution (km^2)	Maximum coverage (km^2)	Location of early-bloom occurrence (Jiangsu coastline)	Date of bloom formation	Date of bloom disappearance	Bloom duration (days)
2008	25,000	540	Yancheng	May	September	110
2009	58,000	2100	Yancheng	May	August	94
2010	29,800	530	Yancheng	May	August	76
2011	26,400	560	Yancheng	May	August	82
2012	19,160	267	Yancheng	May	August	106
2013	29,733	790	Yancheng	May	August	96
2014	50,000	540	Yancheng	May	August	95
2015	52,700	594	Yancheng	May	August	93

Data sources are from the Bulletin of Marine Environmental Status of China (2008–2015) published by the State Oceanic Administration People's Republic of China

Although the initial search for the cause of the green tide during these events focused on the coastal eutrophication of the Qingdao environment, and the action of tides and winds in bringing the algae ashore (e.g., Sun et al. 2008), a series of satellite images clearly demonstrated that the massive green tide actually formed in a broad regional area across the southern Yellow Sea (Fig. 16.2a–e). Small floating green algal patches were found to have initiated in the coast of Jiangsu province

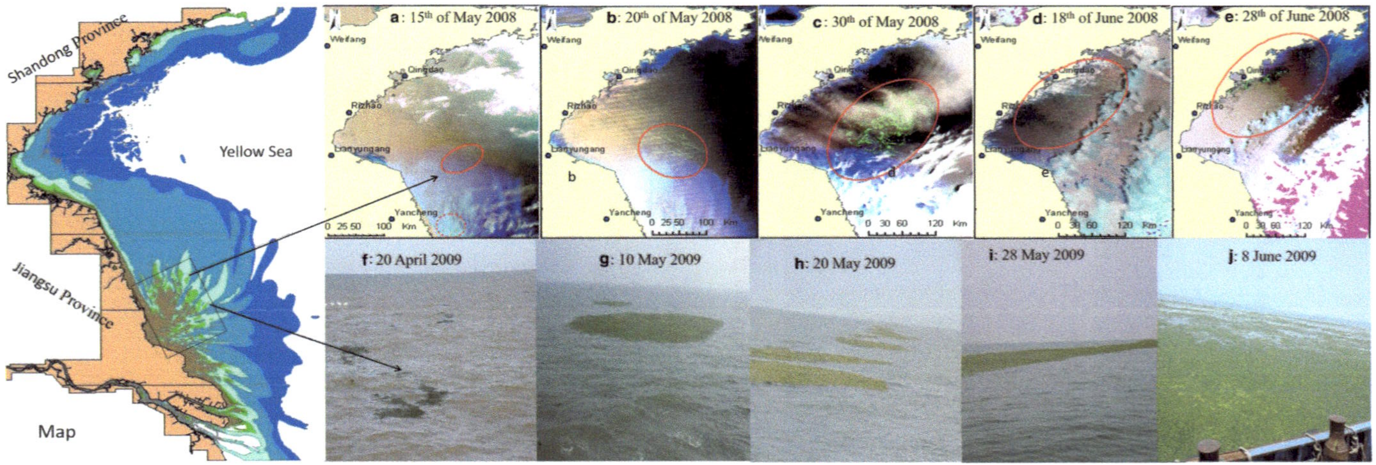

Fig. 16.2 Satellite and field observations of green tide formation in the Yellow Sea (Map, showing the initial location of *Ulva* source (black square: Subei Shoal) along the Jiangsu coastline; (**a–e**), satellite images during May to June of 2008 showing the trajectory of the green tides in the Yellow Sea initially originated from the Jiangsu coast and drifted to Shandong Province over time (the images are cited from Liu et al. 2009); (**f–j**), field observations during April–June of 2009 showing the development and aggregation of free-floating algal mats drifting from the coast of Subei Shoal to the offshore area (from Liu et al. 2010)

near to Subei Shoal (Yancheng) in early May (Table 16.1; Fig. 16.2f–h). In the process of drifting during May to July of 2008, these small patches aggregated and grew rapidly (Fig. 16.2i–j), producing extraordinary amounts of algal biomass, which were eventually scattered across an area of coastal sea of about 84,109 km^2, with a maximum algal mat coverage of 3489 km^2 (Keesing et al. 2011). Similar bloom processes in the Yellow Sea have repeatedly occurred since 2008, and the bloom duration in each summer can last approximately 3 months, but the first location of bloom formation always started from the Subei Shoal (Yancheng) (Table 16.1). Driven by surface currents and southwest and southeast winds, these floating green algae are transported more than 200 km northward in the Yellow Sea, from the Jiangsu coast to the Shandong coast, with most of the biomasses landing in the southern coast of Shandong resulting in severe ecological and environmental damage (Liu et al. 2009; Keesing et al. 2011).

A few species have been identified from the floating green algal canopies based on morphological and genetic analysis, including *Ulva compressa*, *U. flexuosa*, *U. intestinalis*, *U. linza*, *U. pertusa*, and *U. prolifera*, although *U. prolifera* was confirmed to be the dominant species (e.g., Liu et al. 2010; Wang et al. 2010; Duan et al. 2012; Zhang et al. 2015). Evidence from experimental and physiological ecology showed that *U. prolifera* has a number of adaptive physiological traits, including efficient photosynthesis, rapid growth rates, high capacity for nutrient uptake, and diverse reproductive systems which allow it to form impressive biomass within 2 months, when weather in the Yellow Sea is optimum in summer. Xu et al. (2012) found photosynthesis genes of C_3 and C_4 in *U. prolifera* and the key enzymes of C_4 metabolism which can enhance the algal capacity for carbon (C) fixation, biomass accumulation, and environmental adaptation. *U. prolifera* has a diverse reproductive system, including sexual, asexual, and vegetative propagation (Lin et al. 2008); 1 square centimetre of blade can release up to 6 million spores or 27 million gametes, and 92–97% of the spores can germinate (Zhang et al. 2013). The growth rate of this species can reach to 10–37% per day in the field depending on the weather conditions (Liang et al. 2008; Li et al. 2009; Tian et al. 2010). The important reproductive routes to guarantee growth rate of *U. prolifera* during green tide formation are propagation of vegetative fragments and asexual zoospores (Zhang et al. 2011). Moreover, *U. prolifera* displays a high capacity for nutrient uptake, its V_{max} has been shown to increase with increased NH_4^+ concentrations, and it can reach a maximum of 421 mmol g^{-1} DW h^{-1} (Tian et al. 2010). These physiological advantages are important for sufficient proliferation to generate a massive green tide.

16.3 Source of Green Algae in the Yellow Sea

Satellite images in 2008 clearly indicated that the trajectory of the green tides in the Yellow Sea originated from the Jiangsu coast. Liu et al. (2009, 2010) proposed that the biomass source came from the cleaning of fouling green algae at facilities used for more than 20,000 ha of *Porphyra* aquaculture along the Jiangsu coastline. These mariculture activities have expanded nearly 10,000 ha since 2006 in Subei Shoal (Fig. 16.3a, b), a region characterized by large-scale coastal sand ridges (Fig. 16.2). These fouling green algae, including *U. prolifera*, grow on the bamboo poles and ropes used for *Porphyra* aquaculture (Fig. 16.3c), and they are routinely scraped off the poles and ropes after the harvest of *P. yezoensis* in mid-April. The dates of routine removal of green algae coincided with satellite observations of the first occurrence of green patches in the Yellow Sea in late April or early May, just 2 weeks after the *Porphyra* harvest. Although there were different theories about the initial source of propagules for green tides of the Yellow Sea, e.g., Pang et al. (2010) proposed that the propagule source of green tides might have been microscopic germlings of *U. prolifera* produced in coastal crab and shrimp aquaculture ponds situated along the northern coast of Jiangsu province, most satellite evidence and field surveys over the last 5 years point to the *Porphyra* culture rafts in Subei Shoal as an important nursery source for green tides (Zhou et al. 2015; Wang et al. 2015; Fan et al. 2015).

Subei Shoal is the largest intertidal mudflat in China, with an approximate area of 22,740 km^2. It is about 200 km long and 100 km wide and has a pinwheel shape (Fig. 16.2). The unique radial geomorphology of the sand shoals affects the tidal current and results in eddies forming in the deep channels between the sand shoals (Du 2012); tidal residual currents, combined with dominant southeast wind-driven currents and resultant upwelling between the Jiangsu coast and the western Yellow Sea during May to July, appear to play important roles in transporting the floating algae from coast to offshore (Keesing et al. 2011; Liu et al. 2013; Bao et al. 2015). Sea surface temperatures (SST) in the Yellow Sea during May to June generally ranged from 10 to 24 °C, which is optimal for *U. prolifera* (Keesing et al. 2011, 2016). Meanwhile, high dissolved inorganic nitrogen (DIN) concentrations in coastal waters of the Yellow Sea support the growth demand of green algae. Monitoring data shows that DIN concentrations in more than 50% of the coastal areas have exceeded 14 μM since 2003 (State Oceanic Administration 2008–2012). Li et al. (2015) reported that DIN concentrations in the summer survey of 2012 were generally more than 23 μM in the adjacent sea of Subei Shoal. Isotopic N signatures of samples of green tide thalli confirmed sources of nutrients present in the Yellow Sea were available to the macroalgae (Liu et al. 2013; Keesing et al. 2016). δ^{15}N signatures in the thalli of green algae attached to the mariculture rafts ranged from 14 to 25‰, indicating the significant impact of aquaculture, agriculture, and wastewater discharges on coastal water quality in the Yellow Sea (Keesing et al. 2016).

These findings clearly support the conclusion that the extraordinary macroalgal blooms in the Yellow Sea are triggered by a chain of complex events, with human activities (*Porphyra* aquaculture, nutrient-enriched seawaters) interacting with natural geohydrodynamic and climatic conditions (sand shoals, currents, temperature, wind). This has allowed *U. prolifera*, with distinct physiological traits (efficient photosynthesis, rapid growth, high nutrient uptake rates, and diverse reproductive strategies), to proliferate sufficiently to generate massive green tides.

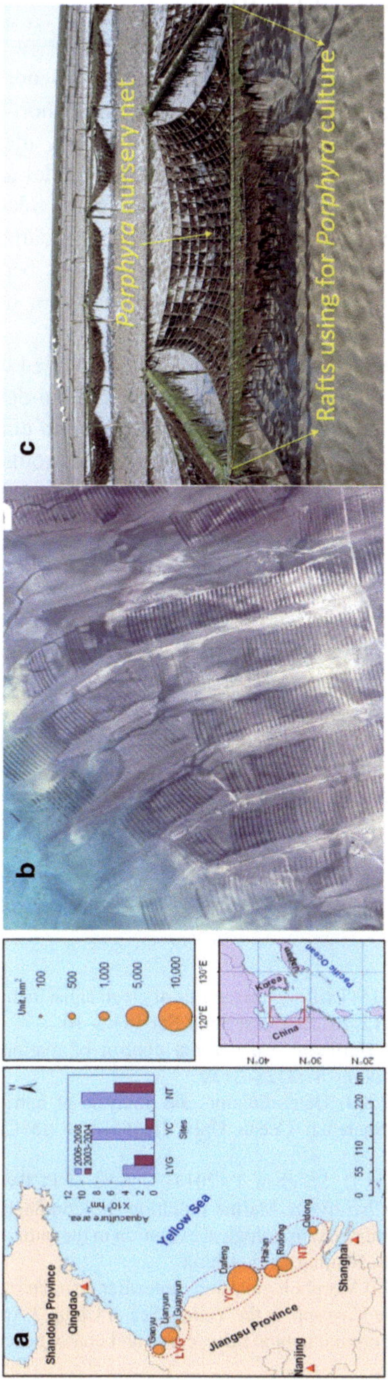

Fig. 16.3 The links between *Porphyra* aquaculture and fouling green *Ulva* fronds. (**a**) Expansion of *Porphyra* aquaculture area (hm^2) along the Jiangsu coastline (cited from Liu et al. 2009); (**b**) Google Map showing the *Porphyra* aquaculture facilities on sandy shoal; (**c**) green *Ulva* fronds growing on support facilities used in *Porphyra* culture (cited from Liu et al. 2010)

16.4 Implications and Future Research

Massive green tides are challenging for management and science. Although reducing eutrophication for long-term benefits is required, a short-term strategy for managing these blooms is also necessary, and it might include controls on macroalgal-related processes and a predictable warning model to mitigate ecological risk. Alternative uses of biomass to profit from the green tide events might be a possible way to partly offset the bill for the environmental damage. For example, *U. prolifera* can be used as food or for medical purposes, because it is rich in polysaccharides, proteins, and essential mineral elements for human health (Cai et al. 2009).

Research in basic knowledge about green tides has provided useful information for understanding the tides of the Yellow Sea. However, an in-depth understanding of these mechanisms in massive green tides is still needed to unravel the complex biological–chemical–physical interactions in coastal ecosystems. For example, in order to reduce nutrient inputs, we need to know the major sources of nutrients from, e.g., river inputs, atmospheric deposition, or others. Regarding the fate of the macroalgal biomass, we need to discover how it is transported and what the consequence is of unchecked growth at sea. In the future, developing a scientific network and interdisciplinary research programme at an international level might be helpful for solving the problem of these massive free-floating seaweed blooms.

Acknowledgments The authors appreciated the comments and manuscript revision from Dr. Patricia Glibert. This work was jointly funded by the State's Key Project of Research and Development Plan (2016YFC1402106) and the Natural Basic Research Priority Project (2010CB428700), supported by the Ministry of Science and Technology in China.

References

Bao M, Guan W, Yang Y et al (2015) Drifting trajectories of green algae in the western Yellow Sea during spring and summer 2012. Estuar Coast Shelf Sci 163:9–16

Burkepile DE, Hay ME (2006) Herbivore vs. nutrient control of marine primary producers: context-dependent effects. Ecology 87:3128–3139

Cai C, Yao B, Shen W et al (2009) Determination and analysis of nutrition compositions in *Enteromorpha clathrata*. J Shanghai Ocean Univ 18:155–159 (in Chinese with English abstract)

Cheney D, Rajic L, Sly E, Meric D, Sheahan T (2014) Uptake of PCBs contained in marine sediments by the green alga *Ulva rigida*. Marine Pollution Bulletin 88(1–2): 207–214.

Du J (2012) Sediment transport and geomorphological evolution in the radial sand ridges, southern Yellow Sea. PhD Dissertation, Nanjing Univ. China

Duan W, Guo L, Sun D et al (2012) Morphological and molecular characterization of free-floating and attached green macroalgae *Ulva* spp. in the Yellow Sea of China. J Appl Phycol 24:97–108

Duarte CM (2009) Global loss of coastal habitat rates, cause and consequence. Fundación BBVA, CSIC Publication, USA

Fan S, Fu M, Wang Z et al (2015) Temporal variation of green macroalgal assemblage on *Porphyra* aquaculture rafts in the Subei Shoal, China. Estuar Coast Shelf Sci 163:23–28

Fletcher RT (1996) The occurrence of 'green tide'. In: Schramm W, Nienhuis PH (eds) Marine Benthic Vegetation-recent changes and the effects of eutrophication. Springer, Berlin, pp 7–43

Gower JFR, King SA (2011) Distribution of floating *Sargassum* in the Gulf of Mexico and the Atlantic Ocean mapped using MERIS. Int J Remote Sens 32:1917–1929

Keesing JK, Liu D, Fearns P et al (2011) Inter- and intra-annual patterns of *Ulva prolifera* green tides in the Yellow Sea during 2007–2009, their origin and relationship to the expansion of coastal seaweed aquaculture in China. Mar Pollut Bull 62:1169–1182

Keesing JK, Liu D, Shi Y et al (2016) Abiotic factors influencing biomass accumulation of green tide causing *Ulva* spp. on *Porphyra* culture rafts in the Yellow Sea, China. Mar Pollut Bull 105:88–97

Lapointe BE (1997) Nutrient thresholds for bottom-up control of macroalgal blooms on coral reefs in Jamaica and southeast Florida. Limnol Oceanogr 42:1119–1131

Li H, Zhang C, Han X et al (2015) Changes in concentrations of oxygen, dissolved nitrogen, phosphate, and silicate in the southern Yellow Sea, 1980–2012: sources and seaward gradients. Estuar Coast Shelf Sci 163:44–55

Li R, Wu X, Wei Q et al (2009) Growth of *Enteromorpha prolifera* under different nutrient conditions. Adv Mar Sci 27(2):211–216 (in Chinese with English abstract)

Liang Z, Lin X, Ma M et al (2008) A preliminary study of the *Enteromorpha prolifera* drift gathering causing the green tide phenomenon. Period Ocean Univ China 38(4):601–604 (in Chinese with English abstract)

Lin A, Shen S, Wang J et al (2008) Reproduction diversity of *Enteromorpha prolifera*. J Integr Plant Biol 50:622–629

Liu D, Keesing JK, Dong Z et al (2010) Recurrence of the world's largest green-tide in 2009 in Yellow Sea, China: *Porphyra yezoensis* aquaculture rafts confirmed as nursery for macroalgal blooms. Mar Pollut Bull 60:1423–1432

Liu D, Keesing JK, He P et al (2013) The world's largest macroalgal bloom in the Yellow Sea, China: formation and implications. Estuar Coast Shelf Sci 129:2–10

Liu D, Keesing JK, Xing Q et al (2009) World's largest macroalgal bloom caused by expansion of seaweed aquaculture in China. Mar Pollut Bull 58:888–895

Nelson TA, Haberlin K, Nelson AV et al (2008) Ecological and physiological controls of species composition in green macroalgal blooms. Ecology 89:1287–1298

Pang S, Liu F, Shan T et al (2010) Tracking the algal origin of the *Ulva* bloom in the Yellow Sea by a combination of molecular, morphological and physiological analyses. Mar Environ Res 69:207–215

Raffaelli DG, Raven JA, Poole LJ (1998) Ecological impact of green macroalgal blooms. Oceanogr Mar Biol Annu Rev 36:97–125

Smetacek V, Zingone A (2013) Green and golden seaweed tides on the rise. Nature 505:84–88

State Oceanic Administration People's Republic of China (SOA) (2008–2012) The national bulletins of marine environment quality status. SOA Publication, Beijing

Sun S, Wang F, Li C et al (2008) Emerging challenges: massive green algae blooms in the Yellow Sea. Nat Proced 2266:1–5

Tian Q, Huo Y, Zhang H et al (2010) Preliminary study on growth and NH_4^+-N uptake kinetics of *Enteromorpha prolifera* and *Enteromorpha clathrata*. J Shanghai Univ 19(2):252–258 (in Chinese with English abstract)

Valiela I, Foreman K, LaMontagne M et al (1992) Couplings of watersheds and coastal waters: sources and consequences of nutrient enrichment in Waquoit Bay, Massachusetts. Estuaries 15:443–457

Valiela I, McClelland J, Hauxwell J et al (1997) Macroalgal blooms in shallow estuaries: controls and ecophysiological and ecosystem consequences. Limnol Oceanogr 42:1105–1118

Wang J, Jiang P, Cui Y et al (2010) Molecular analysis of green-tide-forming macroalgae in the Yellow Sea. Aquat Bot 93:25–31

Wang Z, Xiao J, Fan S et al (2015) Who made the world's largest green tide in China? – an integrated study on the initiation and early development of the green tide in Yellow Sea. Limnol Oceanogr 60:1105–1117

Xu J, Fan X, Zhang X et al (2012) Evidence of coexistence of C_3 and C_4 photosynthetic pathways in a green-tide-forming alga, *Ulva prolifera*. PLoS One 7(5):e37438. https://doi.org/10.1371/journal.pone.0037438

Ye N, Zhang X, Mao Y et al (2011) "Green tides" are overwhelming the coastline of our blue planet: taking the world's largest example. Ecol Res 26:477–485

Zhang J, Huo Y, Yu K et al (2013) Growth characteristics and reproductive capability of green tide algae in Rudong coast, China. J Appl Phycol 25:795–803

Zhang Q, Liu Q, Kang Z et al (2015) Development of a fluorescence in situ hybridization method for rapid detection of *Ulva* (*Enteromorpha*) *prolifera*. Estuar Coast Shelf Sci 163:103–111

Zhang X, Xu D, Mao Y et al (2011) Settlement of vegetative fragments of *Ulva prolifera* confirmed as an important seed source for succession of a large-scale green tide bloom. Limnol Oceanogr 56:233–242

Zhou M, Liu D, Anderson D et al (2015) Introduction to the special issue on green tides in the Yellow Sea. Estuar Coast Shelf Sci 163:1–7

Chapter 17
Ecological Drivers of Green *Noctiluca* Blooms in Two Monsoonal-Driven Ecosystems

Joaquim I. Goes, Helga do R. Gomes, Khalid Al-Hashimi, and Anukul Buranapratheprat

17.1 Introduction

The earliest documented reports of green *Noctiluca* blooms date back to the 1950s (Subrahmanyan 1954; Charernphol 1958), and recent observations in many parts of the world have made it abundantly clear that these blooms are expanding rapidly in terms of their intensity and their longevity, posing serious challenges for local fisheries and coastal activities. What is particularly alarming is that *Noctiluca* blooms have begun to appear offshore and in deeper waters as is the case in the Arabian Sea. Along the coast of Oman, the impacts of *Noctiluca* are not just ecological but economic. Effects of these blooms are felt far beyond fisheries, to tourism and recreation; aquaculture, oil refineries, ship repair, and a host of other coastal industries, including desalination plants threatening the supply of freshwater to major cities of Oman.

In the offshore Arabian Sea, the emergence of *Noctiluca* as the dominant wintertime species in place of diatoms is a recent phenomenon. Ever since *Noctiluca* blooms were first detected in 2000, they have been occurring with predictable regularity and increasing intensity. In the winter of 2017, for instance, the green *Noctiluca* bloom was one of the most widespread and long-lasting blooms of recent years. It appeared off the coast of Oman in November of 2016, and by February 2017, it spread westward, stretching across the entire Arabian Sea into the coastal waters of Pakistan (Fig. 17.1). For the first time ever, the bloom was found

J.I. Goes (✉) • H.d.R. Gomes
Lamont Doherty Earth Observatory at Columbia University, Palisades, NY, USA
e-mail: jig@ldeo.columbia.edu

K. Al-Hashimi
Department of Marine Sciences and Fisheries, Sultan Qaboos University, Muscat, Oman

A. Buranapratheprat
Faculty of Science, Department of Aquatic Science, Burapha University, Chonburi, Thailand

P.M. Glibert et al. (eds.), *Global Ecology and Oceanography of Harmful Algal Blooms*, Ecological Studies 232, https://doi.org/10.1007/978-3-319-70069-4_17

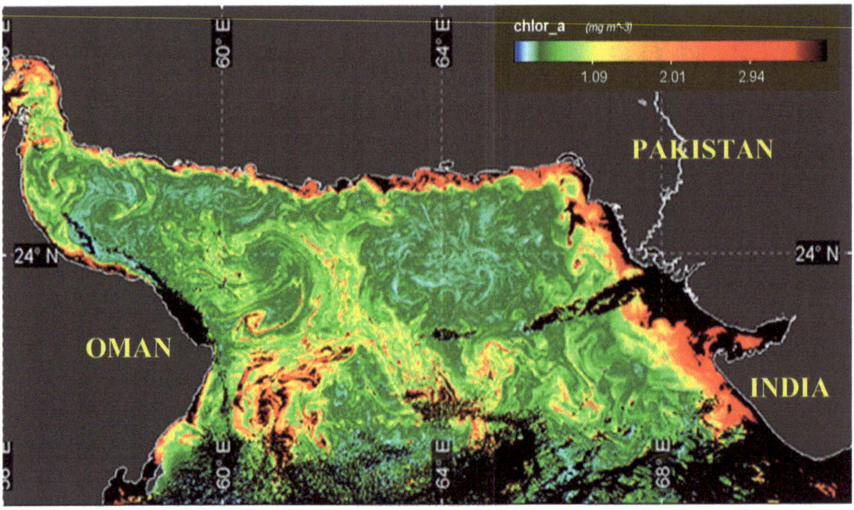

Fig. 17.1 Green *Noctiluca* blooms in the Arabian Sea in the winter of 2017 as captured by NASA MODIS-Aqua image of February 5, 2017

as far south as ~17°N off the coast of Ratnagiri, India. At the peak of its growth, the bloom occupied an area that was roughly three times the size of the state of Texas, USA. Although the bloom disappeared over most of the Arabian Sea by end of April, it persisted along the coasts of Pakistan and Oman even after the commencement of the summer monsoon season in June, causing significant loss of water and air quality, disruption to normal coastal services, and massive losses to fishermen due to fish mortality. Although all coastal regions around the Northern Arabian Sea have been experiencing large *Noctiluca* blooms, the bloom of 2017 was extraordinary because of the socioeconomic losses that it left in its wake, leading to renewed international interest in the rapidly changing state of the Arabian Sea ecosystem.

Outbreaks of *Noctiluca* are not exclusive to the Arabian Sea. They have been frequently reported in the coastal waters of many Southeast Asian countries [Harrison et al. 2011; see also Chap. 14, Furuya et al. (2018)], most often in the Manila Bay, Philippines (Furuya et al. 2006a); the upper Gulf of Thailand (Sriwoon et al. 2008; Lirdwitayaprasit et al. 2006); Jakarta and Ambon Bays, Indonesia (Nurdjaman and Yanagi 2002; Sidharta 2013); and the coastal waters of Vietnam (Hai et al. 2010).

Here, we briefly summarize some of the distinct physiological characteristics of *Noctiluca*, including some common environmental characteristics associated with bloom outbreaks of this organism, in two ecosystems, i.e., the Arabian Sea and the Gulf of Thailand, both of which are monsoonally driven.

17.2 Physiology of Green *Noctiluca*: Mixotrophy

Green *Noctiluca* is a mixotroph (Stoecker et al. 2017). It acquires its colour from the hundreds of green, free-swimming symbionts *Protoeuglena noctilucae* belonging to the class Pedinophyceae (Wang et al. 2016) living within its central symbiosome (Fig. 17.2a). It can sustain itself through carbon (C) fixation by its endosymbionts and by ingestion of exogenous prey (Gomes et al. 2014; Furuya et al. 2006a, b). This dual mechanism of obtaining energy presents *Noctiluca* with several advantages, the most significant being that it affords it the ability to out compete both its prey and predators allowing it to grow as thick green blooms (Fig. 17.2b). This mixotrophic behaviour of green *Noctiluca* makes it very different from the more extensively studied red *Noctiluca* found in temperate coastal waters which is devoid of endosymbionts and exclusively heterotrophic (Harrison et al. 2011).

In the Arabian Sea, surface populations of *Noctiluca* typically contain large quantities of diatoms in their food vacuoles (Fig. 17.2c; Gomes et al. 2014). However, endosymbiont photosynthetic rates are very low because the typical wintertime incident irradiances of 1500–1800 μE m^{-2} s^{-1} are far in excess of the ~200 μE m^{-2} s^{-1} at which light becomes photo-inhibiting for the endosymbionts (Goes and Gomes 2016). In contrast, subsurface populations of *Noctiluca* are generally devoid of ingested diatoms, and endosymbiont photosynthetic rates are unusually high. Despite these stark differences, growth rates of surface and subsurface populations are comparable, suggesting that actively growing cells of *Noctiluca* have a remarkable capacity to transition to a greater dependence on heterotrophy either when conditions for photosynthesis are not favourable to its endosymbionts or when an external source of food is available to the host cells.

In controlled feeding experiments, *Noctiluca* exhibited preferential feeding tendencies, growing best when the external food source was a dinoflagellate, followed by single-celled diatoms (Mile et al. 2017). Growth rates of *Noctiluca* were much slower when the only available extraneous food was a chain-forming diatom and slowest when no food was available. Additional experiments showed

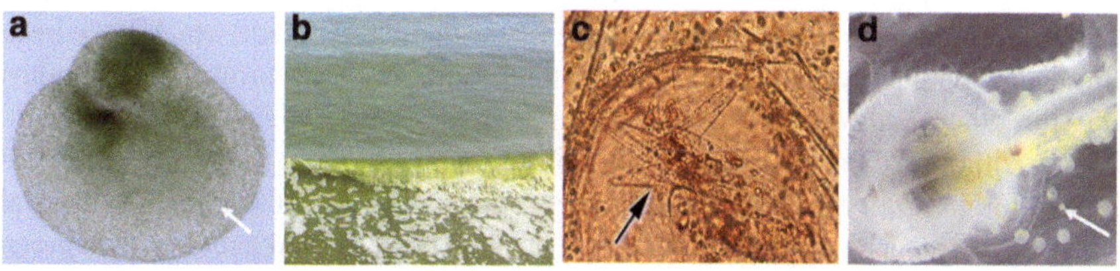

Fig. 17.2 (Panel **a**) green *Noctiluca* cell (~800 μm diameter) with endosymbionts (*arrows*) viewed under a microscope; (Panel **b**) blooms of *Noctiluca* along the coast of Muscat, Oman; (Panel **c**) diatoms (*arrow*) in the food vacuole of *Noctiluca;* (Panel **d**) close-up of a salp with intact cells of Noctiluca inside its gut

that *Noctiluca* grew best in the presence of an extraneous source of food and when incident irradiance light conditions were optimal (~200 μmol m^{-2} s^{-1}, 12L:12D) (Mile et al. 2017). Under light-limiting conditions (30 μmol m^{-2} s^{-1}), even when an external source of food was available, growth of *Noctiluca* was almost negligible, suggesting co-dependency on heterotrophy and autotrophy for growth.

Grazing experiments with *Noctiluca* as the sole food source revealed that it is not a preferred food for most micro- and meso-zooplankton (Gomes et al. 2014). Instead, it was largely preyed upon by salps (Fig. 17.2d) and jellyfish, indicative of a shorter food chain. These observations are consistent with the idea that in mixotroph-dominated systems, the food chain is much shorter and the trophic structure fundamentally different from the traditional planktonic food web [Mitra et al. 2014; see also Chap. 7, Flynn et al. (2018)].

Noctiluca's dependence on inorganic nutrients and its ability to graze on other phytoplankton implies that it competes for resources with both its prey and predators. As with most mixotrophs, the emergence of *Noctiluca* represents a challenge for ecosystem modelling studies because of the complexity of its behaviour [Stickney et al. 2000; Flynn and Mitra 2009; see also Chap. 7, Flynn et al. (2018)].

17.3 Physiology of *Noctiluca*: Nitrogen Sources

Despite evidence that *Noctiluca* blooms are expanding their spatial and temporal range, and becoming more pervasive and intense worldwide, mechanisms that trigger these blooms are still unclear. Global distribution maps of green *Noctiluca* (Harrison et al. 2011) afford us some potential clues and insights about conditions that promote the growth of this organism to bloom proportions. Based on these maps, it is apparent that most, if not all, countries that are being impacted by green *Noctiluca* blooms come under the influence of the Indian monsoon system. Furthermore, these blooms typically occur in regions that experience significant influxes of nutrients due to upward shoaling of nutrient-rich subsurface waters and/or land runoff (Hai et al. 2010; Sriwoon et al. 2008; Buranapratheprat et al. 2008, 2009; Gomes et al. 2014; Swaney et al. 2015). With the exception of Oman, all these countries experiencing *Noctiluca* blooms are predominantly agricultural economies and have a long history of use of synthetic fertilizers, which have been implicated as the predominant cause for the proliferation of harmful algal blooms (HABs) other than *Noctiluca* in the coastal waters of many countries (Glibert et al. 2006). Although there is strong evidence in support of the idea that many HABs are related to fertilizer use and excessive nitrogenous (N) nutrient loading into coastal ecosystems [see also Chaps. 4, 12, Glibert et al. (2018a, b)], linkages between green *Noctiluca* blooms and N nutrient loading at present seem at best tenuous, particularly because Oman, a country which presently experiences the worst outbreaks of *Noctiluca* blooms, has no long history of N fertilizer use, and blooms are most intense and extensive during winter, a period when land runoff is low. In the case of blooms off the coast of Oman, it seems mostly likely that the N-nutrient-rich source

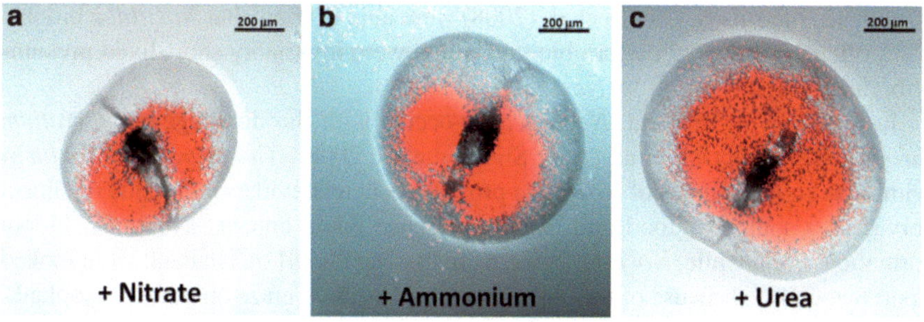

Fig. 17.3 Confocal microscopic images showing differences in cell sizes and endosymbiont content of *Noctiluca* grown in seawater medium enriched with (**a**) nitrate, (**b**) ammonium, and (**c**) urea

waters for *Noctiluca* blooms are from depth. In winter, cyclonic eddies that typically begin to populate the coast of Oman around the beginning of the winter monsoon season have been implicated in bringing subsurface nutrient-rich waters into the euphotic zone and growth of *Noctiluca* blooms (Gomes et al. 2009, 2014).

On account of the uncertainty as to the primary N-nutrient driver for *Noctiluca* blooms, we compared its growth rates in the presence of NO_3^-, NH_4^+, and urea under controlled experimental conditions. The results revealed that urea, a common ingredient in commercially available fertilizers, is the most preferred N source for *Noctiluca* (Tan et al. 2016). Cells of *Noctiluca* grown in seawater enriched with urea were also the largest in size (>1000 μm), had the highest content of endosymbionts, and exhibited the highest levels of photosynthetic competency as compared to other N nutrients tested. Although *Noctiluca* grew well in seawater enriched with NH_4^+, the cells were much smaller, and endosymbiont content much lower as compared to those grown with urea. Inorganic NO_3^- appeared to be the least preferred N source of the three N-nutrient sources, and the cells were by far the smallest and had the lowest biomass of endosymbionts (Fig. 17.3a–c).

In separate experiments, it was observed that when extraneous N nutrients were low, *Noctiluca* was still capable of accumulating a large quantity of N nutrients, particularly NH_4^+, when grown in the presence of an extraneous phytoplankton prey (Mile et al. 2017).

17.4 Comparative Bloom Regions

Although *Noctiluca* blooms occur typically in regions with significant influx of N nutrients, field data as well as onboard experiments suggest that bloom outbreaks typically occur when the upwelled waters are hypoxic, as hypoxia is especially conducive for photosynthesis by *Noctiluca's* population of endosymbionts. In an earlier investigation, which was based on microscopic data on phytoplankton from a sequence of cruises, Gomes et al. (2008) were able to show that *Noctiluca* blooms were often preceded by diatom blooms which were invariably short-lived presumably due to rapid grazing by *Noctiluca*.

In the Gulf of Thailand, *Noctiluca* co-occurs with the dinoflagellate *Ceratium furca* (Lirdwitayaprasit et al. 2006; Sriwoon et al. 2008). The presence of *C. furca* within the food vacuoles of *Noctiluca* provides concrete evidence that this organism serves as a source of food and that its presence could enhance *Noctiluca* bloom formation. For a while, *Noctiluca* blooms in the upper Gulf of Thailand were looked upon favourably, because of the belief that their ability to graze on other phytoplankton, kept toxin-forming HABs in check, especially in an environment that experiences excessive anthropogenic N loading. With *Noctiluca* blooms becoming more intense, this perception is slowly changing, because the increased frequency of *Noctiluca* bloom events and their increasing intensity are posing serious threats to fisheries, tourism, and a host of other coastal industries on which these coastal communities depend.

The predictable regularity with which *Noctiluca* blooms occur in the Arabian Sea and the upper Gulf of Thailand and commonalities in their ocean circulation processes resulting from seasonally reversing monsoonal winds allow us to examine these blooms in the context of regional hydrographic conditions that develop seasonally at both these locations. Both the Arabian Sea and the Gulf of Thailand are semi-enclosed systems that receive considerable amounts of freshwater during the two monsoon seasons. Circulation is anticyclonic during the summer monsoon and cyclonic during the winter monsoon (Fig. 17.4a–d).

In the Arabian Sea, massive basin-wide *Noctiluca* blooms are a winter monsoon phenomenon, facilitated in large measure by the uplift of cold, nutrient-rich hypoxic waters from depth to the surface by large, long-lived cyclonic mesoscale eddies (Gomes et al. 2009). *Noctiluca* blooms have also been recorded during the summer monsoon, along the west coasts of India and Pakistan, but rather sporadically especially during periods when coastal upwelling is strong and there is significant intrusion of hypoxic waters onto the continental shelf (Devassy and Sreekumaran Nair 1987). In recent years, summertime *Noctiluca* blooms have become more frequent in coastal embayments along the coasts of Pakistan and Oman, both of which experience upsloping of hypoxic waters due to wind-driven upwelling (Goes et al. 2005; Munir et al. 2013). These embayments are sheltered from strong monsoonal winds and are much calmer (Al-Azri et al. 2007; Al-Hashmi et al. 2015) as compared to more wind-exposed regions along the coast and offshore, where mixing is stronger and possibly a deterrent to *Noctiluca* blooms. In the upper Gulf of Thailand, the data of Sriwoon et al. (2008) also suggest that *Noctiluca* blooms survive best in waters that are highly stratified and calm. Active mixing within the water column in all likelihood provides an explanation as to why *Noctiluca* blooms are not found offshore during the summer monsoon despite significant influx of nutrient-rich and hypoxic waters into the euphotic column. During winter monsoon, winds coming off the nutrient-rich Indian subcontinent (e.g., Swaney et al. 2015) are comparatively weaker and the water column more stratified which creates ideal conditions for *Noctiluca* to grow to bloom proportions.

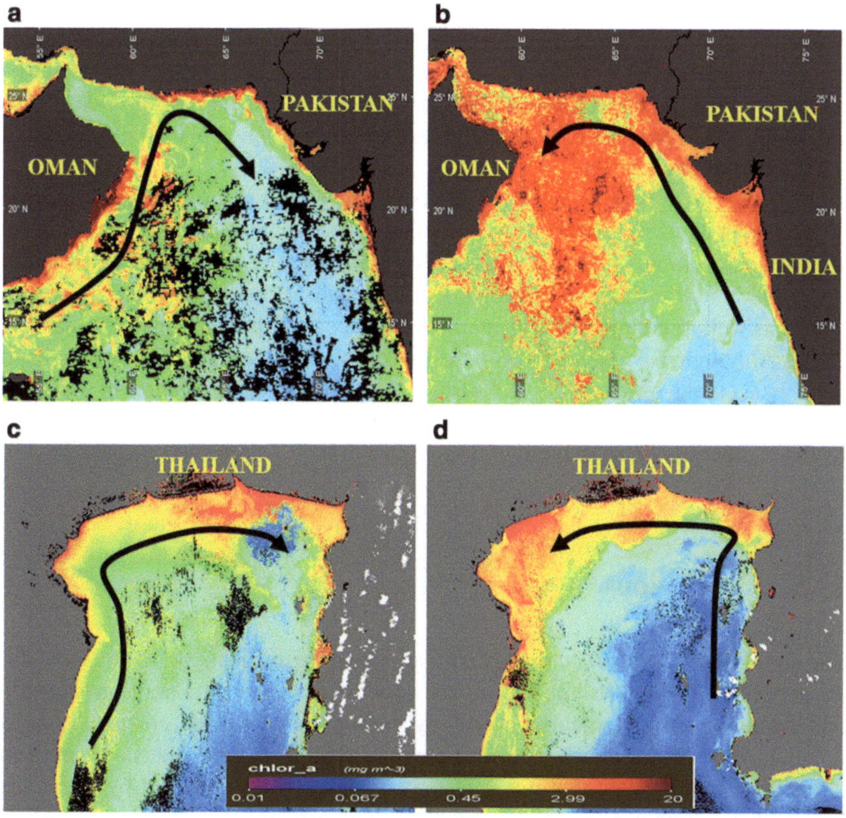

Fig. 17.4 Distribution of phytoplankton blooms in the Arabian Sea during the summer monsoon (Panel **a**) and winter monsoon (Panel **b**) and in the upper Gulf of Thailand during the summer monsoon (Panel **c**) and winter monsoon (Panel **d**). Direction of current flow has been indicated by thick black lines

The upper Gulf of Thailand is much smaller in size (100 km^2) and is extremely shallow (max. depth 25 m, avg. depth ~15 m) as compared to the Arabian Sea which occupies an area of 3.86×10^6 km^2 and is much deeper (max. depth ~4500 m, avg. depth ~2730 m). Four rivers empty into the upper Gulf of Thailand, and because of the disproportionate amount of freshwater that it receives relative to its size, its waters are more estuarine and more stratified than the oceanic waters of the Arabian Sea for most part of the year. This may be the reason why *Noctiluca* blooms occur twice a year in the upper Gulf of Thailand when there is a significant influx of N nutrients either due to land runoff or due to coastal upwelling. In all likelihood, *Noctiluca* blooms in the upper Gulf of Thailand are initiated during the summer upwelling season and in the western half of the Gulf, where there is up-shoaling of subsurface hypoxic nutrient-rich waters. However, because of the prevailing direction of flow of the currents, which is clockwise along the northern coastline, *Noctiluca* blooms accumulate downstream in the eastern half of the Gulf, where they benefit from a huge influx of N nutrients from land runoff during the rainy season (Fig. 17.4c, d). When the currents reverse their direction and become anticlockwise during the winter monsoon, *Noctiluca* begins to accumulate in the western half of the upper Gulf of Thailand. Growth at this time of the year is also sustained by excess nutrients and the presence of prey phytoplankton throughout the water column.

Although there is no indication of a systematic rise in anthropogenic N loading in the upper Gulf of Thailand (Wiriwutikorn 1996), the increased frequency of super-intense blooms of *Noctiluca* witnessed in recent years has been ascribed to excessive N-nutrient influxes from agricultural lands, urban waste water treatment plants, as well as shrimp farms located along the coast (Cheevaporn and Menasveta 2003). Although *Noctiluca* are nontoxic, thick blooms in the upper Gulf of Thailand often exacerbate oxygen loss and also cause massive accumulation of NH_4^+ in water column, both of which have been blamed for massive fish kills that follow *Noctiluca* blooms (Wattayakorn 2006).

17.5 Conclusions

The emergence of *Noctiluca* as the dominant player in the planktonic food web, and its portended disruptive impacts on the food web, demands a more systematic understanding of this organism's ecology, biology, growth, and grazing vis-à-vis its reliance on photosynthesis by its endosymbionts. What lends special urgency to this situation is the fact that many of the countries being impacted by *Noctiluca* blooms have large coastal populations that are dependent on fisheries and several other coastal industries for their livelihoods. Although prior research has provided us with important clues as to what likely triggers the growth of this mixotrophic organism, more detailed ecological studies are urgently needed to address the recent and rapid advent of this organism at the base of the food chain and its implications for the food web and biogeochemical cycles of regions that were once dominated by autotrophs. This ecophysiological information would also be essential to build a holistic perspective of how complex phytoplankton communities evolve in response to human activities and climate-induced changes. Further international collaboration on the ecology, oceanography, and socioeconomics of HABs, as emphasized in the new global HAB Programme, GlobalHAB [see Chap. 22, Berdalet et al. (2018)], will help to advance our understanding and management of these large *Noctiluca* blooms.

Acknowledgments This work is partially supported by grants NNX13AI29A and NNX17AG66G from the National Aeronautical and Space Agency, USA, and grants from the Gordon and Betty Moore Foundation and the Sultan Qaboos Cultural Center, USA, to J.I. Goes and H.do R. Gomes. Al-Hashimi is supported by Sultan Qaboos University and the Ministry of Agriculture and Fisheries Wealth, Sultanate of Oman, and A. Buranapratheprat is supported by the National Research Council of Thailand, Thailand, and the Japan International Cooperation Agency, Japan.

References

Al-Azri A, Al-Hashmi K, Goes JI et al (2007) Seasonality of the bloom-forming heterotrophic dinoflagellate *Noctiluca scintillans* in the Gulf of Oman in relation to environmental conditions. Int J Oceans Oceanogr 2:51–60

Al-Hashmi K, Claereboudt M, Piontkovski S et al (2015) Seasonal variability of size-classes of phytoplankton biomass in a sub-tropical embayment, Muscat, Sea of Oman. Iran J Fish Sci 14:136–149

Berdalet E, Kudela R, Banas NS et al (2018) GlobalHAB: fostering international coordination on harmful algal bloom research in aquatic systems. In: Glibert PM, Berdalet E, Burford M et al (eds) Global ecology and oceanography of harmful algal blooms. Springer, Cham, pp 425–447

Buranapratheprat A, Niemann KO, Matsumura S et al (2009) MERIS imageries to investigate surface chlorophyll in the upper Gulf of Thailand. Coast Mar Sci 33:22–28

Buranapratheprat A, Yanagi T, Niemann KO et al (2008) Surface chlorophyll-a dynamics in the upper Gulf of Thailand revealed by a coupled hydrodynamic-ecosystem model. J Oceanogr 64:639–656

Charernphol S (1958) Preliminary study of discoloration of seawater in the Gulf of Thailand. In: Proceedings of the ninth Pacific congress. pp 131–134

Cheevaporn V, Menasveta P (2003) Water pollution and habitat degradation in the Gulf of Thailand. Mar Pollut Bull 47:43–51

Devassy VP, Sreekumaran Nair SR (1987) Discolouration of water and its effect on fisheries along the Goa coast. Mahasagar 20:121–128

Flynn KJ, Mitra A (2009) Building the "perfect beast": modelling mixotrophic plankton. J Plankton Res 31:965–992. https://doi.org/10.1093/plankt/fbp044

Flynn KJ, Mitra A, Glibert PM et al (2018) Mixotrophy by HABs: by whom, on whom, when, why and what next. In: Glibert PM, Berdalet E, Burford M et al (eds) Global ecology and oceanography of harmful algal blooms. Springer, Cham, pp 113–132

Furuya K, Iwataki M, Lim PT et al (2018) Overview of harmful algal blooms in Asia. In: Glibert PM, Berdalet E, Burford M et al (eds) Global ecology and oceanography of harmful algal blooms. Springer, Cham, pp 289–308

Furuya K, Saito H, Rujinard S et al (2006a) Persistent whole-bay red tide of *Noctiluca scintillans* in Manila Bay, Philippines. Coast Mar Sci 30:74–79

Furuya K, Saito H, Sriwoon R et al (2006b) Vegetative growth of *Noctiluca scintillans* containing the endosymbiont *Pedinomonas noctilucae*. Afr J Mar Sci 28:305–308

Glibert PM, Al-Azri A, Allen JI et al (2018a) Key questions and recent research advances on harmful algal blooms in relation to nutrients and eutrophication. In: Glibert PM, Berdalet E, Burford M et al (eds) Global ecology and oceanography of harmful algal blooms. Springer, Cham, pp 229–259

Glibert PM, Beusen AHW, Harrison JA et al (2018b) Changing land, sea- and airscapes: sources of nutrient pollution affecting habitat suitability for harmful algae. In: Glibert PM, Berdalet E, Burford M et al (eds) Global ecology and oceanography of harmful algal blooms. Springer, Cham, pp 53–76

Glibert PM, Harrison J, Heil C et al (2006) Escalating worldwide use of urea – a global change contributing to coastal eutrophication. Biogeochemistry 77:441–463

Goes JI, Gomes HR (2016) An ecosystem in transition: the emergence of mixotrophy in the Arabian Sea. In: Glibert PM, Kana T (eds) Aquatic microbial ecology and biogeochemistry: a dual perspective. Springer International Publishing, Geneva, pp 155–185

Goes JI, Prasad TG, Gomes HR et al (2005) Warming of the Eurasian landmass is making the Arabian Sea more productive. Science 308:545–547

Gomes HR, Goes JI, Matondkar SGP et al (2008) Blooms of *Noctiluca miliaris* in the Arabian Sea – an *in situ* and satellite study. Deep Sea Res 55:751–765

Gomes HR, Goes JI, Matondkar SGP et al (2009) Unusual blooms of the green *Noctiluca Miliaris* (Dinophyceae) in the Arabian Sea during the winter monsoon. In: Wiggert JD, Hood RR,

Naqvi SWA et al (eds) Indian Ocean: biogeochemical processes and ecological variability. American Geophysical Union, Washington, DC, pp 347–363
Gomes HR, Goes JI, Matondkar SGP et al (2014) Massive outbreaks of *Noctiluca scintillans* blooms in the Arabian Sea due to spread of hypoxia. Nat Commun 5. https://doi.org/10.1038/ncomms5862
Hai DN, Lam NN, Dippner JW (2010) Development of *Phaeocystis globosa* blooms in the upwelling waters of the South Central coast of Viet Nam. J Mar Syst 83:253–261
Harrison PJ, Furuya K, Glibert PM et al (2011) Geographical distribution of red and green *Noctiluca scintillans*. Chin J Oceanol Limnol 29:807–883
Lirdwitayaprasit T, Meksumpun S, Rungsupa S et al (2006) Seasonal variations in cell abundance of *Noctiluca scintillans* in the coastal waters off Chonburi Province, the upper Gulf of Thailand. Coast Mar Sci 30:80–84
Mile A, Thondapu S, McKee K et al (2017) The rise of the mixotroph *Noctiluca scintillans* in the Arabian Sea: disruptive impact on the food web in a warmer world? Assoc Sci Limnol Oceanogr, Annual meeting, Hawaii, February 2017 (abstract only)
Mitra A, Flynn KJ, Burkholder JM et al (2014) The role of mixotrophic protists in the biological carbon pump. Biogeosciences 11:995–1005
Munir S, Naz T, Burhan Z et al (2013) Seasonal abundance, biovolume and growth rate of the heteotrophic dinoflagellate (*Noctiluca scintillans*) from coastal waters of Pakistan. Pak J Bot 45:1109–1113
Nurdjaman S, Yanagi T (2002) Lower trophic level ecosystem in Jakarta Bay, Indonesia. La Mer 20:161–170
Sidharta RB (2013) The current status of research on harmful algal bloom (HAB) in Indonesia. J Coast Dev 8:75–88
Sriwoon R, Pholpunthin P, Lirdwitayaprasit T et al (2008) Population dynamics of green *Noctiluca scintillans* (Dinophyceae) associated with the monsoon cycle in the upper Gulf of Thailand. J Phycol 44:605–615
Stickney HL, Hood RR, Stoecker DK (2000) The impact of mixotrophy on planktonic marine ecosystems. Ecol Model 125:203–230
Stoecker DK, Hansen PJ, Caron DA et al (2017) Mixotrophy in the marine plankton. Annu Rev Mar Sci 9:311–335
Subrahmanyan R (1954) A new member of Euglenineae, *Protoeuglena noctilucae* gen. et. sp. n., occurring in *Noctiluca miliaris* Suriray causing green discoloration off the Sea of Calicut. Proc Indian Acad Sci 39:120–127
Swaney DP, Hong B, Paneer Selvam A et al (2015) Net anthropogenic nitrogen fluxes from Indian watersheds: an initial assessment. J Mar Syst 141:45–58
Tan YS, Ang A, Gomes HdoR et al (2016) Could aerosol dust plume-derived trace metals and inorganic nutrients be fueling the recent growth and proliferation of *Noctiluca scintillans* blooms in the Arabian Sea? Am Geophys Union, Annual meeting, San Francisco, December 2016 (abstract only)
Wang L, Lin X, Goes JI et al (2016) Phylogenetic analyses of three genes of *Pedinomonas noctilucae*, the green endosymbiont of the marine dinoflagellate *Noctiluca scintillans*, reveal its affiliation to the order Marsupiomonadales (Chlorophyta, Pedinophyceae) under the reinstated Name *Protoeuglena noctilucae*. Protist 167:205–216. https://doi.org/10.1016/j.protis.2016.02.005
Wattayakorn G (2006) Environmental issues in the Gulf of Thailand. In: Wolanski E (ed) The environment in Asia Pacific Harbours. Springer, The Netherlands, pp 249–259
Wiriwutikorn T (1996) Long-term variations of nutrients in the upper Gulf of Thailand. M.Sc. Thesis, Inter-Departmentof Environmental Science, Chulalongkorn University, Bangkok

Part VI
Observing and Predicting Harmful Algal Blooms: Tools and Predictive Approaches

Chapter 18
Advancements and Continuing Challenges of Emerging Technologies and Tools for Detecting Harmful Algal Blooms, Their Antecedent Conditions and Toxins, and Applications in Predictive Models

Patricia M. Glibert, Grant C. Pitcher, Stewart Bernard, and Ming Li

18.1 Introduction

Safe beaches, fishing, recreation and drinking water supplies, protection of aquaculture regions, and understanding of mass mammal strandings or die-off events, as well as measuring progress towards water quality targets (e.g., total maximum daily loads, TMDLs, or water quality directives), are just some of the reasons that improved monitoring and prediction of harmful algal blooms (HABs) have become necessary in coastal and freshwaters worldwide. With HABs increasing in frequency, duration, geographic extent, and human and ecological impacts, this need is ever increasing [see Chap. 2, Glibert et al. (2018b)].

When HABs occur, the traditional, and still among the most common, methods for assessing their extent and potential impacts are water sampling (via boats or ships), collection of samples from selected stations, identification of cells with microscopy, followed by toxin extraction and analysis, often referenced relative to mouse bioassays. When HABs or mass die-offs, such as fish kills, or mass strandings of marine mammals occur, the response teams may not have the tools or skills to sample the requisite data on the right time scale to determine exactly what happened. Boats or personnel may not be available. Additionally, insufficient information may be available for teams to sample the most appropriate

P.M. Glibert (✉) • M. Li
University of Maryland Center for Environmental Science, Horn Point Laboratory, Cambridge, MD, USA
e-mail: glibert@umces.edu

G.C. Pitcher
Fisheries Research and Development, Cape Town, South Africa

S. Bernard
Earth Systems Earth Observation, CSIR—NRE Centre for High Performance Computing, Rosebank, Cape Town, South Africa

P.M. Glibert et al. (eds.), *Global Ecology and Oceanography of Harmful Algal Blooms*, Ecological Studies 232, https://doi.org/10.1007/978-3-319-70069-4_18

or impacted sites. Therefore, reconstruction of events is based on best available data, but such data are often inadequate to fully understand the causative factors and the impacts. For protection of human health and seafood safety, many regions have routine monitoring, but an impacted site may not necessarily be part of the sampling grid, or samples may not necessarily be taken on temporal scales needed to fully resolve bloom duration.

Accordingly, many regions now routinely close sites for shellfish collection at certain times of year in a precautionary mode, an approach that may be cost effective from the sampling perspective but may be very costly in terms of lost income potential for harvesters. Needs for improved sampling and prediction of HABs and their impacts are great.

The scientific goal of the Global Ecology and Oceanography of Harmful Algal Blooms (GEOHAB) Programme was to "improve prediction of HABs by determining the ecological and oceanographic mechanisms underlying their population dynamics, integrating biological, chemical, and physical studies supported by enhanced observation and modeling systems" [GEOHAB 2001; see also Chap. 3, Kudela et al. (2018)]. Without question, considerable advances have been made in the tools and technologies for measuring, monitoring, and predicting HABs over the past nearly two decades since GEOHAB was conceived. The goal of this chapter is to highlight some of these advances, in terms of tools and technologies for detecting HAB abundance, their associated environmental conditions, their toxins, and their impacts. Advances in each of these categories have contributed over the past decade to new data for incorporation into new types of models allowing for nowcasts and forecasts. There is great promise for continued advancements although many challenges will remain. It is hoped that in coming years, these challenges can be reduced as approaches advance from conceptual and experimental to operational.

18.2 Advances in Detecting Blooms

18.2.1 In Situ Imaging and Detection Systems

While it is without question that the microscope will remain a critically important tool in identifying HABs, there are various instruments that are available to detect cell abundance remotely. These instruments take advantage of either unique genetic signatures (DNA, RNA) targeted by molecular probes, imaging technology that can take pictures of cells in rapid succession, allowing cell identification through automated image analysis software, or use of other cell signatures, such as fluorescent parameters or other characteristics, and these approaches have allowed automation of cell detection. The Environmental Sample Processor (ESP, Fig. 18.1a,b), either as a fixed instrument on a sampling platform or, in its newest configuration, as an underwater robot, autonomously samples the water, filters the particles, and, using DNA analysis, determines the presence of a particular HAB, and conveys that information to shore (Babin et al. 2005; Seltenrich 2014). An array of the first

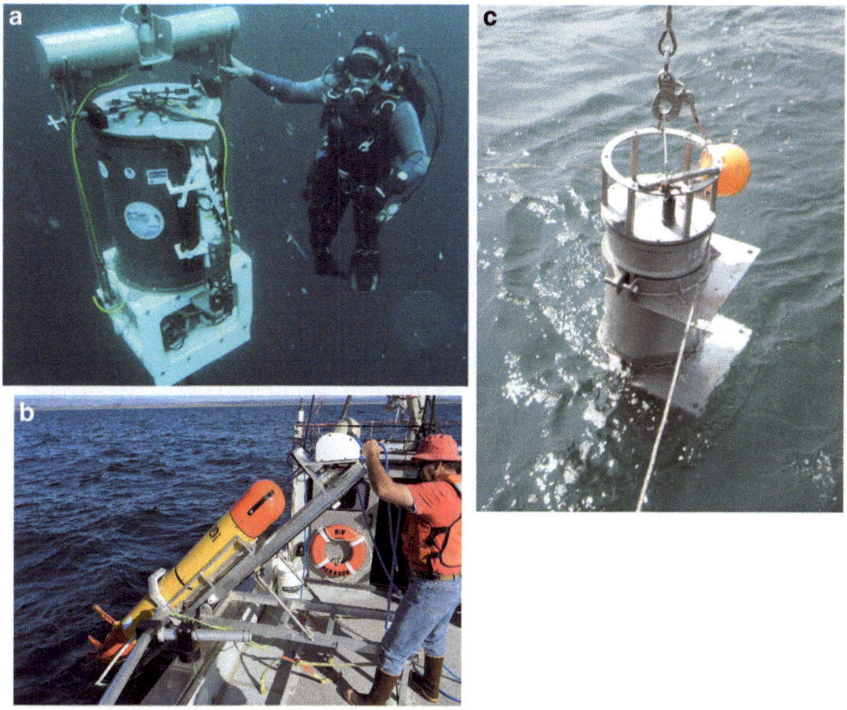

Fig. 18.1 Photographs of the Environmental Sample Processor (ESP) on a mooring and a glider (panels **a**, **b**, photo courtesy of Monterey Bay Aquarium Research Institute) and the Imaging Flow CytoBot (IFCB, panel **c**, photo courtesy of T. Crockford, Woods Hole Oceanographic Institution)

generation of this instrument has been used in the Gulf of Maine providing real-time estimates of *Alexandrium* and *Pseudo-nitzschia* cell abundances that are subsequently incorporated into numerical forecasts to help managers and industry know where and when shellfish may become toxic, enabling proactive management decisions (Anderson 2009). In addition, some ESPs have been outfitted with automated sandwich hybridization assay as well as archival and storage of samples on filters for later toxin assay performed in a laboratory. Some ESPs have been configured to directly perform competitive enzyme-linked immunosorbent assays (ELISA) for shellfish toxins such as domoic acid (DA, Doucette et al. 2009).

Another approach to remote analysis of cells is the Imaging FlowCytobot (IFCB; Fig. 18.1c), an automated underwater flow cytometer and image analysis system capable of high-resolution images that are classified automatically. This instrument, with its high throughput, has the ability to detect multiple species or to resolve multiple life stages of different types of cells. It has been used extensively to provide early warning of HABs along the Texas coast (Seltenrich 2014). Originally developed as an underwater automated flow cytometer (Olson et al. 2003), the instrument now also takes advantage of imaging technology that allows cells to be identified, enumerated, and classified based on imaging software and appropriate training of the software. Imaging of cells is rapidly advancing, especially with the development of the FlowCam, an instrument that uses a microscope objective and can perform high-quality, automated image analysis based on a user-developed library of images and image-processing software (e.g., Camoying and Yñiguez 2016). This instrument, historically laboratory-based, is now being coupled to in-water or on-water instruments, as is the case with the IFCB. Compared to the ESP, which relies on molecular probes, image-processing instruments can detect cells based on image or, depending on configuration, fluorescent properties, allowing these instruments to "see" organisms that may not have been the original target of interest, as is the case with molecular probes which are specific to the organisms for which the probe was designed. This capability proved to be important when the dinoflagellate *Dinophysis* was first detected in the Gulf of Mexico using an IFCB where the main objective had originally been to detect species of the dinoflagellate, *Karenia* (Campbell et al. 2010). Depending on the application, the IFCB can generate many thousands of images per hour, and successful deployments of the instruments have been on the order of many months. New imaging approaches, such as holography, are also rapidly advancing and can provide 3-D images at resolutions not currently available in existing instruments (Erickson et al. 2012).

Another similar in situ instrument, the Optical Plankton Discriminator (OPD), or more fondly termed the BreveBuster for its sensors targeting *Karenia brevis*, optically detects *K. brevis* blooms by comparing light absorption by particles in ambient water to the light absorption fingerprint that is characteristic of this HAB species. The coloured dissolved organic matter (CDOM) absorption spectra and correlation coefficient between the particulate spectra and known phytoplankton classes are resolved in real time. This comparison yields a "similarity index" (SI) that is related to the fraction of the phytoplankton community biomass that is contributed by *K. brevis*. Values of SI < 0.5 indicate $< 10\%$ *K. brevis* biomass, while values > 0.8 indicate $> 90\%$ *K. brevis* biomass (Kirkpatrick et al. 2011). This instrument, also, is now currently deployed on an autonomous underwater vehicle (AUV).

As powerful as the ESP and IFCB, in their conventional operational state, they are fixed onto moored platforms—and they are expensive. Consequently, while they provide highly resolved temporal data, and may even provide highly resolved vertically integrated data, fixed platform instruments are not able to provide spatial information. It is for this reason that these technologies are being transitioned to AUVs, and second-generation ESP and IFCB are being developed for deployment on gliders or other remote vehicles. This is not an easy transition, as AUVs tend to be comparatively small in terms of payload volume, and power is much more restricted than on a moored buoy where solar power can be more readily used. With advancing miniaturization of optics, electronics, and fluidic systems, these tools will be increasingly operational.

18.2.2 Ocean Colour and Optical Sensors

Satellite remote sensing has been widely applied in coastal, estuarine, and inland waters as a monitoring tool to detect HABs (e.g., Stumpf 2001; Stumpf et al. 2003; Tang et al. 2004; Michalak et al. 2013). Satellite remote sensing, specifically ocean colour radiometry, provides the systematic spatial coverage that mooring instruments and ship surveys are unable to achieve—and are subsequently often widely used both for operational bloom monitoring and analysis of bloom phenology.

While satellite ocean colour data are extremely effective at determining gross phytoplankton biomass, they have many limitations in resolving individual HABs. Most critically, ocean colour satellites only resolve the signal from the surface layers, are unlikely to resolve subsurface structure, and are only capable of yielding limited information on phytoplankton community structure at relatively high biomass. Ocean colour has little application when HAB abundance is low relative to the broader phytoplankton assemblage or when HABs are in thin layers or other subsurface aggregations [see Chap. 9, Raine et al. (2018)]. It is also widely recognized that the accuracy of ocean colour remote-sensing products, most typically chlorophyll *a* concentrations, is diminished in coastal and estuarine environments where there is significant presence of non-phytoplankton material (Dall'Olmo et al. 2005; Gitelson et al. 2007). The use of ocean colour to detect phytoplankton in coastal waters is often limited by the fact that other seawater constituents, e.g., sediment, or CDOM, are present in such high quantities that they severely reduce the ability to derive the spectrum of the light leaving the water or to subsequently derive phytoplankton biomass or community structure. It is important that these constraints on ocean colour applications for HABs are clearly communicated to potential users. Nevertheless, ocean colour is very cost effective and valuable for the routine observation of high-biomass HABs, and it can provide some ability to distinguish broad taxonomy—particularly in waters where phytoplankton dominate the optical signal and where bloom ecology is generally well known. It should also be realized that ocean colour and other satellite data, such as sea surface temperature and surface wind, have considerable value for indirect assessment of ecological variability and probabilistic bloom formation models (e.g., Oliveira et al. 2009).

Ocean colour does offer the capability to detect large changes in phytoplankton community size (Evers-King et al. 2014) or large changes in accessory pigments (Bernard et al. 2014)—but only at a substantial phytoplankton biomass, when there is a low presence of any other non-phytoplankton constituents and when the atmospheric correction is sufficiently good. A wide variety of approaches and algorithms have been developed to try and provide improved taxonomic information directly from ocean colour, for both oceanic and HAB applications (IOCCG 2014). Many of the approaches developed for oligotrophic and mesotrophic oceanic environments (e.g., Brewin et al. 2011) are not designed for HAB applications in eutrophic or optically complex coastal waters, and care should be taken with HAB application of statistical or abundance-based phytoplankton functional types (PFT)

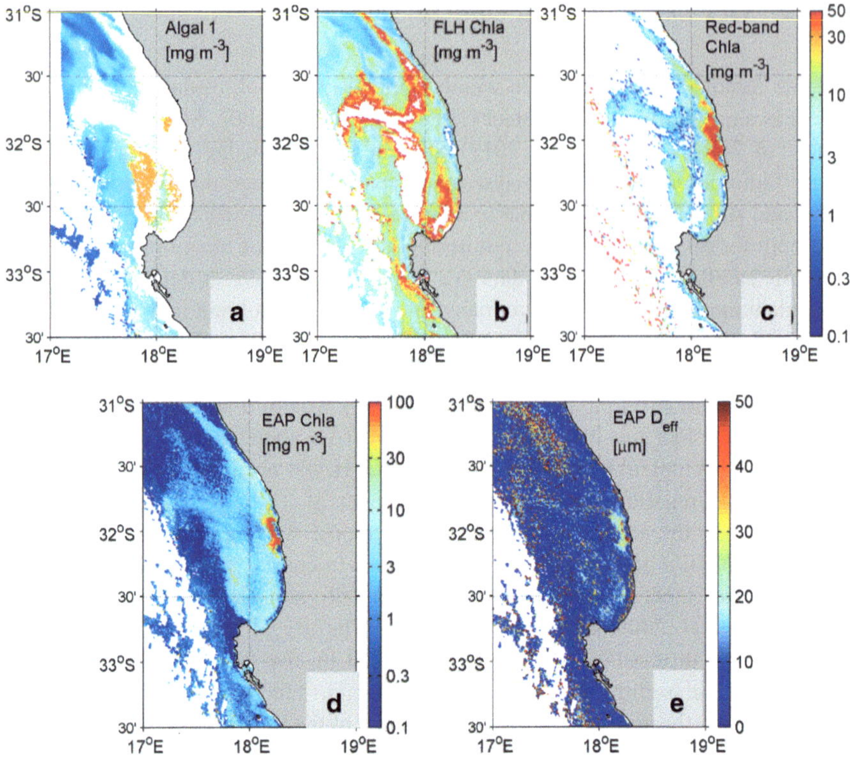

Fig. 18.2 MERIS reduced resolution images from 25 October 2002, showing a bloom of *Alexandrium catenella* in the southern Benguela. Images show chlorophyll *a* calculated from (**a**) the standard chlorophyll *a* algorithm, Algal 1, (**b**) the fluorescence line height (FLH) algorithm, (**c**) an empirical red-near-infrared 2-band ratio algorithm, and (**d**) the equivalent algal population (EAP) algorithm (Evers-King et al. 2014). Panel (**e**) shows the effective cell diameter also calculated from the EAP algorithm. The significant differences between algorithm outputs highlight the need for appropriate algorithm choices for HAB applications

approaches. Such conditions are rather exclusive but are met in upwelling systems. Examples of various algorithmic approaches to derive chlorophyll and effective cell diameter for the Benguela upwelling system (Fig. 18.2) illustrate the need for appropriate algorithm choices for HAB applications.

Aside from atmospheric correction and Case 2 "signal" issues, many challenges lie in the fundamental ability to quantitatively measure, identify, and characterize the phytoplankton community and subcommunities—an issue of relevance to both the PFT and HAB communities. Collaboration between the ocean colour and phycological research communities is needed, to develop better routine means of measuring and characterizing phytoplankton community structure, e.g., imaging flow cytometry, if HAB or PFT capabilities from ocean colour are to be fully realized—within the unavoidable constraints of the biomass dependent contribution that phytoplankton community structure makes to the gross ocean colour signal.

Ocean colour observations are most suited to the observation of relatively high-biomass surface blooms—where they can provide very high-value and low-cost products to a wide variety of research and operational users. These observations provide considerable value to the relatively simple routine detection of high-biomass blooms, providing that there is disambiguation from sediment-dominated turbid waters (e.g., Park et al. 2010) and that algorithms capable of operating at high phytoplankton biomass, e.g., >30 mg m^{-3} chlorophyll *a*, are used (Gurlin et al. 2011). Such approaches can provide first-order assessments of synoptic bloom and eutrophication risks, even without more detailed taxonomic information. In ecosystems where high-biomass blooms are typically associated with a single species, such as *Karenia brevis* in the Gulf of Mexico, a variety of ocean colour approaches, ranging from biomass anomaly detection through to more sophisticated bio-optical modelling, have been successfully used to identify bloom occurrence and type (Blondeau-Patissier et al. 2014). Cyanobacterial blooms are relatively easily identified with ocean colour, due to their typically high biomass, their frequently highly scattering properties associated with intracellular vacuoles (Matthews and Odermatt 2015), and their distinctive spectral signatures associated with phycobiliproteins (Kudela et al. 2015). These capabilities have been used both for extensive decadal studies of cyanobacterial occurrence (e.g., Matthews 2014; Shi et al. 2017) and for the development of operational monitoring systems (e.g., Schaeffer et al. 2015).

Even though there are some constraints in the use of ocean colour for HAB observations, the outlook from a sensor perspective is extremely positive. The Ocean and Land Colour Imagers (OLCI) on the Sentinel 3 satellite constellation are extremely well configured for HAB applications from a spectral and radiometric perspective, and there will be a series of eight satellites in fully operational mode for the next 2 decades (Donlon et al. 2012). In addition, the Sentinel 2 and Landsat 8 satellite series provide free and useable ocean colour data at a 20–60 m spatial resolution, opening up new scales of HAB observation for near coastal and inland waters (Kutser et al. 2016). The availability of space-based hyperspectral sensors, such as NASA's PACE mission in 2022, will further improve the capability to detect HABs, i.e., provide the ability to make broad taxonomic distinctions at somewhat lower phytoplankton biomass than with multispectral sensors, but will still not overcome the inherent signal-related constraints outlined above.

18.3 Advances in Detecting Environmental Conditions

Tools for detecting HABs, via genetic, imaging, or optical signatures, are clearly advancing our knowledge of blooms once they do occur. Yet, unless the conditions that lead to these events are better understood, our ability to predict their occurrence will remain limited. Fortunately, there are also advances in monitoring and resolving the complex environmental conditions that precede, are concurrent with, or follow a HAB event. Resolving antecedent conditions of blooms is necessary to understanding how and why a bloom occurred; resolving environmental conditions following such an event may help to understand environmental impacts, such as development of hypoxia.

The tools available for continuous monitoring of many water quality parameters have become very affordable. Instruments such as the Xylem (YSI), Hach, Eureka, and others can measure water temperature, salinity, pH, chlorophyll, turbidity, and dissolved oxygen in one small package and are in use in many locations. In the Chesapeake Bay, USA, for example, a network of such instruments has been developed. Yet, while such data are useful for understanding general water quality conditions, these parameters are insufficient to resolve conditions that may lead to outbreaks of certain HABs. They are especially useful, though, in characterizing impacts such as low dissolved oxygen conditions. Extensive hypoxic or dead zones do not need to develop for severe impacts to occur; in many waters, low dissolved oxygen can occur on a nightly basis and without in situ water sampling, such a condition might not be readily observed with traditional water sampling.

Understanding nutrient conditions before, during, and after HAB events is especially helpful, as nutrient enrichment and changes in nutrient quality and proportion can profoundly impact susceptibility of a system to certain types of HAB outbreaks [e.g., Glibert et al. 2005, 2016; see also Chap. 12, Glibert et al. (2018a)]. However, in situ nutrient sensors have witnessed variable success and application, depending on the particular nutrient.

Nitrate (NO_3^-) sensors have the longest history. Ion-selective electrodes (ISE) for NO_3^- were developed in the 1970s and have experienced widespread use, especially in fresh waters where salinity interference is minimal. While there is also an ISE for ammonium (NH_4^+), it has less applicability, as its ionic interference is high and its resolution low (Table 18.1; Pellerin et al. 2016). There are also UV sensors for NO_3^-, but not for other nutrients, but while they are relatively easy to deploy, they lack sensitivity and suffer interferences in salt water. The only technology that is capable of sensing all forms of nutrients is that of wet chemistry, i.e., the spectrophotometric methods that have long been applied in benchtop analysis. In the in situ applications of wet chemistry, the reagents and detectors are housed in a chamber that either sits on a buoy with water pumped through it or it is placed underwater where it sips samples at preset intervals, interspersed with sipping of a standard solution. In situ wet chemical methods provide the most complete suite of data relevant for HABs, and a high resolution and accuracy, but are expensive in initial investment and maintenance costs (e.g., Glibert et al. 2005, 2008). With proper attention, the wet chemistry instruments can overcome interferences from suspended material, dissolved humics, and rapid fouling, not all of which can be overcome in the optical or ISE sensors (Table 18.1; Pellerin et al. 2016). A recent federal-state-industry partnership incentivized the development of a comparatively inexpensive NO_3^- and PO_4^{3-} sensor (www.nutrients-challenge.org/), but NH_4^+ sensor development lags behind. Absence of NH_4^+ instruments will limit applications for HAB interpretation and ultimately predictive capability. While ISE sensors have had some success on ROVs or other nonstationary

Table 18.1 Advantages and disadvantages of commercially available nutrient sensor technologies

Type	Principle	Advantages	Disadvantages
Ion-selective electrodes (ISE)	Direct potentiometry between a sensing electrode and a reference electrode	Inexpensive (<\$1000) and easy to use Fast response times Not influenced by colour or turbidity Available for NO_3^- and NH_4^+	Low resolution, accuracy, and precision Subject to ionic interferences High instrument drift Limited shelf life
Optical (UV) sensors	Spectral absorption by a photometer	High resolution, accuracy, and precision Chemical-free Fast response time Additional optical information in spectra	Expensive (>\$15,000) High power requirement and maintenance costs Available for NO_3^- only
Wet chemical sensors	Wet chemical colourimetric reaction with detection by photometry	High resolution, accuracy, and precision Potential for in situ calibrations Relatively fast response times Available for NH_4^+, NO_3^-, PO_4^{3-}, $Si(OH)_4$ (limited availability for urea)	Expensive (>\$10,000) High power requirement and maintenance costs High potential for fouling Requires reagents and generates wastes

Reproduced and modified from Pellerin et al. (2016) under Creative Commons licence

platforms, the wet chemistry instruments have not achieved the same success on these platforms owing to their larger footprint as well as high power demands.

Deployments of wet chemistry nutrient monitors in Chesapeake Bay and East China Sea illustrate the type of nutrient variability that may be relevant to HABs. Both of these examples highlight the facts that (1) variability in one nutrient form is not necessarily the same as variability in another nutrient and (2) excursions from background levels can be short-lived but large, injecting sufficient nutrient for a biotic response.

Numerous deployments of continuous wet chemistry nutrient monitors have been undertaken in Chesapeake Bay, USA (Glibert et al. 2005, 2008). A deployment in the Pocomoke River, one of the heavily agricultural tributaries of Chesapeake Bay, shows how rainfall events were associated with increases in PO_4^{3-}. These increases had two components. Coincident with the rainfall events, PO_4^{3-} "spikes" were recorded which lasted only hours, after which concentrations remained elevated for several days (Fig. 18.3a; Glibert et al. 2005). High PO_4^{3-} levels were implicated in major outbreaks of toxic dinoflagellates in this region. Local rainfall in this area is likely intense enough to cause erosion of PO_4^{3-} laden soils, with ultimate release of PO_4^{3-} to the water. A different pattern was found for NO_3^- following rainfall events. Maximum NO_3^-+NO_2^- concentrations lagged rainfall events by several days, suggesting either a groundwater discharge following the rain or microbial transformation of reduced N (e.g., NH_4^+ and urea)

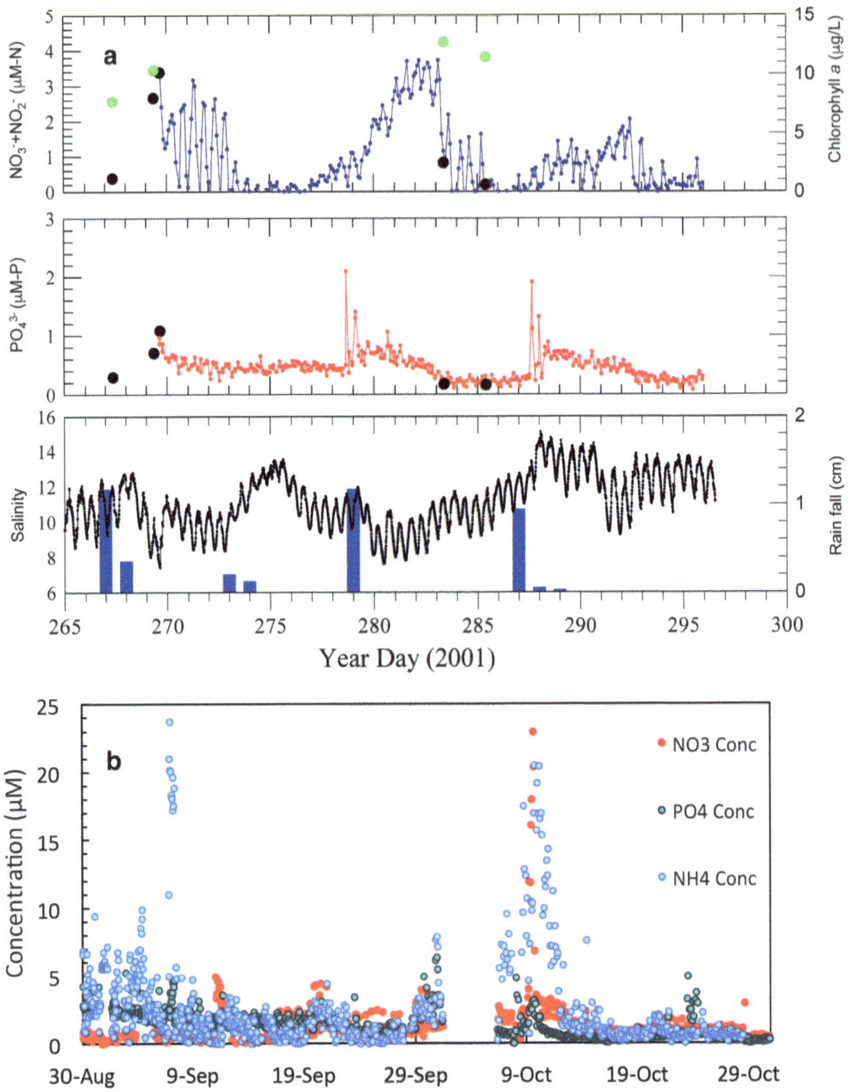

Fig. 18.3 Time series of environmental parameters based on in situ measurements. Panel (**a**): nutrients (*upper panels*) from in situ nutrient sensors deployed in the mesohaline Pocomoke River (tributary of Chesapeake Bay) in late summer of 2001 and corresponding rainfall and salinity (*lower panels*). The *solid green* points in the *upper panels* are chlorophyll *a* taken from small boat sampling. Reproduced from Glibert et al. (2005) with permission of the publisher. Panel (**b**) Time series of nutrients in the Tred Avon River (tributary of Chesapeake Bay) in 2016. Note the ephemeral but highly elevated pulses of NH_4^+ on two occasions, but only one of which had a simultaneous elevation of NO_3^-. Data from V. Kelly (pers. comm)

following its delivery. Similar lags in NO_3^- concentrations following rain events were found in deployments in the Corsica River, another Chesapeake Bay tributary (Glibert et al. 2008), and were also reported by Caffrey et al. (2007) following rainfall in a California estuary. In deployments in the Corsica River, increases in chlorophyll *a* concentrations, dominated by flagellates including mixotrophic dinoflagellates, were observed within days of increases in NH_4^+, which also lagged rainfall events. In a different deployment in yet a different Chesapeake Bay tributary, the mesohaline Tred Avon River, two large pulses of NH_4^+ were observed (Fig. 18.3b; V. Kelly, Green Eyes LLC, pers. comm.). These pulses, with concentrations exceeding 20 μM, deviations from background levels that are normally <5 μM, occurred coincident with very moderate rainfall events and lasted only a period of days or less. Note that the second-nutrient excursion from baseline included both NO_3^- and NH_4^+, whereas the first excursion was only NH_4^+. While they were short-lived, they represented substantial changes in the nutrients available for algal growth. Fewer data are available for continuous wet chemistry measurements of urea, but in the same Corsica River deployment noted above, a several μM-N excursion in urea was also coincident with rainfall concentration on a several-day scale (Glibert et al. 2008).

A month-long deployment of wet chemistry nutrient sensors in East China Sea also revealed two periods of abrupt elevation in nutrients, and, as in the Chesapeake Bay example, NO_3^-+NO_2^- variations were not coupled with those of PO_4^{3-}, but both periods when nutrient elevations were observed were associated with chlorophyll *a* responses (Wang et al. 2017). The East China Sea study also included in situ wet chemistry detection of SiO_3^{2-}, and thus draw-down ratios of different nutrients could be calculated. The change in nutrients with the increase in chlorophyll *a* differed by nutrient. The ratio of the daily averaged $NO_3^- + NO_2^-$ to PO_4^{3-} uptake, $\Delta N{:}\Delta P$, was 18.9 ± 8.8 during the latter part of this first bloom period, but during the second period there was also a significant draw-down of SiO_3^{2-}, and the salinity-normalized $\Delta N{:}\Delta Si$ was 0.59 ± 0.45, indicating that a large amount of SiO_3^{2-} was consumed. From a comparison of the $\Delta N{:}\Delta P$ ratios between the two events, it can be deduced that the two algal blooms were composed largely of different phytoplankton taxa. The first chlorophyll *a* peak was attributed to an accumulation of dinoflagellates (or other non-diatom taxa), while the second peak was attributed to an accumulation of diatoms. These data underscore several important aspects of the relationships between nutrients and HABs, phenomena well reported but not always well appreciated. That is, the dynamics of different nutrients and different nutrient forms are uniquely variable, and reliance on infrequent nutrient measurements to infer relationships with HABs is simply inadequate, as would be reliance on only a single nutrient measured with high frequency (e.g., NO_3^-). These types of data are furthering our understanding of the antecedent conditions of blooms, and these methodologies need to continue to advance.

18.4 Advances in Detecting Toxins

In parallel with advancements in instrumentation to detect cell abundances and blooms, many new approaches have been developed to detect their toxins. With new methodologies has also come knowledge of many new, previously unidentified toxins. The structural diversity of toxins, in some cases their multiple congeners with different properties, increases the challenges for analytical separation, and analysis. This brief review touches on some of the many advances of these new methodologies in achieving many of the goals of increased sensitivity and reliability and of reducing the time for analysis and the costs relative to historical methods.

One of the most important advances is the application of passive samplers for algal toxins. These devices, termed solid phase adsorption toxin trackers (SPATT), incorporate a resin in a mesh bag that can be suspended in water and, after a period of time, can be retrieved, and the toxin is extracted and analysed (MacKenzie et al. 2004). The benefits of these devices are manyfold. As a sampler, they are inexpensive and can be deployed in multiple places and for varying periods of time, yielding a library of spatial coverage and time-integrated results that cannot be easily obtained using classic samples of water or shellfish. These samplers may find application as early warning systems for conditions in which toxin presence and cell presence are uncoupled and, relatedly, may be useful for characterizing or quantifying chronic toxin exposure. They may be also useful in situations where direct sampling of shellfish or other biotic resources is difficult. As passive samplers, these devices have also been used in AUVs or other nonstationary platforms, depending on the period of time for which they are submerged. On the other hand, regulatory application of this technique is limited and awaits further development and validation against standard biotoxin measures (MacKenzie et al. 2004).

The effectiveness of SPATT sampling is a function of several aspects of its implementation. There is much scope for continued investigation and optimization of sorbents for various toxins (Zhang and Zhang 2015). The polymeric resins initially used were found to be effective in detecting lipophilic toxins, such as DA (e.g., MacKenzie et al. 2004; Pizzaro et al. 2013). More recently, using different absorption substrates, the application of this approach to water soluble toxins such as anatoxin-a and homoanatoxin-a (Wood et al. 2011) and hydrophilic toxins, including DA and saxitoxin (STX), has shown to be an effective semiquantitative tool for biotoxin detection. Extraction and analytical methods do vary by toxin. The toxin must be extracted from the resin, and percent of toxin recovered can be quite variable (e.g., Lane et al. 2010). Extracts can be analysed using various approaches, ranging from ELISA using commercially available kits, or by more analytically sensitive (but perhaps more costly) approaches such as liquid chromatography–mass spectrometry (LC–MS) or tandem MS approaches (Zhang and Zhang 2015). In fact, LC–MS/MS has become the preferred methodology for extremely low-level detection and identification of new toxins. ELISAs can be subject to matrix effects (false positives due to compounds in particular tissues or fluids) and must be used at proper dilutions in order to eliminate the possibility of false positive readings. In addition, ELISA results do not always correlate with total toxicity but correlate with affinity of an antibody to a particular toxin structure. For example, a STX ELISA will not recognize all toxin isoforms equally (e.g., STX, neo-STX, gonyautoxins). In contrast, because DA is the major isomer produced by the various species of the diatom, *Pseudo-nitzschia*, the DA ELISA assay is effective in estimating total toxicity. Other methods for toxin testing are the receptor-binding assay (most typically used for STX analysis) that measures total toxicity and is based upon the recognition of the toxin by a purified receptor, usually from rat or pig brain. The LC–MS method measures individual toxins with high molecular specificity and detection sensitivity. The choice of method(s) to be used depends upon equipment available, needs for screening large numbers of samples, and the need for absolute quantification or estimation of toxins in a particular sample.

Beyond monitoring for toxins, there are also advancing technologies for assessing the extent to which algae are synthesizing toxins. There are a number of both housekeeping genes and genes associated with synthesis of specific toxins that can be detected using molecular probes (Zhang and Zhang 2015). Microarrays and quantitative polymerase chain reaction (qPCR) techniques are not only useful in detecting and estimating, quantitatively, specific HAB species, but qPCR has now been used successfully in probing the gene coding for the production of the toxin microcystin (Zhang and Zhang 2015) and a range of other algal toxins. Many cyanobacteria genomes have now been sequenced, revealing toxin biosynthetic gene clusters that may lend themselves to understanding and monitoring bloom toxicity, and progress is advancing with other HABs.

Additionally, progress is also advancing with in vitro assays that may help with understanding impacts and mode of action on targeted organisms affected by toxins. While the classical mouse bioassay is fading from common use, bioassays of fish or oyster larvae continue to be widely applied, and large numbers of assays can be performed without sacrificing large animals. The use of -omics or systems biology approaches, including genomics, transcriptomics, proteomics, and metabolomics, to investigate biotoxin exposures is limited to date, but these methods hold promise for the development of biomarkers that can inform us of the role of biotoxins in unexplained mortality events or the extent of chronic exposures within marine mammal populations. -Omics have evolved over the past two decades into powerful tools to aid in understanding how biological networks respond to perturbations (Veldhoen et al. 2012).

The advancement in instrumentation (sequencing, mass spectrometry), bioinformatic algorithms (for analysing genomic and proteomic data), and public databases of DNA sequences, peptide profiles, and metabolites continues at a rapid pace that will make -omics approaches increasingly applied in both organismal and toxin detection.

18.5 Monitoring to Modelling: Nowcasts and Forecasts

The tools and technologies briefly reviewed are one piece of the puzzle. Linking improved understanding of antecedent conditions, with understanding of cell behaviour (diel cycles, encystment-excystment, thin layer formation, etc.), with physical processes that may aggregate or disperse cells, and along with knowledge of coastal typology and extent of retention of nutrients, for example, requires continued advancement and refinement of algorithms and models.

Many regions in the USA and elsewhere have moved predictive models into operational status. Different types of HABs have different suites of models and measuring/monitoring tools and data that populate these models. The key to operational status is that predictions are made on a regular, uninterrupted basis in a manner that citizens and managers can rely upon. In addition to nowcast models, there is also a need for a different class of models that allow forecasting of future conditions [see also Chap. 19, Franks (2018)], particularly in light of proposed management actions. In the USA, the National Oceanic and Atmospheric Administration (NOAA) supports a HAB operational forecast system for several regions (https://tidesandcurrents.noaa.gov/hab/faq.html). In the case of the Gulf of Mexico, nowcasts and short-term (days scale) forecasts of *Karenia brevis* blooms are based on satellite imagery, field observations, models, public health reports, and buoy data to provide the large spatial scale and high frequency of observations. Information is conveyed regarding potential development, intensification, transport, and associated impacts of *K. brevis* blooms. Similar operational forecasting models have been developed or are a priority for advancement in other regions, including Lake Erie (cyanobacteria), Gulf of Maine (*Alexandrium fundyense*), Washington Coast (*Pseudo-nitzschia*), California coast, and Chesapeake Bay. A challenge for many of these regions and the development of operational models is that they routinely experience multiple different types of HABs.

Managers have other needs for models that may allow interrogation of how a system may respond with future changes—whether it be climate change, a targeted nutrient reduction action, or risk to a developing aquaculture region. Such models build on robust, spatially explicit platforms [i.e., Regional Ocean Modeling System (ROMS) models] and integrate physics, nutrient cycling, physical factors, and nutrient physiology to better predict the long-term changes. There is mounting evidence that climate variability and a higher frequency of extreme weather events will escalate with climate change (Solomon et al. 2007). Climate warming not only directly affects rates of growth of harmful species but also affects the environment indirectly through increased stratification, nutrient recycling, and other changes in biogeochemistry [e.g., Edwards et al. 2006; Glibert et al. 2014; see also Chap. 5, Wells and Karlson (2018)].

As an example, for the Chesapeake Bay, a recent synthesis by Najjar et al. (2009) of Mid-Atlantic climate projections shows that the projected warming will exceed the current interannual variability, that winter/spring precipitation increases are likely, and that there is significant uncertainty in summer and annual projections of precipitation. To project the impacts of climate change on HAB species, simulations were applied from the North American Climate Change Assessment Program (NARCCAP), which are based on dynamical downscaling—embedding a regional climate model (RCM) into a global climate model (GCM)—at 50 km resolution (Mearns et al. 2009). The NARCCAP simulations afford unprecedented resolution in projected air temperature and precipitation over the Chesapeake Bay watershed. Results suggest that despite large model-to-model variability, the projected change is a substantial increase in mean river flow during the winter—spring but decreases in summer. Such changes in river flow have large implications for nutrient runoff and salinity, both of which will impact habitat suitability for specific HABs.

Through scenario testing, it is expected that the effect(s) of specific management actions (e.g., specific targeted nutrient reduction efforts) will be quantifiable in terms of effects on the plankton community composition, including HABs. Such models can therefore be used to provide scenario outcome to such questions as: If nutrients are reduced by x amount, what will be the PFT and HAB species response? What will be the comparative effect of reducing N versus P? What will be the impact of implementation of TMDL criteria? How will such changes in the stoichiometry of nutrient loads alter the likelihood for proliferation of these HABs? How will effects of such reductions vary if temperatures rise or if flow regimes change, i.e., the synergy of nutrient changes and climate related changes? The end result will be more cost effective nutrient reductions that should have significant effects on reducing the numbers of HABs, their frequency or their duration, or a better understanding of changes to be expected with climate change. Only through such models can the relationships between climate forcing and the timing of nutrient loads relative to seasonal warming be understood. Without doubt, such models require robust data for development and calibration along with open communication and collaboration with water quality managers whose needs are being modelled.

18.6 Conclusions and Current/Future Challenges

The overarching needs for HAB detection and ultimately prediction are to have tools available that are affordable, responsive in real time, and reliable. Most of the technologies discussed herein are expensive, requiring investments of many tens of thousands of dollars for a single instrument, and require trained personnel for upkeep and management of the data quality. Given the costs of some of the single instruments now available, spatial coverage is generally minimal at best. Agencies or institutes may have a single instrument that is deployed at a sentinel site—or more likely, may be deployed at a site that is simply the most convenient in terms of existing infrastructure or routine access. The most powerful approaches in interpretation of blooms and their associated environmental conditions come from the synergy of methodologies applied. As an example, during the recent *Pseudo-nitzschia* bloom that blanketed the west coast of the USA during 2015–2016, AUV surveys were conducted during contrasting upwelling conditions and supplemented

ESP instruments moored in the mixed layer of Monterey Bay, while satellite remote sensing was used for regional context (Ryan et al. 2016). Based on such data, the AUV was able to target specific bloom patches and conduct high-resolution surveys. Moreover, long-term time series data, including nutrients, allowed interpretation of macronutrient anomalies during the bloom, and the potential role of altered nutrient stoichiometry in growth and toxin production as the bloom progressed.

Beyond the expense of instrumentation and other challenges described herein, there are also large issues related to the engineering of physical platforms or gliders in terms of power, physical space to accommodate instruments, telecommunications, drift, fouling, and boat strikes or theft. The revolution in imaging and probe techniques, miniaturization, in situ continuous data streams, and -omics data is helping to advance our knowledge on new organisms, pathways, and their variation across species and ecosystems, in space and time. However, this revolution in techniques is creating new challenges in bioinformatics and data interpretation. Huge data streams, the so-called data deluge (Baraniuk 2011; Pellerin et al. 2016), must be calibrated against appropriate standards, processed, and analysed in near real time, and there must be data management tools and appropriate communication tools to allow these data to be available for use by management communities on relevant time scales. Even more daunting is the challenge of incorporation of these new data streams into ecosystem models—most of which are now based on bulk biogeochemical properties and rates (Coles and Hood 2016). Concepts of microbial diversity, mismatches between model structure and data, and the linking of biogeochemical processes to gene expression are difficult enough, even without the computational challenges of increasingly complicated models with enormously complex data (Coles and Hood 2016).

The needs going forward will continue to be great. Investments in infrastructure (equipment and tools) and in conceptual understanding of the antecedent conditions promoting each different type of HAB, as well as improved data management and modelling at all levels, will be required. The avalanche or deluge of data, whether from continuous streams of standard water quality parameters or new types of data from the -omics analyses, can and will be overwhelming but will present exciting opportunities for obtaining data on the temporal, spatial, and organismal diversity levels that will lead to a new generation of tools and models and new understanding of the dynamics of blooms. Nowcasts are urgently needed for many parts of the world, while longer-term planning requires a different suite of modelling approaches. Managers need models that are user-friendly. Development of cell phone apps, for example, is a goal whereby the public, public health officials, or those making their livelihood on the water can be notified if beaches, lakes, or other recreational areas are safe for swimming or fishing or if managers are meeting targets in terms of total nutrient loads. It is the hope that investments at all levels, from local municipalities to countries, will be made to advance these needs, and through global exchanges of knowledge, such as that through GlobalHAB and other scientific networks, the goal of protection of water resources and human health can be achieved.

Acknowledgments This is a contribution of the GEOHAB Core Research Project on HABs in Eutrophic Systems and HABs in Upwelling Systems. We thank V. Kelly of Greens Eyes, Inc., Easton, MD, USA, for supplying unpublished nutrient data and for helpful comments. Support for the preparation of this paper was provided in part by NOAA award NA17NOS4780180 to ML and PMG. This is contribution number 5408 from the University of Maryland Center for Environmental Science and NOAA ECOHAB contribution number 919.

References

Anderson DM (2009) Approaches to monitoring, control and management of harmful algal blooms (HABs). Ocean Coast Manag 52(7):342. https://doi.org/10.1016/j.ocecoaman.2009.04.006

Babin M, Cullen JJ, Roesler CS et al (2005) New approaches and technologies for observing harmful algal blooms. Oceanography 18(2):210–227

Baraniuk RG (2011) More is less: signal process and the data deluge. Science 331:717–719

Bernard S, Pitcher G, Evers-King H et al (2014) Ocean colour remote sensing of harmful algal blooms in the Benguela System. In: Barale V, Gade M (eds) Remote sensing of the African seas. Springer, Dordrecht, pp 185–203

Blondeau-Patissier D, Gower JFR, Dekker AG et al (2014) A review of ocean color remote sensing methods and statistical techniques for the detection, mapping and analysis of phytoplankton blooms in coastal and open oceans. Prog Oceanogr 123:123–144

Brewin RJW, Hardman-Mountford NJ, Lavender SJ et al (2011) An inter-comparison of bio-optical techniques for detecting dominant phytoplankton size class from satellite remote sensing. Remote Sens Environ 115:325–339

Caffrey JM, Chapin TP, Jannasch HW et al (2007) High nutrient pulses, tidal mixing and biological response in a small California estuary: variability in nutrient concentrations from decadal to hourly time scales. Estuar Coast Shelf Sci 71:368–380

Camoying MG, Yñiguez AT (2016) FlowCAM optimization: attaining good quality images for higher taxonomic classification resolution of natural phytoplankton samples. Limnol Oceanogr Methods 14:305–314

Campbell L, Olson RJ, Sosik HM et al (2010) First harmful *Dinophysis* (Dinophyceae, Dinophysiales) bloom in the U.S. is revealed by automated imaging flow cytometry. J Phycol 46:66–75

Coles VJ, Hood RR (2016) Approaches and challenges for linking marine biogeochemical models with the "omics" revolution. In: Glibert PM, Kama TM (eds) Aquatic microbial ecology and biogeochemistry: a dual perspective. Springer, Geneva, pp 45–66

Dall'Olmo G, Gitelson AA, Rundquist DC et al (2005) Assessing the potential of SeaWiFS and MODIS for estimating chlorophyll concentration in turbid productive waters using red and near-infrared bands. Remote Sens Environ 96:176–187

Donlon C, Berruti B, Buongiorno A et al (2012) The global monitoring for environment and security (GMES) sentinel-3 mission. Remote Sens Environ 120:37–57

Doucette GJ, Mikulski CM, Jones KL et al (2009) Remote, subsurface detection of the algal toxin domoic acid onboard the Environmental Sample Processor: assay development and field trials. Harmful Algae 8:880–888

Edwards M, Johns DG, Leterme SC et al (2006) Regional climate change and harmful algal blooms in the northeast Atlantic. Limnol Oceanogr 51:820–829

Erickson JS, Hashemi N, Sullivan JM et al (2012) In situ phytoplankton analysis: there's plenty of room at the bottom. Anal Chem 84:839–850. https://doi.org/10.1021/ac201623k

Evers-King H, Bernard S, Robertson Lain L et al (2014) Sensitivity in reflectance attributed to phytoplankton cell size: forward and inverse modelling approaches. Opt Express 22:11536–11551

Franks PJS (2017) Recent advances in modelling of harmful algal blooms. In: Glibert PM, Berdalet E, Burford M et al (eds) Global ecology and oceanography of harmful algal blooms. Springer, Cham, pp 359–377

GEOHAB (2001) Global ecology and oceanography of harmful algal blooms, science plan. Glibert P, Pitcher G (eds), SCOR and IOC, Baltimore and Paris, 86 pp

Gitelson AA, Schalles JF, Hladik CM (2007) Remote chlorophyll-a retrieval in turbid, productive estuaries: Chesapeake Bay case study. Remote Sens Environ 109:464–472

Glibert PM, Al-Azri A, Allen JI et al (2018a) Key questions and recent research advances on harmful algal blooms in relation to nutrients and eutrophication. In: Glibert PM, Berdalet E, Burford M et al (eds) Global ecology and oceanography of harmful algal blooms. Springer, Cham, pp 229–259

Glibert PM, Allen JI, Artioli Y et al (2014) Vulnerability of coastal ecosystems to changes in harmful algal bloom distribution in response to climate change: projections based on model analysis. Global Change Biol. https://doi.org/10.1111/gcb.12662

Glibert PM, Berdalet E, Burford M et al (2018b) Harmful algal blooms and the importance of understanding their ecology and oceanography. In: Glibert PM, Berdalet E, Burford M et al (eds) Global ecology and oceanography of harmful algal blooms. Springer, Cham, pp 9–25

Glibert PM, Kelly V, Alexander J et al (2008) In situ nutrient monitoring: a tool for capturing nutrient variability and the antecedent conditions that support algal blooms. Harmful Algae 8:175–181

Glibert PM, Seitzinger S, Heil CA et al (2005) The role of eutrophication in the global proliferation of harmful algal blooms: new perspectives and new approaches. Oceanography 18(2):198–209

Glibert PM, Wilkerson FP, Dugdale RC et al (2016) Pluses and minuses of ammonium and nitrate uptake and assimilation by phytoplankton and implications for productivity and community composition, with emphasis on nitrogen-enriched conditions. Limnol Oceanogr 61:165–197

Gurlin D, Gitelson AA, Moses WJ (2011) Remote estimation of chl-*a* concentration in turbid productive waters—return to a simple two-band NIR-red model? Remote Sens Environ 115:3479–3490

International Ocean Colour Coordinating Group (IOCCG) (2014) Phytoplankton functional types from space. In: Sathyendranath S (ed) Reports of the International Ocean-Colour Coordinating Group, No. 15. IOCCG, Dartmouth, Canada

Kirkpatrick GJ, Millie DF, Lohrenz SE et al (2011) Automated, in-water determination of colored dissolved organic material and phytoplankton community structure using the optical phytoplankton discriminator. In: Proceedings SPIE 8030, Ocean Sensing and Monitoring III 8030, 80300E. https://doi.org/10.1117/12.884250

Kudela RM, Palacios SL, Austerberry DC et al (2015) Application of hyperspectral remote sensing to cyanobacterial blooms in inland waters. Remote Sens Environ 167:196–205

Kudela RM, Raine R, Pitcher G et al (2018) Establishment, goals, and legacy of the Global Ecology and Oceanography of Harmful Algal Blooms (GEOHAB) Programme. In: Glibert PM, Berdalet E, Burford M et al (eds) Global ecology and oceanography of harmful algal blooms. Springer, Cham, pp 27–49

Kutser T, Paavel B, Verpoorter C et al (2016) Remote sensing of black lakes and using 810 nm reflectance peak for retrieving water quality parameters of optically complex waters. Remote Sens 8:497

Lane JQ, Roddam CM, Langlois GW et al (2010) Application of solid phase adsorption toxin tracking (SPATT) for field detection of the hydrophilic phycotoxins domoic acid and saxitoxin in coastal California. Limnol Oceanogr Methods 8:645–660

MacKenzie L, White DA, Sim PG et al (2004) Solid phase adsorption toxin tracking (SPATT): a new monitoring tool that simulates the biotoxin contamination of filter feeding bivalves. Toxicon 44:901–918. https://doi.org/10.1016/j.toxicon.2004.08.020.

Matthews MW (2014) Eutrophication and cyanobacterial blooms in South African inland waters: 10 years of MERIS observations. Remote Sens Environ 155:161–177

Matthews MW, Odermatt D (2015) Improved algorithm for routine monitoring of cyanobacteria and eutrophication in inland and near-coastal waters. Remote Sens Environ 156:374–382

Mearns LO, Gutowski WJ, Jones R et al (2009) A regional climate change assessment program for North America. EOS Trans Am Geophys Union 90:311–312

Michalak AM, Anderson EJ, Beletsky D et al (2013) Record-setting algal bloom in Lake Erie caused by agricultural and meteorological trends consistent with expected future conditions. Proc Natl Acad Sci USA 110:6448–6452

Najjar RG, Patterson L, Graham S (2009) Climate simulations of major estuarine watersheds in the Mid-Atlantic region of the United States. Clim Change 95:139–168

Oliveira PB, Moita T, Silva A et al (2009) Summer diatom and dinoflagellate blooms in Lisbon Bay from 2002 to 2005: pre-conditions inferred from wind and satellite data. Prog Oceanogr 83:270–277

Olson RJ, Shalapyonok AA, Sosik HM (2003) An automated submersible flow cytometer for pico- and nanophytoplankton: FlowCytobot. Deep Sea Res I Oceanogr Res Pap 50:301–315

Park Y-J, Ruddick K, Lacroix G (2010) Detection of algal blooms in European waters based on satellite chlorophyll data from MERIS and MODIS. Int J Remote Sens 31:6567–6583

Pellerin BA, Stauffer BA, Young DA et al (2016) Emerging tools for continuous nutrient monitoring networks: sensors advancing science and water resources protection. J Am Wat Res Assoc 52:993–1008

Pizzaro G, Moroño Á, Paz B et al (2013) Evaluation of passive samplers as a monitoring tool for early warning of *Dinophysis* toxins in shellfish. Mar Drugs 11:3823–3845. https://doi.org/10.3390/md11103823

Raine R, Berdalet E, Yamazaki H et al (2018) Key questions and recent research advances on harmful algal blooms in stratified systems. In: Glibert PM, Berdalet E, Burford M et al (eds) Global ecology and oceanography of harmful algal blooms. Springer, Cham, pp 165–186

Ryan JP, Kudela RM, Birch JM et al (2016) Causality of an extreme harmful algal bloom in Monterey Bay, California, during the 2014–2016 northeast Pacific warm anomaly. Geophys Res Letts 44:5571–5579. https://doi.org/10.1002/2017GL072637

Schaeffer BA, Loftin K, Stumpf RP et al (2015) Agencies collaborate, develop a cyanobacteria assessment network. EOS 96. https://doi.org/10.1029/2015EO038809

Seltenrich N (2014) Keeping tabs on HABs: new tools for detecting, monitoring, and preventing harmful algal blooms. Environ Health Perspect 122:A206–A213

Shi K, Zhang Y, Zhou Y et al (2017) Long-term MODIS observations of cyanobacterial dynamics in Lake Taihu: responses to nutrient enrichment and meteorological factors. Sci Rep 7:40326. https://doi.org/10.1038/srep40326

Solomon S, Qin D, Manning M et al (eds) (2007) Climate Change 2007: the physical science basis. Cambridge University Press, Cambridge, p 996

Stumpf RP (2001) Applications of satellite ocean color sensors for monitoring and predicting harmful algal blooms. J Human Ecol Risk Assess 7:1363–1368

Stumpf RP, Culver ME, Tester PA et al (2003) Monitoring *Karenia brevis* blooms in the Gulf of Mexico using satellite ocean color imagery and other data. Harmful Algae 2:147–160

Tang DL, Kawamura H, Doan-Nhu H et al (2004) Remote sensing oceanography of a harmful algal bloom off the coast of southeastern Vietnam. J Geophys Res 109:C03014. https://doi.org/10.1029/2003JC002045

Veldhoen N, Ikonomou MG, Helbing CC (2012) Molecular profiling of marine fauna: integration of omics with environmental assessment of the world's oceans. Ecotoxicol Environ Safety 76:23–38

Wang K, Chen J, Ni X et al (2017) Real-time monitoring of nutrients in the Changjiang Estuary reveals short-term nutrient-algal bloom dynamics. J Geophs Res 122. https://doi.org/10.1002/2016JC012450

Wells ML, Karlson B (2018) Harmful algal blooms in a changing ocean. In: Glibert PM, Berdalet E, Burford M et al (eds) Global ecology and oceanography of harmful algal blooms. Springer, Cham, pp 77–90

Wood SA, Holland PT, MacKenzie L (2011) Development of solid phase adsorption toxin tracking (SPATT) for monitoring anatoxin-a and homoanatoxin-a in river water. Chemosphere 82:888–894

Zhang C, Zhang J (2015) Current techniques for detecting and monitoring algal toxins and causative harmful algal blooms. Environ Anal Chem 2:article 1. https://doi.org/10.4172/JREAC.1000123

Chapter 19
Recent Advances in Modelling of Harmful Algal Blooms

Peter J.S. Franks

19.1 Introduction

Models have proven to be extremely useful tools in HAB research, from synthesizing knowledge into a conceptual framework to testing hypotheses that would be otherwise intractable. Models have been used to tease apart the dynamics underlying observations, to simulating and predicting bloom events. And, as models improve in predictive ability, they are being increasingly used as management tools.

One of the goals of the Global Ecology and Oceanography of Harmful Algal Blooms Programme (GEOHAB) was to improve prediction of HABs by determining the ecological and oceanographic mechanisms underlying the organism's population dynamics. The approach was to integrate biological, chemical, and physical studies employing enhanced observational and modelling systems (GEOHAB 2001). GEOHAB has supported modelling through its many core projects and activities and in June 2009 supported a workshop on GEOHAB Modelling and Linking Observations to Predictions (GEOHAB 2011). One product of this workshop was a special issue of the *Journal of Marine Systems* containing papers covering a wide variety of modelling approaches—and their challenges.

Since the GEOHAB modelling workshop, I recently reviewed a decade of the modelling of harmful algal blooms (HABs) covering the period 1997–2012 (Franks 2014), building on an earlier review of HAB modelling up to 1997 (Franks 1997). Here I will synthesize advances in the modelling of HABs over the last half decade, 2012–2016, following the same organizational structure I introduced in Franks (2014).

To identify HAB models for this review, I searched *ISI Web of Science* using combinations of the key words "harmful algal bloom HAB model." Sifting through

P.J.S. Franks (✉)
Scripps Institution of Oceanography, University of California, San Diego, La Jolla, CA, USA
e-mail: pfranks@ucsd.edu

P.M. Glibert et al. (eds.), *Global Ecology and Oceanography of Harmful Algal Blooms*, Ecological Studies 232, https://doi.org/10.1007/978-3-319-70069-4_19

the thousands of papers that appeared in this search between 2012 and 2016, I reduced them to just over 100 papers, based on my access to the journals and the paper's subject matter. Selections from these papers form the basis of this synthesis.

Rather than concentrate on the scientific results of the models, I explore the models from a modeller's perspective: how the models were constructed and how they were used to gain insights into the system. Here I organize models into six categories: conceptual, empirical-statistical, process, diagnostic, predictive-simulation, and management.

Conceptual models are syntheses of ideas concerning the dynamics of a system. Conceptual models usually precede mathematical models, though many conceptual models have yet to be formulated as mathematical models.

Empirical-statistical models are often models built up through statistical analyses of observations. They may not include any cause-effect dynamics, but rather describe the system through statistical correlations, lags, and other relationships. Such models can be used in process, diagnostic, predictive, or management modes but are most often used as predictive or management tools. This is due to their relative tractability compared to the more numerically intensive dynamic models. Empirical-statistical models are only interpretable within the limits of the data used to create them; this presents a strong constraint to prediction in the face of changing conditions such as those driven by anthropogenic climate change. Such models are often used to predict one variable (e.g., HAB concentration) given measurements of other variables.

Process models are mathematical models designed to test particular hypotheses concerning the dynamics of an organism or system. They are often used to answer "what if?" questions—what if the organisms had no toxin? What if the organisms had geotaxis rather than phototaxis? What if the nutrient levels were doubled?

Diagnostic models tend to be more intimately tied to data than process models. Typically a diagnostic model is used to identify and quantify the dominant dynamics underlying a set of observations. A set of equations describes the system, and the model is run forward (or sometimes backward) in time, based on initial conditions (often observations) and external forcings (wind, light, nutrients, etc.) to attempt to reproduce the observations. Once an acceptable hindcast is produced, the model can be pulled apart to quantify the processes that were important in leading to the observed patterns. Diagnostic models tend to be interpolative—they help us understand the dynamics that occurred between sets of observations and the dynamics of properties that were not measured.

Predictive-simulation models go a step beyond diagnostic models in attempting to predict the occurrence (timing, location, spatial extent, intensity) of events, rather than hindcasting previous events. Predictive models are extrapolative—they give predictions or forecasts beyond the limits of existing data. Predictive models are a logical increment to diagnostic models—a good diagnostic model should contain the essential ingredients to reproduce future events. Predictive models require a thorough understanding of the essential forcings, conditions, and responses of the system; this knowledge is usually obtained through a progression of process and diagnostic models.

Management models are operational forecast systems. They are typically products of multiple tests of predictive models, to the point that the predictive models have some statistical power. Management models are often run as ensembles of different models or cases, to give some statistical robustness to the predictions or forecasts. State or federal agencies—rather than academic scientists—typically run such operational models, as they attempt to protect the public and inform fishers and those involved in coastal commerce.

In the following sections, I will review recent advances in each of these types of model. I will focus mainly on how they were used, rather than the scientific results they generated. I hope that this review will educate and inspire researchers to develop their own models of HAB dynamics and use them in fruitful ways to answer their questions. In particular, those who make the effort to gather dense sets of laboratory or field data can often benefit from a close collaboration with modellers: the models can reveal surprising facets of the dynamics underlying the observations that might not be obtainable in other ways.

19.2 The Models

19.2.1 Conceptual Models

Every model begins as a conceptual model. Before regressions can be run or equations can be written, the researcher has to have some sense of how to reduce an impossibly complex system (reality) down to what they perceive to be its essential elements. This includes the state variables (often the organisms and their constituents), the transfer functions that describe the interactions among the variables, and the forcings such as light, temperature, or nutrients. Conceptual models are the first step in all subsequent models; however, it is often useful to pause at the conceptual model stage to assess what is known and how the pieces might fit together.

Margalef's mandala (Margalef 1978; Margalef et al. 1979) is a useful conceptual model describing the types of organisms that might arise under different environmental conditions. This mandala was recently thoroughly reviewed by Wyatt (2014), who updated the mandala in light of contemporary knowledge of bloom-forming plankton, their life histories, their growth conditions, and their sources of mortality. Glibert (2016) has gone several steps farther in updating the mandala to include 12 dimensions: (1) preference for oxidation state of nitrogen, (2) inorganic nitrogen and phosphorus availability, (3) preference for high light and the ability to be mixotrophic, (4) motility, (5) ambient turbulence, (6) pigmentation, (7) temperature, (8) size, (9) relative growth rate, (10) toxin or reactive oxygen species production, (11) *r* vs. *K* strategy, and (12) grazing mortality. Glibert's analysis leverages recent advances in trait-based modelling of plankton (e.g., Litchman et al. 2013; Flynn et al. 2015), which invokes the existence of traits (such as size or nutrient preference) inherent to a species and the tradeoffs an organism must make in exhibiting a given trait. Glibert's synthesis presents a conceptual framework for subsequent mathematical modelling of HABs.

An interesting conceptual synthesis of the various physiological and ecological mechanisms leading to HABs is presented in Jeong et al. (2015). Using the dinoflagellate, *Cochlodinium (Margalefidinium) polykrikoides* as a model organism, Jeong et al. develop a hierarchy of four generation mechanisms (GM) for HABs. The most basic GM includes nutrients and light; GM2 adds the effects of swimming, while GM3 further includes mixotrophy. GM4 then puts these dynamics into the context of the ecosystem by more explicitly including competition and mortality due to other organisms. These GMs were then synthesized into a "biological interaction map" indicating the types of interaction (grazing, inhibition), the known species, and the direction of the interaction. Simple, general equations were used to describe each GM. These equations could not be solved as presented—the functional forms of the interactions would need to be specified, and the models parameterized with data. This approach, however, laid down the bones for a suite of mathematical models that could serve as testable hypotheses for *C. polykrikoides* HABs.

Conceptual models are particularly useful in synthesizing observations into a coherent description of the system. Accoroni et al. (2015) brought together temperature and nutrient data to understand the occurrence of *Ostreopsis* cf. *ovata* blooms in the Adriatic Sea. By exploring fluctuations and thresholds over several bloom cycles, they were able to construct a conceptual model of bloom dynamics (see Chap. 13, Berdalet and Tester 2018). Tobin et al. (2013) used a similar approach, combining physiological and behavioural observations of different strains of *Heterosigma akashiwo* to develop a strain-specific model explaining how different growth strategies result in different bloom time scales. Gobler and Sunda (2012) summarized knowledge of brown-tide species (*Aureococcus anophagefferens* and *Aureoumbra lagunensis*) to explore conceptual models of bloom formation in response to nutrients, grazing, and the ecological conditions preceding the blooms. A particularly visionary conceptual model was constructed by Visser et al. (2016) to generate predictions concerning the occurrence of cyanobacterial blooms in response to increased atmospheric CO_2 and global warming. Their model was built up through a synthesis of a diverse literature, ultimately helping to focus where future research efforts could be most effectively applied to improve our understanding and predictive ability.

Conceptual models thus serve as a simplified structure for synthesizing information about a complex system. This structure can then be used to motivate further research, as well as the basis for formulating mathematical models of the system.

19.2.2 Empirical-Statistical Models

Empirical-statistical models are used to quantify relationships among observations. Unlike dynamical models that use equations to evolve initial conditions forward in time, empirical-statistical models are often used to predict the value of a variable given measurements of other variables. In that sense, they are often static (no time dependence), revealing correlations and relationships rather than underlying dynamics. Such models were probably introduced to biological oceanography by Gordon Riley (e.g., Riley 1941) in his seminal studies of plankton fluctuations on Georges Bank and the adjacent continental shelf. Since that time, statistical techniques have improved markedly; much of the growth in HAB modelling studies over the past 5 years has been in empirical-statistical models.

Riley made (for that time) extensive measurements of temperature, nutrients, phytoplankton, and zooplankton and used those measurements to derive correlations among the variables. For a given temperature, phosphate, and nitrate concentration, for example, Riley could predict the phytoplankton concentration with some statistical accuracy.

Modest advances on Riley's multiple linear regressions can be found in the use of various forms of generalized linear models (GLM). The GLM differs from ordinary linear regression in allowing the variables to have non-normal distributions. Singh et al. (2014) used a logistic (or logit) form of GLM in which *Dinophysis* concentrations in the coastal Arabian Sea were categorized as being above or below a threshold concentration. A wide range of predictor variables was considered, with the aim of understanding the effects of climate change on *Dinophysis* bloom dynamics. The model was evaluated based on metrics such as the probability of detection, false detections, and an overall skill metric. The models showed that the N:P ratio was significantly correlated with *Dinophysis* abundance; blooms could be predicted based on linear relationships with sea-surface temperature and salinity, total suspended solids, and N:P. Given predictions of changes in these explanatory variables—and given the assumption that these relationships are unchanging – blooms of *Dinophysis* are predicted to become more frequent.

Feki et al. (2013) used a similar approach but employed a generalized linear mixed-effect model (GLMM) that includes both fixed predictors (the environmental data) and a random component (the sampling sites) for understanding the dinoflagellate *Karenia selliformis* blooms in the southwestern Mediterranean Sea. Particularly notable in this study was the careful approach to processing and reducing the data prior to application of the GLMM, including the use of a generalized additive model (GAM) to reduce redundancy among the different observations. In a GAM, the relationship between the observations and the predicted variable does not have to be linear—the GAM can be used non-parametrically to estimate a smoothed function relating the noisy observations to the predicted variable. Multiple predictors can be incorporated, each with different functional forms. Díaz et al. (2016) used GAMs to explore blooms of the dinoflagellate, *Dinophysis acuta* in an upwelling-influenced ria in Galicia, finding that a combination of a cumulative upwelling index, sea-surface temperature, and tidal range explained 77.3% of the deviance of *D. acuta* blooms. Interestingly, the model showed that the bloom relationship to the upwelling index and tidal range was nonlinear.

Among the many applications of empirical-statistical models to HAB research, I noticed two relatively new approaches that appeared frequently: Bayesian approaches and machine learning approaches. Bayesian statistics differ from "frequentist" approaches in assigning prior probabilities (with potentially non-normal distributions) to parameters which are then updated to a posterior probability given new data. This approach was used by Obenour et al. (2014) to improve forecasts of cyanobacterial blooms in Lake Erie; by testing different forms of the nutrient-load/nutrient-bloom relationship, each with a different statistical error distribution, they obtained a model that had both high-predictive skill and a statistical characterization of the prediction uncertainty.

One of the problems inherent to statistical modelling of HABs is that most of the time there is no HAB—there are many zero measurements. These often lead to Poisson or negative binomial distributions, though the abundance of zeros can lead to departures from these distributions. To accommodate such zero-heavy data, Cusack et al. (2015) employed a zero-inflated negative binomial (ZINB) model, a specific version of the more general Bayesian hurdle Poisson model used by Cha et al. (2014). In ZINB models the zeros are considered to arise from a mixture of the underlying distribution (negative binomial) and some second process that generates zeros. Hurdle models, on the other hand, separate the zeros (which arise from one process) from the non-zeros that arise from a different process that does not generate zeros. The goal of both models was to predict HABs: cyanobacteria (Cha et al. 2014) or the diatom genus, *Pseudo-nitzschia* (Cusack et al. 2015). Based on these models, Cha et al. were able to define nonlinear functions (and their confidence limits) relating cyanobacterial presence to water temperature and mean weekly freshwater outflow (flushing time) in Lake Paldang (Korea). Cusack et al. were able to use wind and sea-surface temperature to predict *Pseudo-nitzschia* in bays along the southwest coast of Ireland. The key to these statistically successful hindcasts was careful consideration of the non-bloom observations.

Machine-learning approaches to HAB prediction have been employed for several decades, though they appear to be more common in the last 5 years. Bayesian networks (Rigosi et al. 2015; Moe et al. 2016), neural networks (Gokaraju et al. 2012; Millie et al. 2014), and support vector machines (González Vilas et al. 2014; Nieto et al. 2015) have all been employed to understand and predict HABs. Machine-learning techniques take observations and use various algorithms to learn from them to make predictions.

Bayesian network modelling treats observations as "nodes," which are linked by probabilities, the whole network ultimately generating the probability of an outcome. Because of the two-way probabilistic linking of nodes, the complexity of Bayesian network models increases exponentially with the number of nodes. It is thus desirable to use a conceptual model to limit the complexity of the system. Rigosi et al. (2015) used the model itself to inform the ideal number of nodes by testing models of three, four, and nine nodes—each node being a different data source—with the objective of deriving a model to predict cyanobacterial blooms in lakes. Moe et al. (2016) used a nested approach, using separate modules for (1) the Bayesian network which fed into (2) a process model that was then linked to (3) monitoring data, finally generating (4) a classification system for cyanobacterial blooms in lakes. Both models were used to predict the changes in bloom occurrence under future climate warming scenarios.

Warming was found to increase bloom probability, though this depended on the lake's trophic status and coincident management strategies.

Neural networks (NNs) use layers of nodes to connect observations through different weightings. After being trained with a data set of known inputs and outputs, the NN can be used with new data to predict the outputs. Millie et al. (2014) used 31 hydrological and meteorological predictors (observation data types) to predict presence/absence of freshwater cyanobacterial blooms. To better understand the drivers of the blooms, Millie et al. whittled the predictors down to total phosphorus, temperature, and wind speed as the main drivers controlling bloom occurrence. From this they were able to derive conditions (particularly total phosphorus levels) that would help to mitigate cyanobacterial blooms.

The use of support vector machines (SVMs) to predict HABs is relatively new, even though this technique is particularly applicable to such problems. SVMs are commonly used to classify data; this is often a difficult task, as data may not be easily separable. Data are mapped onto "hyperplanes" on which the data points cluster in obvious ways and are therefore easily separated and classified. The hyperplanes are chosen to maximize the "margin"—the distance between a separating surface and the closest points to that surface on either side. Typically the margin is maximized while the overall error is minimized. As with NNs, the SVM is trained on a known data set; this trained system is then used to classify subsequent data. During the training, optimal parameter values are chosen for the SVM kernel to allow it to optimally classify the data. The end result is a system that can take observations of environmental and biological variables and classify them into, say, bloom and non-bloom conditions. Like other statistical regression techniques, the SVM will additionally give information concerning the relative importance of the data type in determining the output. This approach was used to both predict and understand the drivers of cyanobacterial blooms in a reservoir (Nieto et al. 2015) and *Pseudo-nitzschia* blooms in rias of Spain (González Vilas et al. 2014).

Empirical-statistical models are powerful tools when sufficient data are available. Though they are static in the sense that they do not include dynamic relationships among variables, with sufficient data (which usually means many bloom/non-bloom cycles), they can reveal relationships among environmental and ecological forcings and HAB responses. If conditions remain stationary, they are useful predictive tools and form the basis for most management models.

19.2.3 Process Models

The following three types of model—process, diagnostic, and predictive—tend to be numerical, rather than statistical models. These models contain "state variables" (e.g., nitrate, cell concentration, zooplankton concentration, etc.) connected by "transfer functions" (e.g., grazing rate, nutrient uptake rate, etc.). The dynamics of the model are determined by the choice of state variable, transfer function, and parameters that are used to describe the transfer function.

I have found it useful to distinguish three types of process model: applied mathematics (math), hypothesis testing, and sensitivity analysis. Applied math models tend to be fairly divorced from data, often using seemingly simple nonlinear systems (e.g., phytoplankton-toxic phytoplankton-zooplankton) to explore the model dynamics in relation to the model parameter space: bifurcations, chaotic behaviour, stability, sensitivity, etc. There have always been a substantial number of such models in the literature, and over the last 5 years, such models were the most numerous category. However, I will not be reviewing them here, as they tend not to be incorporated into laboratory or field research in any substantial way, so they remain largely mathematical curiosities for the HAB community.

Using models for hypothesis testing and sensitivity analysis has a long history in HAB research; over the past 5 years, the increase in the amount and availability of data has made such models extremely powerful research tools. Hypothesis-testing models span a wide range of physical and ecological dynamics. Butman et al. (2014) and Waters et al. (2015) both explored the importance of benthic or near-bottom seed populations for HABs, combining models and data in novel ways. Butman et al. used a sequence of physical models to assess sediment resuspension to predict the injection of benthic cysts of *Alexandrium fundyense* into the water column. The models gave spatially and temporally resolved measures of resuspension, including interannual variability from decades-long regional model runs. Exploring the relationship of resuspension to wind stress led to the hypothesis that strong northeast winds in spring could determine subsequent HAB intensity. This hypothesis is an excellent candidate for testing in the field.

Waters et al. (2015) used a novel combination of laboratory and field data, buoyancy-controlled drifters, and a STELLA-based model of organism behaviour to investigate the behaviour of *Karenia brevis* in benthic boundary layers. A particularly novel aspect of the model was that it was incorporated into the software controlling the drifters (in a quarry), so that they behaved similar to the plankton—their vertical movements were determined by their internal N and C stores, calculated from time of day, temperature, and light sensed by the drifters. The combination of data and models supported the hypothesis that near-bottom populations of *K. brevis* could be the inoculum for subsequent dense blooms.

Process models are particularly good tools for exploring how systems work and for testing particular mechanisms controlling system behaviour. Sunda and Shertzer (2014), Yamaguchi and Sai (2015), and Alves-de-Souza et al. (2015) all used models to explore top-down and/or bottom-up controls on HABS, combining laboratory and field data to parameterize, drive, and test their models. The Sunda and Shertzer and Alves-de-Souza et al. models were run without explicit physical forcing, while the Yamaguchi and Sai model was coupled to a one-dimensional (vertical) model to drive nutrient inputs from deep waters. Based on extensive laboratory data (including videos of *Heterosigma akashiwo* swimming), Harvey and Menden-Deuer (2012) developed a model to test the hypothesis that a predator-avoidance behaviour would lead to the formation of dense surface aggregations—HABs. Parameterizing a model with the observed behaviours (persistent upward swimming in the presence of predators but random swimming in their absence), they showed that dense layers of *H. akashiwo* formed at the surface as the organisms fled their predators, supporting laboratory observations.

These combinations of data and models are powerful tools for understanding the potential for certain dynamics to lead to HAB formation. Models are particularly good for experimenting with such dynamics: their intensity can be changed in a model to explore the outcome given different forcings. Warns et al. (2013) published what may be the first model of the complete life cycle of the dinoflagellate, *Biecheleria baltica*—complete with encystment/excystment dynamics and vegetative and sexual phases. The model was particularly useful in several respects: (1) it served as a platform to synthesize understanding of the *B. baltica* life cycle and physiological responses to environmental forcing, (2) it could be used to understand annual cycles in *B. baltica* abundances, and (3) the environmental forcing could be varied to quantify the sensitivity to changes in temperature. As with Moore et al. (2015) (discussed later), increased temperature was found to cause earlier, more intense blooms and the formation of more cysts.

Coupling HAB models to three-dimensional (3D) circulation models is still a relatively nascent field, but several recent papers demonstrate the utility of 3D models in elucidating transport pathways of HAB organisms. Pinto et al. (2016) used passive numerical drifters in a 3D circulation model of the Iberian coast, showing the possibility of local HAB presence based on transport of toxic cells from distant point sources. Hickey et al. (2013) used a similar approach employing virtual drifters in a well-constrained model of the Oregon-Washington coastal currents to evaluate the influences of source location, wind forcing, and presence and movements of freshwater plumes (Columbia and Fraser Rivers) on the occurrence of *Pseudo-nitzschia* blooms in the region. The model was used to test hypotheses formulated through examination of the field data: the model showed that the Columbia River plume could be both a barrier and a conduit for toxic cells reaching certain shellfish beds. Lai and Yin (2014) performed extensive experiments using their 3D physical model, to test whether physical convergence zones could account for locally dense HABs. They used both neutrally buoyant and vertically migrating numerical drifters in their model of a bay in northeast Hong Kong, testing the effects of stratification and wind on drifter accumulation. They showed that intense HAB patches could form through accumulation, a process not often included in HAB analyses.

Gillibrand et al. (2016) went a step further and included biological dynamics following the numerical drifters—the biology being determined by an individual-based model of *Karenia mikimotoi*. This approach allowed them to assess not only drift pathways but the likelihood of growth and mortality along those pathways. They tested hypotheses of the spatial structure of initial conditions (how the cells were distributed prior to blooms) and the importance of growth and mortality, vertical migration, wind forcing, advection, and temperature—each of these latter processes could be turned on or off in a model run. This is a particularly good example of the use of a relatively complex model for hypothesis testing. The process was made more efficient by using archived flow fields: only the biological models needed to be rerun for each experiment, saving considerable time over fully coupled models.

Ryan et al. (2014) took an interesting approach in modelling *Pseudo-nitzschia* in Monterey Bay: using a physical model that assimilated data such as satellite altimeter, satellite temperature, and in situ CTD data, they used an adjoint model to "predict" where HAB patches may have come from 3.5 days earlier. The adjoint technique runs the model backward and forward to obtain a statistically optimal solution of initial conditions (and sometimes forcings) that would lead to the final (observed) condition. Combining these model solutions with an extensive in situ data set allowed them to assemble a very complete story concerning the physical and biological dynamics governing the appearance of HABs in the bay.

Moore et al. (2015) used global climate model ensemble simulations to force a regional 3D model of Puget Sound. Their goal was to predict the occurrence of the dinoflagellate, *Alexandrium,* under the influence of climate change, and to use the model to understand the contributions of potential effects on HABs. Predicting out to the year 2069, Moore et al. were able to quantify how changes in heat flux, river flow, and coastal upwelling would affect the growth and proliferation of *Alexandrium*. The model experiments were interesting: upwelling, for instance, was assessed through a sensitivity analysis using atmospheric forcing from particular years chosen to represent long-term changes in wind stress. Through a suite of model experiments, heating was found to be particularly important in both increasing growth rates and prolonging the bloom season.

Process models remain one of the most important and useful tools in the HAB modelling repertoire. They are particularly useful for testing hypotheses and exploring the dynamics of HAB systems. Hierarchies of models can be used to investigate the influence of increasingly complex combinations of forcings and dynamics on HABs. Process models are typically the next logical step after a conceptual model and a necessary step to achieving a dynamic, predictive model.

19.2.4 Diagnostic Models

Once a model has been well-calibrated for a system, it can be used to diagnose the dynamics that led to a particular set of observations. In this way, a complex system can be reduced to a simpler one, or the importance of nonlinear events (a particular wind speed and direction at a certain time, a confluence of tides and wind, etc.) can be assessed. "Systems" can be simple laboratory experiments or complex field observations.

Lim et al. (2014) explored the interactions between *Cochlodinium polykrikoides* and the diatoms *Chaetoceros danicus*, *Skeletonema costatum*, and *Thalassiosira decipiens*. Experiments with different combinations of *C. polykrikoides* and the diatoms showed that the diatoms (and their exudates) decreased growth rates and swimming speeds of *C. polykrikoides*. Simple models were used to test hypotheses concerning the nature of the interactions (contact or chemical). Once the models were fit to the data, the parameters revealed threshold concentrations of diatoms that would inhibit *C. polykrikoides* growth. The model fits – obtained using an objective statistical criterion—also generated information on the unknown model parameters, such as the interaction strengths among the species. The model could thus be used as a diagnostic tool to recover missing or difficult-to-measure information.

Terseleer et al. (2013) took a similar approach, fitting a model to experimental data investigating domoic acid (DA) production in *Pseudo-nitzschia*. With the model parameterized to reproduce time-series experiments with different nutrient limitations, it could then be used to better understand the factors controlling DA synthesis and release. Numerical experiments thus supplemented laboratory experiments to gain understanding about this complex system.

The simplest form of diagnostic model using a 3D circulation model is particle-tracking. Aoki et al. (2012) followed particles representing the raphidophyte, *Chattonella antiqua*, in the Yatsushiro Sea (Japan); the particles were seeded in a bay that was presumed to be a source of blooms in the sea. Curiously, the model showed that the bay was unlikely to have been a source of the blooms. Instead, they likely originated from a coastal location in the sea itself. The model was then used to diagnose the dominant forcings for a sequence of blooms observed in the bay; different combinations of wind and river flow accounted for each of the blooms, showing that a single conceptual model was an inadequate explanation for the various HABs.

Li et al. (2014) went a step further than Aoki et al. in their examination of the advection of *Alexandrium fundyense* in the Gulf of Maine. They released particles every 5 days at 7 sites that had been previously identified as potential source regions and tracked the particles as they were moved about the gulf. The particle locations were used to produce Lagrangian probability density functions for giving the probability of particles released from a given site arriving at another location; these were used to construct a Gulf of Maine coastal connectivity matrix—the probability of a particle released from a given point arriving at another point in a certain amount of time. Through this process Li et al. could quantify the relative strength of source regions, and "attractiveness" of destination regions, as well as assess the interannual variability of transport pathways for *A. fundyense* HABs.

Giddings et al. (2014) took a similar approach to Li et al. (2014) in their study of *Pseudo-nitzschia* sources and advection pathways in the Pacific Northwest of the United States, releasing particles into a model semi-continuously at suspected HAB source regions. They also gave their particles behaviour—in particular a tendency to float. Before the model was used to test HAB pathways, the model currents and hydrography were tested against an extensive data set using skill metrics. The model was then teased apart to diagnose the influence of wind, the Columbia River plume, and source region on the probability of coastal HABs. This is a particularly thorough study embracing numerical modelling, statistical analyses, and thoughtful analysis of the model results to gain understanding of an extremely complex system.

Henrichs et al. (2015) added biological dynamics to their particle-tracking model of *Karenia brevis* in the Gulf of Mexico. In addition to a physiological model for each particle, the organisms (particles) were allowed to migrate vertically—interacting with the surrounding flows—based on their internal nutrient stores.

The interesting twist in this modelling study was that the model was run both forward and backward in time. Forward runs showed where cells from particular source regions might go, while backward runs showed where cells forming observed blooms might have originated. This combination of data, biological model, and physical model proved to be a uniquely powerful tool for gaining understanding of the dynamics underlying observed HABs in the region.

Diagnostic models have been applied to simple laboratory systems and complex natural systems to gain understanding of sources, pathways, and dynamics of HABs. The combination of models and data is synergistic, giving a tool that is more powerful than either approach individually. It is likely that with the advent of new observing systems and enhanced data streams, we will be seeing an impressive increase in the number and utility of diagnostic models of HABs.

19.2.5 Predictive Models

A well-calibrated diagnostic model can be transitioned to a mode in which it is used for out-of-sample forecasts: prediction. Accurate model predictions require that the model remains an accurate depiction of the system in the future, that is, there are no substantial changes in the ecosystem parameters or structure that could not be captured by the model. Predictive models are often used to forecast HABs, both in the near term and increasingly over climatological time scales.

Brown et al. (2013) coupled a detailed and well-calibrated 3D circulation model with empirical-statistical models of a wide range of potentially harmful species, to enable forecasts of—among other things—HABs in the Chesapeake Bay. For example, the model for *Karlodinium veneficum* was an artificial neural network (ANN), while the dinoflagellate, *Prorocentrum minimum*, model was a general linear logistic regression. The circulation model assimilated measurements and was forced with observations (meteorological, river inputs), generating habitat maps that provided the conditions for the HAB predictions. The authors made a significant effort to make the model solutions and predictions available in real time to interested parties. The coupling of the circulation model with empirical models of the various species—rather than a full ecosystem simulation including the target species—was made for a very efficient and statistically robust forecast system.

As useful as the Brown et al. (2013) model is for forecasting, it is not a dynamic model; it does not solve equations giving the rate of change of state variables based upon its sources and sinks. Walsh et al. (2016) developed a dynamic model of the West Florida shelf that includes 29 explicit (and 7 implicit) state variables, including HAB species such as *Karenia brevis* and *Cochlodinium polykrikoides*. The model was run in a 2D (cross-shore section) mode, requiring careful parameterization of the boundary conditions for the biological variables, such as zooplankton and bacteria. This model represents one of the most ecologically complex models formulated to date, presenting significant challenges in parameterization, implementation, and analysis. Such models, however, provide one of the few means to assess the interaction of HAB species with their ecosystem, rather than treating them as isolated species. This may be the key to gaining a deeper understanding of HAB dynamics.

Kibler et al. (2015) took an intermediate approach between Brown et al.'s use of empirical models for their HAB species and Walsh et al.'s dynamic models: they used empirical data to parameterize temperature-growth rate curves for five different ciguatera-producing species in the Gulf of Mexico and Caribbean Sea. Ensemble averages from a suite of 11 global climate models (GCM) were used to provide temperature data over decadal time scales to hindcast and predict interannual changes in the growth rates of the target species, using those empirical models. Given the GCM predictions of regional temperature changes, three species were predicted to become increasingly dominant, creating higher local risks of ciguatera poisoning, while the other two species were expected to expand their geographic ranges, creating new risks in areas of the western Atlantic.

Note that the Kibler et al. model did not include an ecosystem in which the target species were embedded; there was no mortality or possibility of changes in ecosystem structure in response to the temperature increases. Glibert et al. (2014) addressed this issue by modelling the habitat changes of HAB species, rather than the ecological dynamics of the species. Using a GCM forced with projected climate characteristics, Glibert et al. predicted changes in the HAB genera *Karenia* and *Prorocentrum* driven by nutrient loading and climate change in three regions of the globe: the Baltic Sea and Northeast Atlantic, the Northeast Asian coast, and Southeast Asia. Different patterns of range expansion/contraction were predicted in different regions by the end of this century. Again, this model did not include any HAB dynamics—only changes in the available habitat. Thus, changes in ecosystems or species invasions could not be accommodated in the predictions.

The models discussed here in the context of predicting HABs represent an interesting spectrum of approaches. They all used a well-calibrated physical model—regional or GCM—coupled to biological models varying from empirical to extraordinarily detailed ecosystem representations. The simpler biological models do not allow understanding of the potential for changes in the HAB organism's interactions with its ecosystem, while the detailed ecosystem model has the usual problems associated with poorly constrained parameterizations of physiological and ecological processes. Still, long-term forecasts suggest that some HAB species will become more prevalent or intense and/or will expand their geographic extent as newly favourable habitat opens up due to climate change.

19.2.6 Management Models

Management models are those used to help minimize the economic and health effects of HABs. They typically will take inputs of environmental conditions and HAB cell counts and use them to forecast HAB intensity in a given area. The models differ in how information is processed and used, though a key element is making the information easily accessible to relevant parties.

Hu et al. (2016) developed a system for information delivery, as opposed to forecasting of *Karenia brevis* blooms in the Gulf of Mexico. The system integrated satellite imagery, *K. brevis* cell counts from various stations, and ocean currents from a well-calibrated 3D model. The integrated product is made available to the community in keyhole markup language, which can be plotted with other products in Google Earth. The model will produce maps of *K. brevis* concentrations, and the model can be used to predict bloom trajectories over the next few days. The combination of data and model is particularly useful to extend the effective reach of the field samples, which are comparatively sparse in time and space.

A particularly impressive effort was undertaken in Europe as part of the EU FP7 programme: Applied Simulations and Integrated Modelling for the Understanding of Toxic and Harmful Algal Blooms—"ASIMUTH" (e.g., Davidson et al. 2016). In this effort, data from local and regional shellfish monitoring programmes (acquiring both shellfish toxicity data and HAB species cell counts) were integrated in near-real time with 3D circulation models forced with measured (in situ and remote sensing) meteorological variables. The models were run forward in time to generate 3–5-day forecasts of HAB location, intensity, and toxicity. Silva et al. (2016) deployed the system for the coast of Portugal, predicting *Pseudo-nitzschia* HABs on the coast and in the lagoons and estuaries. For the 2013–2014 period, they achieved a forecast skill of 85% of predictions correct, 12% fail (an unpredicted bloom occurred), and 3% of the predictions were for a bloom that did not occur. Prediction skill depended on the area being considered, as well as biological and physical factors that could not be taken into account in the models. The ASIMUTH system predictions were published in a HAB warning bulletin that could be used by the community to mitigate the economic, health, and societal effects of the HABs.

Ruiz-Villarreal et al. (2016) deployed the ASIMUTH system to forecast *Dinophysis* spp. HABs along the Northwest Iberian coast. In addition to the 3D circulation model and field data on HABs, they included an offline Lagrangian particle-tracking model that took archived model output to predict the transport of HAB patches. The model was tested with data from the 2005 and 2013 diarrhetic shellfish poisoning (DSP) outbreaks and was found capable of forecasting HABs based on predictions of upwelling and downwelling and alongshore advection. A more rigorous statistical test of the model would reinforce its utility as a HAB warning tool.

Maguire et al. (2016) describe the use of the ASIMUTH system to provide forecasts and bulletins concerning HAB species around Ireland and the British Isles. A key element in the development of these warnings was the integration of regional physical models (e.g., Dabrowski et al. 2016; Cusack et al. 2016) with larger-scale models that provided boundary conditions. As in the other ASIMUTH projects, the physical models were integrated with in situ monitoring data to provide the forecasts. In all cases, the models provided not only forecasts, but a synthetic platform for a wide range of data types—from HAB cell counts to wind stresses—that could then be used in a diagnostic sense to better understand the dynamics that led to the blooms and their transport and dissipation. These models thus go well beyond the relatively straightforward statistical forecasts and include a dynamic component that provides a gateway to improved insights into local and regional HAB dynamics.

Achieving the ASIMUTH goals required the participation of multiple agencies and institutions in multiple countries; this significant effort appears to have paid off in a system that can serve as a global model for regional and large-scale HAB forecasting.

19.3 Conclusions

Our understanding of the dynamics of HABs, in the context of both their ecosystem and their environment, has increased tremendously through the combination of models and data. The engagement of statisticians, physical oceanographers, and modellers with the biologists investigating HABs has been a powerful boon to progress in this field. New statistical techniques have been deployed to understand and predict HABs in response to measured forcings, and new dynamic models of HABs and their ecosystems are enabling a deeper understanding of the processes governing the initiation, growth, transport, and decline of toxic phytoplankton species and their toxins.

Though process models tend to be the most often published, the numbers of empirical-statistical and diagnostic models are increasing every year. It is particularly pleasing to see such impressive progress in the development and deployment of management models; it is likely that their utility will help to justify the decades of public investment in research aimed at understanding HABs.

The great variety of approaches in formulating and analysing HAB models reflects the goals of the researchers and the data and tools/expertise available. Some models are all physics, some are all biology, and most are some fruitful combination of the two. However, it seems clear that the significant advances in the field of HAB modelling over the last half decade have been fueled by a remarkable increase in the frequency, geographic coverage, and range of types of observational data. Future advances in modelling will occur at this junction of models and data, using data to conceptualize models and using models to understand data.

Acknowledgments There are far more published model studies than I am able to review here. I apologize to those whose work I have not included—there just wasn't enough room!

References

Accoroni S, Glibert PM, Pichierri S et al (2015) A conceptual model of annual *Ostreopsis* cf. *ovata* blooms in the northern Adriatic Sea based on the synergic effects of hydrodynamics, temperature, and the N:P ratio of water column nutrients. Harmful Algae 45:14–25. https://doi.org/10.1016/j.hal.2015.04.002

Alves-de-Souza C, Pecqueur D, Le Floc'h E et al (2015) Significance of plankton community structure and nutrient availability for the control of dinoflagellate blooms by parasites: a modeling approach. PLoS One 10(6):e0127623. https://doi.org/10.1371/journal.pone.0127623

Aoki K, Onitsuka G, Shimizu M et al (2012) Factors controlling the spatio-temporal distribution of the 2009 *Chattonella antiqua* bloom in the Yatsushiro Sea, Japan. Estuar Coast Shelf Sci 114:148–155. https://doi.org/10.1016/j.ecss.2012.08.028

Berdalet E, Tester PA (2018) Key questions and recent research advances on harmful algal blooms in benthic systems. In: Glibert PM, Berdalet E, Burford M et al (eds) Global ecology and oceanography of harmful algal blooms. Springer, Cham, pp 261–286

Brown CW, Hood RR, Long W et al (2013) Ecological forecasting in Chesapeake Bay: using a mechanistic-empirical modeling approach. J Mar Syst 125:113–125. https://doi.org/10.1016/j.jmarsys.2012.12.007

Butman B, Aretxabaleta AL, Dickhudt PJ et al (2014) Investigating the importance of sediment resuspension in *Alexandrium fundyense* cyst population dynamics in the Gulf of Maine. Deep Sea Res Part II Top Stud Oceanogr 2(103):79–95. https://doi.org/10.1016/j.dsr2.2013.10.011

Cha Y, Park SS, Kim K et al (2014) Probabilistic prediction of cyanobacteria abundance in a Korean reservoir using a Bayesian Poisson model. Water Resour Res 50(3):2518–2532. https://doi.org/10.1002/2013WR014372

Cusack C, Dabrowski T, Lyons K et al (2016) Harmful algal bloom forecast system for SW Ireland. Part II: are operational oceanographic models useful in a HAB warning system? Harmful Algae 53:86–101. https://doi.org/10.1016/j.hal.2015.11.013

Cusack C, Mouriño H, Moita MT et al (2015) Modelling *Pseudo-nitzschia* events off southwest Ireland. J Sea Res 105:30–41. https://doi.org/10.1016/j.seares.2015.06.012

Dabrowski T, Lyons K, Nolan G et al (2016) Harmful algal bloom forecast system for SW Ireland. Part I: description and validation of an operational forecasting model. Harmful Algae 53:64–76. https://doi.org/10.1016/j.hal.2015.11.015

Davidson K, Anderson DM, Mateus M et al (2016) Forecasting the risk of harmful algal blooms. Harmful Algae 53:1–7. https://doi.org/10.1016/j.hal.2015.11.005

Díaz PA, Ruiz-Villarreal M, Pazos Y et al (2016) Climate variability and *Dinophysis acuta* blooms in an upwelling system. Harmful Algae 53:145–159. https://doi.org/10.1016/j.hal.2015.11.007

Feki W, Hamza A, Frossard V et al (2013) What are the potential drivers of blooms of the toxic dinoflagellate *Karenia selliformis*? A 10-year study in the Gulf of Gabes, Tunisia, southwestern Mediterranean Sea. Harmful Algae 23:8–18. https://doi.org/10.1016/j.hal.2012.12.001

Flynn KJ, St John M, Raven JA et al (2015) Acclimation, adaptation, traits and trade-offs in plankton functional models: reconciling terminology for biology and modeling. J Plankton Res 37:683–691

Franks PJS (1997) Models of harmful algal blooms. Limnol Oceanogr 42:1273–1282

Franks PJS (2014) Modeling of harmful algal blooms: advances in the last decade. In: Rossini GP (ed) Toxins and biologically active compound from microalgae, Biological effects and risk management, vol 2. CRC Press, Boca Raton, FL, pp 538–565

GEOHAB (2001) Global ecology and oceanography of harmful algal blooms. Glibert P, Pitcher G (eds) Science plan. SCOR, IOC, Baltimore, Paris. 86 pp

GEOHAB (2011) GEOHAB modelling: linking observations to predictions, a workshop report. McGillicuddy DJ Jr, Glibert PM, Berdalet E et al (eds). IOC, SCOR, Paris, Newark, Delaware, 85 pp

Giddings SN, MacCready P, Hickey BM et al (2014) Hindcasts of potential harmful algal bloom transport pathways on the Pacific Northwest coast. J Geophys Res Oceans 119(4):2439–2461. https://doi.org/10.1002/2013JC009622

Gillibrand PA, Siemering B, Miller PI et al (2016) Individual-based modelling of the development and transport of a *Karenia mikimotoi* bloom on the North-west European continental shelf. Harmful Algae 53:118–134. https://doi.org/10.1016/j.hal.2015.11.011

Glibert PM (2016) Margalef revisited: a new phytoplankton mandala incorporating twelve dimensions, including nutritional physiology. Harmful Algae 55:25–30. https://doi.org/10.1016/j.hal.2016.01.008

Glibert PM, Allen JI, Artioli Y et al (2014) Vulnerability of coastal ecosystems to changes in harmful algal bloom distribution in response to climate change: projections based on model analysis. Glob Change Biol 20(12):3845–3858. https://doi.org/10.1111/gcb.12662

Gobler CJ, Sunda WG (2012) Ecosystem disruptive algal blooms of the brown tide species, *Aureococcus anophagefferens* and *Aureoumbra lagunensis*. Harmful Algae 14:36–45. https://doi.org/10.1016/j.hal.2011.10.013

Gokaraju B, Durbha SS, King RL et al (2012) Ensemble methodology using multistage learning for improved detection of harmful algal blooms. IEEE Geosci Remote Sens Lett 9(5):827–831. http://ieeexplore.ieee.org/xpls/abs_all.jsp?arnumber=6153049

González Vilas L, Spyrakos E, Palenzuela T et al (2014) Support vector machine-based method for predicting *Pseudo-nitzschia* spp. blooms in coastal waters (Galician rías, NW Spain). Prog Oceanogr 124:66–77. https://doi.org/10.1016/j.pocean.2014.03.003

Harvey EL, Menden-Deuer S (2012) Predator-induced fleeing behaviors in phytoplankton: a new mechanism for harmful algal bloom formation? PLoS One 7(9):e46438. https://doi.org/10.1371/journal.pone.0046438

Henrichs DW, Hetland RD, Campbell L (2015) Identifying bloom origins of the toxic dinoflagellate *Karenia brevis* in the western Gulf of Mexico using a spatially explicit individual-based model. Ecol Model 313:251–258. https://doi.org/10.1016/j.ecolmodel.2015.06.038

Hickey BM, Trainer VL, Kosro M et al (2013) A springtime source of toxic *Pseudo-nitzschia* cells on razor clam beaches in the pacific northwest. Harmful Algae 25:1–14. https://doi.org/10.1016/j.hal.2013.01.006

Hu C, Murch B, Corcoran AA et al (2016) Developing a smart semantic web with linked data and models for near-real-time monitoring of red tides in the eastern Gulf of Mexico. IEEE Syst J 10 (3):1282–1290. https://doi.org/10.1109/JSYST.2015.2440782

Jeong HJ, Lim AS, Franks PJS et al (2015) A hierarchy of conceptual models of red-tide generation: nutrition, behavior, and biological interactions. Harmful Algae 47:97–115. https://doi.org/10.1016/j.hal.2015.06.004

Kibler SR, Tester PA, Kunkel KE et al (2015) Effects of ocean warming on growth and distribution of dinoflagellates associated with ciguatera fish poisoning in the Caribbean. Ecol Model 316:194–210. https://doi.org/10.1016/j.ecolmodel.2015.08.020

Lai Z, Yin K (2014) Physical-biological coupling induced aggregation mechanism for the formation of high biomass red tides in low nutrient waters. Harmful Algae 31:66–75. https://doi.org/10.1016/j.hal.2013.09.011

Li Y, He R, Manning JP (2014) Coastal connectivity in the Gulf of Maine in spring and summer of 2004-2009. Deep Sea Res Part II Top Stud Oceanogr 103:199–209. https://doi.org/10.1016/j.dsr2.2013.01.037

Lim AS, Jeong HJ, Jang TY et al (2014) Inhibition of growth rate and swimming speed of the harmful dinoflagellate *Cochlodinium polykrikoides* by diatoms: Implications for red tide formation. Harmful Algae 37:53–63. https://doi.org/10.1016/j.hal.2014.05.003

Litchman E, Ohman MD, Kiørboe T (2013) Trait-based approaches to zooplankton communities. J Plankton Res 35:473–484

Maguire J, Cusack C, Ruiz-Villarreal M et al (2016) Applied simulations and integrated modelling for the understanding of toxic and harmful algal blooms (ASIMUTH): integrated HAB forecast systems for Europe's Atlantic arc. Harmful Algae 53:160–166. https://doi.org/10.1016/j.hal.2015.11.006

Margalef R (1978) Life forms of phytoplankton as survival alternatives in an unstable environment. Oceanol Acta 1:493–509

Margalef R, Estrada M, Blasco D (1979) Functional morphology of organisms involved in red tides, as adapted to decaying turbulence. In: Taylor DL, Seliger HH (eds) Toxic dinoflagellate blooms. Elsevier, North Holland, pp 89–94

Millie DF, Weckman GR, Fahnenstiel GL et al (2014) Using artificial intelligence for CyanoHAB niche modeling: discovery and visualization of Microcystis – environmental associations within western Lake Erie. Can J Fish Aquat Sci 71(11):1642–1654. https://doi.org/10.1139/cjfas-2013-0654

Moe SJ, Haande S, Couture RM (2016) Climate change, cyanobacteria blooms and ecological status of lakes: a Bayesian network approach. Ecol Model 337:330–347. https://doi.org/10.1016/j.ecolmodel.2016.07.004

Moore SK, Johnstone JA, Banas NS et al (2015) Present-day and future climate pathways affecting *Alexandrium* blooms in Puget Sound, WA, USA. Harmful Algae 48:1–11. https://doi.org/10.1016/j.hal.2015.06.008

Nieto PJG, Fernández JRA, Suárez VMG et al (2015) A hybrid PSO optimized SVM-based method for predicting of the cyanotoxin content from experimental cyanobacteria concentrations in the Trasona reservoir: a case study in Northern Spain. Appl Math Comput 260:170–187. https://doi.org/10.1016/j.amc.2015.03.075

Obenour D, Gronewold A, Stow CA et al (2014) Using a Bayesian hierarchical model to improve Lake Erie cyanobacteria bloom forecasts. Water Resour Res 50:7847–7860. https://doi.org/10.1002/2014WR015616

Pinto L, Mateus M, Silva A (2016) Modeling the transport pathways of harmful algal blooms in the Iberian coast. Harmful Algae 53:8–16. https://doi.org/10.1016/j.hal.2015.12.001

Rigosi A, Hanson P, Hamilton DP et al (2015) Determining the probability of cyanobacterial blooms: the application of Bayesian networks in multiple lake systems. Ecol Appl 25(1):186–199. https://doi.org/10.1890/13-1677.1

Riley GA (1941) Plankton studies. IV. Georges Bank. Bull Bingham Oceanogr Coll 7:1–73

Ruiz-Villarreal M, García-García LM, Cobas M et al (2016) Modelling the hydrodynamic conditions associated with *Dinophysis* blooms in Galicia (NW Spain). Harmful Algae 53:40–52. https://doi.org/10.1016/j.hal.2015.12.003

Ryan JP, McManus MA, Kudela RM et al (2014) Boundary influences on HAB phytoplankton ecology in a stratification-enhanced upwelling shadow. Deep Sea Res Part II Top Stud Oceanogr 101:63–79. https://doi.org/10.1016/j.dsr2.2013.01.017

Silva A, Pinto L, Rodrigues SM et al (2016) A HAB warning system for shellfish harvesting in Portugal. Harmful Algae 53:33–39. https://doi.org/10.1016/j.hal.2015.11.017

Singh A, Hårding K, Reddy HRV, Godhe A (2014) An assessment of *Dinophysis* blooms in the coastal Arabian sea. Harmful Algae 34:29–35. https://doi.org/10.1016/j.hal.2014.02.006

Sunda WG, Shertzer KW (2014) Positive feedbacks between bottom-up and top-down controls promote the formation and toxicity of ecosystem disruptive algal blooms: a modeling study. Harmful Algae 39:342–356. https://doi.org/10.1016/j.hal.2014.09.005

Terseleer N, Gypens N, Lancelot C (2013) Factors controlling the production of domoic acid by *Pseudo-nitzschia* (Bacillariophyceae): a model study. Harmful Algae 24:45–53. https://doi.org/10.1016/j.hal.2013.01.004

Tobin ED, Grünbaum D, Patterson J et al (2013) Behavioral and physiological changes during benthic-pelagic transition in the harmful alga, *Heterosigma akashiwo*: potential for rapid bloom formation. PLoS One 8(10):e76663. https://doi.org/10.1371/journal.pone.0076663

Visser PM, Verspagen JMH, Sandrini G et al (2016) How rising CO_2 and global warming may stimulate harmful cyanobacterial blooms. Harmful Algae 54:145–159. https://doi.org/10.1016/j.hal.2015.12.006

Walsh JJ, Lenes JM, Darrow B et al (2016) Impacts of combined overfishing and oil spills on the plankton trophodynamics of the West Florida shelf over the last half century of 1965-2011: a two-dimensional simulation analysis of biotic state transitions, from a zooplankton- to a bacterioplankton-modulated ecosystem. Cont Shelf Res 116:54–73. https://doi.org/10.1016/j.csr.2016.01.007

Warns A, Hense I, Kremp A (2013) Modelling the life cycle of dinoflagellates: a case study with *Biecheleria baltica*. J Plankton Res 35(2):379–392. https://doi.org/10.1093/plankt/fbs095

Waters LG, Wolcott TG, Kamykowski D et al (2015) Deep-water seed populations for red tide blooms in the Gulf of Mexico. Mar Ecol Prog Ser 529:1–16. https://doi.org/10.3354/meps11272

Wyatt T (2014) Margalef's mandala and phytoplankton bloom strategies. Deep Sea Res Part II Top Stud Oceanogr 101:32–49. https://doi.org/10.1016/j.dsr2.2012.12.006

Yamaguchi H, Sai K (2015) Simulating the vertical dynamics of phosphate and their effects on the growth of harmful algae. Estuar Coast Shelf Sci 164:425–432. https://doi.org/10.1016/j.ecss.2015.07.043

Part VII
Moving Forward: Emerging Issues and a New Global Programme

Chapter 20
Emerging HAB Research Issues in Freshwater Environments

Michele A. Burford, David P. Hamilton, and Susanna A. Wood

20.1 Introduction

Harmful algal blooms (HABs) are a common feature of many freshwater systems, particularly those impacted by human activity (Fig. 20.1). The dominant group of causal organisms is cyanobacteria (CyanoHABs). These prokaryotic organisms share some features in common with eukaryotic algae, such as the ability to photosynthesize and produce toxins. However, they have fundamental structural differences and vary in their physiological traits.

CyanoHABs can have major economic, social and environmental impacts (O'Neil et al. 2012). CyanoHABs can cause hypoxic conditions, resulting in fish kills and changes in the nutrient biogeochemistry of a waterbody, as well as affecting the diversity and abundance of other species. The economic costs of CyanoHABs in the USA have been estimated to be US$2.2 billion p.a., via the effects on water quality, recreation use, fishing and property values (Dodds et al. 2008). Additional water treatment costs are the most costly consequences of CyanoHABs. These can amount to several million dollars for a single treatment plant alone, with overall annualized figures of US$813 million in the USA, £19 million in the UK and AUD$65 million in Australia (see review by Hamilton et al.

M.A. Burford (✉)
Australian Rivers Institute, Griffith University, Nathan, QLD, Australia
e-mail: m.burford@griffith.edu.au

D.P. Hamilton
Australian Rivers Institute, Griffith University, Nathan, QLD, Australia

Environmental Research Institute, University of Waikato, Hamilton, New Zealand

S.A. Wood
Environmental Research Institute, University of Waikato, Hamilton, New Zealand

Cawthron Institute, Nelson, New Zealand

P.M. Glibert et al. (eds.), *Global Ecology and Oceanography of Harmful Algal Blooms*, Ecological Studies 232, https://doi.org/10.1007/978-3-319-70069-4_20

Fig. 20.1 *Microcystis* blooms in Lake Rotorua, South Island (*left*), and Lake Rotoehu, North Island (*right*), New Zealand. Photos: David Hamilton (*left*), Hamish Lass, Bay of Plenty Regional Council (*right*)

2014a). Globally, these economic costs are predicted to increase over time due to increasing incidences of algal blooms from a warming climate (Wagner and Adrian 2009; Carey et al. 2012).

The most commonly studied cyanobacterial HAB genera are the pelagic species producing toxins harmful to human and animal health. The genera include *Cylindrospermopsis*, *Microcystis*, *Dolichospermum* (basionym *Anabaena*), *Aphanizomenon*, *Planktothrix* and *Nodularia*, but there are a number of less well-studied toxic species (Table 20.1). These species, as well as new and emerging cyanotoxins produced by CyanoHABs, have been examined in detail in a series of reviews in *Harmful Algae* (Vol. 54). Many of these cyanobacterial species are found throughout the world, typically blooming in mesotrophic or eutrophic lentic waterbodies where water residence times are sufficiently long to support a high biomass. Biomass accumulation leading to a CyanoHAB therefore reflects the antecedent conditions, including water temperature, water column stratification and mixing dynamics, light climate and nutrient supply (O'Neil et al. 2012). Over time, the number of reports of CyanoHAB incidences is increasing worldwide, including in developing countries (see review by Mowe et al. 2015) and large lake systems, e.g., Lake Erie, USA (Bullerjahn et al. 2016).

The objective of this chapter is to briefly summarize some key areas where there have been significant advances in knowledge and outline some of the major challenges for future research on freshwater CyanoHABs. We focus on new insights into CyanoHAB responses to nutrients and measurement technologies applicable at molecular or cellular levels through to landscape-scale monitoring. Some emerging challenges, such as improved analytical methods and quantifying the toxicity of cyanotoxins, e.g., beta-*N*-methylamino-L-alanine (BMAA) with a proposed link to neurological diseases (Lage et al. 2016), are not addressed. We also examine some of the challenges and opportunities for deterministic models to utilize the complex data generated from these studies and integrate key ecological and physiological processes.

Table 20.1 Summary of traits of the most common freshwater cyanobacterial bloom-forming genera

Genera	Taxonomic group	Toxins that may be produced	Morphological features	Distinctive physiological characteristics	References
Microcystis	*Chroococcales*	Microcystins	Mucilaginous colonies of cells	Positively buoyant	Harke et al. (2016)
Cylindrospermopsis raciborskii	*Nostocales*	Cylindrospermopsins, Saxitoxins	Solitary coiled and straight trichomes, heterocysts, akinetes	N-fixing, akinete formation, high P affinity, optimal growth in low to moderate light intensities, neutrally buoyant	Burford et al. (2016)
Dolichospermum (previously *Anabaena*)	*Nostocales*	Anatoxins Microcystins Saxitoxins Cylindrospermopsins	Solitary coiled and straight trichomes, heterocysts	N-fixing, akinete formation, positively buoyant, taste and odour producer	Li et al. (2016)
Aphanizomenon	*Nostocales*	Anatoxins Cylindrospermopsins Saxitoxins	Solitary straight trichomes, heterocysts	N-fixing, optimal growth in low to moderate light intensities,	Cires and Ballot (2016)
Planktothrix	*Oscillatoriales*	Anatoxins, Aplysiatoxins Microcystins Saxitoxins	Solitary straight or slightly curved trichomes	Neutrally buoyant, efficient light harvesters	Kurmayer et al. (2016)
Nodularia	*Nostocales*	Nodularins	Solitary straight trichomes, heterocysts	Brackish waters, N-fixing, akinete formation, buoyancy regulation	O'Neil et al. (2012)

These advances in knowledge are consistent with the scientific goal of GEOHAB to improve our prediction of HABs by understanding the mechanisms driving population dynamics, despite the fact that CyanoHABs were not explicitly included in the GEOHAB Programme [see also Chap. 3, Kudela et al. (2018)]. The GEOHAB Programme included programmes on eutrophication and adaptive strategies as well as observation, modelling and prediction, all of which are as relevant to CyanoHABs, as they are to eukaryotic HABs. GEOHAB has recently been replaced with a new global Programme, GlobalHAB [see Chap. 22, Berdalet et al. (2018)]. It has an expanded brief beyond GEOHAB to include freshwater CyanoHABs, with a focus on advancing the science and management, via a global perspective that encourages communication across the scientific community, as well as with water managers and the public.

20.2 Recent Scientific Insights

20.2.1 Strain Differences or Intraspecific Variability

Strain (= ecotype) variability, and its implications for physiological and ecological studies, is an ongoing challenge for studies of all microalgae, including cyanobacteria (see Burkholder and Glibert 2009 *Harmful Algae*, special issue). In recent years there has been an increased focus on the extent of strain variability within and between populations, as well as over time, and the implications for doing culture studies (see review by Lakeman et al. 2009). Understanding the variability in environmental responses between toxic CyanoHAB strains is a particularly pertinent question for water resource managers attempting to predict and manage blooms. This is because blooms of CyanoHAB toxin producers and toxin levels in waterbodies are often not strongly correlated (Wood et al. 2011). This creates challenges when managing the risks to humans and other animals. It has been established for many years that environmental conditions can affect toxin cell quotas in some cyanobacterial species (e.g., Oh et al. 2000). However, it is increasingly clear that strains, or isolates from within or between waterbodies, can vary substantially in their production of toxins under the same environmental conditions. For example, Willis et al. (2016a) found that 24 isolates of *C. raciborskii* from one bottle of lake water, grown under the same environmental conditions, had a 2.5-fold difference in their toxin cell quotas. This helps explain why there is so much variation in published studies of responses of different isolates to the same physicochemical variables.

Nutrients, for example, can affect the relative abundance of toxin and non-toxin-producing strains (Glibert and Burford 2017). Studies have shown that non-toxin-producing strains of *Microcystis* require lower nutrient concentrations to achieve maximal growth rates compared with toxin-producing strains, whereas toxin-producing strains yield higher growth rates than non-toxin-producing strains at higher nutrient concentrations (Vezie et al. 1998; Davis et al. 2009). As a result, a study in a North American lake showed that the proportion of toxic to non-toxic strains of *M. aeruginosa* changed in response to nitrogen (N) and phosphorus (P) concentrations, possibly due to the energetic costs of toxin production and the fact that microcystins contain N (Davis et al. 2010). A study by Puddick et al. (2016a) also showed that N played a key role in modulating toxicity of *Microcystis*. A recent study of competition between *M. aeruginosa* and *C. raciborskii* showed that strain, rather than species, determined which species would displace the other based on light intensity and P concentration (Marinho et al. 2013). A mesocosm study found that an increase in P, but not N availability, led to dominance of toxic strains of *C. raciborskii* (Burford et al. 2014). It is not clear if this occurred because the toxic strains require more P for growth or that the balanced N/P ratio favoured these strains.

The physiological characteristics of blooms may also vary over time based on the dominance of particular strains. Kardinaal et al. (2007) found seasonal patterns in strain dominance with a shift from a higher proportion of toxic strains early in a *Microcystis* bloom, to mostly non-toxic strains towards the end of the bloom. Welker et al. (2007) found a reduction of diversity of *Microcystis* strains, based on distinct peptides, and a decline in the toxin-producing colonies from summer to autumn. Studies of strain variability in response to environmental conditions highlight the need to capture this variability in laboratory studies and to design field studies with this variability in mind.

20.2.2 Nutrient Utilization Strategies

Increased nutrient loads to aquatic systems are a well-established driver of CyanoHAB development (Paerl 2014; Hamilton et al. 2016). Despite this knowledge, our ability to predict the magnitude, duration and frequency of blooms, and the relative dominance of CyanoHABs compared with other algal species, can be improved as we learn more about how nutrient concentrations, ratios and forms affect CyanoHABs (Oliver et al. 2012; Glibert et al. 2016; Glibert and Burford 2017). Downing et al. (2001), in a study of 99 lakes around the world, showed that occurrence of cyanobacteria blooms was most closely correlated with variations in total N and P, or standing algal biomass, rather than N to P ratio. The importance of lower N/P ratios in stimulating cyanobacterial (particularly diazotrophic species) dominance and blooms in lakes and reservoirs is contentious, with some studies showing that lower ratios appear to benefit cyanobacteria (e.g., Harris et al. 2016; Smith et al. 2016), while others show no benefit (e.g., Dolman et al. 2012) or negative effects (Lv et al. 2011). The benefit of low N/P ratios for N-fixing cyanobacteria is based on the premise that they can outcompete other algal species under these conditions (Orihel et al. 2015). Despite this, Scott and McCarthy (2010) showed that N fixation rates in lakes may be insufficient to offset N demands, and therefore low N/P ratios may not be the main cause of blooms of N-fixing cyanobacteria in some systems.

A range of physiological mechanisms for utilizing N by CyanoHABs has been identified. *C. raciborskii* has been shown to only produce heterocysts, and fix N, when internal N stores were depleted (Willis et al. 2016b). However, *Nodularia spumigena* from the Baltic Sea maintained heterocysts without any N fixation (Vintila and El-Shehawy 2007, 2010). For *Microcystis*, Harke and Gobler (2015) demonstrated daily changes in the expression of a suite of genes for N uptake under fluctuating N conditions, with different, complex responses according to the form of N involved.

Forms of N have been shown to affect dominance of specific microalgal groups. High levels of ammonium (NH_4^+) potentially repress nitrate (NO_3^-) uptake but with strong interactions amongst the microalgal assemblage and affected by the relative concentrations of the inorganic N species (Glibert et al. 2016). Harris et al. (2016), in an analysis of reservoir water quality, found that cyanobacterial biovolume, as well as the production of toxins and secondary metabolites, was inversely related to the ratio of NO_3^-/NH_4^+ in the water. Domingues et al. (2011) found that NH_4^+ promoted cyanobacteria and chlorophytes in a freshwater tidal estuary, while NO_3^- promoted diatom species. Urea is also acknowledged as an important N source that may promote cyanobacterial blooms (Solomon and Glibert 2008; Finlay et al. 2010) although long-term studies are lacking in both freshwater and marine waters (Davidson et al. 2014) to allow conclusions to be made about its stimulatory effect [but see also Chap. 12, Glibert et al. (2018a)].

Availability of N can also affect toxin production by some cyanobacterial species, with much of our understanding based on studies of *Microcystis* (see review by Gobler et al. 2016). N may either be a component of the toxin molecules or be required for the synthesis pathway. N also affects the proportion of congeners of toxins. For example, a culture study showed that the toxicity of *Microcystis* was linked to N availability by affecting the relative abundance of more toxic congeners of microcystin with higher arginine and hence N content within molecules (Puddick et al. 2016a). However, not all species have the ability to regulate their toxin production based on N availability, e.g., *C. raciborskii* has constitutive production (Willis et al. 2015).

Laboratory studies of nutrient physiology are also identifying adaptation strategies of cyanobacteria to low P conditions. For example, Australian strains of *C. raciborskii* have been shown to have the highest recorded phosphate (PO_4^{3-}) uptake affinity of a freshwater cyanobacteria (and eukaryotic algae), as well as a high storage capacity, and an enhanced ability for active P uptake below defined concentration thresholds (Fig. 20.2, Prentice et al. 2015; Willis et al. in 2017). Additionally, Amaral et al. (2014), in a study of *C. raciborskii* growth, concluded that frequency and duration of PO_4^{3-} addition is a more important factor promoting growth of this species than concentrations of PO_4^{3-} per se.

A study of marine cyanobacteria has also identified the ability of some species to switch from using phospholipids in the cell membrane to sulpholipids in response to

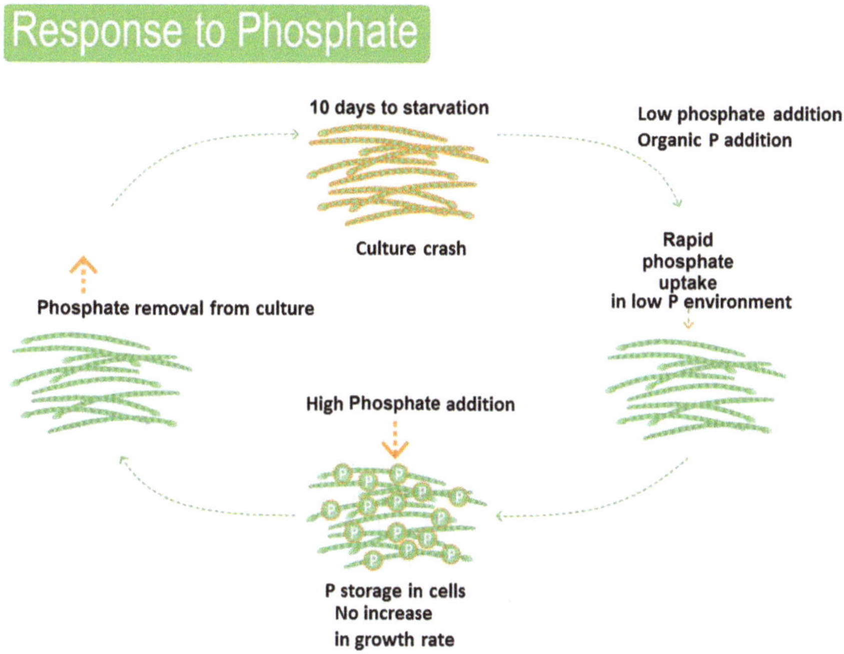

Fig. 20.2 Diagram of the range of responses of the toxic cyanobacterium, *Cylindrospermopsis raciborskii* to fluctuating P availability. The diagram shows high phosphate addition to cells resulting in storage of P with no increase in growth, followed by removal of available phosphate and continued growth of cells at the same rate until culture crash. Addition of low concentrations of phosphate or organic P stimulates rapid uptake of P, with sustained inputs resulting in P storage in cells

low P availability (van Mooy et al. 2009). Although there is less information for freshwater cyanobacteria, a recent study has shown that *Microcystis* upregulates production of sulphate-binding and permease proteins under low external P concentrations (Harke and Gobler 2013). This suggests that *Microcystis* has the capacity to utilize sulphur for physiological processes in place of P. There are also numerous studies showing that the organic form of P, i.e. phosphoesters, is utilized by CyanoHABs when inorganic P availability is low (e.g., Wu et al. 2011; Bai et al. 2014). Additionally, genes responsible for upregulating enzymes capable of cleaving other organic P sources, such as phosphonates, have been identified in some cyanobacteria (Stucken et al. 2010).

20.2.3 *Interactions Between Nutrients and Climate Change*

Eutrophication is likely to act synergistically with climate change to exacerbate CyanoHAB blooms and toxin production [Wood et al. 2015; see also Chaps. 4 and 5,

Glibert et al. (2018b), Wells and Karlson (2018)]. Researchers identified the potential for climate change to promote cyanobacterial blooms nearly 10 years ago (Paerl and Huisman 2008; see review by Visser et al. 2016) but soon after realized the potential for bloom control under the influence of a warming climate by strictly controlling nutrient inputs to aquatic ecosystems (Brookes and Cayelan 2011). There are now a number of studies showing the interplay between nutrients and climate change variables, e.g., van de Waal et al. (2009) showed that *Microcystis* growth was enhanced by increased CO_2 and N levels. Cyanobacteria possess a number of physiological traits that enable them to be selectively advantaged over eukaryotic algae under climate change (Carey et al. 2012). Some of these are direct responses to temperature, including exponential growth rates that increase more rapidly with increasing temperature (i.e. high Q_{10} values) than those of eukaryotic algae. For example, Reynolds (2006) gives a Q_{10} value for *M. aeruginosa* that is higher than that recorded for any eukaryotic algae as well as being higher than any other cyanobacteria. However, Lurling et al. (2012) found little difference in temperature optima between a group of eight cyanobacteria species versus a group of eight chlorophyte species. Other responses are indirect and may relate to the extended periods of thermal stratification further promoting cyanobacterial species with buoyancy regulation which allows migration to well-lit surface waters (Carey et al. 2012).

20.3 New Technologies for Monitoring and Management

Many planktonic cyanobacterial species are positively buoyant and therefore have a tendency to accumulate at the water surface, with aggregations often reinforced by light winds and flow-mediated processes (Hunter et al. 2008). This can make sampling and monitoring of cyanobacteria and cyanotoxins problematic. Cell concentrations can vary rapidly with changing meteorological conditions which lead to shifts in water temperature and mixing and transport processes. Traditional sampling practices (e.g., grab samples taken at a regular monitoring site) provide only a snapshot of cyanobacteria present at that one point in time and may miss areas or times of highest risk. Analysis of these samples relies on microscopy which can be time consuming and is dependent upon expert taxonomic identification. Techniques based on detection of specific pigments using satellites (Hunter et al. 2008) and data loggers or molecular markers have advanced cyanobacterial monitoring and management and are providing greater insight into the dynamics of CyanoHABs (Fig. 20.3).

Fig. 20.3 Schematic showing a range of new technologies available for cyanobacterial bloom monitoring and the different scales at which they function. (1) Satellite remote sensing can detect and track surface cyanobacterial blooms at lake or regional scales. (2) Unmanned aerial vehicles (or 'drones') have the potential to determine the extent of surface cyanobacterial blooms within a lake. (3) Observational platforms equipped with sensors can measure physicochemical variables and cyanobacterial biomass (using proxies such as phycocyanin) at high frequencies within a lake. (4) Water samples can now be rapidly analysed using molecular techniques to determine species composition and abundance of toxin genes at a specific site

20.3.1 Probes for Quantification of Cyanobacteria

Cyanobacteria contain many pigments that serve different functions including light-harvesting, structural elements of thylakoid membranes, N storage and photoprotection (Oliver et al. 2012). Several of these pigments are carotenoids such as β-carotene, zeaxanthin, echinenone, canthaxanthin and myxoxanthophyll (Schagerl and Müller 2006). The light absorption spectra of cyanobacteria are largely determined by pigments specific to this taxon, including phycocyanin, phycoerythrin and allophycocyanin. Phycocyanin is a phycobiliprotein which absorbs red and orange light at wavelengths between 610 and 630 nm and emits fluorescence from 600 to 700 nm. This fluorescence signature has been utilized to enable cyanobacteria to be detected by sensors from within a mixed assemblage of phytoplankton and to make a biomass estimation (Bastien et al. 2011). Recent field and laboratory studies have demonstrated strong relationships between phycocyanin concentrations and cyanobacterial cell concentrations or biovolumes (e.g., Kong et al. 2014). Field data loggers equipped with phycocyanin-specific sensors have been developed to monitor cyanobacterial populations and can provide early warning signals, for example, in drinking water supplies (Izydorczyk et al. 2005).

Many benthic cyanobacteria (see Sect. 4) contain higher amounts of phycoerythrin, a phycobiliprotein, which can also be measured fluorometrically (emission peak = 573 nm) with probes. Some commercially available probes are also able to measure fluorescence emitted at different wavelengths (commonly known as fluoroprobes), and these sensors therefore purport to be able to distinguish different algal taxa. However, it has been shown that benthic cyanobacteria may be underestimated by these sensors because some do not measure phycoerythrin, which is an important constituent of these species (Echenique-Subiabre et al. 2016).

Studies have highlighted caveats when using pigment sensor data, including the effects of temperature, light, species composition and colony or filament morphology on phycocyanin concentrations (e.g., Kasinak et al. 2015). Additionally, most sensors only perform accurately within limited concentration ranges (Kong et al. 2014). Despite these limitations, phycocyanin sensors provide a rapid and extremely cost-effective method of obtaining data on total cyanobacterial biomass. The technology is now routinely utilized in drinking water supplies worldwide (e.g., Zamyadi et al. 2012), and sensors have been installed on water quality monitoring buoys to provide real-time data on cyanobacterial biomass (e.g., Hamilton et al. 2014b).

20.3.2 Remote Sensing

Tracking cyanobacterial blooms by satellite remote sensing has the potential for monitoring over large spatial scales. However, obtaining data at sufficiently fine spatial and spectral resolution to identify phycocyanin (and therefore distinguish cyanobacteria from other phytoplankton) has proved challenging [Kudela et al. 2015; see also Chap. 18, Glibert et al. (2018c)]. The use of hyperspectral sensors to enable spectral shape to be delineated has improved knowledge of the spectral reflectance spectra of cyanobacteria in situ (Hunter et al. 2010). Wynne et al. (2008) developed a cyanobacterial index that relies on changes in the shape of the spectrum between the wavelengths of 665, 681 and 709 nm (caused by the strong scattering associated with cyanobacteria at around 709 nm) and successfully applied this index to the detection of blooms in the Laurentian Great Lakes using the MEdium Resolution Imaging Spectrometer (MERIS). A different maximum peak height algorithm was applied to cyanobacterial detection in inland and coastal waters in South Africa by Matthews et al. (2012), also using MERIS data. Kudela et al. (2015) made a significant advance in the ability to distinguish cyanobacteria genera using remote sensing, by using data from a HyspIRI Airborne Campaign (undertaken as part of a NASA Student Airborne Research Program). When high spectral resolution data were available, with the appropriate atmospheric correction, these authors were able to separate the cyanobacterial genera, *Aphanizomenon* and *Microcystis*.

Future satellite missions, such as the Ocean Land Colour Instrument (OLCI, European Space Agency), HyspIRI, Pre-Aerosol Clouds and Ocean Ecosystem (NASA) and Sentinel-2 MSI (Toming et al. 2016), will provide opportunities to significantly reduce pixel size. This will lead to much higher spatial resolution, as well as greater spectral resolution which is essential for distinguishing cyanobacteria amongst other optically active agents of the water column, including eukaryotic algae (Kudela et al. 2015).

At local lake or river scales, the use of unmanned aerial vehicles (UAVs or 'drones') is expanding rapidly and may enable data collection at a high spatial and temporal resolution. This will improve the ability to manage water quality risks and provide new knowledge on bloom dynamics, particularly the spatial extent and temporal variability of blooms during daytime. UAVs, equipped with cameras, hyperspectral sensors and even samplers, could be deployed to coincide with regular monitoring or target specific sites (i.e. areas with high recreational use). This provides a means to reduce the interval between data gathering and generation of results, thus markedly improving risk management. Promising results have been shown for monitoring both planktonic and benthic cyanobacteria (Hemple et al. 2014; Van der Merwe and Price 2015; Aguirre-Gómez et al. 2016).

20.3.3 Molecular-Based Identification and Enumeration

Molecular-based techniques are evolving rapidly, and there are now many methods for identifying and, in some cases, enumerating target species or genes in environmental samples (Humbert et al. 2010; Kurmayer et al. 2017). These techniques often allow for high throughput of samples, require no taxonomic expertise and enable detection of functional genes (e.g., genes involved in toxin production), thus providing additional information not obtainable from morphological identification.

Quantitative PCR (qPCR) is now widely used in cyanobacterial research and monitoring programmes for identification and enumeration of specific species or genes involved in toxin synthesis (e.g., Vaitomaa et al. 2003; Al-Tebrineh et al. 2012). Commercial testing kits are available for simultaneous detection and quantification of total cyanobacteria (16S ribosomal RNA; 16S rRNA) and a range of genes responsible for toxin production. Such tools are increasingly being incorporated into monitoring programmes for initial screening of samples, whereby positive results can be followed up with confirmatory chemical analysis.

The advent of high-throughput sequencing (HTS) has fundamentally changed the ability to characterize entire microbial communities in environmental samples (Shokralla et al. 2012). It allows enormous volumes of sequence (DNA or RNA) data to be produced cost-effectively, and assays can be developed to target specific organisms or groups, i.e. metabarcoding (Taberlet et al. 2012). Cyanobacterial-specific assays have been tested, and although a significant proportion of the sequences are general to bacteria (Kleinteich et al. 2014), the data provide a wealth of information on cyanobacterial species diversity. Generic assays targeting all bacteria (16S rRNA) are now commonly used to characterize freshwater samples. These provide valuable additional information on patterns and shifts in bacteria associated with cyanobacterial communities and have the potential to rapidly advance knowledge on microbial interactions (Brasell et al. 2015). A well-acknowledged limitation of HTS is the need for more comprehensive and robust reference databases which are usually generated from morphologically verified specimens (Nilsson et al. 2006). Data generated using HTS is considered semi-quantitative, and therefore its application for routine monitoring is limited but likely to improve with new technologies, e.g., gene enrichment coupled with HTS (e.g., Dowle et al. 2016).

20.4 Benthic CyanoHABs

Benthic cyanobacteria are found in all freshwater environments across the globe [see also Chap. 13, Berdalet and Tester (2018)]. An increasing number of species have been shown to produce microcystins, nodularins, anatoxins, cylindrospermopsins and saxitoxins (Quiblier et al. 2013), and ingestion of benthic biofilms has been associated with animal poisonings (e.g., Mez et al. 1998; Gugger et al. 2005; Wood et al. 2007). Many benthic species (like some phytoplankton species) also produce non-toxic taste and odour compounds, geosmin and 2-methylisoborneol (MIB), and these can also be problematic when the water is used for human consumption (e.g., Izaguirre and Taylor 2007).

Recent surveys have highlighted the widespread distribution of benthic cyanobacteria and associated cyanotoxins in multiple countries. For example, in a survey of wadeable rivers in California (USA), microcystins were detected in benthic samples at over one third ($n = 368$) of sites surveyed (Fetscher et al. 2015), and in New Zealand, proliferations (defined as covering greater than 20% of the river bed) of potentially toxic *Phormidium* have been identified in more than 100 rivers (Fig. 20.4, McAllister et al. 2016). A notable feature of these problem species is their occurrence in habitats which are typically thought to be 'free' of

Fig. 20.4 Benthic bloom of *Phormidium* (*dark brown* patches) in the Motueka River, South Island, New Zealand (*left*), and underwater image of a *Phormidium* mat in the Wai-iti River, South Island, New Zealand (*right*) (Photos: S. Wood, Cawthron)

toxic cyanobacteria, such as in the metaphyton of oligotrophic lakes (Smith et al. 2012) and the benthos of rivers with high water quality (McAllister et al. 2016). Lack of historic monitoring data makes it difficult to ascertain if there has been a recent increase in the prevalence of benthic freshwater cyanobacteria due to changing environmental conditions or as a result of more targeted surveys and more frequent reporting (Quiblier et al. 2013; McAllister et al. 2016). Many knowledge gaps remain, in particular regarding interactions amongst organisms and internal nutrient cycling within biofilms and the nutrient and flow preferences of species in lotic habitats (Quiblier et al. 2013).

20.5 Challenges and Opportunities for Modelling Cyanobacterial Dynamics

In this section we have focused primarily on deterministic models for CyanoHAB assessment and prediction because they allow for incremental developments using improved or new algorithms of processes (e.g., toxin production, N fixation or gene expression) that could potentially parallel advances in understanding of physiological processes relevant to CyanoHAB formation and relevant monitoring technologies (e.g., remote sensing, PCR and HTS). Process-based CyanoHAB models may serve a range of purposes including predictive assessments of the spatial and temporal dynamics of blooms (Carraro et al. 2012), toxin production (Islam et al. 2012) and simulation of specific processes associated with cyanobacteria including N fixation, buoyancy regulation and heterocyst production.

The selection of an appropriate CyanoHAB model involves careful consideration of its purpose, relevant time and space scales of prediction, levels of complexity and process representation, availability of model input data and the degree of familiarity of the user with the model, which is particularly relevant to complex deterministic models [see also Chap. 19, Franks (2018)]. Alignment of scales of model and observational data is critical to enable a model to be appropriately calibrated and validated (Kreft et al. 2013). For example, CyanoHABs and toxin production may vary at scales from submillimetre in surface scums (Puddick et al. 2016b) to blooms that span many hundreds of kilometres (Bowling and Baker 1996). Models must therefore be adapted to be 'fit for purpose,' and an important first step in this process is conceptualization involving aspects of problem definition or articulation, development of a hypothesis or testing framework and construction of a system dynamics diagram. This essential first step can help to identify whether a suitable model exists or needs to be constructed or whether incremental improvements and extensions are required using an existing model.

20.5.1 Integrating Models with New Monitoring Technologies

While observational technologies have advanced rapidly in recent years, there has been considerably less effort devoted to advancing or developing new models for CyanoHABs [see Glibert et al. 2010; Oliver et al. 2012; see also Chaps. 18 and 19, Franks (2018), Glibert et al. (2018c)]. Hellweger (2015), for example, considers that ecological modellers have been slow to incorporate new knowledge of nutrient uptake dynamics into microbial models, instead resorting to Monod kinetics to explain nutrient uptake dynamics. He provides other examples relevant to CyanoHABs to support this perceived lack of progress, including the inability to include microbial attack and defence mechanisms (e.g., microcystins), and lack of quantification of heterogeneity at a population level (e.g., colony size). Cyanobacteria have an exceptional level of cellular heterogeneity amongst the bacterial kingdom, including structural attributes that have significant physiological roles. Few workers have attempted to include aspects of morphogenesis or cellular differentiation in CyanoHAB models, so functional attributes associated with structural attributes such as hormogonia, akinetes or heterocysts are only infrequently encapsulated in simple empirical expressions (see Oliver et al. 2012).

Gene sequencing and expression have provided a way forward to address this predicament, but molecular- and biogeochemical-level understandings of cell division and the implications of morphological diversity are still poorly resolved (e.g., Garcia-Pichel and Wojciechowski 2009) and place limitations on the process descriptions that are possible in CyanoHAB models. Coles and Hood (2016), using examples of marine models, provide recommendations for improving incorporation of genomics findings into three-dimensional, coupled physical–biogeochemical models. These include improved scientific exchanges and collaborations amongst disciplines (i.e. 'omics' and modelling), techniques to compare genomic data with model outputs and explicitly capturing the diversity of microbial community composition in model outputs. Meeting this challenge may, however, involve increasing the complexity and parameterization of these physical–biogeochemical models in order to address additional functionalities.

In terms of spatial and/or temporal resolution of CyanoHAB models, models need to align with the information ('big data') collected by rapidly evolving technologies for cyanobacteria monitoring (e.g., in situ phycocyanin sensors) and toxin detection. Additionally, new theories of CyanoHAB dynamics need to be tested and adapted via models at an ecosystem scale. Not only can new technologies assist algorithm development, but they have the potential to revolutionize the calibration and validation of CyanoHAB models (e.g., Hamilton et al. 2014b). For example, spectral reflectance spectra from remote sensing images can be processed to provide in situ estimates of surface cyanobacterial biomass (see Sect. 3.1). These data would provide a modeller interested in spatial heterogeneity of CyanoHABs in surface waters with large amounts of data with which to validate output from a two-dimensional (vertically integrated) or three-dimensional model (Ibelings et al. 2003; Rose et al. 2016).

20.5.2 Models of Adaptive Strategies in Cyanobacteria

Cyanobacteria have a range of adaptive strategies that operate differently amongst species or strains and over different timescales. This necessitates careful consideration of the level of model detail and complexity required to address the questions of interest, as well as model alignment to the time and space scales of interest. For example, relatively simple models of buoyancy regulation were a major focus of efforts for the last two decades to simulate diurnal CyanoHAB events (Kromkamp and Walsby 1990). The models were used to simulate rates of vertical migration of cells, trichomes or colonies of individual species that were measured in laboratory experiments. These processes underpin whether colonies or filaments are entrained in the turbulence or disentrain from it to form blooms (Huisman et al. 2004) and have been applied to natural water bodies by coupling with hydrodynamic models (Wallace and Hamilton 2000). The impetus for more recent models of *Microcystis* blooms has been to demonstrate the importance of colony size on diurnal migrations (Aparicio Medrano et al. 2013). Additionally, three-dimensional advection models coupled with Lagrangian particle tracking have been used to capture CyanoHAB distributions in Lake Erie (Rowe et al. 2016). However, these models have little or no representation of a range of other physiological processes that affect buoyancy regulation, such as variations in internal nutrient stores, gas vacuole formation or extracellular polysaccharides. Therefore, the models may be suitable for applications over short timescales relevant to advection and mixing processes, but not over longer timescales corresponding to changes in biomass or cell morphology.

Cyanobacteria also have a number of reproductive strategies to optimize survival. These include benthic overwintering, resting stages (akinetes) and morphological changes (e.g., Borges et al. 2016). Hellweger et al. (2008) used an individual-based model of akinete formation in *Dolichospermum flos-aquae* to track its life cycle stages and demonstrate the importance of the benthic (akinete) stage in allowing nutrient uptake from sediment porewaters to support subsequent nutritional demand for vegetative growth in the water column. Similarly, Hense and Burchard (2010) demonstrated the importance of a heterocystous stage amongst three other stages (vegetative cells, akinetes and 'recruiting cells') in *N. spumigena*, to enable it to meet its N requirements for growth. These types of morphological features and adaptations are encoded for genetically and raise the question of whether cyanobacteria models should attempt to simulate individual life cycle stages. Here models may bifurcate to operate mechanistically at a molecular or cellular level, based on gene expression responses (e.g., for N fixation in *Dolichospermum*; Hellweger et al. 2016), or empirically using simplifying assumptions that relate the life cycle process directly to environmental drivers. Kreft et al.

(2013) developed a case for individual-based mechanistic modelling because of the rapid rise of microbial technologies for single cell investigations. A hybrid approach may even be possible that uses mechanistic functions to describe processes that are well understood and described and an empirical approach for other processes that are less well understood. Kreft et al. (2013) argue that models can be developed which reproduce observed patterns at the individual and population level. These models have direct application to CyanoHABs for which it may be possible to constrain the number of species or strains that are modelled. Extrapolation of this approach to the community or ecosystem level with multiple interacting populations is likely to remain a major challenge for the foreseeable future. Another key challenge involves the strong interdependence of physical, chemical and biological variables related to CyanoHABs. For example, heat trapping by cyanobacteria in dense surface layers influences light attenuation which feeds back to thermal stratification, mixing dynamics and growth responses of the cyanobacteria (Kumagai et al. 2000). In order to capture these interrelationships, highly integrated interdisciplinary models are needed with teams that bridge disciplines.

20.6 Conclusions

Research into freshwater CyanoHABs has made substantial inroads into understanding factors affecting bloom development, assisted by new technologies, such as molecular methods and remote sensing. These new technologies also have great potential for more effective monitoring of blooms and toxins. There have also been substantial advances into our understanding of the complexity of responses to environmental conditions. This includes both the acknowledgement of strain variability within species and the physiological- and community-level responses of CyanoHABs to nutrient regimes. Increasingly, benthic CyanoHABs are being recognized as toxin producers with the potential for serious risks to humans and other animals. Improvements in predictive models have lagged the other technological advances, but there is an acknowledgement in the modelling community that closer links with researchers studying the physiology and ecology of CyanoHABs are needed. Integrating data from traditional and emerging techniques into deterministic models has the potential to rapidly improve knowledge of CyanoHAB processes and ultimately enable more accurate risk assessment and mitigation strategies. The new global Programme, GlobalHAB, provides a framework for sharing new international research on cyanobacteria, with a focus on advancing the science and management, via a global perspective that encourages communication across the scientific community, water managers and the public.

References

Aguirre-Gómez R, Salmerón-García O, Gómez-Rodríguez G et al (2016) Use of unmanned aerial vehicles and remote sensors in urban lakes studies in Mexico. Int J Remote Sens. https://doi.org/10.1080/01431161.2016.1264031

Al-Tebrineh J, Pearson LA, Yasar SA et al (2012) A multiplex qPCR targeting hepato- and neurotoxigenic cyanobacteria of global significance. Harmful Algae 15:19–25

Amaral V, Bonilla S, Aubriot L (2014) Growth optimization of the invasive cyanobacterium *Cylindrospermopsis raciborskii* in response to phosphate fluctuations. Eur J Phycol 49:134–141

Aparicio Medrano E, Uittenbogaard RE, Dionisio Pires LM et al (2013) Coupling hydrodynamics and buoyancy regulation in *Microcystis aeruginosa* for its vertical distribution in lakes. Ecol Model 248:41–56

Bai F, Liu R, Yang Y et al (2014) Dissolved organic P use by the invasive freshwater diazotroph cyanobacterium. *Cylindrospermopsis raciborskii*. Harmful Algae 39:112–120

Bastien C, Cardin R, Veilleux E et al (2011) Performance evaluation of phycocyanin probes for monitoring of cyanobacteria. J Environ Monit 13:110–118

Berdalet E, Kudela R, Banas NS et al (2018) GlobalHAB: fostering international coordination on harmful algal bloom research in aquatic systems. In: Glibert PM, Berdalet E, Burford M et al (eds) Global ecology and oceanography of harmful algal blooms. Springer, Cham, pp 425–447

Berdalet E, Tester PA (2018) Key questions and recent research advances on harmful algal blooms in benthic systems. In: Glibert PM, Berdalet E, Burford M et al (eds) Global ecology and oceanography of harmful algal blooms. Springer, Cham, pp 261–286

Borges H, Wood SA, Puddick J et al (2016) Intracellular, environmental and biotic interactions influence recruitment of benthic *Microcystis* (Cyanophyceae) in a shallow eutrophic lake. J Plankton Res. https://doi.org/10.1093/plankt/fbw046

Bowling LC, Baker PD (1996) Major cyanobacterial bloom in the Barwon-Darling River, Australia, in 1991, and underlying limnological conditions. Mar Freshw Res 47:643–657

Brasell K, Heath M, Ryan K et al (2015) Successional change in microbial communities of benthic *Phormidium*-dominated biofilms. Microb Ecol 69:254–266

Brookes JD, Cayelan CC (2011) Resilience to blooms. Science 334:46–47

Bullerjahn GS, Robert M, McKay A et al (2016) Global solutions to regional problems: Collecting global expertise to address the problem of harmful cyanobacterial blooms. A Lake Erie case study. Harmful Algae 54:223–238

Burford MA, Beardall J, Willis A et al (2016) Understanding the winning strategies used by the bloom forming cyanobacterium *Cylindrospermopsis raciborskii*. Harmful Algae 54:44–53

Burford MA, Davis TW, Orr PT et al (2014) Nutrient-related changes in the toxicity of field blooms of the cyanobacterium *Cylindrospermopsis raciborskii*. FEMS Microbiol Ecol 89:135–148

Burkholder JM, Glibert PM (2009) The importance of intraspecific variability in harmful algae – preface to a collection of topical papers. Harmful Algae 8:744–745

Carey CC, Ibelings BW, Hoffman EP et al (2012) Eco-physiological adaptations that favour freshwater cyanobacteria in a changing climate. Water Res 46:1394–1407

Carraro E, Guyennon N, Hamilton D et al (2012) Coupling high-resolution measurements to a three-dimensional lake model to assess the spatial and temporal dynamics of the cyanobacterium *Planktothrix rubescens* in a medium-sized lake. Hydrobiologia 698:77–95

Cires S, Ballot A (2016) A review of the phylogeny, ecology and toxin production of bloom-forming *Aphanizomenon* spp. and related species within the Nostocales (cyanobacteria). Harmful Algae 54:21–43

Coles VJ, Hood RR (2016) Approaches and challenges for linking marine biogeochemical models with the "omics" revolution. In: Glibert PM, Kana TM (eds) Aquatic microbial ecology and biogeochemistry: a dual perspective. Springer, Cham, pp 45–63

Davidson K, Gowen R, Harrison P, Fleming L, Hoagland P, Moschonas G (2014) Anthropogenic nutrients and harmful algae in coastal waters. J Environ Manage 146:206–216

Davis TW, Berry DL, Boyer GL et al (2009) The effects of temperature and nutrients on the growth and dynamics of toxic and non-toxic strains of *Microcystis* during cyanobacteria blooms. Harmful Algae 8:715–725

Davis TW, Harke MJ, Marcoval MA et al (2010) Effects of nitrogenous compounds and phosphorus on the growth of toxic and non-toxic strains of *Microcystis* during cyanobacterial blooms. Aquat Microb Ecol 61:149–162

Dodds WK, Bouska WW, Eitzmann JI et al (2008) Eutrophication of U.S. freshwaters: analysis of potential economic damages. Environ Sci Technol 43:12–19

Dolman AM, Rücker J, Pick FR et al (2012) Cyanobacteria and cyanotoxins: the influence of N versus P. PLoS One 7(6):e38757. https://doi.org/10.1371/journal.pone.0038757

Domingues RB, Barbosa AB, Sommer U et al (2011) Ammonium, nitrate and phytoplankton interactions in a freshwater tidal estuarine zone: potential effects of cultural eutrophication. Aquat Sci 73:331–343

Dowle E, Pochon X, Banks J et al (2016) Targeted gene enrichment and high throughput sequencing for environmental biomonitoring: a case study using freshwater macroinvertebrates. Mol Ecol Resour 16:1240–1254

Downing JA, Watson SB, McCauley E (2001) Predicting cyanobacteria dominance in lakes. Can J Fish Aquat Sci 58:1905–1908

Echenique-Subiabre I, Dalle C et al (2016) Application of a spectrofluorimetric tool (BBE BenthoTorch) for the monitoring of potentially toxic benthic cyanobacteria in rivers. Water Res 101:341–350

Fetscher AE, Howard MDA, Stancheva R et al (2015) Wadeable streams as widespread sources of benthic cyanotoxins in California, USA. Harmful Algae 49:105–116

Finlay K, Patoine A, Donald D et al (2010) Experimental evidence that pollution with urea can degrade water quality in phosphorus-rich lakes of the Northern Great Plains. Limnol Oceanogr 55:1213–1230

Franks PJS (2018) Recent advances in modelling of harmful algal blooms. In: Glibert PM, Berdalet E, Burford M et al (eds) Global ecology and oceanography of harmful algal blooms. Springer, Cham, pp 359–377

Garcia-Pichel F, Wojciechowski MF (2009) The evolution of a capacity to build supra-cellular ropes enabled filamentous cyanobacteria to colonize highly erodible substrates. PLoS One 4(11):e7801. https://doi.org/10.1371/journal.pone.0007801

Glibert PM, Al-Azri A, Allen JI et al (2018a) Key questions and recent research advances on harmful algal blooms in relation to nutrients and eutrophication. In: Glibert PM, Berdalet E, Burford M et al (eds) Global ecology and oceanography of harmful algal blooms. Springer, Cham, pp 229–259

Glibert PM, Allen JI, Bouwman AF et al (2010) Modeling of HABs and eutrophication: status, advances, challenges. J Mar Syst 83:262–275

Glibert PM, Beusen AHW, Harrison JA et al (2018b) Changing land, sea- and airscapes: sources of nutrient pollution affecting habitat suitability for harmful algae. In: Glibert PM, Berdalet E, Burford M et al (eds) Global ecology and oceanography of harmful algal blooms. Springer, Cham, pp 53–76

Glibert PM, Burford MA (2017) Globally changing nutrient loads and harmful algal blooms: recent advances, new paradigms, and continuing challenges. Oceanography 30(1):58–69

Glibert PM, Pitcher GC, Bernard S et al (2018c) Advancements and continuing challenges of emerging technologies and tools for detecting harmful algal blooms, their antecedent conditions and toxins, and applications in predictive models. In: Glibert PM, Berdalet E, Burford M et al (eds) Global ecology and oceanography of harmful algal blooms. Springer, Cham, pp 339–357

Glibert PM, Wilkerson FP, Dugdale RC et al (2016) Pluses and minuses of ammonium and nitrate uptake and assimilation by phytoplankton and implications for productivity and community composition, with emphasis on nitrogen-enriched conditions. Limnol Oceanogr 61:165–197

Gobler CJ, Burkholder JM, Davis TW et al (2016) The dual role of N supply in controlling the growth and toxicity of cyanobacterial blooms. Harmful Algae 54:87–97

Gugger M, Lenoir S, Berger C et al (2005) First report in a river in France of the benthic cyanobacterium *Phormidium favosum* producing anatoxin-a associated with dog neurotoxicosis. Toxicon 45(7):919–928

Hamilton DP, Carey CC, Arvola L et al (2014a) A Global Lake Ecological Observatory Network (GLEON) for synthesising high–frequency sensor data for validation of deterministic ecological models. Inland Waters 5:49–56

Hamilton DP, Salmaso N, Paerl HW (2016) Mitigating harmful cyanobacterial blooms: strategies for control of nitrogen and phosphorus loads. Aquat Ecol 50(3):351–366

Hamilton DP, Wood SA, Dietrich DR et al (2014b) Costs of harmful blooms of freshwater cyanobacteria. In: Sharma NK, Rai AK, Stal LJ (eds) Cyanobacteria: an economic perspective, 1st edn. Wiley, New York, pp 245–256

Harke MJ, Gobler CJ (2013) Global transcriptional responses of the toxic cyanobacterium, *Microcystis aeruginosa*, to N stress, P stress, and growth on organic matter. PLoS One 8(7): e69834. https://doi.org/10.1371/journal.pone.0069834

Harke MJ, Gobler CJ (2015) Daily transcriptome changes reveal the role of N in controlling microcystin synthesis and nutrient transport in the toxic cyanobacterium, *Microcystis aeruginosa*. BMC Genomics 16:1068

Harke MJ, Steffen MM, Gobler CJ et al (2016) A review of the global ecology, genomics, and biogeography of the toxic cyanobacterium, *Microcystis* spp. Harmful Algae 54:4–20

Harris TD, Smith VH, Graham JL et al (2016) Combined effects of N and P and nitrate to ammonia ratios on cyanobacterial metabolite concentrations in eutrophic Midwestern USA reservoirs. Inland Waters 6:199–210

Hellweger FL (2015) 100 Years since Streeter and Phelps: it is time to update the biology in our water quality models. Environ Sci Technol 49:6372–6373

Hellweger FL, Fredrick ND, McCarthy MJ et al (2016) Dynamic, mechanistic, molecular-level modelling of cyanobacteria: *Anabaena* and N interaction. Environ Microbiol 18:2721–2731

Hellweger FL, Kravchuk ES, Novotny V et al (2008) Agent-based modeling of the complex life cycle of a cyanobacterium (*Anabaena*) in a shallow reservoir. Limnol Oceanogr 53:1227–1241

Hemple SME, Heath MW, Olds J et al (2014) Monitoring benthic *Phormidium* blooms using imagery gathered with an unmanned aerial vehicle. Greater Wellington Regional Council. Victoria University Report. June 2014, 26 pp

Hense I, Burchard H (2010) Modelling cyanobacteria in shallow coastal seas. Ecol Model 221:238–244

Huisman J, Sharples J, Stroom J et al (2004) Changes in turbulent mixing shift competition for light between phytoplankton species. Ecology 85:2960–2970

Humbert JF, Quiblier C, Gugger M (2010) Molecular approaches for monitoring potential toxic marine and freshwater phytoplankton species. Anal Bioanal Chem 397:1723–1732

Hunter PD, Tyler AN, Willby NJ et al (2008) The spatial dynamics of vertical migration by *Microcystis aeruginosa* in a eutrophic shallow lake: a case study using high spatial resolution time-series airborne remote sensing. Limnol Oceanogr 53:2391–2406

Hunter PD, Tyler AN, Carvalho L et al (2010) Hyperspectral remote sensing of cyanobacterial pigments as indicators for cell populations and toxins in eutrophic lakes. Remote Sens Environ 114:2705–2718

Ibelings BW, Vonk M, Los FH et al (2003) Fuzzy modeling of cyanobacterial surface waterblooms: validation with NOAA-AVHRR satellite images. Ecol Appl 13:1456–1472

Islam M, Kitazawa D, Park H (2012) Numerical modeling on toxin produced by predominant species of cyanobacteria within the ecosystem of Lake Kasumigaura, Japan. Procedia Environ Sci 13:166–193

Izaguirre G, Taylor WD (2007) Geosmin and MIB events in a new reservoir in southern California. Water Sci Technol 55:9–14

Izydorczyk K, Tarczynska M, Jurczak T et al (2005) Measurement of phycocyanin fluorescenceas an online early warning system for cyanobacteria in reservoir intake water. Environ Toxicol 20:425–430
Kardinaal WEA, Janse I, Kamst-van Agterveld MP et al (2007) *Microcystis* genotype succession in relation to microcystin concentrations in freshwater lakes. Aquat Microb Ecol 2(48):1–12
Kasinak J, Holt BM, Chrislock ME et al (2015) Benchtop fluorometry of phycocyanin as a rapid approach for estimating cyanobacterial biovolume. J Plankton Res 37:248–257
Kleinteich J, Hildebrand F, Wood SA et al (2014) Diversity of toxin and non-toxin containing cyanobacterial mats of meltwater ponds on the Antarctic Peninsula: a pyrosequencing approach. Antarct Sci 26:521–532
Kong Y, Lou I, Zhang Y et al (2014) Using an online fluorescence probe for rapid monitoring of cyanobacteria in Macau freshwater reservoir. Hydrobiologia 741:33–49
Kreft J-U, Plugge CM, Grimm V et al (2013) Mighty small: observing and modeling individual microbes becomes big science. Proc Natl Acad Sci USA 110:18027–18028
Kromkamp J, Walsby AE (1990) A computer model of buoyancy and vertical migration in cyanobacteria. J Plankton Res 12:161–183
Kudela R, Palacios SL, Austerberry DC (2015) Application of hyperspectral remote sensing to cyanobacterial blooms in inland waters. Remote Sens Environ 167:196–205
Kudela RM, Raine R, Pitcher G et al (2018) Establishment, goals, and legacy of the Global Ecology and Oceanography of Harmful Algal Blooms (GEOHAB) Program. In: Glibert PM, Berdalet E, Burford M et al (eds) Global ecology and oceanography of harmful algal blooms. Springer, Cham, pp 27–49
Kumagai M, Nakano S, Jiao C et al (2000) Effect of cyanobacterial blooms on thermal stratification. Limnology 1:191–195
Kurmayer R, Deng L, Entfellner E (2016) Role of toxic and bioactive secondary metabolites in colonization and bloom formation by filamentous cyanobacteria. Harmful Algae 54:69–86
Kurmayer R, Sivonen K, Wilmotte A et al (2017) Molecular tools for the detection and quantification of toxigenic cyanobacteria. Wiley, Hoboken, NJ, 276 p
Lage S, Burian A, Rasmussne U et al (2016) BMAA extraction of cyanobacteria samples: which method to choose? Environ Sci Pollut Res 23(1):338–350
Lakeman MB, von Dassow P, Cattolico RA (2009) The strain concept in phytoplankton ecology. Harmful Algae 8:746–758
Li X, Dreher TW, Li R (2016) An overview of diversity, occurrence, genetics and toxin production of bloom-forming *Dolichospermum* (Anabaena) species. Harmful Algae 54:54–68
Lurling M, Eshetu F, Faassen EJ et al (2012) Comparison of cyanobacterial and green algal growth rates at different temperatures. Freshw Biol 58:552–559
Lv J, Wu H, Chen M (2011) Effects of nitrogen and phosphorus on phytoplankton composition and biomass in 15 subtropical, urban shallow lakes in Wuhan, China. Limnologica 41:48–56
Marinho MM, Gonçalves Souza MV, Lürling M (2013) Light and phosphate competition between *Cylindrospermopsis raciborskii* and *Microcystis aeruginosa* is strain dependent. Microb Ecol 66:479–488
Matthews MW, Bernard S, Robertson L (2012) A new algorithm for detecting trophic status (chlorophyll-a), cyanobacterial-dominance, surface scums and floating vegetation in coastal and inland waters from MERIS. Remote Sens Environ 124:637–652
McAllister TG, Wood SA, Hawes I (2016) The rise of toxic benthic *Phormidium* proliferations: a review of their taxonomy, distribution, toxin content and factors regulating prevalence and increased severity. Harmful Algae 55:282–294
Mez K, Hanselmann K, Preisig HR (1998) Environmental conditions in high mountain lakes containing toxic benthic cyanobacteria. Hydrobiologia 368(1–3):1–15
Mowe MAD, Mitrovic SM, Lim RP et al (2015) Tropical cyanobacterial blooms: a review of prevalence, problem taxa, toxins and influencing environmental factors. J Limnol 74:205–224
Nilsson RH, Ryberg M, Kristiansson E et al (2006) Taxonomic reliability of DNA sequences in public sequence databases: a fungal perspective. PLoS One 1(1):e59. https://doi.org/10.1371/journal.pone.0000059

O'Neil JM, Davis TW, Burford MA et al (2012) The rise of harmful cyanobacteria blooms: the potential roles of eutrophication and climate change. Harmful Algae 14:313–334

Oh H-M, Lee SJ, Jang M-H, Yoon B-D (2000) Microcystin production by *Microcystis aeruginosa* in a P-limited chemostat. Appl Environ Microbiol 66:176–179

Oliver R, Hamilton DP, Brookes J et al (2012) Physiology, blooms and prediction of planktonic cyanobacteria. In: Whitton BA (ed) Ecology of cyanobacteria II: their diversity in space and time. Springer, New York, pp 155–194

Orihel DM, Schindler DW, Ballard NC et al (2015) The "nutrient pump:" Iron-poor sediments fuel low nitrogen-to-phosphorus ratios and cyanobacterial blooms in polymictic lakes. Limnol Oceanogr 60:856–871

Paerl HW (2014) Mitigating harmful cyanobacterial blooms in a human- and climatically-impacted world. Life 4(4):988–1012

Paerl HW, Huisman J (2008) Climate - blooms like it hot. Science 320:57–58

Prentice MJ, O'Brien KR, Hamilton DP et al (2015) High-and low-affinity phosphate uptake and its effect on phytoplankton dominance in a phosphate-depauperate lake. Aquat Microb Ecol 75:139–153

Puddick J, Prinsep MR, Wood SA et al (2016a) Modulation of microcystin congener abundance following N depletion of a *Microcystis* batch culture. Aquat Ecol 50:235–246

Puddick J, Wood SA, Hawes I et al (2016b) Fine-scale cryogenic sampling of planktonic microbial communities: application to toxic cyanobacterial blooms. Limnol Oceanogr Methods 14:600–609

Quiblier C, Wood SA, Echenique-Subiabre I et al (2013) A review of current knowledge on toxic benthic freshwater cyanobacteria–ecology, toxin production and risk management. Water Res 47:5464–5479

Reynolds CS (2006) Ecology of phytoplankton. Cambridge University Press, Cambridge, p 535. https://doi.org/10.1017/CBO9780511542145

Rose KC, Weathers KC, Hetherington AL et al (2016) Insights from the Global Lake Ecological Observatory Network (GLEON). Inland Waters 6:476–482

Rowe MD, Anderson EJ, Wynne TT et al (2016) Vertical distribution of buoyant *Microcystis* blooms in a Lagrangian particle tracking model for short-term forecasts in Lake Erie. J Geophys Res Oceans 121:5296–5314

Schagerl M, Müller B (2006) Acclimation of chlorophyll a and carotenoid levels to different irradiances in four freshwater cyanobacteria. J Plant Physiol 163:709–716

Scott JT, McCarthy MJ (2010) N fixation may not balance the N pool of lakes over timescales relevant to eutrophication management. Limnol Oceanogr 55:1265–1270

Shokralla S, Spall JL, Gibson JF et al (2012) Next generation sequencing technologies for environmental DNA research. Mol Ecol 21:1794–1805

Smith F, Wood SA, Wilks T et al (2012) Survey of *Scytonema* (Cyanobacteria) and associated saxitoxins in the littoral zone of recreational lakes in Canterbury (New Zealand). Phycologia 51:542–551

Smith VH, Wood SA, McBride CG et al (2016) P and N loading restraints are essential for successful eutrophication control of Lake Rotorua, New Zealand. Inland Waters 6(2):273–283

Solomon C, Glibert P (2008) Urease activity in five phytoplankton species. Aquat Microb Ecol 52:149–157

Stucken K, John U, Cembella A et al (2010) The smallest known genomes of multicellular and toxic cyanobacteria: comparison, minimal gene sets for linked traits and the evolutionary implications. PLoS One 5:e9235. https://doi.org/10.1371/journal.pone.0009235

Taberlet P, Coissac E, Pompanon F et al (2012) Towards next-generation biodiversity assessment using DNA metabarcoding. Mol Ecol 21:2045–2050

Toming K, Kutser T, Laas A et al (2016) First experiences in mapping lake water quality parameters with Sentinel-2 MSI Imagery. Remote Sens 8:640

Vaitomaa J, Rantala A, Halinen K et al (2003) Quantitative real-time PCR for determination of microcystin synthetase E copy numbers for *Microcystis* and *Anabaena* in lakes. Appl Environ Microb 69:7289–7297
Van de Waal DB, Verspagen JM, Lürling M et al (2009) The ecological stoichiometry of toxins produced by harmful cyanobacteria: an experimental test of the carbon-nutrient balance hypothesis. Ecol Lett 12:1326–1335
Van der Merwe D, Price KP (2015) Harmful algal bloom characterization at ultra-high spatial and temporal resolution using small unmanned aircraft systems. Toxins 7:1065–1078
Van Mooy BAS, Fredricks HF, Pedler BE et al (2009) Phytoplankton in the ocean substitute lipids in response to P scarcity. Nature 458:69–72
Vezie C, Brient L, Sivonen K et al (1998) Variation of microcystin content of cyanobacterial blooms and isolated strains in Grand-Lieu lake (France). Microb Ecol 35:126–135
Vintila S, El-Shehawy R (2007) Ammonium ions inhibit N fixation but do not affect heterocyst frequency in the bloom-forming cyanobacterium *Nodularia spumigena* strain AV1. Microbiology 153:3704–3712
Vintila S, Jonasson S, Wadensten H et al (2010) Proteomic profiling of the Baltic Sea cyanobacterium *Nodularia spumigena* strain AV1 during ammonium supplementation. J Proteomics 73 (9):1670–1679
Visser PM, Verspagen JMH, Sandrini G et al (2016) How rising CO_2 and global warming may stimulate harmful cyanobacterial blooms. Harmful Algae 54:145–159
Wagner C, Adrian R (2009) Cyanobacteria dominance: quantifying the effects of climate change. Limnol Oceanogr 54:2460–2468
Wallace BB, Hamilton DP (2000) Simulation of water bloom formation in the cyanobacterium *Microcystis aeruginosa*. J Plankton Res 22:1127–1138
Welker M, Sejnohova D, von Dohren H et al (2007) Seasonal shifts in chemotype composition of *Microcystis* sp. Communities in the pelagial and the sediment of a shallow reservoir. Limnol Oceanogr 52:609–619
Wells ML, Karlson B (2018) Harmful algal blooms in a changing ocean. In: Glibert PM, Berdalet E, Burford M et al (eds) Global ecology and oceanography of harmful algal blooms. Springer, Cham, pp 77–90
Willis A, Adams MP, Chuang AW et al (2015) Constitutive toxin production and different optimal N concentrations leads to variation in toxin yield with three ecotypes of *Cylindrospermopsis raciborskii* ((Woloszynska) Seenayya et Subba Raju). Harmful Algae 47:27–34
Willis A, Chuang AW, Burford MA (2016a) N fixation by the reluctant diazotroph *Cylindrospermopsis raciborskii* (Cyanophyceae). J Phycol 52:854–862
Willis A, Chuang AW, Woodhouse JN et al (2016b) Intraspecific variation in growth, morphology and toxin quotas for the cyanobacterium *Cylindrospermopsis raciborskii*. Toxicon 119:307–310
Willis A, Posselt AJ, Burford MA (2017) *Cylindrospermopsis raciborskii* is highly adapted to very low P. Eur J Phycol. https://doi.org/10.1080/09670262.2017.1286524
Wood SA, Puddick J, Borges H et al (2015) Potential effects of climate change on cyanobacterial toxin production. In: Botana LM, Louzao C, Vilariño N et al (eds) Climate change and marine and freshwater toxins. Springer, Berlin, pp 155–180
Wood SA, Rueckert A, Hamilton DP et al (2011) Switching toxin production on and off: intermittent microcystin synthesis in a *Microcystis* bloom. Environ Microbiol Rep 3:118–124
Wood SA, Selwood AI, Rueckert A et al (2007) First report of homoanatoxin-a and associated dog *neurotoxicosis* in New Zealand. Toxicon 50:292–301
Wu Z, Zeng B, Li R et al (2011) Physiological regulation of *Cylindrospermopsis raciborskii* (Nostocales, Cyanobacteria) in response to inorganic P limitation. Harmful Algae 15:53–58
Wynne TT, Stumpf RP, Tomlinson MC et al (2008) Relating spectral shape to cyanobacterial blooms in the Laurentian Great Lakes. Int J Remote Sens 29:3665–3672
Zamyadi A, Mcquaid N, Prevost M et al (2012) Monitoring of potentially toxic cyanobacteria using an online multi-probe in drinking water sources. J Environ Monit 14:579–588

Chapter 21
Mitigation and Control of Harmful Algal Blooms

Zhiming Yu, Xiuxian Song, Xihua Cao, and Yang Liu

21.1 Introduction

Harmful algal blooms (HABs) are a natural phenomenon of ecosystems, with records as far back as in Ancient Egypt (Hallegraeff 1993). However, coastal water eutrophication caused by human activities since the 1950s has promoted their occurrence and spread worldwide (Glibert et al. 2006). The frequency, intensity, and geographic extent of HABs have increased worldwide in recent years (Glibert et al. 2005; Anderson et al. 2012; Liu et al. 2013; Al Shehhi et al. 2014; Teen et al. 2012), resulting in profound and deleterious effects on aquatic ecosystems, aquaculture, tourism, public health (Hallegraeff 1993; Kim 2006; Jiang et al. 2009; Anderson et al. 2012), and even the safety of nuclear power plants (Guangxi Academy of Sciences 2015). Now it has become one of the most serious aquatic disasters in the world.

Although many studies have been conducted to understand this ecological phenomenon, the mechanisms of bloom formation are still not fully understood, and accurate predication of HABs is still a highly challenging task (Anderson 2014).

Z. Yu (✉) • X. Song • Y. Liu
Key Laboratory of Marine Ecology and Environmental Sciences, Institute of Oceanology, Chinese Academy of Sciences, Qingdao, China

University of Chinese Academy of Sciences, Beijing, China

Laboratory for Marine Ecology and Environmental Science, Qingdao National Laboratory for Marine Science and Technology, Qingdao, China
e-mail: zyu@qdio.ac.cn

X. Cao
Key Laboratory of Marine Ecology and Environmental Sciences, Institute of Oceanology, Chinese Academy of Sciences, Qingdao, China

Laboratory for Marine Ecology and Environmental Science, Qingdao National Laboratory for Marine Science and Technology, Qingdao, China

P.M. Glibert et al. (eds.), *Global Ecology and Oceanography of Harmful Algal Blooms*, Ecological Studies 232, https://doi.org/10.1007/978-3-319-70069-4_21

HABs affect the production and activity of human beings, and people need appropriate methods and technologies to manage blooms. These methods and technologies can be summarized into two groups: preventative methods and mitigating methods.

Preventative methods aim to address the primary causes of HABs to prevent their occurrence. Those methods include nutrient control (Okaichi and Yanagi 1997; Anderson et al. 2001, 2004; Glibert et al. 2006; Heisler et al. 2008), ecological aquaculture strategies and biological control strategies (Zhang et al. 2005a, b; Tang and Gobler 2011; Tang et al. 2015; Yang et al. 2015), and control of exotic species via ships' ballast water treatment (Maranda et al. 2013; Guilbaud et al. 2015; Olsen et al. 2016). The purpose of these preventative methods is to create acceptable water quality and ecosystem conditions to reduce the possibilities of HAB occurrence. In order to achieve this, some international nutrient control plans have been implemented, such as Total Maximum Daily Loads (TMDL) targets and other nutrient reduction measures.

However, once HABs are established, adequate control and mitigation technologies and methods are needed to limit damage, especially in critical locations, such as cooling water sites of nuclear power plants, aquaculture sites, and waters for large sporting events. These blooms can result in enormous economic losses and need to be mitigated and controlled rapidly. Effective and environmental-friendly HAB control and mitigation methods are needed by enterprises and governments. As a result, studies on effective and environment-friendly controlling methods have become a priority in this field, particularly the methods to mitigate the HABs occurred in waters with important social and economic value, which is the major concern in this chapter.

21.2 Methods that Could Be Used to Control and Mitigate HABs

The principles of controlling HABs are to decrease the algal cell densities, stop its rapid proliferation, and mitigate its harmful effects. Methods can be divided into physical methods, chemical methods, and biological methods based on mechanisms of actions (Table 21.1; Yu et al. 1993; Kim 2006; Sengco 2009; Secher 2009).

21.2.1 Physical Methods

Physical methods are those that rely on physical mechanisms to directly remove or isolate the cells from water, and these techniques include skimming, isolation, and disruption (Shirota 1989; Kim 2006). Sometimes these methods are coupled with the use of some coagulants or flocculants. Physical methods are usually simple and do not cause secondary pollution. They have been successful in controlling HABs

Table 21.1 The control strategies of HABs (modified from Kim 2006)

Methods	Principle/methods	Advantages/disadvantages	References
Physical methods	– Skimming algae flocs using boat after air flotation – Isolation by fence and oil spill collector – Lysis by ultrasonic and electromagnetic waves – Electrolysis	– Advantages: simple, easy to operate and environmental-friendly – Disadvantages: high cost, only can be used in small scale waters, lysis of toxic algae can release toxin to waters	Shirota (1989), Kim (2006), Xu et al. (2007)
Biological methods	– Predation by, e.g., zooplankton and filter-feeding shellfish – Enzymolysis or parasitism by bacteria, virus and/or parasitic dinoflagellates – Nutrient competition or allelopathy	– Advantages: environmental friendly, some virus and parasitic dinoflagellates are specific to algae cells – Disadvantages: may cause invasion of alien species, problems of cultivation, storage, transportation and timeliness, not suitable to control field HABs, low predation rate of toxin algae by filter-feeding animals, lysis of toxic algae can release toxin to waters	Doucette (1995), Kodama et al. (2006), Secher (2009), Marcoval et al. (2013), Yang et al. (2015), Tang et al. (2015)
Chemical methods	– Killing by, e.g., inorganic killing reagent (cupric sulphate, hydrogen peroxide, chlorine, sodium hypochlorite, sodium percarbonate, and ozone) and organic algicide (such as HDTMA, sophorolipid, and aponin) – Flocculating by, e.g., flocculants setting, including inorganic, organic and microbial flocculants; clay mineral setting, such as kaolinite and montmorillonite	– Advantages: takes effects fast, easy to store and transport – Disadvantages: potential secondary pollution, nonspecific, may cause harm to non-HABs organisms, lysis of toxic algae can release toxin to waters	Rounsefell and Evans (1958), Divakaran and Pillai (2002), Ma and Liu (2002), Secher (2009), Li et al. (2014)

organisms in small volume ballast water and cyanobacterial blooms that form surface scums, especially in the later stages of a bloom in some field studies. However, physical methods usually show little effect on low density and benthic HABs organisms and are usually expensive when applied in large volume of water (Shirota 1989).

21.2.2 Biological Methods

Biological methods to control HABs usually utilize enzymolysis or parasitism effects on microorganisms (Shao et al. 2013; Zheng et al. 2013; Park et al. 2013a; Tilney et al. 2014; Yang et al. 2014; Mu et al. 2014), nutrient competition or allelopathic effects of macroalgae or seagrass (Tang and Gobler 2011; Yang et al. 2015), or grazing by marine protozoa, zooplankton, and filter-feeding shellfish (Jeong et al. 2008; Marcoval et al. 2013; Park et al. 2013b). Biological methods usually do not cause secondary pollution and can be species-specific, without impacts on non-HAB species. However, there are possibilities of introducing invasive species into the ecosystem, resulting in ecosystem disruption. Furthermore, filtration of toxic algae by shellfish may cause harm to human beings via toxin bioaccumulation. Additionally, there are great difficulties in cultivation, transportation, and timeliness when these methods are applied in the field, which makes it difficult to use biological methods to control HABs when rapid response is necessary. The feasibility of controlling HABs using zooplankton (*Acartia clausi*) and bivalves (clam and oyster) was estimated and showed that the method was completely infeasible, considering the cost, space, and facilities (Shirota 1989; Anderson et al. 2001). Nevertheless, filter-feeding shellfish do contribute to prevention of HABs occurrence (Anderson et al. 2001). Thus far, biological methods have only studied in the laboratory, and few field applications have been reported. Biological control in many regions also requires extensive permits, again limiting the timeliness of such approaches.

21.2.3 Chemical Methods

Chemical methods employ either chemicals toxic to HAB organisms (killing them directly) or flocculants that precipitate algal cells to the water bottom.

Reagents used in the direct killing methods can be an inorganic agent, including cupric sulphate, hydrogen peroxide, chlorine, sodium hypochlorite, sodium percarbonate and ozone, or organic algicides, such as hexadecyl trimethyl ammonium bromide (HDTMA) (Cao and Yu 2003b), sophorolipid (Baek et al. 2003), and aponin (Secher 2009). Direct killing methods that take effect rapidly are currently prohibited in most regions (e.g., cupric sulphate) due to their toxic effects on the ecosystem. In recent years, researchers have started to consider using more environmental-friendly compounds, such as those that can rapidly decompose, including hypochloric acid, sodium hypochlorite, sodium percarbonate and ozone (Yu et al. 1993; Jeong et al. 2002; Burson et al. 2014; Randhawa et al. 2013), organic algicides (e.g., carboxylic acid, organic amine, and ketone) (Wu et al. 2014; Baek et al. 2014), and plant extracts (Barani et al. 2015). After considering the potential environment problems and costs, however, these chemicals are usually not recommended to be applied in the field, except perhaps in some limited aquaculture or localized regions, such as lakes.

Flocculation methods involve physically aggregating algal cells using flocculants which then settle to the bottom of the waterbody, ultimately resulting in the death of HAB organisms. The flocculants can be inorganic, organic, as well as natural minerals. The inorganic flocculants are mainly compounds of ferrum or aluminium, such as aluminium sulphate, ferric chloride, and their polycompounds (Sukenik et al. 1988). The organic flocculants are mainly surfactant and polymer flocculants (Tenney et al. 1969; Tilton et al. 1972; Sukenik et al. 1988; Pang et al. 2013). Flocculation functions like water purification in settling tanks in water treatment plants and takes effect rapidly in the control of HABs. However, this method is too expensive to be applied in the field (Shirota 1989; Anderson et al. 2001).

Natural minerals, such as clay and loess, can be used to flocculate HAB organisms. Natural minerals are abundant, cheap, and environment-friendly. The method has been intensively studied since the 1980s. Up to now, controlling HABs using natural minerals is one of the few methods that can be applied widely in the field.

21.2.4 Flocculation Using Natural Minerals

Technically, many methods seem to be able to inhibit the growth of HABs. Taking the requirements of field application for efficiency, safety, low cost, and operability into consideration, however, many methods are feasible only in laboratory, and only a few can be applied in the field, among which natural clays are promising (Shirota 1989; Yu et al. 1994a, b, c, d; Anderson 1997; Sengco et al. 2001; Lewis et al. 2003; Sengco and Anderson 2004; Beaulieu et al. 2005; Kim 2006; Park et al. 2013b).

Studies on controlling HABs using clay started in Japan. In the 1960s, Japanese scientists found that natural clay minerals could flocculate cells of HABs (Yu et al. 1993), and from the late 1970s to early 1980s, the Japanese used this method to control HABs in coastal aquaculture sites in Kagoshima (Shirota 1989; Yu et al. 1993; Imai et al. 2006). Kim (1987) studied mitigating *Prorocentrum triestinum* using loess in Korea, and in the late 1990s, Koreans mitigated HABs in aquaculture sites using loess on many occasions (Kim 2006). Yu et al. (1994b, c, d, 1995, 1999) conducted studies of HAB mitigation using clay in China in the early 1990s and proposed the clay modification theory and methods to improve removal efficiency. Since 2005 (Zhang 2006), modified clays have been applied widely in the field in China.

Low flocculating efficiency is the most serious drawback with natural mineral methods, which often leads to a requirement of huge amount of clay to meet an effective efficiency in the field. For instance, it was reported that the clay dosage in aquaculture sites in Japan was 110–400 t/km^2 (Shirota 1989), and the loess dosage used in control of *Cochlodinium* (*Margalefidinium*) *polykrikoides* blooms in Korea was 384 t/km^2 (Anderson et al. 2001). However, the low removal efficiency, high dosage, and heavy sediments serve as the bottleneck to the use of natural mineral methods when applied in the field. Subsequently, numerous studies have been carried out on how to improve the removal efficiency of natural minerals. For example, Shirota (1989) thought that among all natural minerals employed in the field, montmorillonite exhibited the highest efficiency. Yu et al. (1994a, 1995) found that the surface property of clay minerals was the main factor affecting efficiency, and some types of kaolinites showed better flocculation methods. In addition, studies also demonstrated that the removal efficiency of clay was algal-species-dependent and, in general, was lower for species with higher mobility. Screening for suitable clays in controlling specific HAB species became an important research topic (Sengco et al. 2001; Padilla et al. 2010).

To further increase the removal efficiency of natural clay minerals, some scientists proposed to pretreat clay minerals, such as by acidification (Maruyama et al. 1987; Miao et al. 2014) or adding other components (Yu et al. 1994b, 1999; Sengco et al. 2001; Sun et al. 2004a, b; Pierce et al. 2004; Lee et al. 2008; Liu et al. 2010). Among these pretreatments, modified clay, prepared under the directions of clay surface modification theory and methods, greatly improved the removal efficiency (Yu et al. 1994b, 1999; Anderson 1997) and has been applied successfully to mitigating HABs of different species and water bodies (both marine and freshwater) in the field (Zhang 2006).

21.3 The Theory and Method of Clay Surface Modification

21.3.1 The Theory of Clay Surface Modification

Yu et al. (1994a, b, c, d, 1995) studied the removal efficiency of different types of clay and proposed that the key mechanism of controlling HABs using clay is flocculation. Based on the traditional flocculation theory (DLVO), the model of interaction between algal cells and clay particles could be mathematically expressed as follows:

$$V = V_{\mathrm{R}} + V_{\mathrm{A}} \tag{21.1}$$

$$V_{\mathrm{R}} = \frac{\varepsilon r_a r_b\left(\psi_a^2 + \psi_b^2\right)}{4(r_a + r_b)}\left[\frac{2\psi_a\psi_b}{\psi_a^2 + \psi_b^2}\,\mathrm{In}\left(\frac{1 + \exp(-\kappa H_0)}{1 - \exp(-\kappa H_0)}\right) + \mathrm{In}(1 - \exp(-2\kappa H_0))\right] \tag{21.2}$$

$$V_{\mathrm{A}} = A\int\int\frac{d\upsilon_1 d\upsilon_2}{L^6} \tag{21.3}$$

where V denotes the flocculation interaction between algal cells and clay particles, which can be divided into electrostatic interaction (V_{R}) and Van der Waals interaction (V_A); a and b denote the clay particles and algal cells, respectively; r_i and ψ_i denote the radius and surface potential of particle "i"; ε denotes dielectric permittivity; κ denotes the reciprocal of Debye–Hückel length; H_0 denotes the distance between the interacting particles; L denotes the distance between particles; and v denotes the volume of the clay particles or algal cells.

Through this model, it was found that the surface potential and effective interaction radius of clay particles were the main factors that influenced the removal efficiency of clays. This model further explained the result that the removal efficiency of the kaolinite was higher than that of montmorillonite (Yu et al. 1994a, 1995) and also provided scientific guidelines on how to further improve the capability of clays to remove HAB organisms, i.e., changing the surface potential of clays and increasing the effective radius of interaction between clay particles and HAB organisms (Yu et al. 1994d, 1995, 1999). Based on the model and findings described above, the theory of clay surface modification was proposed in order to improve the removal efficiency of clays.

The fact that the surfaces of clay particles and algal cells are both electronegative in seawater provided the fundamental basis for the development of the clay surface modification theory, which aimed to increase the positive surface potential of the clay particles, so as to decrease electrostatic repulsion and ultimately convert the interaction from repulsion to electrostatic attraction between clay particles and algal cells. According to the theory (Yu et al. 1994b), the flocculation between algal cells and clay particles could be improved if the clay surfaces are modified with positively charged compound M^{Z+} (Fig. 21.1): the higher the M^{Z+} concentration and the bigger

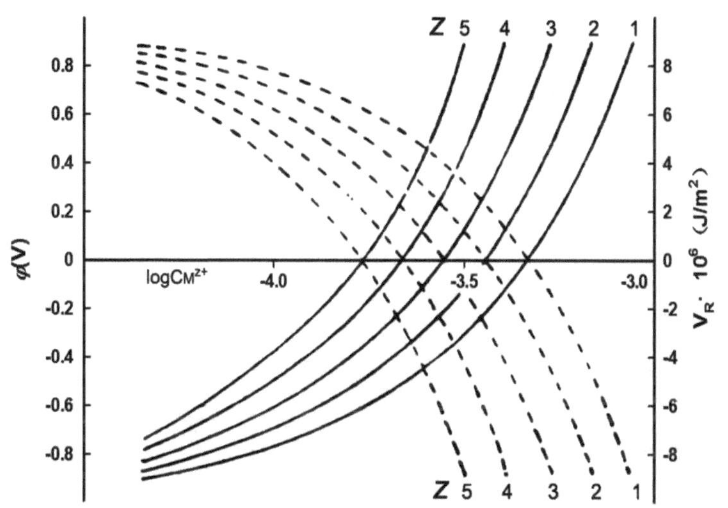

Fig. 21.1 The theoretical analog curves showing the importance of the concentration of the second component C_M^{Z+} affecting clay surface potential (φ) and electrostatic interaction (V_R) between clay particles and red tide organisms (Reproduced from Yu et al. 1994b with permission of this publisher)

the charge Z, the stronger the flocculation should be. Besides, according to Van der Waals interaction (V_A), the longer the M^{Z+} molecular chain, the stronger the "bridging" effects are and the stronger the flocculation should be as well (Yu et al. 1995).

The theory of clay surface modification provided a guiding basis to further improve the flocculation between clay particles and algal cells. Since then, the flocculation of clay particles and algal cells was studied through fractal theory (Wang et al. 2011) and adsorption theory (Lin 2013), respectively. Both approaches were highly consistent with the theory of clay surface modification.

21.3.2 Techniques and Methods to Modify Clay Surfaces

According to the theory of clay surface modification, the bigger the charge Z in M^{Z+} and the longer the molecular chain, the stronger the flocculation should be. Considering the high dispersity of modified clay and decreasing self-aggregation in seawater, the molecular chain of the modifier M^{Z+} should not be too long. Yu et al. (1999) used compounds of aluminium, magnesium, and others with appropriate molecular weights as the modifiers and prepared modified clays through adsorption or insertion methods. The results showed that the electronegative potential of clay surface decreased with an increase in modifier concentrations and finally changed from negative to positive regardless of the type of clay (Fig. 21.2). The removal efficiency could be increased several dozen times after clay modification. The extent of the increase was well correlated with changes of the clay surface potential (Fig. 21.3), which further proved the validity of the clay surface modification theory (Yu et al. 1994b, 1999). Afterward, Cao and Yu (2003a) used the positively charged organic compound HDTMA to modify clays, and the results showed that removal efficiencies were increased by about 100 times due to the algicide effects of HDTMA.

Further studies showed that the improvement in removal efficiency of modified clay was related to clay type (Yu et al. 1994a, 1995; Sengco et al. 2001; Anderson

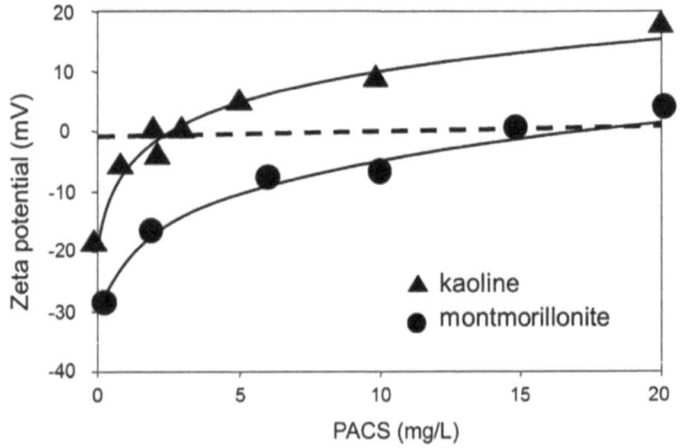

Fig. 21.2 The effect of polyhydroxy aluminium chloride (PACS) concentration on the surface potential of clay particles (Reproduced from Yu et al. 1994b with permission of the publisher)

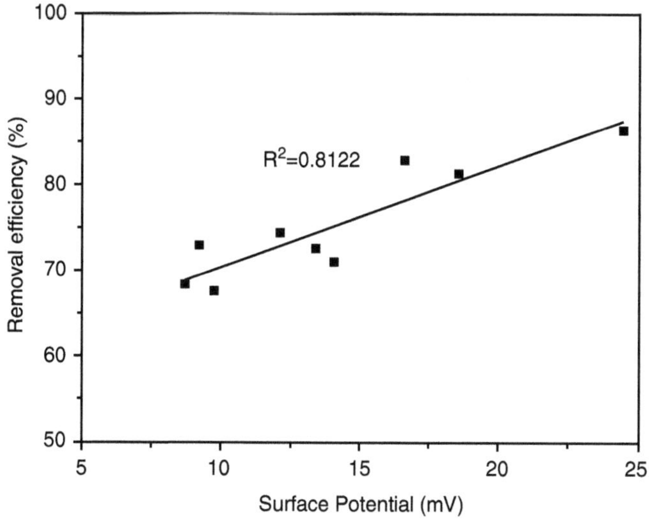

Fig. 21.3 A positive correlation between removal efficiency of *Heterosigma akashiwo* and surface charge property of the montmorillonite (Reproduced from Yu et al. 1999 with permission of the publisher)

2004; Pierce et al. 2004; Padilla et al. 2010; Seger et al. 2015), clay size (Han and Kim 2001; Yu et al. 2004), surface morphology (Yu et al. 1995), modifier type (Maruyama et al. 1987; Yu et al. 1994b, 1999; Liu et al. 2016), algal features (Padilla et al. 2010), environmental factors (Avnimelech et al. 1982; Sengco and Anderson 2004), and the properties of modified clay suspension (Liu et al. 2015).

21.3.3 Assessment of Possible Ecological Effects of Modified Clay (Clay)

A critical precondition of HAB control or mitigation approach is that there is no negative impact on the ecosystem, including harm to other aquatic organisms. Hence, the possible ecological effects of the modified clay have been examined.

Numerous studies have been carried out to examine the effects of modified clay on aquatic organisms and benthos (Table 21.2). Results showed that at the effective clay dosage, the clay or modified clay did not show any negative effects on the survival of aquatic organisms and benthos, and the modified clay even improved the survival rates of some aquaculture animals (Archambault et al. 2004). Besides, the modified clay improved water quality: nutrient concentrations were significantly decreased, and consequently the regrowth of HAB organisms was suppressed (Lu 2014).

Table 21.2 Test on the potential effects of clay or modified clay on marine organisms

Clay materials	Experimental organisms	Effects	References
Clay	*Pagrosomus major* and *Ocyurus chrysurus*	200–400 g/m^{2}clay showed no effects to fishes	Shirota (1989)
Loess	Benthic fauna	Tongyeong City, Korea, 96–99, sprayed every year, showed no changes in the species composition, diversity, and abundance of benthos	Kim (2006)
Loess	Fish and shellfish, including abalone and flatfish	No significant impacts at a clay concentration of 20 g/L within 24 h	Kim (2006)
Anti-HAB reagents (Clay, $Ca(ClO)_2$, $AlCl_3$, $FeCl_3$)	*Penaeus chinensis*	Clay no toxicity, $Ca(ClO)_2$, $AlCl_3$ and $FeCl_3$ can be used to modify the clay minerals	Sun et al. (2001)
Clay-complex including kaolinite, sodium salt, and polyaluminium salt	*Penaeus japonicus* (length between 1.0 cm and 1.5 cm)	Harmless	Song et al. (2003)
Yellow clay	Seven benthic, filter-feeding invertebrates	Species-specific, some suffer negative impacts	Shumway et al. (2003)
Phosphatic clay mixed with PAC	Infaunal amphipods (*Leptocheirus plumulosus* and *Ampelisca abdita*), grass shrimp embryos (*Palaemonetes pugio*) and larval sheepshead minnows (*Cyprinodon variegatus*)	Not acutely toxic in most cases, neither increase nor decrease toxicity to benthic organisms compared with untreated bloom	Lewis et al. (2003)
Phosphatic clay	Juvenile hard clams, *Mercenaria mercenaria*	No clam mortalities occurred, particle-suspension trials grows better than no-clay controls	Archambault et al. (2004)
Hexadecyltrimethylammonium bromide(HDTMA) modified clay	*Penaeus japonicus* larvae	Survival rates 100% at 0.03 g/L and 0.09 g/L	Cao et al. (2004)
Gemini surfactant trialkyl-polyoxyethenyl quaternary ammonium compound (TPQAC) modified clay	*Neomysis awatschensis*	No apparent influences to the mariculture	Wu and Yu (2007)

(continued)

Table 21.2 (continued)

Clay materials	Experimental organisms	Effects	References
HDTMA modified clay, polyaluminium chloride modified clay	Infant oyster (*Crassostrea gigas*, about 0.2 cm in size)	No obvious influences to survival rate, no negative effects on the gill and digestive gland super microstructure, increased survival rates apparently compared with control	Gao et al. (2007)
Sophorolipid-loess mixture	Bacterioplankton, heterotrophic protists, and zooplankton	No variation in bacterial abundance, heterotrophic protist and zooplankton abundance decreased, and the mixture had less adverse effects than the yellow clay	Lee et al. (2008)
Chitosan-modified soils	Morphological composition and genetic diversity of 44 zooplankton taxa	No considerable difference in morphological composition and genetic diversity of zooplankton, and zooplankton community could recover original condition after half year	Ni et al. (2010)
Aminoclay	Non-harmful phytoplankton, zooplankton and farmed fish, *Pagrus major* and *Paralichthys olivaceus*	Phytoplankton unchanged, negligible impacts under recommended concentrations	Lee et al. (2013)
Ball clay	Non-harmful microalgae (*Isochrysis galbana* Parke), diatom (*Skeletonema tropicum*), mussel (*Perna viridis*) and fishes such as milkfish (*Chanos chanos*), sea bass (*Lates calcarifer*), and rabbitfish (*Siganus guttatus*)	Not significantly affected by ball clay addition under the optimum concentration	Orizar et al. (2013)
Polyaluminium chloride modified clay	Infant of *Apostichopus japonicus Selenka*, infant of *Patinopecten yessoensis*	No effects under the optimum clay dosage in chronic and acute toxicity tests	Wang (2014)

For some toxic algal blooms, modified clay not only inhibited the cell growth and suppressed toxin production (Yu and Rao 1998), it also decreased the breakdown of organic nitrogen (N) and organic phosphorus (P) in sediment, hence effectively suppressing the release of nutrients into the ambient water. Studies also found that modified clay could promote the transformation of paralytic shellfish toxins (PSTs) from highly toxic GTX1 and GTX4 to less toxic GTX2 and dcGTX3 (Lu et al. 2015a, b). Some HAB species have the ability to form cysts under adverse conditions, and the formed cysts may act as the "seeds" of another bloom (Joyce et al. 2005). Wang (2014) studied the effects of modified clay on the cyst formation of *Scrippsiella trochoidea*. Results showed that, although modified clay promoted the formation of cysts, the formed cysts were mainly abnormal cysts with very low germination percentage and no thorn on the cell surface. Therefore, modified clay also has the additional positive effect of inhibiting cyst germination.

As stated above, modified clays (clays) could control HABs, improve water quality, decrease the risk of algal toxicity, and improve the survival of aquaculture animals, with no negative environmental impact. These benefits and advantages, compared with other methods, lay an important foundation for the large-scale application of modified clay (clay) in the field.

21.4 Application Cases of Modified Clays and Clays in In Situ Control of HABs

To date, controlling HABs using modified clays or clays has been fully implemented in the field in Japan, South Korea, and China. Many economic and environmental losses have been avoided (Kim 2010; Kim et al. 2010).

21.4.1 Field Applications in Japan

The first field application using clay to control HABs was in Japan. In the late 1970s and early 1980s, large-scale field experiments were conducted at aquaculture sites, and results demonstrated that, for example, montmorillonite sprayed at 110–400 g/m^2 could effectively "kill" *Cochlodinium* cells without any observed negative impact on fish (Shirota 1989). In 1979, a *Cochlodinium* bloom occurred in Yatsushiro Sea, and fishermen used clay to mitigate the blooms. An investigation was done, in which 70% of the fishermen thought that the clay method was effective and no negative effects were reported. Japanese scientists also sprayed clay powders from airplanes, with a reported effective dosage of 200 g/m^2 (Shirota 1989).

21.4.2 Field Applications in Korea

Korea is the country where natural minerals are thought to be the most effective and operable method, and clays have been widely used to mitigate HABs in aquaculture sites, and therefore massive economic damage has been avoided with no observed negative impacts (Kim 2010; Kim et al. 2010).

Loess is the main natural mineral that has been used in HAB mitigation in Korea because of its availability and low cost. Ever since *Prorocentrum triestinum* removal experiments using loess were conducted (Kim 1987, 2006) in 1986, several natural minerals were used to control *Cochlodinium polykrikoides* blooms in fishery sites (Kim 2006). In 1996, 1×10^5 t of loess was used to control a 260 km^2-sized bloom of *C. polykrikoides* bloom, and the effective dosage was reported to be 384 g m^2. In 1997, another *C. polykrikoides* bloom was controlled using 1×10^5 t loess (Anderson et al. 2001). Korea also published the guidebook *HABs and Clay Dispersion* to help the government and fishermen to master the techniques and methods (Kim 2006). In this book, the suitable particle size of minerals (less than 50 μm), the dosage (100–400 g/m^2), and the best spraying time (noon) were recommended (Kim 2006). This guide also gives words of caution and instructions when used in aquaculture sites (Kim 2006). Furthermore, special loess dispensers (since 1997), including the electrolytic loess dispenser (ECD), the third generation dispenser were also designed (Kim 2010; Kim et al. 2010; Park et al. 2013b).

21.4.3 Field Applications in China

As described above, Yu et al. (1994b, c, d, 1995, 1999) proposed the theory and method of clay surface modification to improve control efficiency. In 2005, modified clay was first applied in the field to control a cyanobacterial bloom of *Microcystis aeruginosa* in Xuanwu Lake in Nanjing, China (Zhang 2006) (Fig. 21.4). The results showed that modified clay could control the cyanobacteria bloom effectively (Zhang 2006; Mei et al. 2010). After implementation of the modified clay method, the biomass of *M. aeruginosa* decreased by 82%, while the dominance of *M. aeruginosa* decreased from 95% to 11%. Water quality significantly improved. Chemical oxygen demand (COD) and total P also greatly decreased. Cyanotoxin concentrations dropped from 0.03–0.62 μg/L to 0.01 μg/L. The number of zooplankton increased by 47%, while the number of species of benthos was unchanged. Overall, the number of bacteria decreased one or two orders of magnitude. There were no observable impacts on higher organisms, including plants and fish (Zhang 2006). In 2006–2008, monitoring results showed clear lake water with 20–30% macrophyte coverage (*Potamogeton crispus L.*) at the bottom. The aquatic community structure of the lake was thus transformed from microalgae-dominated to macrophyte-dominated. *Microcystis aeruginosa* blooms have not been observed since then.

Up to now, modified clay methods have been widely used to mitigate and control of HABs from the north to the south in Chinese coastal waters. Thousands of square

Fig. 21.4 Control of *Microcystis aeruginosa* bloom using modified clay in Xuanwu Lake in Nanjing, China. Comparison before (**a**) and after implement (**b**); Spraying modified clay in the field (**c**) and (**d**); Next year after modified clay implement, the grass (*Potamogeton crispus*) grew vigorously (**e**) and the *Microcystis aeruginosa* bloom was not observed from then on (**f**)

kilometres of different HAB blooms (e.g., *Heterosigma akashiwo*, *Chattonella* sp., *Thalassios* sp., *Skeletonema costatum*, and *Aureococcus anophagefferens*) have been controlled in the past 10 years (Fig. 21.5). Some important sites, i.e., aquaculture facilities, tourism sites, areas of beach use, and aquatic sports (e.g., the Olympic games, the Asian Games, Universiade, etc.) were successfully prevented

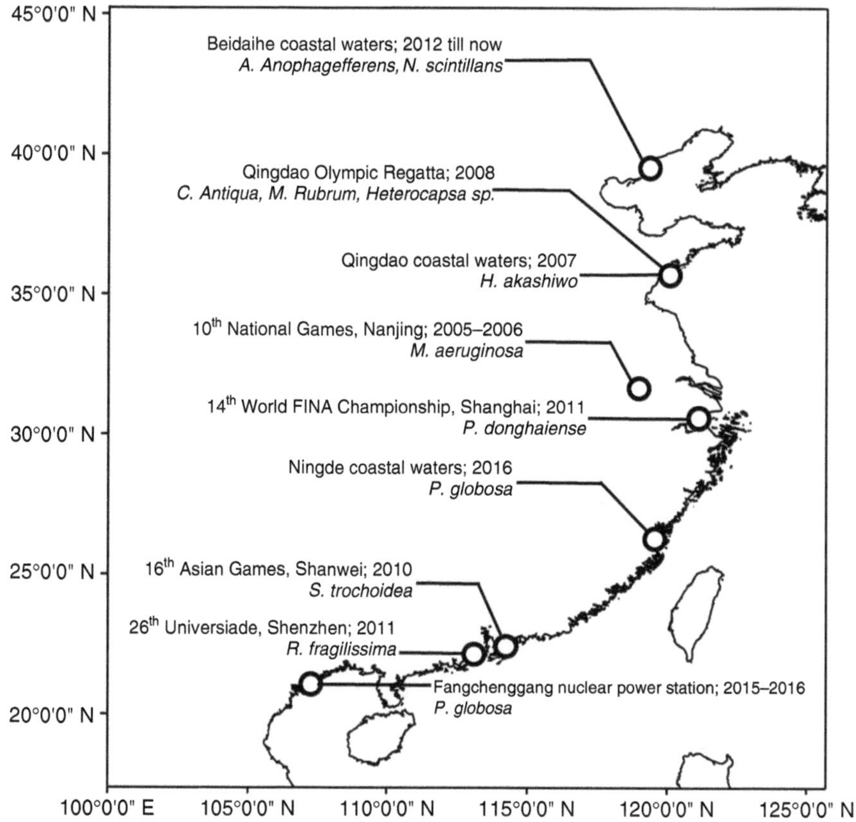

Fig. 21.5 Application of modified clay in control of HABs in China

from HAB outbreaks (Fig. 21.6) (Aweb 2006; People's Network 2006; Qingdao News 2007a, b; Public Network 2008; Hebnews 2012; China Daily 2015). The average dosage of modified clays was 4–10 t km^2, which is much lower than that of original (natural) clays (CNC News 2016). In 2015, modified clay methods were used to mitigate *Phaeocystis globosa* blooms that threatened the cooling water system of a nuclear power plant in China and effectively guaranteed the safety of the nuclear power plant (Guangxi Academy of Sciences 2015). In 2014, the modified clay technology was listed in the "Technical guidelines for treatment with red tide disaster" (GB/T 30743-2014) as the method of national standards in China. Now, it is preferred in mitigation and control of HABs in China.

Fig. 21.6 Mitigation HABs by using modified clay in different waters of China (**a**) Aquaculture sites; (**b**) 2008 Olympic Regatta in Qingdao; (**c**) Beidaihe coastal waters; (**d**) Cooling water area of Fangchenggang nuclear power station

References

Al Shehhi MR, Gherboudj I, Ghedira H (2014) An overview of historical harmful algae blooms outbreaks in the Arabian Seas. Mar Pollut Bull 86(1–2):314–324

Anderson DM (1997) Turning back the harmful red tide. Nature 388(6642):513–514

Anderson DM (2004) Prevention, control, and mitigation of harmful algal blooms: multiple approaches to HAB management. In: Hall S, Etheridge S, Anderson D et al (eds) Harmful algae management and mitigation. Asia-Pacific Economic Cooperation, APEC Publication 204-MR-04.2, Singapore, pp 123–130

Anderson DM (2014) HABs in a changing world: a perspective on harmful algal blooms, their impacts, and research and management in a dynamic era of climactic and environmental change. In: Kim HG, Reguera B, Hallegraeff GM et al (eds) Harmful algae 2012: proceedings of the 15th International conference on Harmful Algae. International Society for the Study of Harmful Algae, pp 3–17

Anderson DM, Andersen P, Bricelj VM et al (2001) Monitoring and management strategies for harmful algal blooms in coastal waters Asia-Pacific Economic Program. Singapore, and Intergovernmetal Oceanographic Commission, Paris, 268 pp

Anderson DM, Cembella AD, Hallegraeff GM (2012) Progress in understanding harmful algal blooms: paradigm shifts and new technologies for research, monitoring, and management. Annu Rev Mar Sci 4:143–176

Archambault M-C, Bricelj VM, Grant J et al (2004) Effects of suspended and sedimented clays on juvenile hard clams, *Mercenaria mercenaria*, within the context of harmful algal bloom mitigation. Mar Biol 144(3):553–565
Avnimelech Y, Troeger BW, Reed LW (1982) Mutual flocculation of algae and clay: evidence and implications. Science 216(4541):63–65
Aweb (2006) http://news.aweb.com.cn/2006/4/15/8373997.htm (in Chinese)
Baek SH, Son M, Jung SW et al (2014) Enhanced species-specific chemical control of harmful and non-harmful algal bloom species by the thiazolidinedione derivative TD49. J Appl Phycol 26(1):311–321
Baek SH, Sun XX, Lee YJ et al (2003) Mitigation of harmful algal blooms by sophorolipid. J Microbiol Biotech 13(5):651–659
Barani M, Yousefzadi M, Moezi M (2015) Essential oils, new source of algicidal compounds. J Appl Phycol 27(1):267–273
Beaulieu SE, Sengco MR, Anderson DM (2005) Using clay to control harmful algal blooms: deposition and resuspension of clay/algal flocs. Harmful Algae 4(1):123–138
Burson A, Matthijs HC, de Bruijne W et al (2014) Termination of a toxic *Alexandrium* bloom with hydrogen peroxide. Harmful Algae 31:125–135
Cao X, Song X, Yu Z (2004) Removal efficiency of red tide organisms by modified clay and its impacts on cultured organisms. Environ Sci 25(5):148–152 (in Chinese)
Cao X, Yu Z (2003a) Extinguishment of harmful algae by organo-clay. Chin J Appl Ecol 14(7):1169–1172 (in Chinese)
Cao X, Yu Z (2003b) Mechanism of quaternary ammonium compounds extinguishing *Heterosigma Akashiwo*. Oceanol Limnol Sin 34(2):201–207 (in Chinese)
China Daily (2015) http://www.chinadaily.com.cn/interface/toutiao/1120781/2015-7-3/cd_21170934.html?tt_from=mobile_qq&utm_campaign=client_share&app=news_article&utm_source=mobile_qq&iid=2832028472&utm_medium=toutiao_ios&tt_group_id=4642294177 (in Chinese)
CNC News (2016) http://news.cncnews.cn/2016-05-19/124418660.html (in Chinese)
Divakaran R, Pillai VNS (2002) Flocculation of algae using chitosan. J Appl Phycol 14(5):419–422
Doucette GJ (1995) Interactions between bacteria and harmful algae: a review. Nat Toxins 3(2):65–74
Gao Y, Yu Z, Song X et al (2007) Impact of modified clays on the infant oyster (*Crassostrea gigas*). Mar Sci Bull 26(3):53–60 (in Chinese)
GB/T 30743-2014. Technical guidelines for treatment with red tide disaster. China's National Standards. China Standard Press, Beijing (in Chinese)
Glibert PM, Anderson DM, Gentien P et al (2005) The global, complex phenomena of harmful algal blooms. Oceanography 18(2):136–147
Glibert PM, Harrison J, Heil C et al (2006) Escalating worldwide use of urea – a global change contributing to coastal eutrophication. Biogeochem 77(3):441–463
Guangxi Academy of Sciences (2015) http://www.gxas.cn/Gxas/news.aspx?id=3099 (in Chinese)
Guilbaud J, Masse A, Wolff FC et al (2015) Porous membranes for ballast water treatment from microalgae-rich seawater. Mar Pollut Bull 101(2):612–617
Hallegraeff GM (1993) A review of harmful algal blooms and their apparent global increase. Phycologia 32(2):79–99
Han MY, Kim W (2001) A theoretical consideration of algae removal with clays. Microchem J 68(2–3):157–161
Hebnews (2012) http://hebei.hebnews.cn/2012-06/28/content_2744957.htm (in Chinese)
Heisler J, Glibert PM, Burkholder JM et al (2008) Eutrophication and harmful algal blooms: a scientific consensus. Harmful Algae 8(1):3–13
Imai I, Yamaguchi M, Hori Y (2006) Eutrophication and occurrences of harmful algal blooms in the Seto Inland Sea, Japan. Plankton Benthos Res 1(2):71–84
Jeong HJ, Kim HR, Kim KI et al (2002) NaOCl produced by electrolysis of natural seawater as a potential method to control marine red-tide dinoflagellates. Phycologia 41(6):643–656

Jeong HJ, Kim JS, Yoo YD et al (2008) Control of the harmful alga *Cochlodinium polykrikoides* by the naked ciliate *Strombidinopsis jeokjo* in mesocosm enclosures. Harmful Algae 7(3):368–377
Jiang X, Tang Y, Lonsdale DJ et al (2009) Deleterious consequences of a red tide dinoflagellate *Cochlodinium polykrikoides* for the calanoid copepod *Acartia tonsa*. Mar Ecol Prog Ser 390(10):105–116
Joyce LB, Pitcher GC, Randt AD et al (2005) Dinoflagellate cysts from surface sediments of Saldanha Bay, South Africa: an indication of the potential risk of harmful algal blooms. Harmful Algae 4(2):309–318
Kim CH, Park TG, Lee C (2010) Harmful dinoflagellates and mitigation strategies in Korea. Philipp J Sci 139(2):139–147
Kim HG (1987) Ecological study of dinoflagellates responsible for red tide. 1. The population growth and control of *Prorocentrum triestinum* Schiller. Bull Nat Fish Res Dev Instit 39:1–6
Kim HG (2006) Mitigation and controls of HABs. In: Granéli E, Turner J (eds) Ecology of harmful algae. Springer, Berlin, pp 327–338
Kim HG (2010) An overview on the occurrences of harmful algal blooms (HABs) and mitigation strategies in Korean coastal waters. In: Ishimatsu A, Lie H-J (eds) Coastal environmental and ecosystem issues of the East China Sea. Terrapub and Nagasaki University Publication, Tokyo, pp 121–131
Kodama M, Doucette G, Green D (2006) Relationships between bacteria and harmful algae. In: Granéli E, Turner J (eds) Ecology of harmful algae. Springer, Berlin, pp 243–255
Lee YC, Jin E, Jung SW et al (2013) Utilizing the algicidal activity of aminoclay as a practical treatment for toxic red tides. Scientific Rep 3:1292. https://doi.org/10.1038/srep01292
Lee YJ, Choi JK, Kim EK et al (2008) Field experiments on mitigation of harmful algal blooms using a sophorolipid—yellow clay mixture and effects on marine plankton. Harmful Algae 7(2):154–162
Lewis MA, Dantin DD, Walker CC et al (2003) Toxicity of clay flocculation of the toxic dinoflagellate, *Karenia brevis*, to estuarine invertebrates and fish. Harmful Algae 2(4):235–246
Li YH, Wu T, Yang WD et al (2014) The effectiveness of five natural products against three species of harmful algae. Water Environ J 28(2):270–276
Lin Y (2013) Influence of aggregates morphology and algal organic materials on the efficiency of modified clays deployed in mitigation of HAB organisms. PhD Dissertation, University of Chinese Academy of Sciences (in Chinese)
Liu G, Fan C, Zhong J et al (2010) Using hexadecyl trimethyl ammonium bromide (CTAB) modified clays to clean the *Microcystis aeruginosa* blooms in Lake Taihu, China. Harmful Algae 9(4):413–418
Liu L, Zhou J, Zheng B et al (2013) Temporal and spatial distribution of red tide outbreaks in the Yangtze River Estuary and adjacent waters, China. Mar Pollut Bull 72(1):213–221
Liu Y, Cao X, Yu Z et al (2015) Flocculation of harmful algal cells using modified clay: effects of the properties of the clay suspension. J Appl Phycol:1–11
Liu Y, Cao X, Yu Z et al (2016) Controlling harmful algae blooms using aluminum-modified clay. Mar Pollut Bull 103(1–2):211–219
Lu G (2014) Effects of HABs mitigation by modified clay on major nutrient cycling and algal toxins. PhD Dissertation, University of Chinese Academy of Sciences (in Chinese)
Lu G, Song X, Yu Z et al (2015a) Effects of modified clay flocculation on major nutrients and diatom aggregation during *Skeletonema costatum* blooms in the laboratory. Chin J Oceanol Limnol 33:1007–1019
Lu G, Song X, Yu Z et al (2015b) Environmental effects of modified clay flocculation on *Alexandrium tamarense* and paralytic shellfish poisoning toxins (PSTs). Chemosphere 127: 188–194
Ma J, Liu W (2002) Effectiveness and mechanism of potassium ferrate (VI) preoxidation for algae removal by coagulation. Water Res 36(4):871–878
Maranda L, Cox AM, Campbell RG et al (2013) Chlorine dioxide as a treatment for ballast water to control invasive species: shipboard testing. Mar Pollut Bull 75(1–2):76–89

Marcoval MA, Pan J, Tang Y et al (2013) The ability of the branchiopod, *Artemia salina*, to graze upon harmful algal blooms caused by *Alexandrium fundyense*, *Aureococcus anophagefferens*, and *Cochlodinium polykrikoides*. Estuar Coast Shelf Sci 131:235–244
Maruyama T, Yamada R, Usui K et al (1987) Removal of marine red tide planktons with acid treated clay. Nippon Suisan Gakkaishi 53(10):1811–1819
Mei Z, Zhang Z, Zhao C et al (2010) Dynamics of phytoplankton and water quality with control of cyanobacterial bloom in Lake Xuanwu, Nanjing. J Lake Sci 22(1):44–48 (in Chinese)
Miao C, Tang Y, Zhang H et al (2014) Harmful algae blooms removal from fresh water with modified vermiculite. Environ Technol 35(3):340–346
Mu R, Ma G, Sun C et al (2014) Inhibition of *Microcystis aeruginosa* and microcystin-LR with one algicidal bacterium isolated from a eutrophic lake. Afr J Biotech 11(2):460–465
Ni J, Yu Y, Feng W et al (2010) Impacts of algal blooms removal by chitosan-modified soils on zooplankton community in Taihu Lake, China. J Environ Sci 22(10):1500–1507
Okaichi T, Yanagi T (eds) (1997) Sustainable development in the Seto Inland Sea, Japan: from the viewpoint of fisheries. Terra Scientific Publishing, Tokyo. 329 pp
Olsen RO, Hoffmann F, Hess-Erga OK et al (2016) Ultraviolet radiation as a ballast water treatment strategy: Inactivation of phytoplankton measured with flow cytometry. Mar Pollut Bull 103(1–2):270–275
Orizar IS, Rivera PPL, Azanza RV (2013) Harmful algal bloom (HAB) mitigation using ball clay: effect on non-target organisms. J Environ Sci Manag 5(2):36–43
Padilla L, San Diego-McGlone M, Azanza R (2010) Exploring the potential of clay in mitigating *Pyrodinium bahamense* var. compressum and other harmful algal species in the Philippines. J Appl Phycol 22(6):761–768
Pang Y, Ding Y, Sun B (2013) Removal of red tide organism by a novel cationic polymeric flocculant. Procedia Environ Sci 18:602–609
Park MG, Kim S, Shin E-Y et al (2013a) Parasitism of harmful dinoflagellates in Korean coastal waters. Harmful Algae 30:S62–S74
Park TG, Lim WA, Park YT et al (2013b) Economic impact, management and mitigation of red tides in Korea. Harmful Algae 30:S131–S143
People's Network (2006) http://news.163.com/06/0106/15/26PUQJHP0001124R.html (in Chinese)
Pierce RH, Henry MS, Higham CJ et al (2004) Removal of harmful algal cells (*Karenia brevis*) and toxins from seawater culture by clay flocculation. Harmful Algae 3(2):141–148
Public Network (2008) http://news.163.com/08/0118/01/42F2Q7H00001124J.html (in Chinese)
Qingdao News (2007a) http://news.sina.com.cn/o/2007-09-27/030612641122s.shtml (in Chinese)
Qingdao News (2007b) http://www.sd.xinhuanet.com/news/2007-10/28/content_11516498.htm (in Chinese)
Randhawa V, Thakkar M, Wei L (2013) Effect of algal growth phase on *Aureococcus anophagefferens* susceptibility to hydrogen peroxide. Aquat Toxicol 142:230–238
Rounsefell GA, Evans JE (1958) Large-scale experimental test of copper sulfate as a control for the Florida red tide. Special Scientific Report Fisheries No. 270. US Department of the Interior, Fish and Wildlife Service, Washington, DC
Secher S (2009) Measures to control harmful algal blooms. The Plymouth Stud Sci 2(1):212–227
Seger A, Dorantes-Aranda JJ, Müller MN et al (2015) Mitigating fish-killing *Prymnesium parvum* algal blooms in aquaculture ponds with clay: the importance of pH and clay type. J Mar Sci Eng 3(2):154–174
Sengco MR (2009) Prevention and control of *Karenia brevis* blooms. Harmful Algae 8(4):623–628
Sengco MR, Anderson DM (2004) Controlling harmful algal blooms through clay flocculation. J Eukaryot Microbiol 51(2):169–172
Sengco MR, Li A, Tugend K et al (2001) Removal of red- and brown-tide cells using clay flocculation. I. Laboratory culture experiments with *Gymnodinium breve* and *Aureococcus anophagefferens*. Mar Ecol Prog Ser 210:41–53

Shao J, Li R, Lepo JE et al (2013) Potential for control of harmful cyanobacterial blooms using biologically derived substances: problems and prospects. J Environ Manage 125:149–155
Shirota A (1989) Red tide problem and countermeasures (2). Int J Aquat Fish Technol 1:195–223
Shumway SE, Frank DM, Ewart LM (2003) Effect of yellow yellow clay on clearance rate in seven species of benthic, filter-feeding invertebrates. Aquac Res 34(15):1391–1402
Song X, Yu Z, Gao Y (2003) Removal of different species of red tide organisms with an effective clay-complex system. Chin J Appl Ecol 14(7):1165–1168 (in Chinese)
Sukenik A, Bilanovic D, Shelef G (1988) Flocculation of microalgae in brackish and sea waters. Biomass 15(3):187–199
Sun X, Zhang B, Yu Z (2001) Toxicity study of anti-HABs agents on *Penaeus chinensis*. Mar Sci Bull 3(1):51–54 (in Chinese)
Sun X-X, Han K-N, Choi J-K et al (2004a) Screening of surfactants for harmful algal blooms mitigation. Mar Pollut Bull 48(9–10):937–945
Sun X-X, Lee Y-J, Choi J-K et al (2004b) Synergistic effect of sophorolipid and yellow clay combination in harmful algal blooms mitigation. Mar Pollut Bull 48(9–10):863–872
Tang YZ, Gobler CJ (2011) The green macroalga, *Ulva lactuca*, inhibits the growth of seven common harmful algal bloom species via allelopathy. Harmful Algae 10(5):480–488
Tang YZ, Kang Y, Berry D et al (2015) The ability of the red macroalga, *Porphyra purpurea* (Rhodophyceae) to inhibit the proliferation of seven common harmful microalgae. J Appl Phycol 27(1):531–544
Teen LP, Gires U, Pin LC (2012) Harmful algal blooms in Malaysian waters. Sains Malays 41(12): 1509–1515
Tenney MW, Echelberger WF Jr, Schuessler RG et al (1969) Algal flocculation with synthetic organic polyelectrolytes. Appl Microb 18(6):965–971
Tilney CL, Pokrzywinski KL, Coyne KJ et al (2014) Growth, death, and photobiology of dinoflagellates (Dinophyceae) under bacterial-algicide control. J Appl Phycol 26(5):2117–2127
Tilton RC, Murphy J, Dixon JK (1972) The flocculation of algae with synthetic polymeric flocculants. Water Res 6(2):155–164
Wang H, Yu Z, Cao X et al (2011) Fractal dimensions of flocs between clay particles and HAB organisms. Chin J Oceanol Limnol 29(3):656–663
Wang Z (2014) The effects of modified clay on benthonic organisms in mitigation of harmful algal blooms (HABs). PhD Dissertation, University of Chinese Academy of Sciences (in Chinese)
Wu Y, Lee Y, Jung S-G et al (2014) A novel thiazolidinedione derivative TD118 showing selective algicidal effects for red tide control. World J Microb Biotech 30(5):1603–1614
Wu P, Yu Z (2007) Extinguishment of harmful algae by organo—clay modified by Gemini surfactant. Environ Sci 28(1):80–86 (in Chinese)
Xu Y, Yang J, Ou M et al (2007) Study of *Microcystis aeruginosa* inhibition by electrochemical method. Biochem Eng J 36(3):215–220
Yang X, Li X, Zhou Y et al (2014) Novel insights into the algicidal bacterium DH77-1 killing the toxic dinoflagellate *Alexandrium tamarense*. Sci Total Environ 482:116–124
Yang Y, Liu Q, Chai Z et al (2015) Inhibition of marine coastal bloom-forming phytoplankton by commercially cultivated *Gracilaria lemaneiformis* (Rhodophyta). J Appl Phycol 27(6):2341–2352
Yu Z, Rao DS (1998) Impact of halloysite on growth of *Psuedonitzschia pungens* f. multiseries and production of algal toxin. Oceanol Limnol Sin 29(1):47–52 (in Chinese)
Yu Z, Zou J, Ma X (1993) The chemical means of controlling red tides. Oceanol Limnol Sin 24(3):314–318 (in Chinese)
Yu Z, Zou J, Ma X (1994a) A more effective clay for removing red tide organisms. J Nat Dis 3(2):105–109 (in Chinese)
Yu Z, Zou J, Ma X (1994b) A new method to improve the capability of clays for removing red tide organisms. Oceanol Limnol Sin 25(2):226–232 (in Chinese)
Yu Z, Zou J, Ma X (1994c) Application of clays to removal of red tide organisms I. Coagulation of red tide organisms with clays. Chin J Oceanol Limnol 12(3):193–200

Yu Z, Zou J, Ma X (1994d) Application of clays to removal of red tide organisms II. Coagulation of different species of red tide organisms with montmorillonite and effect of clay pretreatment. Chin J Oceanol Limnol 12(4):316–324

Yu Z, Zou J, Ma X (1995) Application of clays to removal of red tide organisms III. The coagulation of kaolin on red tide organisms. Chin J Oceanol Limnol 13(1):62–70

Yu Z, Sun X, Song X (1999) Clay surface modification and its coagulation of red tide organisms. Chin Sci Bull 44(7):617–620

Yu Z, Sengco M, Anderson D (2004) Flocculation and removal of the brown tide organism, *Aureococcus anophagefferens* (Chrysophyceae), using clays. J Appl Phycol 16(2):101–110

Zhang S, Song X, Wang Y et al (2005a) Competition on nutrients between *Gracilaria lemaneiformis* and *Scrippsiella trochoidea* (Stein) Loeblich. Oceanol Limnol Sin 36(6):556–561 (in Chinese)

Zhang S, Yu Z, Song X et al (2005b) Competition about nutrients between *Gracilaria lemaneiformis* and *Prorocentrum donghaiens*. Acta Ecol Sin 25(10):2676–2680 (in Chinese)

Zhang Z (2006) Analysis of emergency control effectiveness of blue-green alge in Xuanwu Lake. Pollut Control Technol 19(5):56–59 (in Chinese)

Zheng X, Zhang B, Zhang J et al (2013) A marine algicidal actinomycete and its active substance against the harmful algal bloom species *Phaeocystis globosa*. Appl Microbiol Biotechnol 97(20):9207–9215

Chapter 22
GlobalHAB: Fostering International Coordination on Harmful Algal Bloom Research in Aquatic Systems

Elisa Berdalet, Raphael M. Kudela, Neil S. Banas, Eileen Bresnan, Michele A. Burford, Keith Davidson, Christopher J. Gobler, Bengt Karlson, Po Teen Lim, Lincoln Mackenzie, Marina Montresor, Vera L. Trainer, Gires Usup, Kedong Yin, Henrik Enevoldsen, and Ed Urban

E. Berdalet (✉)
Institute of Marine Sciences (CSIC), Barcelona, Catalonia, Spain
e-mail: berdalet@icm.csic.es

R. M. Kudela
University of California, Santa Cruz, CA, USA

N.S. Banas
University of Strathclyde, Glasgow, UK

E. Bresnan
Marine Scotland Science, Aberdeen, UK

M. A. Burford
Australian Rivers Institute, Griffith University, Nathan, QLD, Australia

K. Davidson
The Scottish Association for Marine Science, Scottish Marine Institute, Oban, UK

C.J. Gobler
Stony Brook University, Southampton, NY, USA

B. Karlson
Swedish Meteorological and Hydrological Institute, Västra Frölunda, Sweden

P.T. Lim
Institute of Ocean and Earth Sciences, University of Malaya, Kuala Lumpur, Malaysia

L. Mackenzie
Cawthron Institute, Nelson, New Zealand

M. Montresor
Stazione Zoologica Anton Dohrn, Napoli, Italy

V.L. Trainer
Northwest Fisheries Science Center, National Marine Fisheries Service, National Oceanic and Atmospheric Administration, Seattle, WA, USA

G. Usup
School of Environmental and Resource Science, Faculty of Science and Technology, Universiti Kebangsaan Malaysia, Bangi, Selangor, Malaysia

P.M. Glibert et al. (eds.), *Global Ecology and Oceanography of Harmful Algal Blooms*, Ecological Studies 232, https://doi.org/10.1007/978-3-319-70069-4_22

22.1 Genesis of GlobalHAB

The idea for establishment of the GlobalHAB Programme arose at the final Synthesis Open Science Meeting (OSM) of the Global Ecology and Oceanography of Harmful Algal Blooms (GEOHAB) Programme in Paris, in April 2013 (GEOHAB 2014). Participants at the meeting evaluated and synthesized the outcomes of GEOHAB and agreed that international coordination of HAB science was still needed to increase understanding of these events and how to mitigate their impacts on humans and aquatic ecosystems. In 2015, the GlobalHAB Programme was formally adopted by the two cosponsors, the Intergovernmental Oceanographic Commission (IOC) of UNESCO [through the Intergovernmental Panel on Harmful Algal Blooms (IPHAB)] and the Scientific Committee on Oceanic Research (SCOR), and was also endorsed by the International Atomic Energy Agency (IAEA). IOC and SCOR appointed a Scientific Steering Committee (SSC) for the new GlobalHAB Programme in 2015, which met for the first time in March 2016 to initiate the development of the GlobalHAB Science and Implementation Plan that is summarized in this chapter.

The GlobalHAB Science and Implementation Plan builds on the GEOHAB Science Plan [GEOHAB 2001; Kudela et al. 2017; see also Chap. 3, Kudela et al. (2018)] and integrates suggestions from members of the international HAB community that participated in the GEOHAB Synthesis OSM (GEOHAB 2014). For example, while GEOHAB focused on marine HABs, GlobalHAB incorporates research on freshwaters and brackish waters. Also, while GEOHAB studies were centred on unicellular planktonic microalgae and some marine cyanobacteria only, GlobalHAB will foster research on harmful marine benthic taxa, macroalgae, and freshwater cyanobacteria as well.

The GlobalHAB Scientific and Implementation Plan (Fig. 22.1) includes the five programme elements that structured the GEOHAB Implementation Plan (GEOHAB 2003), namely, Biodiversity and Biogeography, Nutrients and Eutrophication, Adaptive Strategies, Comparative Ecosystems, and Observation, Modelling, and Predictions. Significant progress was made on these programme elements in GEOHAB [Kudela et al. 2017; see also Chap. 3, Kudela et al. (2018)]; these programme elements provide a sound and useful scientific framework to be incorporated and continued under GlobalHAB with an updated perspective. To this end, the SSC reviewed the accomplishments attained within each programme element, revisited the open questions, and analysed the next-generation tools available to advance knowledge with respect to these original programme elements.

K. Yin
Sun Yat-Sen (Zhongshan) University, Guangzhou, China

H. Enevoldsen
IOC UNESCO/University of Copenhagen, Nørregade, Denmark

E. Urban
Scientific Committee on Oceanic Research, Newark, DE, USA

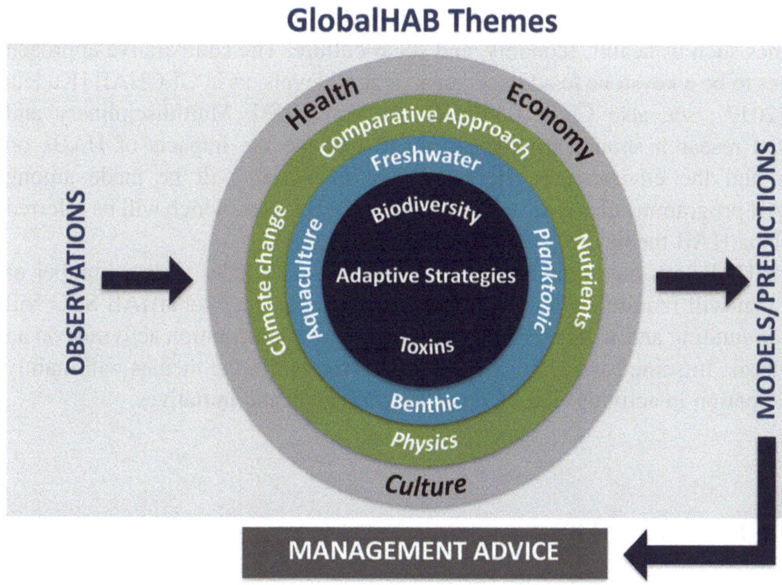

Fig. 22.1 The themes integrated in GlobalHAB cover from small, cell scale subjects (e.g., biodiversity, adaptive strategies) to studies at ecosystem scale and climate change-related processes. GlobalHAB perspective is multidisciplinary, integrating physics, chemistry, and biology and addressing the impacts of HABs on the environment and human societies (including health, economy, and sociocultural aspects). Improved observation within and across the different themes should result in better predictions of the HAB events and contribute to the management of their impacts

Furthermore, seven new topics were added: Benthic HABs, Freshwater HABs and cyanobacteria HABs, Aquaculture and HABs, Toxins, Health, Economy, and Climate Change. Although these terms were mentioned in the GEOHAB Science Plan, the topics were either deliberately omitted in order to focus on oceanography and ecology of marine microalgae or could not be examined fully during GEOHAB because they were not yet sufficiently mature. Exceptionally, research on Benthic HABs was fostered through the creation of a core research project (CRP) towards the end of GEOHAB Programme (GEOHAB 2012), given the increasing demand for techniques and knowledge synthesis on the causative agents for ciguatera fish poisoning (CFP). The relevance of the problem and the rapid progresses achieved in this research area [Berdalet et al. 2017a; see also Chap. 13, Berdalet and Tester (2018)] justifies the inclusion of this topic in GlobalHAB with objectives that reach beyond CFP to encompass other benthic HAB issues. In the conceptual model, the different GEOHAB programme elements and the new topics form a comprehensive and ambitious structure. GlobalHAB's remit covers processes occurring from cellular to ecosystem scales, concerning benthic and planktonic, unicellular and multicellular organisms and integrating physical, chemical, and biological processes, and allows interaction between HABs and (traditionally) non-HAB disciplines such as health, economy, and socio-culture. The comparative approach continues to be a keystone to address research at all levels, as in GEOHAB [Kudela et al. (2017); see also Chap. 3, Kudela et al. (2018)]. Multidisciplinary and integrated research should provide advice to manage the impacts of HABs on humans and the environment. Hereafter, no distinction will be made among GEOHAB programme elements and new GlobalHAB topics, which will be referred to as GlobalHAB themes.

GlobalHAB lays out an ambitious programme and includes a large number of themes that will require significant funding to address. The GlobalHAB SSC will seek such funding and will determine priorities for implementation activities on an annual basis. In some cases, GlobalHAB's involvement in the themes will mainly be participation in activities led by other organizations and initiatives.

22.2 GlobalHAB Themes

GlobalHAB includes 12 themes that cover the full spectrum of HAB science. Many of these themes intersect and these intersections will be considered when planning the science activities described.

22.2.1 Biodiversity and Biogeography

The overall objective of this theme is to identify the factors that determine the changing biogeographical distribution of HAB species (including potential trends associated with climate change), their genetic variability, and the biodiversity of HAB-associated communities, including diversity of toxins. Understanding the biodiversity and biogeography of harmful organisms and their toxins is fundamental to identify trends in HAB occurrence, enable their prediction and management, and, when possible, mitigate their impacts.

Correct taxonomic identification of harmful species is crucial to achieve this objective. New molecular technologies with progressively reduced cost [e.g., qPCR, fluorescent in situ hybridization, high-throughput sequencing (HTS)] are helping to refine species descriptions and to identify cryptic and pseudo-cryptic diversity among HAB species formerly defined based on morphological characters (Lelong et al. 2012; Kremp et al. 2014). In this sense, a priority is educating a future generation of taxonomists worldwide who can bridge both morphological and molecular identification techniques. Other related challenges are establishing reference sequences to interpret molecular data and achieve identification at the species level and defining standardized protocols for physiological investigations (including e.g., toxin composition, physiology, life cycle).

To implement the objectives for this theme, GlobalHAB will collaborate with other projects and organizations to maintain and reinforce taxonomic training initiatives for the identification of microalgae and support the establishment of a database for reference sequences of HAB species. Furthermore, GlobalHAB will align with the GlobalHAB Status Report and harmful algal events database and map (HAEDAT-HABMAP) initiatives to create a baseline of biogeographic species distribution and will support long time series, such as the ones considered within the recently finished SCOR Working Group (WG), Phytoplankton Time Series and Ocean Observing Sites (SCOR WG 137 http://wg137.net/time-series/time-series-map), and the new IOC WG TrendsPO, where biological data are collected.

22.2.2 Adaptive Strategies

The overall objective of this theme is to determine the adaptations of HAB species and how these adaptations help to explain their proliferation or harmful effects. Detecting unique characteristics and adaptations of HAB species in particular environments will contribute to the development of predictive models.

Under GEOHAB, significant effort was devoted to identifying harmful species uniquely associated with particular habitats. However, cosmopolitan rather than unique species thrive in upwelling systems (Smayda 2010a, b), and no unique taxa have been identified in fjords and coastal embayments [see also Chap. 10, Roy et al. (2018)]. Habitat-specific examples are the pelagophytes, *Aureococcus* and *Aureoumbra*, which cause persistent brown tides in coastal eutrophied embayments (Gobler and Sunda 2012). Other habitat-specific examples for eutrophic environments are described in Chap. 12 (Glibert et al. 2018a).

Still, it is likely that particular adaptive strategies play a role in the dynamics of harmful blooms. Understanding and characterizing these strategies in detail and, if possible, parameterizing the processes involved (especially under natural conditions) are main challenges to understand and predict HAB events. Examples include dinoflagellate life history strategies, e.g., encystment and excystment processes, sexual reproduction, vertical migration, which may explain the recurrence of the blooms in certain areas [see also Chap. 8, Azanza et al. (2018)]; the production of allelopathic compounds (biologically active compounds that elicit specific responses in competitors or predators, e.g., Ianora et al. 2011, Selander et al. 2015); and intrinsic chemical or life-form defences against parasites (viruses, protists; e.g., Chambouvet et al. 2008), which can provide competitive advantages under certain conditions. In the case of cyanobacteria, buoyancy regulation to access light and nutrients, cyst formation under adverse conditions, superior nutrient-scavenging physiology, genetic variability among strains within populations and over time, and capacity to fix atmospheric nitrogen and synthesize secondary metabolites appear as some of the particular adaptive strategies to this group (O'Neil et al. 2012). Finally, the intrinsic capacity to attach to surfaces (including plastics; e.g., Masó et al. 2003), exhibited by benthic dinoflagellates and macrophytes, potentially constitutes a major adaptive mechanism to colonize and disperse into new habitats and requires further investigation.

Some of these adaptive strategies can be strain- and species-specific, making it difficult to establish general models. However, trait-based approaches, i.e., identification of common traits across HAB species, would facilitate integration of the outcomes from this theme into new modelling approaches. New in situ observation technologies and cutting-edge methodologies (metabolomics, proteomics, etc.) offer new possibilities for progress.

22.2.3 HAB Toxins

The overall objective of this theme is to characterize the genetic and environmental aspects of toxin production, to determine the mode of action of selected toxins, and to address several limitations in toxin analysis techniques. Toxins are a crosscutting subject and a common concern through all the GlobalHAB research topics.

To obtain a better understanding of toxin biosynthesis pathways, and the genes involved, constitutes a fundamental challenge. Knowledge of the genetic basis of toxin production could reveal the existence of suitable target genes for the identification of toxic species. Furthermore, a better understanding of environmental controls of toxin production is necessary for better prediction of the probable impact of changes in marine habitats on the toxicity of HAB species. Significant progress on the environmental controls on paralytic shellfish poisoning (PSP) toxin (saxitoxins) synthesis (see review in Krüger et al. 2010) stimulates analogous research on other phycotoxins, such as domoic acid (DA, e.g., Jeffery et al. 2004), ciguatoxins, okadaic acid (OA), and emerging toxins.

Identifying and characterizing the modes of action of fish-killing microalgal toxins—such as the ones produced by *Cochlodinium* (*Margalefidinium*) *polykrikoides*, *Chattonella* spp., *Heterosigma akashiwo*, and *Karlodinium australe* (e.g., Black et al. 1991; Van Wagoner et al. 2008; Lim et al. 2014)—and of other cytotoxic species affecting other marine cultured organisms (e.g., shrimp) are still challenging issues. This information is particularly relevant for development of effective strategies to protect the marine aquaculture industry from massive mortality episodes of farmed species as a result of toxic microalgal blooms. Emerging toxins in marine and freshwater environments will require similar approaches.

Sustainability of marine aquaculture (as explained below) and the protection of public health and food security also require sensitive, accurate, fast, and cost-effective means of microalgal toxin detection and analysis. Recent advances in analytical techniques, and cell-based and functional assays, are contributing to this progress (e.g., Yogi et al. 2011; Turner et al. 2015). However, official and reliable regulatory methods that are used for human health protection are still not available for routine analysis of some toxins, with ciguatoxin being the most notable. The role of different (marine and freshwater) toxin analogues and their bio-stability also require further attention (Pearson et al. 2016).

Within this theme, GlobalHAB will foster, for instance, workshops focused on the genetic and environmental basis of toxicity (especially focused on domoic acid and PSP), and the mode of action of fish-killing phycotoxins, and will endorse interlaboratory toxin analysis method validation, such as the ongoing validation for the PSP toxin methods.

22.2.4 Nutrients and Eutrophication

This theme continues to focus on the links between nutrient supply, eutrophication, and on the occurrence of HABs. Nutrient pollution, typically driving eutrophication, is often considered to be a main driver responsible for the increased number of high-biomass HABs in marine, brackish and freshwaters (Paerl et al. 2001; Anderson et al. 2002; Heisler et al. 2008). Important advances were achieved by GEOHAB on this research topic [Glibert and Burford 2017; see also Chap. 12, Glibert et al. (2018a)], which contributed, for instance, to the implementation of policies limiting nutrient (N and P) inputs and proved useful to prevent HABs in certain coastal and freshwater areas. However, the relationship between nutrient pollution and HABs is more complex than previously thought.

New challenges include understanding the role of organic nutrient availability and ratios on some HAB events, as well as in the modulation of HAB toxin production (Glibert and Burford 2017), and unveiling the mixotrophic capacities of HAB species [Burkholder et al. 2008; see also Chap. 7, Flynn et al. (2018)]. Furthermore, modifications in nutrient supply to aquatic systems may accompany climate change and other anthropogenic forcing factors [see also Chap. 4, Glibert et al. (2018b)]. These questions are fundamental for planktonic (e.g., *Pseudo-nitzschia*, *Alexandrium*), benthic (e.g., *Gambierdiscus*, *Ostreopsis*), macroalgal (e.g., *Sargassum*, *Enteromorpha*), and harmful freshwater and cyanobacteria (e.g. *Microcystis*, *Cylindrospermopsis*) species. GlobalHAB will also investigate the links of aquaculture-related nutrients and HAB occurrence and toxicity and deoxygenation and anoxia processes resulting from high-biomass HABs (Fig. 22.2).

22.2.5 HABs in Freshwater Systems and Cyanobacterial HABs (CyanoHABs) from Marine to Freshwater Systems

The overall objective of this theme is to develop a global perspective on the science and management of freshwater HABs and cyanobacterial HABs in marine, brackish, and freshwater habitats.

Fig. 22.2 A typical springtime *Noctiluca scintillans* bloom in Eastsound, Orcas Island, in Puget Sound, Washington State, USA. High-biomass blooms can cause anoxia in the water column when the bloom decays. Photo credit: V. L. Trainer, NOAA

Some eukaryotic groups and a range of cyanobacterial species (CyanoHABs, including *Microcystis*, *Cylindrospermopsis*, *Dolichospermum*, *Aphanizomenon*, *Planktothrix*, *Lyngbya*, *Phormidium*) are involved in harmful events in freshwater and brackish systems, e.g., *Nodularia* in the Baltic Sea [see also Chap. 20, Burford et al. (2018)]. In marine ecosystems, toxic filamentous cyanobacteria such as *Lyngbya* and *Moorea* have been more recently associated with harmful proliferations in tropical regions. Overall, CyanoHABs and other freshwater HABs produce a wide range of toxins, including microcystins, cylindrospermopsins, anatoxins, nodularins, saxitoxins, aplysiatoxins, and lyngbyatoxins.

HABs in freshwater systems (especially drinking water reservoirs) and CyanoHABs have major economic, social, and environmental impacts, with millions of dollars spent annually testing water supplies and mitigating CyanoHAB effects (Hamilton et al. 2014). Under global warming scenarios, these events are expected to increase, although the picture is still not clear. In addition to fostering research to clarify potential trends (in coordination with activities to implement theme 12, Climate Change and HABs), GlobalHAB's distinctive role on this theme will be to bridge the gap between freshwater and marine HAB researchers to share knowledge, techniques, and approaches and to facilitate more effective communication between HAB scientists and policy makers internationally. A final aim will be to determine the current state of knowledge and potential approaches to managing, mitigating, and predicting CyanoHABs in freshwater, brackish, and marine habitats. Crosscutting objectives concerning specific adaptive strategies of CyanoHAB species, genetic and environmental control of toxin production, modulation of the events by nutrient availability, and modelling [see also Chap. 20, Burford et al. (2018)] are also included in this theme and will be taking into account the activities to implement the themes described in Sects. 22.2.2, 22.2.3, 22.2.4, and 22.2.9.

22.2.6 HABs and Marine Aquaculture

The objectives of this theme are to determine the links between marine aquaculture and HABs occurrence and to find efficient methods to protect farmed seafood products from HABs impacts in different regions.

Aquaculture of shellfish and finfish has many benefits, including the production of nutritious high-protein food, reducing the pressure on natural resources, and supporting sustainable economic development and employment. Unfortunately, contamination of seafood by phycotoxins and mass mortalities of cultured animals caused by harmful microalgae (raphidophytes, dictyochophytes, haptophytes, dinoflagellates) have devastating impacts on aquaculture operations in certain areas. Improved awareness and identification of HAB events have coincided with increases in aquaculture development, raising concern about the direct implications of aquaculture on the (increased) incidence and intensity of HABs (Price et al. 2015). Through focused studies and improved observational methods, this theme will seek to provide fresh perspectives for assessing the potential effects of nutrients and shifting nutrient ratios, organic matter, and fish/shellfish transfer from aquaculture activities on the promotion of HABs in different areas of the planet. Progress in modelling (hydrodynamic and coupled biogeochemical models) and technology can notably contribute to improve early warning and mitigation methods, appropriately designed to the specific aquaculture sites.

There is a further need to find efficient methods to protect farmed seafood products from HAB impacts. Basic research issues include clarification of the deleterious effects of some fish-killing algae (as indicated in theme 3), as well as other emergent toxins and mechanisms of harm in cultured seafood (e.g., Borcier et al. 2017), the ecophysiology and ecology of fish-killing algae, and the global extent of these blooms. Advances in technology (e.g., high-tech, autonomous, in situ molecular and imaging flow cytometry methods) are offering new possibilities for early warning and real-time sensing of impending blooms, although some refinement and field trials are still required [see also Chap. 18, Glibert et al. (2018c)]. In turn, detection of organisms and toxins in waters and shellfish through rapid test kits is becoming more available, expecting to cover wider range of targets and to be validated in the near future.

GlobalHAB will endorse and seek to promote activities in this important field of both fundamental and applied research to provide reliable, mature, and cost-effective solutions for the prevention and mitigation of the impacts of HABs on aquaculture.

22.2.7 Benthic HABs (BHABs)

The overall objective of this theme is to achieve a better understanding of BHABs and to provide tools to manage and mitigate the impacts of these events on human health and the environment. As mentioned in Sect. 22.1 above, this theme constitutes a direct continuation of the GEOHAB CRP on HABs in benthic systems (GEOHAB 2012) established in response to more frequent events and geographic expansion of BHABs, especially those caused by the toxic benthic dinoflagellate genera, *Gambierdiscus* and *Ostreopsis*. Ciguatoxins produced by *Gambierdiscus* species are bioaccumulated in reef fish, causing ciguatera fish poisoning (CFP) in humans when eaten. CFP, the most frequent HAB-related illness in the world, often with significant long-term health effects (Friedman et al. 2017), is endemic in tropical areas and is also expanding to temperate latitudes (e.g., Boada et al. 2010). *Ostreopsis* produces palytoxins and analogues, associated with dramatic seafood intoxications in the tropics. *Ostreopsis* blooms have become more frequent and intense, especially in temperate waters, where some outbreaks have been related to sporadic acute respiratory irritations in humans exposed to marine aerosols and massive benthic faunal damage [Berdalet et al. 2017a; see also Chap. 13, Berdalet and Tester (2018)].

The objectives and questions formulated in GEOHAB (2012) are still valid, and significant progress was achieved in a relatively short time (Berdalet et al. 2017a), in part due to the cooperative international research and capacity building fostered by GEOHAB.

GlobalHAB will continue the lines of research indicated in GEOHAB (2012) to improve knowledge of the ecology, physiology, and toxin-transfer mechanisms through marine food webs and to determine fundamental parameters to develop predictive models on BHAB dynamics. This theme is not only focused on *Gambierdiscus* and *Ostreopsis* but also on other BHAB taxa. The impacts of

BHABs in marine organisms and ecosystems and the modulation of BHAB dynamics by climate change should be investigated as well.

GlobalHAB will be directly involved in regional and international initiatives focused on multidisciplinary research on CFP (including ecology, toxicology, medicine, and epidemiology), especially in the most affected tropical and subtropical areas, such as the "Global Ciguatera Strategy" (http://hab.ioc-unesco.org/) adopted by IOC, WHO, IAEA, and FAO in 2015 (see also theme 10). Emphasis will be put in the standardization of sampling methods for BHAB organisms (e.g., Tester et al. 2014; Jauzein et al. 2016; Mangialajo et al. 2017) and to use common databases on organisms' biogeography and harmful events. GlobalHAB will facilitate interactions among public health, social, economy experts, stakeholders, and people directly suffering BHAB impacts, through transdisciplinary meetings and workshops (see also Health and Economy Themes, described 22.2.10 and 22.2.11).

22.2.8 Comparative Approach

The overall objective of this theme is to determine the extent to which HAB species, their population dynamics, and community interactions respond similarly within comparable ecosystem types. The GEOHAB Programme adopted the comparative approach (GEOHAB 2001; Anderson et al. 2005), from the cellular to the ecosystem level [c.f., Kudela et al. 2010a, b; Trainer et al. 2010; see also Chaps. 11 and 12, Pitcher et al. (2018), Glibert et al. (2018c)]. This approach is based on the view that the ecology and oceanography of HABs can be best understood through the study of the causative organisms and affected systems in relation to comparable organisms and systems. Improved generalizations about the causes and consequences of HABs would be particularly useful in modelling HABs and management and mitigation of their effects.

GlobalHAB will also adopt a comparative approach and extend it to comparisons within marine, brackish, and freshwater ecosystem types and, if possible, among the different aquatic systems as well. Case studies where the comparative approach can provide useful insights and progress include, for example, comparisons of *Gambierdiscus* dynamics in the main affected areas (the Pacific Ocean and Caribbean Sea); comparison of the dynamics of *Ostreopsis* and *Gambierdiscus* (the main benthic HAB taxa); comparison of *Pseudo-nitzschia* harmful events in the eastern and western Pacific, to understand their differential impacts (Trainer and Kameneva 2017); and comparisons of the blooms of *Pseudochattonella* species (Dictyochophyceae) that caused fish mortalities in northern Europe at the end of the twentieth century, as well as in Japan (e.g., Dahl and Tangen 1993; Fukuyo et al. 2002) and more recently in South America (Clement et al. 2016; in this case also followed by exceptional *Alexandrium catenella* blooms, Hernández et al. 2016). By studying conditions leading to blooms at different latitudes and under different environmental conditions, an increased understanding in bloom formation and effects of climate change on the cyanobacteria blooms may also be achieved. The comparative ecosystem approach is expected to be central to many of the GlobalHAB implementation activities, as it has been to GEOHAB [see, e.g., Chaps. 8, 12, and 17, Azanza et al. (2018), Glibert et al. (2018a), and Goes et al. (2018)].

22.2.9 Observation, Modelling, and Prediction

The overall objective of this theme is to improve the detection and prediction of HABs by developing new capabilities in observation and modelling (GEOHAB 2011).

Understanding and describing HAB dynamics requires highly resolved observations of biological, chemical, and physical processes that interplay from cell to ecosystem scales. Parametrization of the different processes are fundamental for development and validation of short-term, regional forecasting, as well as long-term or large-scale projections of HABs [see also Chap. 19, Franks (2018)].

Technical progress has favoured improved visualization and/or characterization of small (cellular) and fine-scale processes (e.g., thin layers in stratified systems) (Berdalet et al. 2014, 2017b). Examples include the FlowCytobot and the Imaging FlowCytobot (IFCB, Campbell et al. 2010; Brosnahan et al. 2015), the underwater long working distance microscope (Jaffe et al. 2013), the fine-scale sampler (FSS) (Lunven et al. 2005), and the high-resolution IFREMER particle-size analyzer profiler (IPSAP) (Lunven et al. 2012); tow-body systems such as acrobat (SeaSciences; Timmerman et al. 2014) and autonomous underwater vehicles like the MBARI Dorado (Ryan et al. 2010); and the free-gliding quasi-horizontal profiler (TurboMAP-G, Foloni-Neto et al. 2014) and the Environmental Sample Processor (ESP, e.g., Greenfield et al. 2008). Furthermore, long-term coordinated and multiparameter observation systems constitute invaluable tools for early warnings and predictions. Recent examples of projects effectively merging observation and modelling to inform prediction at a regional scale include the ASIMUTH Project (Davidson et al. 2016) and the development of predictive models for *Alexandrium* (Anderson et al. 2014) and *Pseudo-nitzschia* (Anderson et al. 2016).

GlobalHAB will follow on from GEOHAB (see, e.g., HABWATCH 2004, with lectures and tutorials available at http://HABWatch.obs-vlfr.fr; GEOHAB 2011; Berdalet et al. 2014) cosponsoring initiatives such as specific technical and modelling workshops. GlobalHAB will also coordinate with the Global Ocean Observing System (GOOS), the International Council for the Exploration of the Sea (ICES), and the International Ocean-Colour Coordination Group (IOCCG) to identify and provide justification for long-term HAB sentinel sites. Recently, HABs have been included in the definition of the GOOS "phytoplankton" essential ocean variable (EOV; http://goosocean.org/index.php?option=com_content&view=article&id=14&Itemid=114).

22.2.10 HABs and Human and Environmental Health

The overall objective of this theme is to increase collaborations among HAB scientists with medical, veterinary, public health, and social science expertise to help minimize the risk of HAB impacts to human and animal health.

In the case of CFP, there is a general consensus about the need to integrate medical and public health information, including methods of communicating risk, and HAB research (Friedman et al. 2017). In some of the most affected areas, public health and HAB specialists have begun to establish systems for compiling multidisciplinary environmental and human health data. This is the case both for *One Health Harmful Algal Bloom System* (OHHABS) launched in June 2016 by the US Centres for Disease Control and Prevention (CDC) (http://www.cdc.gov/habs/ohhabs.html) and for the platform at the Institut Louis Malardé (http://www.ciguatera-online.com/index.php/en/) in French Polynesia. The simultaneous collection of environmental and human health data over time will help public health practitioners identify long-term trends in HAB-related diseases in humans and animals. Importantly, at the international level, the coordinated IOC-IAEA-FAO-WHO "Global Ciguatera Strategy" was adopted in 2015 (http://hab.ioc-unesco.org/) for improved research and management. GlobalHAB has been aligned with this initiative since its inception, actively collaborating with the task team established to implement the strategy.

Furthermore, the need for an integrated understanding of the linkages between human health and environmental characteristics of the ocean is progressively getting visibility in Europe (e.g., Moore et al. 2013; www.ecehh.org/events/oceans-human-health/) and the USA. This includes HABs at the core of both ocean and human health. Information on HAB-related illnesses in animals and humans and on certain environmental characteristics of the blooms are collected by OHHABS. Information about HAB-related illnesses is also registered by the CDC through the National Outbreak Reporting System (NORS) (www.cdc.gov/). GlobalHAB will align with these initiatives facilitating, for instance, joint trans-disciplinary meetings among HAB and non-HAB specialists.

22.2.11 Economic Impacts of HABs

The overall objective of this theme is to develop cross-community understanding of the economic impacts of HABs and to define methods and criteria capable of robustly assessing (at both regional and national levels) the economic costs of HABs, the full costs associated with monitoring, predicting, and mitigating HABs, and the economic value of minimizing HAB impacts.

Hoagland et al. (2002) estimated the economic effects of HABs in the USA to be $50 million per year (between 1987 and 1992, at the 2002 value of the dollar). Multiyear losses in farmed fish and shellfish production were estimated to be about US$94.0 million for Korea, Japan, and China from 2006 to 2012 (Trainer and

Fig. 22.3 Girls enjoying the April 2017 razor clam dig on the Pacific coast of Washington State, USA, using a white plastic tube called a "clam gun." These clams can retain phycotoxins, such as domoic acid, for many months. Photo credit: M. O'Connell, Long Live the Kings Salmon Hatchery

Yoshida 2014). In Washington State, USA, a yearlong closure of the recreational razor clam harvest (Fig. 22.3) resulted in almost US$25 million in lost income, including over 400 jobs, as well as lost wages, and spending during trips to the coast on hotels, groceries, camping, and restaurants (Dyson and Huppert 2010). The increasing magnitude of macroalgal (*Ulva*, *Sargassum*) HABs is dramatically impacting the economy of countries highly dependent on tourism and coastal fisheries [e.g., Smetacek and Zingone 2013; see also Chap. 16, Liu and Zhou (2018)]. However, there is a general consensus that many studies on the negative economic impacts of HABs have employed relatively crude measures and methodologies, the results of which are often difficult to compare (Davidson et al. 2014). Revision of regional estimates with more modern environmental economic methods is required, as are more local evaluations in regions of particular concern or economic value. The economic effects of HABs arise from public health costs, commercial fishery closures and fish kills, insurance costs, possible medium- and long-term declines in coastal and marine recreation and tourism, and the costs of monitoring, management, and mitigation. Environmental evaluation of HABs also needs to be extended to ensure that the less tangible benefits of HAB mitigation are included, for example, the value in terms of health and well-being of eating seafood and of maintaining sustainable employment in remote regions to foster sustainable development.

This is a complex and transdisciplinary theme to address. The main contribution of GlobalHAB will be to bring together natural scientists with economists and to serve as a catalyst for further advances between these communities.

22.2.12 Climate Change and HABs

The overall objective of this topic is to understand global patterns in HAB responses to common drivers (thermal windows, stratification, changing levels of CO_2) associated with climate change.

Concern about the potential impacts of long-term climate change on HAB dynamics was already noted in the GEOHAB Science Plan (2001). In 2013, a workshop on HAB's and climate change was arranged and the main conclusions were summarized in Wells et al. (2015), see also Chap. 5, Wells and Karlson (2018). Subsequently, the ICES-PICES-IOC Scientific Symposium on Harmful Algae and Climate Change (https://pices.int/meetings/international_symposia/2015/2015-HAB/scope.aspx) was held in Gothenburg, Sweden, 19–22 May 2015, supported financially by the Swedish Research Council Formas, PICES and SCOR.

Scientific data obtained in the last 20 years suggest that the main factors controlling microalgal populations (surface water temperature, ocean stratification, wind and water circulation patterns, precipitation-linked nutrient inputs, and salinity in marine and brackish waters) are changing in ways that could foster the geographic expansion and increase the impacts of HABs [e.g., Hallegraeff 2010; Wells et al. 2015, see also Chap. 5, Wells and Karlson (2018) and Chap. 12, Glibert et al. (2018a)]. Alteration of marine habitats is another global change that is affecting HAB occurrences, and surface water acidification may have a potential influence as well. A fundamental question is whether the environmental windows of opportunity for HAB species are expanding (e.g., Moore et al. 2015) or simply shifting geographically and seasonally.

GlobalHAB has identified different levels of research to address this theme, from experimental work on climate effects on HAB organisms using laboratory and mesocosm experiments as a basis for predictions of climate change effects on HABs to the development of comprehensive, region-specific studies integrating biological process data with downscaled climate projections. To reach a global comprehension requires adoption of good practices in lab and field approaches to investigate HAB responses to climate-linked drivers (see Wells et al. 2015), a manual that will be fostered by GlobalHAB. In this regard, the efforts of SCOR WG 149, Changing Ocean Biological Systems (COBS): how will biota respond to a changing ocean?, may also be a relevant partner as they are developing protocols for multi-stressor, climate-related experiments.

Interest in this theme is shared by several international organizations and initiatives as indicated earlier (see theme 1). For instance, GlobalHAB will work with GOOS to identify the best sites for time-series observations of HABs and related oceanographic parameters (EOVs, essential ocean variables) to track the potential impact of climate change on HABs. GlobalHAB will also cooperate with other international groups (particularly the IOC Global Ocean Oxygen Network, GO_2NE) to investigate ocean deoxygenation in response to climate change and eutrophication, with a specific focus on HABs.

22.3 Implementation of GlobalHAB

The GlobalHAB Science and Implementation Plan provides detailed implementation activities for the near term (1–3) and medium term (10 years). GlobalHAB will help coordinate research internationally to address specific knowledge and methodological gaps when possible and will focus its efforts on activities to synthesize and communicate existing knowledge as well. Implementation activities arise from recognizing the need for interactions among some of the GlobalHAB themes and the benefits of working with other organizations with interests in these themes. Funding for implementation and prioritization of implementation activities will be evaluated annually, as part of GlobalHAB SSC meetings. Examples of some of the specific activities have been illustrated in the description of the GlobalHAB objectives. General tasks to be undertaken will include:

Review papers—Much can be accomplished in the early years of the programme through scientific syntheses related to the topics that will be the focus of GlobalHAB. These syntheses will contribute important information to advance international HAB research and will provide a foundation for future GlobalHAB activities.

Science workshops—The workshop approach will be used by GlobalHAB to bring together small groups of relevant experts to address specific topics necessary to overcome barriers to HAB research; the outcome of the workshops will usually result in peer-reviewed journal articles, manuals, summaries for policymakers, and other publications.

Open science meetings—OSMs provide opportunities for the international HAB community to gather for discussion of issues relevant to a variety of GlobalHAB themes. OSMs will allow HAB scientists to present current research on specific GlobalHAB themes and to identify and prioritize research topics, debate controversial issues in HAB science, and establish new international research partnerships across disciplines.

Special sessions—In situations where the number of scientists working on a specific topic is insufficient to merit an OSM, GlobalHAB will propose special sessions at international science meetings. Special sessions can be efficient at providing venues for scientists to present their results, not just to other scientists working on the same specific topic but also to the broader HAB community.

Intercalibration activities—Many measurement methods and types of equipment are used by more than one laboratory, but there are no standard reference materials available (in particular for most phycotoxins) for laboratories to analyse and to make their results comparable. In chemical oceanography, it has been found that results from analyses using the same methods may vary widely due to subtle differences in techniques. The establishment of standardized methods for sampling and estimating cell concentrations of BHAB species also requires an international approach. GlobalHAB will facilitate intercalibration activities when it is suspected that differences among laboratories, methods, and/or equipment may be hindering progress in the field.

Training—Standard methods of research and observations are available for some areas of HAB science, but these methods need to be taught to a greater number of scientists and technicians. Sometimes, suitable manuals already exist, but in other cases manuals will need to be developed for use in the training (see below). GlobalHAB will provide mechanisms to organize training materials and activities. GlobalHAB will especially target training and capacity-building initiatives in these areas facing basic problems related to HABs and their effects.

Manuals—GlobalHAB will develop "good practice" manuals to help standardize commonly used methods and compare field data and experimental results and provide instructions for people who want to use the methods. Manuals that would be beneficial include a user-friendly electronic manual on mitigation strategies for freshwater CyanoHABs across the world and a good practices manual for HAB and climate change research.

Databases—GlobalHAB will promote the development of lists and databases helpful for HAB sciences, for example, updating a list of fish-killing harmful algae in the IOC Reference List that in turn feeds into the World Register of Marine Species.

Outreach to policymakers—Scientific results from HAB studies are often not accessible to policymakers in an understandable and attractive format. GEOHAB produced a Scientific Summary for Policy Makers (see Kudela et al. 2015) and GlobalHAB will do the same for those topics for which there are no sources of objective and authoritative information.

22.4 Collaboration of GlobalHAB with Other International Entities

GlobalHAB will serve the international community studying HABs by (1) providing a common umbrella of objectives and (2) facilitating coordination of activities by national, regional, and other international bodies interested in HAB research. Accordingly, GlobalHAB will work to establish strategic partnerships among organizations and projects, and the success of GlobalHAB will also depend on the engagement of the international community. At the beginning of GlobalHAB, synergies with some entities were already established given the links built previously during GEOHAB [see Chap. 3, Kudela et al. (2018)]. Some of the organizations that GlobalHAB will maintain their partnerships or establish new synergies with are listed below:

Global organizations and projects—At the initiation of GlobalHAB, partnerships have been established with the Intergovernmental Panel on Harmful Algal Blooms (IPHAB), the GOOS Biology and Ecosystems Panel, the Group on Earth Observation's Blue Planet initiative, the International Society for the Study of Harmful Algae (ISSHA), IOCCG, and the International Atomic Energy Agency (IAEA). New links will be explored with the Partnership for Observation of the

Global Oceans (POGO), the World Health Organization (WHO), the Food and Agriculture Organization (FAO), the Integrated Marine Biosphere Research (IMBeR) project, and the Intergovernmental Panel on Climate Change (IPCC). Among the listed bodies, only IPHAB and ISSHA have HABs in their terms of reference. The links with other entities not strictly focused on HABs will reinforce addressing the global and transdisciplinary impacts of HABs.

Regional organizations and projects—GlobalHAB success depends on the promotion of its objectives in a variety of regional organizations and projects and on the involvement of regional scientists in GlobalHAB. Accordingly, GlobalHAB is developing partnerships with PICES, ICES, the Mediterranean Science Commission (CIESM), and regional bodies that are part of IOC (e.g., IOC/WESTPAC/HAB, IOC/ANCA, IOC/FANSA, IOC/HANA, UNEP/NOWPAP).

Endorsed national projects—GlobalHAB provides an international structure for planning and coordination of international scientific activities related to HABs in aquatic environments and their effects on the environment and human health and well-being. In addition, GlobalHAB provides a framework for national and regional projects to (1) coordinate their activities and (2) participate in coordinated initiatives of research, observations, and modelling on HABs. The GlobalHAB SSC may endorse already funded projects, proposals for projects prior to submission for funding, or even letters of intent. The primary criterion for endorsement is applicability to any of the themes of the GlobalHAB Science and Implementation Plan, which will be available through IOC and SCOR. Endorsed national HAB projects can provide valuable contributions to GlobalHAB, and GlobalHAB can help national projects work collaboratively around central themes of the GlobalHAB project and provide visibility to national projects.

22.5 International Approach of GlobalHAB

The international approach used by GEOHAB is at the core of the GlobalHAB Programme. As seen from the description of the programme in the previous sections, GlobalHAB recognizes that much remains to be learned about HABs in order to help protect marine ecosystems and human health and that there are advantages to bringing marine and freshwater HAB scientists together on issues of common interest. Therefore, the *general mission* of GlobalHAB is to *foster international cooperative science on HABs,* following in the steps of GEOHAB. Furthermore, the *overall goal* of GlobalHAB is *to improve understanding and prediction of HABs in aquatic ecosystems and management and mitigation of their impacts*. GlobalHAB will conduct the following actions to achieve its mission and goal:

- Address the scientific and societal challenges of HABs, including the environmental, human health, and economic impacts, in a rapidly changing world.
- Consolidate linkages with broader scientific fields and other regional and international initiatives relevant to HABs.

- Foster the development and adoption of advanced, cost-effective technologies.
- Promote training, capacity building, and communication of HAB research to society.
- Serve as a liaison between the HAB-related scientific community, stakeholders, and policymakers, informing science-based decision-making.

The GlobalHAB Science and Implementation Plan is available at www.globalhab.info.

Acknowledgments The GlobalHAB Scientific Steering Committee acknowledges the financial support provided by IOC and by SCOR through Grants OCE 12-1243377 and OCE-1546580 from the US National Science Foundation and from ICES, IPHAB, ISSHA, PICES, and the Scottish Association for Marine Science.

References

Anderson CR, Kudela RM, Kahru M et al (2016) Initial skill assessment of the California Harmful Algae Risk Mapping (C-HARM) system. Harmful Algae 59:1–8

Anderson DM, Glibert PM, Burkholder JM (2002) Harmful algal blooms and eutrophication: nutrient sources, composition, and consequences. Estuaries 25:562–584

Anderson DM, Keafer BA, Kleindinst JL et al (2014) *Alexandrium fundyense* cysts in the Gulf of Maine: long-term time series of abundance and distribution, and linkages to past and future blooms. Deep Sea Res II Top Stud Oceanogr 103:6–26

Anderson DM, Pitcher GC, Estrada M (2005) The comparative "systems" approach to HAB research. Oceanography 18:148–157

Azanza RV, Brosnahan ML, Anderson DM et al (2018) The role of life cycle characteristics in harmful algal bloom dynamics. In: Glibert PM, Berdalet E, Burford M et al (eds) Global ecology and oceanography of harmful algal blooms. Springer, Cham, pp 133–161

Berdalet E, McManus MA, Ross ON et al (2014) Understanding harmful algae in stratified systems: review of progress and future directions. Deep Sea Res II Top Stud Oceanogr 101:4–20

Berdalet E, Tester PA, Chinain M et al (2017a) Harmful algal blooms in benthic systems: recent progresses and future research. Oceanography 30:36–45

Berdalet E, Montresor M, Reguera B et al (2017b) Harmful algal blooms in fjords and coastal embayments and stratified systems: recent progress and future research. Oceanography 30:46–57

Berdalet E, Tester P (2018) Key questions and recent research advances on harmful algal blooms in benthic systems. In: Glibert PM, Berdalet E, Burford M et al (eds) Global ecology and oceanography of harmful algal blooms. Springer, Cham, pp 261–286

Black EA, Whyth JNC, Bagshaw JW et al (1991) The effects of *Heterosigma akashiwo* on juvenile *Oncorhynchus tshawytscha* and its implications for fish culture. J Appl Ichthyol 7:168–175

Boada LD, Zumbado M, Luzardo OP et al (2010) Ciguatera fish poisoning on the West Africa Coast: an emerging risk in the Canary Islands (Spain). Toxicon 56:1516–1519

Borcier AE, Morvezen R, Boudry P et al (2017) Effects of bioactive extracellular compounds and paralytic shellfish toxins produced by *Alexandrium minutum* on growth and behaviour of juvenile great scallops *Pecten maximus*. Aquat Toxicol. https://doi.org/10.1016/j.aquatox.2017.01.009

Brosnahan ML, Velo-Suárez L, Ralston DK et al (2015) Rapid growth and concerted sexual transitions by a bloom of the harmful dinoflagellate *Alexandrium fundyense* (Dinophyceae). Limnol Oceanogr 60:2059–2078

Burford M, Hamilton DP, Wood SA (2018) Emerging HAB research issues in freshwater environments. In: Glibert PM, Berdalet E, Burford M et al (eds) Global ecology and oceanography of harmful algal blooms. Springer, Cham, pp 381–402
Burkholder JM, Glibert PM, Skelton H (2008) Mixotrophy, a major mode of nutrition for harmful algal species in eutrophic waters. Harmful Algae 8:77–93
Campbell L, Olson RJ, Sosik HM et al (2010) First harmful *Dinophysis* (Dinophyceae, Dinophysiales) bloom in the U.S. revealed by automated imaging flow cytometry. J Phycol 46:66–75
Chambouvet A, Morin P, Marie D et al (2008) Control of toxic marine dinoflagellate blooms by serial parasitic killers. Science 322(5905):1254–1257
Clement A, Lincoqueo L, Saldivia M et al (2016) Exceptional summer conditions and HABs of *Pseudochattonella* in Southern Chile create record impacts on salmon farms. Harmful Algae News 53:1–3
Dahl E, Tangen K (1993) 25 years experience with *Gyrodinium aureolum* in Norwegian waters. In: Smayda TJ, Shimizu Y (eds) Toxic phytoplankton blooms in the sea. Elsevier, New York, pp 15–21
Davidson K, Anderson DM, Mateus M et al (2016) Forecasting the risk of harmful algal blooms. Harmful Algae 53:1–7
Davidson K, Gowen RJ, Harrison PJ (2014) Anthropogenic nutrients and harmful algae in coastal waters. J Environ Manage 146:206–216
Dyson K, Huppert DD (2010) Regional economic impacts of razor clam beach closures due to harmful algal blooms (HABs) on the Pacific coast of Washington. Harmful Algae 9:264–271
Flynn KJ, Mitra A, Glibert PM et al (2018) Mixotrophy in HABs: by whom, on whom, when, why, and what next. In: Glibert PM, Berdalet E, Burford M et al (eds) Global ecology and oceanography of harmful algal blooms. Springer, Cham, pp 113–132
Foloni-Neto H, Lueck R, Mabuchi Y et al (2014) A new quasi-horizontal glider to measure biophysical microstructure. J Atmos Ocean Technol 31:2278–2293. https://doi.org/10.1175/JTECH-D-13-00240.1
Franks PJS (2018) Recent advances in modelling of harmful algal blooms. In: Glibert PM, Berdalet E, Burford M et al (eds) Global ecology and oceanography of harmful algal blooms. Springer, Cham, pp 359–377
Friedman MA, Fernandez M, Backer L et al (2017) An updated review of ciguatera fish poisoning: clinical, epidemiological, environmental, and public health management. Mar Drugs 15(72). https://doi.org/10.3390/md15030072
Fukuyo Y, Imai I, Kodama M et al (2002) Red tides and harmful algal blooms in Japan. In: Taylor FJ, Trainer VL (eds) Harmful algal blooms in the PICES region of the North Pacific. PICES, Scientific Report No. 23. North Pacific Marine Science Organization (PICES), Sidney, BC, pp 7–20
GEOHAB (2001) Global ecology and oceanography of harmful algal blooms, science plan. Glibert PM, Pitcher G (eds) SCOR and IOC, Baltimore and Paris 87 pp
GEOHAB (2003) Global ecology and oceanography of harmful algal blooms, implementation plan. Gentien P, Pitcher G, Cembella A et al (eds) SCOR and IOC, Baltimore and Paris, 36 pp
GEOHAB (2011) GEOHAB modelling: a workshop report. McGillicuddy DJ Jr, Glibert PM, Berdalet E et al (eds), IOC and SCOR, Paris and Newark, Delaware 85 pp
GEOHAB (2012) Global ecology and oceanography of harmful algal blooms: core research project – harmful algal blooms in benthic systems. Berdalet E, Tester P, Zingone A (eds) IOC of UNESCO and SCOR, Paris and Newark 64 pp
GEOHAB (2014) Global ecology and oceanography of harmful algal blooms: synthesis open science meeting. Berdalet E, Bernard S, Burford MA et al (eds) IOC and SCOR, Paris and Newark, Delaware, USA 78 pp
Glibert PM, Al-Azri A, Allen JI et al (2018a) Key questions and recent research advances on harmful algal blooms in in relation to nutrients and eutrophication. In: Glibert PM, Berdalet E, Burford M et al (eds) Global ecology and oceanography of harmful algal blooms. Springer, Cham, pp 229–259

Glibert PM, Beusen AHW, Harrison JA et al (2018b) Changing land, sea- and airscapes: sources of nutrient pollution affecting habitat suitability for harmful algae. In: Glibert PM, Berdalet E, Burford M et al (eds) Global ecology and oceanography of harmful algal blooms. Springer, Cham, pp 53–76

Glibert PM, Burford M (2017) Globally changing nutrient loads and harmful algal blooms: recent advances, new paradigms, and continuing challenges. Oceanography 30(1):58–69. https://doi.org/10.5670/oceanog.2017.110

Glibert PM, Pitcher GC, Bernard S et al (2018c) Advancements and continuing challenges of emerging technologies and tools for detecting harmful algal blooms, their antecedent conditions and toxins, and applications in predictive models. In: Glibert PM, Berdalet E, Burford M et al (eds) Global ecology and oceanography of harmful algal blooms. Springer, Cham, pp 339–357

Gobler CJ, Sunda WG (2012) Ecosystem disruptive algal blooms of the brown tide species, *Aureococcus anophagefferens* and *Aureoumbra lagunensis*. Harmful Algae 14:36–45

Goes JI, Gomes HR, Al-Hashimi K et al (2018) Ecological drivers of green *Noctiluca* blooms in two monsoonally driven ecosystems. In: Glibert PM, Berdalet E, Burford M et al (eds) Global ecology and oceanography of harmful algal blooms. Springer, Cham, pp 327–336

Greenfield DI, Marin R III, Doucette GJ et al (2008) Field applications of the second-generation environmental sample processor (ESP) for remote detection of harmful algae, 2006-2007. Limnol Oceanogr Methods 6:667–679

HABWATCH (2004) In: Babin M, Roesler C, Cullen J (eds) Real-time coastal observing systems for marine ecosystem dynamics and harmful algal blooms: theory, instrumentation and modelling. Oceanographic methodology series. UNESCO Printers, p 799

Hallegraeff GM (2010) Ocean climate change, phytoplankton community responses, and harmful algal blooms: a formidable predictive challenge. J Phycol 46:220–235

Hamilton DP, Wood SA, Dietrich DR et al (2014) Costs of harmful blooms of freshwater cyanobacteria. In: Sharma NK, Rai AK, Stal LJ (eds) Cyanobacteria: an economic perspective, 1st edn. Wiley, New York, pp 245–256

Heisler J, Glibert PM, Burkholder JM et al (2008) Eutrophication and harmful algal blooms: a scientific consensus. Harmful Algae 8:3–13

Hernández C, Díaz PA, Molinet C et al (2016) Exceptional climate anomalies and northwards expansion of Paralytic Shellfish Poisoning outbreaks in Southern Chile. Harmful Algae News 54:1–2

Hoagland P, Anderson DM, Kaoru Y et al (2002) The economic effects of harmful algal blooms in the United States: estimates, assessment issues, and information needs. Estuaries 25:677–695

Ianora A, Bentley MG, Caldwell GS et al (2011) The relevance of marine chemical ecology to plankton and ecosystem function: an emerging field. Mar Drugs 9:1625–1648

Jaffe JS, Franks PJS, Briseno C et al (2013) Advances in underwater fluorometry, from bulk fluorescence to planar laser imaging of individuals. In: Watson J, Zielinski O (eds) Subsea optics and imaging. Woodhead Publishing, Cambridge, pp 536–549

Jauzein C, Fricke A, Mangialajo L et al (2016) Sampling of *Ostreopsis* cf. *ovat*a using artificial substrates: optimization of methods for the monitoring of benthic harmful algal blooms. Mar Pollut Bull. https://doi.org/10.1016/j.marpolbul.2016.03.047

Jeffery B, Barlow T, Moizer K, Paul S et al (2004) Amnesic shellfish poison. Food Chem Toxicol 42:545–557

Kremp A, Tahvanainen P, Litaker W et al (2014) Phylogenetic relationships, morphological variation, and toxin patterns in the *Alexandrium ostenfeldii* (Dinophyceae) complex: implications for species boundaries and identities. J Phycol 50:81–100

Krüger T, Mönch B, Oppenhäuser S et al (2010) LC-MS/MS determination of the isomeric neurotoxins BMAA (β-N-methylamino-L-alanine) and DAB (2,4-diaminobutyric acid) in cyanobacteria and seeds of *Cycas revoluta* and *Lathyrus latifolius*. Toxicon 55:547–557

Kudela RM, Berdalet E, Bernard S et al (2015) Harmful algal blooms – a scientific summary for policy makers. IOC/UNESCO, Paris IOC/INF-1320

Kudela RM, Berdalet E, Enevoldsen H et al (2017) GEOHAB–The Global Ecology and Oceanography of Harmful Algal Blooms Program: motivation, goals, and legacy. Oceanography 30:12–21

Kudela RM, Berdalet E, Enevoldsen H et al (2018) Establishment, goals, and legacy of the Global Ecology and Oceanography of Harmful Algal Blooms (GEOHAB) Program. In: Glibert PM, Berdalet E, Burford M et al (eds) Global ecology and oceanography of harmful algal blooms. Springer, Cham, pp 27–49

Kudela RM, Howard MDA, Jenkins BD et al (2010a) Using the molecular toolbox to compare harmful algal blooms in upwelling systems. Prog Oceanogr 85:108–121

Kudela RM, Seeyave S, Cochlan WP (2010b) The role of nutrients in regulation and promotion of harmful algal blooms in upwelling systems. Prog Oceanogr 85:122–135

Lelong A, Hégaret H, Soudant P et al (2012) *Pseudo-nitzschia* (Bacillariophyceae) species, domoic acid and amnesic shellfish poisoning: revisiting previous paradigms. Phycologia 51:168–216

Lim HC, Leaw CP, Tan TH et al (2014) A bloom of *Karlodinium australe* (Gymnodiniales, Dinophyceae) associated with mass mortality of cage cultured fishes in West Johor Strait, Malaysia. Harmful Algae 40:51–62

Liu D, Zhou M (2018) Green tides of the Yellow Sea: massive free-floating blooms of *Ulva prolifera*. In: Glibert PM, Berdalet E, Burford M et al (eds) Global ecology and oceanography of harmful algal blooms. Springer, Cham, pp 317–326

Lunven M, Guillaud JF, Youenou A et al (2005) Nutrient and phytoplankton distribution in the Loire River plume (Bay of Biscay, France) resolved by a new fine scale sampler. Estuar Coast Shelf Sci 65:94–108

Lunven M, Landeira JM, Lehaître M et al (2012) In situ video and fluorescence analysis (VFA) of marine particles: applications to phytoplankton ecological studies. Limnol Oceanogr Methods 10:807–823

Mangialajo L, Fricke A, Perez-Gutierrez G et al (2017) Benthic Dinoflagellate Integrator (BEDI): a new method for the quantification of benthic harmful algal blooms. Harmful Algae 64:1–10

Masó M, Garcés E, Pagès F et al (2003) Drifting plastic debris as a potential vector for dispersing harmful algal bloom (HAB) species. Sci Mar 67:107–111

Moore MN, Depledge MH, Fleming LE et al (2013) Oceans and Human Health (OHH): a European perspective from the Marine Board of the European Science Foundation (Marine Board-ESF). Microb Ecol 65:889–900

Moore SK, Johnstone JA, Banas NS et al (2015) Present-day and future climate pathways affecting *Alexandrium* blooms in Puget Sound, WA, U.S. Harmful Algae 48:1–15

O'Neil JM, Davis TW, Burford MA et al (2012) The rise of harmful cyanobacteria blooms (CHABs): role of eutrophication and climate change in freshwater, estuarine and marine ecosystems. Harmful Algae 14:313–334

Paerl HW, Fluton RS, Moisander PH et al (2001) Harmful freshwater algal blooms. Sci World J 1:76–113

Pearson LA, Dittmann E, Mazmouz R et al (2016) The genetics, biosynthesis and regulation of toxic specialized metabolites of cyanobacteria. Harmful Algae 54:98–111

Pitcher G, Kudela RM, Reguera B et al (2018) Key questions and recent research advances on harmful algal blooms in eastern boundary upwelling systems. In: Glibert PM, Berdalet E, Burford M et al (eds) Global ecology and oceanography of harmful algal blooms. Springer, Cham, pp 205–227

Price C, Black KD, Hargave BT et al (2015) Marine cage culture and the environment: effects on water quality and primary production. Mar Ecol Prog Ser 6:151–174

Roy S, Montresor M, Cembella A (2018) Key questions and recent research advances on harmful algal blooms in fjords and coastal embayments. In: Glibert PM, Berdalet E, Burford M et al (eds) Global ecology and oceanography of harmful algal blooms. Springer, Cham, pp 187–203

Ryan JP, McManus MA, Sullivan JM (2010) Interacting physical, chemical and biological forcing of phytoplankton thin-layer variability in Monterey Bay, California. Cont Shelf Res 30:7–16

Selander S, Kubanek J, Hamberg M et al (2015) Predator lipids induce paralytic shellfish toxins in bloom-forming algae. Proc Natl Acad Sci USA 112:6395–6400

Smayda TJ (2010a) Adaptations and selection of harmful and other dinoflagellate species in upwelling systems 1. Morphology and adaptive polymorphism. Prog Oceanogr 85:53–70

Smayda TJ (2010b) Adaptations and selection of harmful and other dinoflagellate species in upwelling systems 2. Motility and migratory behaviour. Prog Oceanogr 85:71–91

Smetacek V, Zingone A (2013) Green and golden seaweed tides on the rise. Nature 504:84–88

Tester PA, Kibler SA, Holland WC et al (2014) Sampling harmful benthic dinoflagellates: comparison of artificial and natural substrate methods. Harmful Algae 39:8–25

Timmerman AHV, McManus MA, Cheriton OM et al (2014) Hidden thin layers of toxic diatoms in a coastal bay. Deep Sea Res II Top Stud Oceanogr 101:129–140

Trainer VL, Kameneva P (2017) Conditions promoting extreme *Pseudo-nitzschia* events in the eastern but not the western Pacific. PICES Scientific Report, North Pacific Marine Science Organization (PICES), Sidney, BC, Canada

Trainer VL, Pitcher GC, Reguera B et al (2010) The distribution and impacts of harmful algal bloom species in eastern boundary upwelling systems. Prog Oceanogr 85:35–52

Trainer VL, Yoshida T (eds) (2014) Proceedings of the Workshop on Economic impacts of harmful algal blooms on fisheries and aquaculture. PICES Scientific Report No. 47, p 85

Turner AD, McNabb PS, Harwood DT et al (2015) Single laboratory validation of a multi-toxin UPLC-HILIC-MS/MS method for quantitation of paralytic shellfish toxins in bivalve shellfish. J AOAC Int 98:609–621

Van Wagoner RM, Deeds JR, Satake M (2008) Isolation and characterization of karlotoxin 1, a new amphipathic toxin from *Karlodinium veneficum*. Tetrahedron Lett 49:6457–6461

Wells ML, Karlson B (2018) Harmful algal blooms in a changing ocean. In: Glibert PM, Berdalet E, Burford M et al (eds) Global ecology and oceanography of harmful algal blooms. Springer, Cham, pp 77–90

Wells ML, Trainer VL, Smayda TJ et al (2015) Harmful algal blooms and climate change: learning from the past and present to forecast the future. Harmful Algae 49:68–93

Yogi K, Oshiro N, Inafuku Y et al (2011) Detailed LC-MS/MS analysis of ciguatoxins revealing distinct regional and species characteristics in fish and causative algae from the Pacific. Anal Chem 83:8886–8891

Index

P.M. Glibert et al. (eds.), *Global Ecology and Oceanography of Harmful Algal Blooms*, Ecological Studies 232, https://doi.org/10.1007/978-3-319-70069-4

D

E

Q

R

MIX
Papier aus verantwortungsvollen Quellen
Paper from responsible sources
FSC® C105338

If you have any concerns about our products,
you can contact us on
ProductSafety@springernature.com

In case Publisher is established outside the EU,
the EU authorized representative is:
Springer Nature Customer Service Center GmbH
Europaplatz 3, 69115 Heidelberg, Germany

Printed by Libri Plureos GmbH
in Hamburg, Germany